Societal Challenges and Opportunities of Low-Carbon Energy Transformations

Low-carbon energy transformations to support carbon-neutral societies is an important and urgent topic and subject to current national priorities and socio-economic planning. This book addresses the societal challenges and opportunities that come with the transition in EU member states such as energy poverty and inequality, energy security, gender inequality, energy efficiency improvements, climate change mitigation, growth in green investments, etc. It contains a balanced approach of theoretical concepts with an impressive range of relevant case studies to investigate the issues, develop indicator frameworks, identify barriers and drivers, and create policies and measures to unlock opportunities.

Features:

- Addresses systematically and comprehensively the low-carbon energy transition, its barriers, and its societal implications.
- Discusses the main societal challenges and opportunities of low-carbon energy transition from theoretical and practical points of view.
- Provides definitions of concepts, measurement indicators, policies, and a framework for overcoming barriers and enforcing drivers.
- Includes case studies on low-carbon energy transition challenges and opportunities that are developed for EU countries.
- Encourages discussion on policies and measures necessary to overcome identified barriers and proposes how to promote a just and smooth low-carbon energy transition.

This book is a great reference for academics, researchers, graduate students, and professionals such as energy producers, city planners, policymakers, etc. interested in current and emerging trends in sustainable energy.

Societal Challenges and Opportunities of Low-Carbon Energy Transformations

Edited by Dalia Streimikiene

CRC Press
Taylor & Francis Group
Boca Raton London New York

CRC Press is an imprint of the
Taylor & Francis Group, an **informa** business

Designed cover image: Shutterstock

First edition published 2025
by CRC Press
2385 NW Executive Center Drive, Suite 320, Boca Raton FL 33431

and by CRC Press
4 Park Square, Milton Park, Abingdon, Oxon, OX14 4RN

CRC Press is an imprint of Taylor & Francis Group, LLC

ISBN: 978-1-032-85280-5 (hbk)
ISBN: 978-1-003-51749-8 (pbk)
ISBN: 978-1-003-51748-1 (ebk)

DOI: 10.1201/9781003517481

Typeset in Times
by Apex CoVantage, LLC

Contents

Preface......ix
About the Editor......xi
List of Contributors......xiii

Chapter 1 Introduction: Low-Carbon Energy Transition......1

Dalia Streimikiene

1.1 Concept of the Low-Carbon Energy Transition......1
1.2 Policy Documents of Low-Carbon Energy Transition......12
1.2.1 Germany......12
1.2.2 France......12
1.2.3 Sweden......13
1.2.4 United Kingdom......13
1.2.5 Norway......13
1.2.6 Denmark......14
1.2.7 Netherlands......16
1.2.8 Spain......17
1.3 Targets and Indicators of Low-Carbon Transition......18
1.3.1 Limited Financial Resources......21
1.4 Drivers of and Barriers to Low-Carbon Energy Transition......27
1.5 Policy Areas of Low-Carbon Energy Transition......38
1.5.1 Renewable Energy Promotion......39
1.5.2 Energy Efficiency Standards and Regulations......40
1.5.3 Carbon Pricing Mechanisms......42
1.5.4 Emission Reduction Targets......43
1.5.5 Regulatory Framework for Clean Technology Deployment......43
1.5.6 Infrastructure Investment and Modernization......45
1.5.7 Energy Transition Planning and Governance......46
1.5.8 Just Transition Policies......46
1.6 Conclusions......48
References......49

Chapter 2 Energy Poverty......55

Dalia Streimikiene

2.1 Energy Poverty Concepts......55
2.1.1 Energy Inequalities and Justice......55
2.1.2 Energy Vulnerability and Resilience......57
2.1.3 Energy Poverty Concepts Differences in EU Countries......59
2.2 Energy Poverty Indicators......65
2.3 Drivers of Energy Poverty......70
2.4 Policies and Measures to Fight Energy Poverty......77
2.5 Energy Poverty Case Study......78
2.5.1 EU Energy Poverty Indicators......78
2.5.2 Assessment of Energy Poverty Indicators in Lithuania......83
References......90

Chapter 3 Energy Security ... 95

Indre Siksnelyte-Butkiene

3.1 Perception of the Energy Security Concept ... 95
3.2 Energy Security Indicators ... 97
3.3 Policies and Measures to Overcome Energy Security Challenges ... 121
3.4 Comparative Assessment of Energy Security across the EU Countries: A Case Study ... 122
3.4.1 Basis of the Framework ... 123
3.4.2 Development of Indicators Set ... 125
3.4.3 Assessment and Ranking Tool ... 130
3.4.4 Data ... 131
3.4.5 Results and Guidance for Future Research ... 134
3.5 Conclusions ... 135
References ... 136

Chapter 4 Gender Equality ... 139

Lina Volodzkiene

4.1 Gender Equality Concept in Low-Carbon Transition ... 139
4.1.1 Theoretical Frameworks for Gender Equality in the Low-Carbon Transition ... 139
4.1.2 Specific Content and Expression of Gender Equality in Climate Action ... 144
4.2 Indicators and Frameworks of Gender Equality ... 150
4.2.1 Social Indicators of Gender Equality ... 150
4.2.2 Economic Indicators of Gender Equality ... 154
4.2.3 Political and Legal Indicators of Gender Equality ... 160
4.2.4 Cultural and Normative Indicators of Gender Equality ... 164
4.2.5 Environmental and Sustainability Indicators of Gender Equality ... 165
4.2.6 Technological Indicators of Gender Equality ... 169
4.3 Drivers of and Barriers to Gender Equality ... 172
4.4 Policies and Measures to Ensure Gender Equality ... 177
4.5 Conclusions ... 183
References ... 184

Chapter 5 Energy Efficiency ... 191

Mindaugas Butkus and Raimonda Valasevičienė

5.1 The Energy Efficiency Concept ... 191
5.1.1 Energy Conservation ... 192
5.1.2 Energy Intensity ... 193
5.1.3 Energy Efficiency in Consumption ... 193
5.1.4 Efficiency of Energy in Production ... 194
5.1.5 Benefits of Energy Efficiency ... 194
5.1.6 Energy Efficiency Losses ... 195
5.1.7 Energy Efficiency Gap and Rebound Effect ... 195

5.2 Energy Efficiency Indicators ... 198
5.2.1 Comparative Analysis of Energy Efficiency Indices across Different Countries ... 200
5.2.2 Indicators of Energy Efficiency in the Industrial Sector ... 202
5.2.3 Residential Sector Energy Efficiency Indicators ... 202
5.2.4 Challenges in Evaluating the Efficiency of the Economy ... 204
5.3 Drivers of and Barriers to Energy Efficiency ... 206
5.4 Policies and Measures to Increase Energy Efficiency ... 213
5.4.1 Policies Aimed at Improving the Efficiency of Energy Use ... 213
5.4.2 Political Regulations and Standards ... 214
5.4.3 Strategies to Enhance Energy Efficiency ... 215
5.4.4 Causes of Policy Failure ... 216
5.4.5 The Process of Formulating and Executing Energy Efficiency Policies ... 216
5.4.6 Assessment of the Efficacy of Energy Efficiency Initiatives ... 217
5.4.7 Strategies to Minimize Energy Efficiency Deficits ... 218
5.5 Case Study ... 220
5.6 Conclusions ... 230
References ... 230

Chapter 6 Renewable Energy ... 235

Ahmad Bathaei

6.1 Renewable Energy Sources ... 235
6.1.1 Sources ... 235
6.1.2 Benefits and Features ... 242
6.2 Renewable Energy Indicators ... 244
6.3 Drivers of and Barriers to Renewable Energy Usage ... 247
6.3.1 Drivers ... 247
6.3.2 Barriers ... 249
6.3.3 Opportunities ... 251
6.4 Policies and Measures to Promote Renewables ... 253
6.4.1 Feed-in Tariffs (FiTs) ... 255
6.4.2 Tax Incentives and Credits ... 256
6.4.3 Renewable Energy Grants and Subsidies ... 257
6.4.4 Public Awareness Campaigns and Education Programs ... 261
6.5 Case Study ... 262
6.6 Conclusion ... 268
References ... 269

Chapter 7 Green Jobs ... 274

Justas Streimikis

7.1 The Green Jobs Concept ... 274
7.2 Indicators of Green Jobs ... 282
7.3 Drivers of and Barriers to Green Jobs ... 293
7.4 Policies and Measures to Promote Green Employment ... 306
7.5 Case Study ... 325
7.6 Conclusions ... 329
References ... 330

Chapter 8 Green Investments 333

Karolis Andriuskevicius

8.1 The Green Investments Concept 333
8.2 Indicators 342
8.3 Drivers and Barriers 350
8.4 Policies and Measures to Promote Green Investments 356
8.5 Case Study 361
8.6 Conclusions 370
References 372

Index 377

Preface

Low-carbon transition is the main policy agenda in the European Union and other developed and developing nations. Through low-carbon transition the increased use of renewable energy sources in households and a growing number of prosumers in developed economies have many benefits for society, such as reduced GHG emissions, increased energy security, reduced energy costs, new green jobs creation, etc.; however, there are important challenges linked to these processes like market failures and barriers of renewable energy penetration, energy poverty, and inequality, gender inequalities etc. that also should be taken into account during the process of energy transformation.

This book aims to analyze and systematize the main societal challenges and opportunities of low-carbon energy transition comprehensively and systematically based on a scientific literature review, to develop concepts and frameworks of indicators and main barriers and drivers and policies and measures to address them and present the case studies for assessing low-carbon energy transition challenges and opportunities in European Union member states.

The low-carbon energy transition has many societal challenges. It promises many opportunities and benefits for societies and, therefore, needs to be investigated systematically by showing their mutual interrelations and overlapping barriers and drivers. The book deals with a new and very topical issue that will affect everyone. Though there are many studies dealing with various aspects of low-carbon energy transition, this book proposes a new approach to the investigation of this transformation process by introducing a harmonized framework for analysis of the main challenges and opportunities, which is based on the definition of the main concepts, construction of indicators frameworks for measuring the phenomenon, systematization of the main drivers and barriers of the phenomenon and developing the corresponding policies and measures to overcome these barriers and enforce of drivers. The case studies for EU countries allow us to investigate the low-carbon energy transition processes more deeply by providing practical application of the developed theoretical framework. The book is constituted of 8 interconnected parts that sequentially reveal the main challenges and opportunities low-carbon transition.

The introductory chapter introduces the concept of low-carbon energy transition by providing a theory for the transition to a carbon-neutral society. This chapter analyzes the international political support for climate change mitigation and low-carbon transition, including major conventions and international agreements and their targets. A systematic approach to investigating interrelated societal challenges and opportunities of low-carbon energy transition is developed. The main targets and indicators to measure these targets are discussed, and the barriers and drivers to low-carbon transition and policies to promote low-carbon transition and overcome barriers and enforce drivers are scrutinized systematically based on a comprehensive literature review.

Chapter 2 analyzes energy equality and energy poverty issues, which are key challenges to be addressed to ensure just a low-carbon energy transition. As new renewable energy microgeneration technologies are expensive, it is important to ensure that low-income populations will receive support for the implementation of these technologies to reduce energy poverty risk. This chapter applies the framework developed in Chapter 2 for analysis of energy equality and defines the concepts of energy equality and poverty, introduces indicators to measure energy poverty and inequality and systematizes barriers and drivers of energy equality and policies and measures to address them. The case study on the energy poverty challenge in low-carbon energy transition was developed for selected EU countries.

Chapter 3 deals with the challenge of energy security as switching from fossil fuels to 100% renewables in final energy consumption causes problems of security of energy supply due to the high intermittency of renewable energy generation and the high need for energy storage capacities. In this chapter, the same approach was applied and developed in Chapter 2, and the concept of energy security was defined, indicators to measure energy security were constructed and barriers and drivers of energy security and policies and measures to address them were systematized. The case study on the energy security challenge in low-carbon energy transition was developed for selected EU countries.

Chapter 4 investigates the challenge of gender equality in low-carbon energy transition. Gender equality is necessary to address just low-carbon transition as women face many difficulties in green job markets, establishments of green businesses etc. and many other barriers in implementing of renewable energy microgeneration technologies. In this chapter the framework developed in Chapter 2 was applied for analysis of gender equality and concept of gender equality was defined, indicators to measure gender equality were proposed and the main barriers and drivers of equality in low-carbon energy transition were identified by selecting appropriate policies and measures to address them. The case study on the gender equality challenge in low-carbon energy transition was developed for selected EU countries.

Chapter 5 analyzes the problems of energy efficiency improvement and the benefits of low-carbon energy transition. There are many barriers to energy efficiency improvement, though this is the main way to reduce GHG emissions and reduce energy expenses for households, enterprises, and other organizations. This chapter applies the framework developed in Chapter 2 for analysis of energy efficiency and defines the concepts of energy efficiency, presents indicators to measure energy efficiency, and systematizes barriers and drivers of energy efficiency and policies and measures to address them. The case study on the energy efficiency improvement challenges and opportunities in low-carbon energy transition was developed for selected EU countries.

Chapter 6 examines replacing fossil fuels by renewable energy sources as an important challenge and opportunity to reduce GHG emissions and achieve a carbon-neutral society. This chapter applies the framework developed in Chapter 2 for analysis of the replacement of fossil fuels by renewables and defines the concepts of renewable energy fats penetration and replacement of fossil fuels making 100% of renewables in final energy consumption, presents indicators to measure penetration of renewable energy sources and systematizes barriers and drivers of renewables and policies and measures to address them. The case study on the penetration of renewables in low-carbon energy transition was developed for selected EU countries.

Chapter 7 is devoted to the investigation of green jobs. Green jobs are an important challenge, providing a lot of opportunities and benefits of low-carbon energy transition like economic growth and social welfare. However, green job market development also faces many barriers and policies and measures are necessary to overcome them and enforce drivers of green jobs creation. This chapter applies the framework developed in Chapter 2 and defines the green job concepts, develops a framework of indicators to measure green jobs and systematizes barriers to and drivers of green jobs expansion and policies and measures to address them. The case study on green job creation challenges and opportunities in low-carbon energy transition was developed for selected EU countries.

Chapter 8 presents an analysis of green investment challenges and opportunities. Green investments ensure penetration of new energy efficient and renewable energy technologies and green infrastructure development, providing for smooth low-carbon energy transition and economic growth, including new green job creation. However, many barriers to green investments exist, and policies and measures are necessary to support them. The chapter applies the framework developed in Chapter 2 and defines the green investment concepts, develops a framework of indicators to measure green investments and systematizes barriers to and drivers of green investment growth and policies and measures to address them. The case study on green investment growth challenges and opportunities in low-carbon energy transition was developed for selected EU countries.

About the Editor

Dalia Streimikiene is a leading researcher at the Lithuanian Centre for Social Sciences, Institute of Economics and Rural Development. Her main areas of research are sustainable development, sustainability assessment, and corporate social responsibility. She is the author of more than 200 papers in international journals referred to in WoS and Editor-in-Chief of the international journal *Transformations in Business and Economics.*

Contributors

Karolis Andriuskevicius
Senior Researcher
Lithuanian Centre for Social Sciences
Institute of Economics and Rural Development
Vilnius, Lithuania

Ahmad Bathaei
Researcher
Lithuanian Centre for Social Sciences
Institute of Economics and Rural Development
Vilnius, Lithuania

Mindaugas Butkus
Senior Researcher
Lithuanian Centre for Social Sciences
Institute of Economics and Rural Development
Vilnius, Lithuania

Indre Siksnelyte-Butkiene
Senior Researcher
Lithuanian Centre for Social Sciences
Institute of Economics and Rural Development
Vilnius, Lithuania

Dalia Streimikiene
Leading Researcher
Lithuanian Centre for Social Sciences
Institute of Economics and Rural Development
Vilnius, Lithuania

Justas Streimikis
Senior Researcher
Lithuanian Centre for Social Sciences
Institute of Economics and Rural Development
Vilnius, Lithuania

Raimonda Valasevičienė
Engineer
Lithuanian Water Management Company LLC
Vilnius, Lithuania

Lina Volodzkiene
Senior Researcher
Lithuanian Centre for Social Sciences
Institute of Economics and Rural Development
Vilnius, Lithuania

1 Introduction
Low-Carbon Energy Transition

Dalia Streimikiene

1.1 CONCEPT OF THE LOW-CARBON ENERGY TRANSITION

A low-carbon energy transition is the process of moving from conventional carbon sources of energy—fossil fuels, such as coal, oil, and gas—to clean or green sustainable alternatives. The very purpose of the transition is reducing greenhouse gas emissions to moderate change effects and ensure long-term environmental sustainability. It is these drivers that give rise to this transition: hence, addressing climate change, reducing greenhouse gas emissions, and creating an environmentally sustainable energy system.

The low-carbon energy transition is an intricate and multifaceted process (Green & Gambhir, 2020). This transition entails moving from high- to low-carbon energy sources, and it can significantly impact the energy return on investment (EROI) and net energy per capita. The significance of careful planning when retiring aging coal-fired power plants and introducing renewable energy sources is to ensure a seamless transition. Wu et al. (2018) emphasizes the role of market regulation, energy consumption, and energy capacity in facilitating the transition, especially in distributed energy systems. Collectively, these studies underline the necessity for a comprehensive and strategic approach to low-carbon energy transition.

The intricate low-carbon energy transition is influenced by a multitude of factors. Wu et al. (2018) highlight the significance of market regulation, energy consumption, and capacity in facilitating this transition, as well as the necessity for an environmentally, socially equitable, and economically sustainable approach to low-carbon energy and the retirement of aging coal generators, proposing a model for strategically planning their replacement with renewable power plants to achieve benefits in carbon mitigation, cost, and risk. Chen et al. (2021) explore the transition pathway of the power sector with a high penetration of renewable energy, underscoring the importance of assessing the integration costs of flexible generation from thermal power plants. Key elements of the low-carbon energy transition include:

- **Renewable energy sources**. Harnessing and development of renewable sources of energy, including solar, wind, hydro, geothermal, and biomass. These give low or no carbon emissions energy compared to conventional fossil fuels.

Some key renewable energy sources are presented in Figure 1.1

Renewable energy sources present numerous benefits, including the reduction of greenhouse gas emissions, advancements in technological development, and the potential for improved energy efficiency (Lackner, 2021). Additionally, they offer a sustainable and reliable energy supply, contributing to economic growth, job creation, and enhanced energy security. Moreover, the incorporation of renewable energy sources into smart grids has the probability to boost energy efficiency and facilitate real-time information flow. Nevertheless, challenges such as the requirement for substantial quantities of raw materials and energy for production, along with the recycling issue, need to be addressed (Kabeyi & Olanrewaju, 2022).

DOI: 10.1201/9781003517481-1

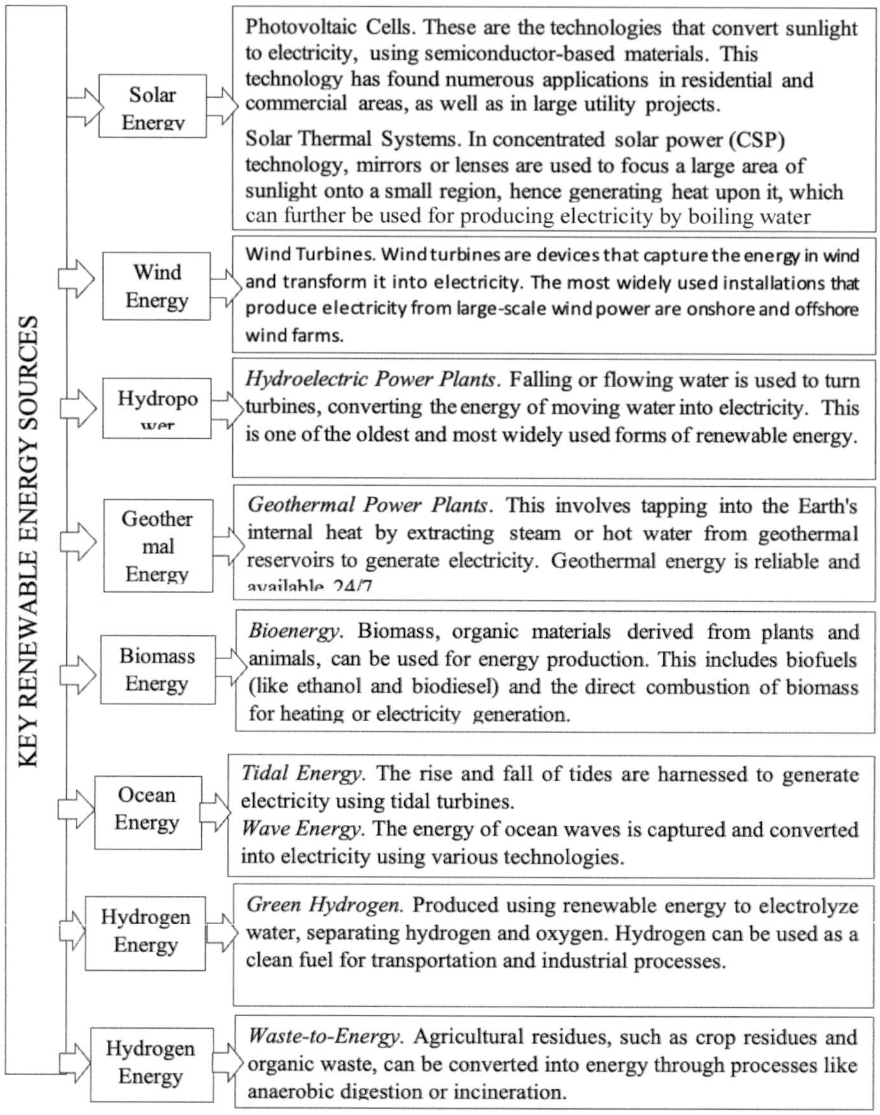

FIGURE 1.1 Some key renewable energy sources.

Source: created by authors.

One very crucial benefit that renewable energy holds is the role it plays in reducing greenhouse gas emissions. Unlike fossil fuels, which in combustion discharge substantial volumes of carbon dioxide and other pollutants into the atmosphere, renewables like solar, wind, and hydropower either have very minimal electricity generation emission or zero such emission at all. Generally, technologies for renewable energy have a limited environmental burden from pollution in air and water. Unlike the conventional power plants that burn fossil fuels, renewable energy sources add to cleaner air and water, hence better human health and ecosystems. Most of the renewable energy sources are virtually inexhaustible at least on human timescales. This includes energy derived from the sun and energy drawn as a result of the movement of the atmosphere in the form of wind. These sources provide an ecological and long-term resolution for the world's energy requirements. Diversification of energy sources enhances energy security. Countries relying heavily on the importation of fossil fuels could reduce their vulnerability to supply disruptions by making investments in abundant locally available renewable resources. The sector provides employment in various fields, such as business, installation, maintenance, and research. This growth will not only stimulate the economy but also breed innovation and skills development by creating jobs. Figure 1.2 presents some key advantages of renewable energy.

Renewable energy sources also carry significant drawbacks, including their reliance on weather conditions, low energy efficiency, and the demand for substantial quantities of raw materials. Additionally, there are concerns about their ability to meet the growing global energy demand, and integrating them into existing supply systems can present engineering, environmental, health, and economic challenges. Despite these disadvantages, the utilization of renewable energy sources can play a key role in mitigating the adverse impacts of conventional energy sources on the environment, society, and economy, as emphasized by Kumar and Rathore (2023). Figure 1.3 presents some key disadvantages of renewable energy.

The scientific evidence overwhelmingly supports the advantages (pros) of renewable energy, particularly in terms of environmental sustainability and mitigating climate change. However, it's significant to note that no energy source is entirely without challenges, and critics may raise concerns about certain aspects of renewable energy technologies. While the majority of scientists and experts

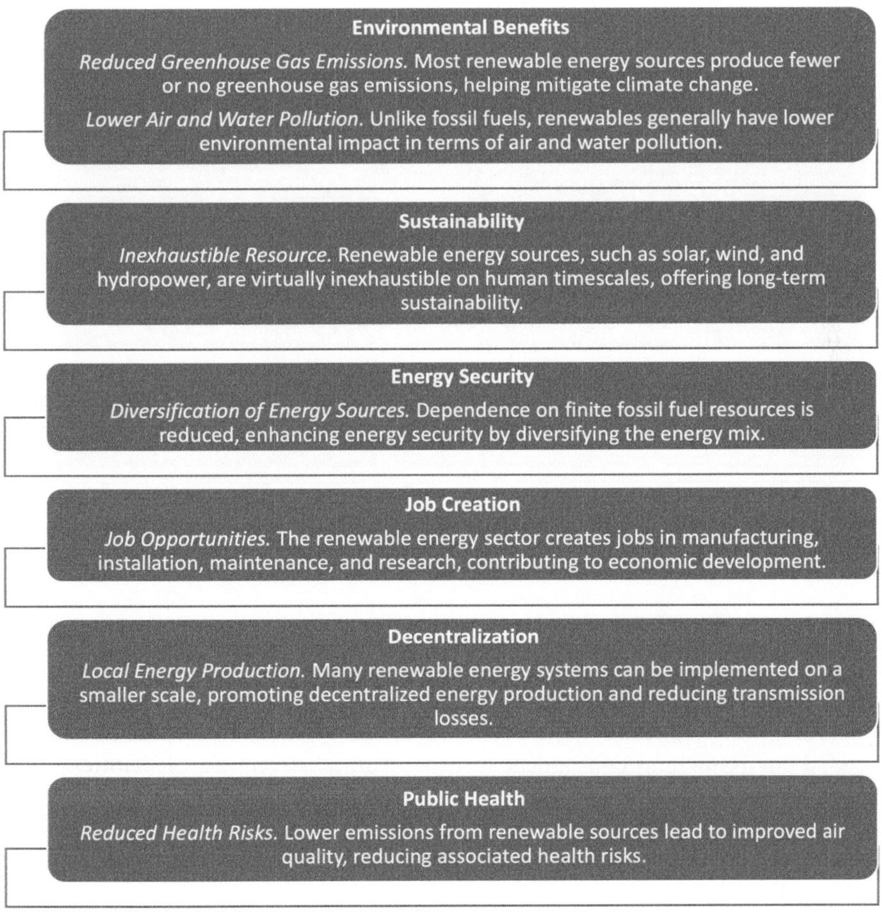

FIGURE 1.2 Some key advantages of renewable energy.

Source: created by authors according to International Renewable Energy Agency (2017), Maradin (2021), Cakir et al., (2022), Baidya and Nandi (2020).

FIGURE 1.3 Some key disadvantages of renewable energy.

Source: created by authors according to Maradin (2021), Nath and Sen (2021), Carpejani et al. (2019).

support the conversion to renewable energy, there are critics and skeptics. Criticisms often revolve around issues such as economic viability, technological readiness, and the potential environmental impact of certain renewable technologies. Renewable energy faces challenges due to its high costs and unintended environmental consequences. Moreover, it encounters difficulties in meeting the increasing global energy demand without causing integration issues in existing supply systems, The creation of supporting facilities for renewable energy sources demands a substantial amount of energy and raw materials, including some that are rare, and it presents challenges in terms of network grids and recycling. It's essential to distinguish between legitimate concerns backed by scientific evidence and unfounded skepticism.

The scientific evidence strongly supports the benefits of renewable energy in addressing climate change and promoting sustainability. However, ongoing research is essential to address challenges and refine technologies for a more efficient and widespread adoption of renewable energy.

Key elements of the low-carbon energy transition include:

- **Energy efficiency.** Improving the efficiency of energy use in several segments, including industries, transportation, and buildings. This involves implementing technologies and practices that reduce energy consumption and waste.

The enhancement of energy efficiency plays a critical role in mitigating the ecological and societal costs of economic growth. This importance is especially pronounced in areas such as buildings, industry, and transportation, where it has the potential to result in substantial energy savings. Efficient energy use is also vital for promoting energy conservation. The significance of energy is end-use efficiency, which can deliver a higher level of desired service per unit of energy consumed and has the potential to markedly decrease energy intensity.

Energy efficiency is one of the significant parts of a low-carbon transition to energy, where it reduces greenhouse gas emissions, minimizes resource consumption, and contributes to a more sustainable society. The definition of energy efficiency is giving a desired output of something by optimum use of input and minimizing waste.

In the perspective of the low-carbon energy transition, several key aspects highlight the significance of energy efficiency:

1. *Reducing energy waste.* Energy efficiency aims to minimize the amount of energy wasted in various processes, systems, and technologies. By improving the efficiency of energy use, a smaller amount energy is required to complete the same tasks, resulting in lower overall energy consumption.

The concept of reducing energy waste is a critical aspect of energy efficiency in the perspective of the low-carbon energy transition. According to Andersson et al. (2017), Ruparathna et al. (2016), Galvão et al. (2017), Ashrafian (2023), Palm and Bryngelsson (2023), Xu et al. (2018), Perrone et al. (2023), Bayendang et al. (2023), Liu et al. (2023), Ali et al. (2023), and Song et al. (2023), the concept is shown in Figure 1.4 and Figure 1.5.

In general, energy efficiency is a multi-faceted approach that includes technological, behavioral, and policy measures in the optimization of energy used in various sectors. It runs under the dogma of the low-carbon energy transition and, as such, is central to the achievement of environmental sustainability, with very high leverage in mitigating the outcomes of climate change.

2. *Improving industrial processes.* The industries consume a huge proportion of the energy. In industrial processes, energy efficiencies are measured through process improvements on the equipment and optimal manufacturing, including technologies for waste heat that are recovered and reused. Technological improvements enable energy savings in an industry's competitiveness. In the low-carbon energy transitions, the importance of improving industrial processes has been major in relation to energy efficiency. Most industrial activities account for a high proportion of global energy and greenhouse gas emissions. Improving efficiency of the industrial processes will mean reduced waste of energy; hence, it is in itself a contribution to sustainable environmental development.

Here are several strategies and aspects involved in improving industrial processes (Najjarpour & Tousi, 2024; Sukhawatthanakun et al., 2023):

- Energy audits and process optimization.
- Advanced process control systems.
- High-efficiency equipment and machinery.
- Cogeneration and combined heat and power (CHP).

- Waste heat recovery.
- Process integration and pinch analysis.
- Energy-efficient lighting and HVAC systems.
- Lean manufacturing and resource efficiency.
- Process heat integration.
- Materials and product innovation.
- Employee training and engagement.
- Lifecycle assessment.
- Collaborative research and development.
- Regulatory compliance and certification.

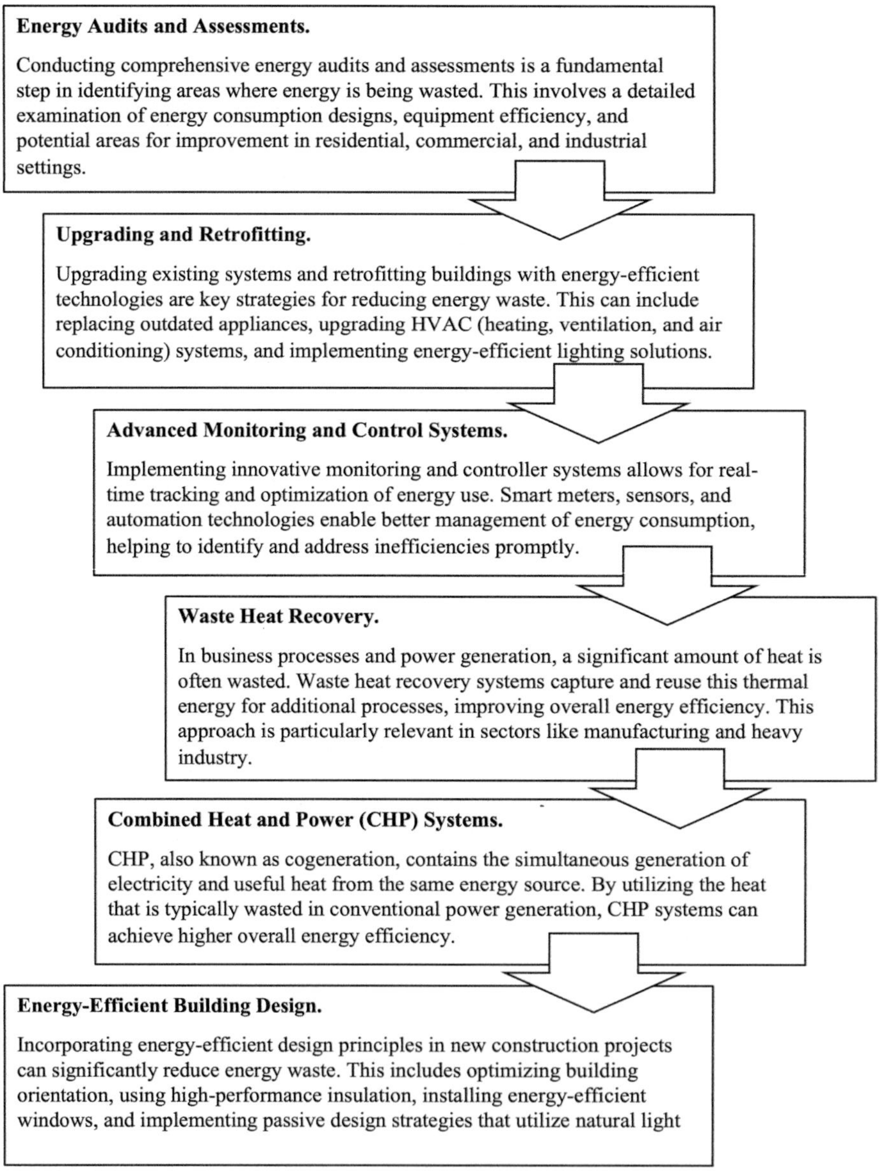

FIGURE 1.4 The concept of reducing energy waste (1).

Source: created by authors.

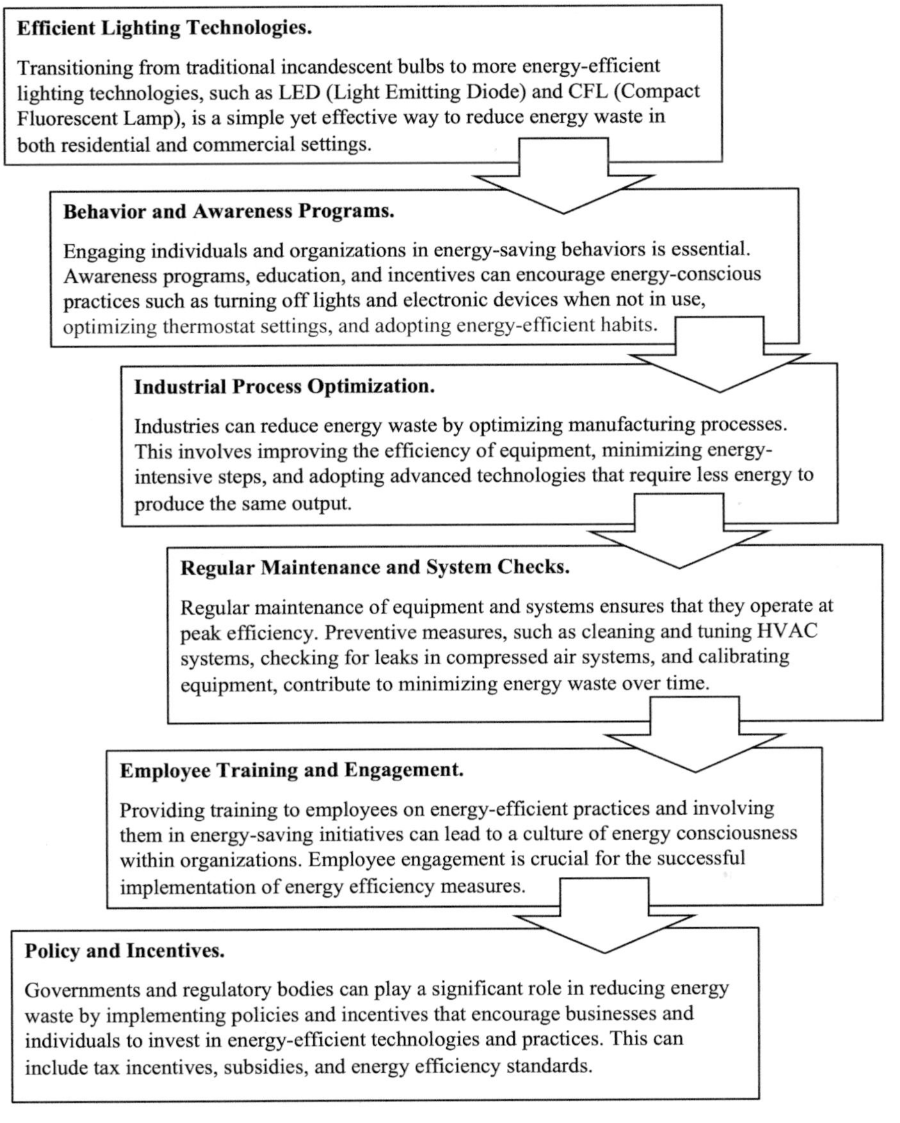

FIGURE 1.5 The concept of reducing energy waste (2)

Source: created by authors.

Conducting detailed energy audits helps industries identify areas of inefficiency. Through process optimization, companies can streamline their operations to reduce energy consumption and improve overall efficiency (Najjarpour & Tousi, 2024). Implementing advanced process control systems enhances the precision and efficiency of industrial processes. Automation technologies, sensors, and real-time monitoring allow for better control of variables, minimizing energy waste and improving production outcomes. Upgrading to high-efficiency equipment, including motors, pumps, compressors, and industrial boilers, reduces energy consumption. Investing in state-of-the-art machinery often pays off through long-term energy savings (Wang et al., 2023). Cogeneration and CHP systems allow industries to simultaneously produce electricity and capture waste heat for use in heating or other industrial processes. This integrated approach increases overall energy efficiency and reduces emissions. Waste heat recovery systems capture and reuse thermal energy generated during industrial processes. This recovered heat can be utilized for additional heating or power generation, minimizing the need for supplemental energy sources. Implementing energy-efficient lighting and heating, ventilation, and air conditioning (HVAC) systems in industrial facilities contributes to energy savings. Technologies like LED lighting and advanced HVAC controls enhance efficiency without compromising productivity. Adopting lean manufacturing principles involves minimizing waste, including energy waste, in production processes (Maware & Parsley, 2022). Resource efficiency measures focus on using raw materials more effectively, reducing waste generation, and optimizing resource consumption. Collaborative efforts among industry stakeholders, research institutions, and government agencies can drive research and development initiatives focused on improving industrial processes. This can lead to the acceptance of innovative tools and best practices. Governments often set energy efficiency standards and regulations for industrial processes. Complying with these standards and obtaining certifications can drive industries to adopt energy-efficient technologies and practices.

3. *Building and Construction Practices.* Building and construction practices in themselves are, therefore, very key to the context of energy efficiency and, further, within low-carbon energy transition. The manner in which buildings are designed, constructed, and operated has significant impacts on energy use, GHG emissions, and thus sustainability. Key aspects of building and construction practices in the context of energy efficiency are shown in Figure 1.6.

Energy efficiency in transportation is critical for reducing carbon emissions. This includes the development and adoption of fuel-efficient vehicles, electrification of transportation, improvements in public transportation systems, and the rise of sustainable urban planning to reduce the need for extensive travel (Popescu, 2022). Governments and regulatory bodies play a key role in promoting energy efficiency through policies and rules. These may include energy efficiency standards for appliances, building codes that require energy-efficient projects, and incentives for businesses and individuals to adopt energy-efficient practices. Educating consumers about the welfares of energy efficiency and encouraging energy-conscious behavior is crucial. This can include initiatives to promote energy-saving practices at home, such as using energy-efficient appliances, turning off lights when not needed, and adopting energy-conscious lifestyles (Soares et al., 2021). Energy efficiency measures often lead to economic savings by reducing energy costs for businesses and individuals. Simultaneously, these measures contribute to environmental conservation by lowering carbon emissions, air and water pollution, and the overall ecological footprint.

- **Carbon capture and storage (CCS)**. Carbon capture and storage (CCS) is an essential technology for the low-carbon energy transition, whereby emissions of carbon dioxide should be captured from industrial processes and power generation. The captured CO_2 produced from industrial processes and power generation is transported and finally stored to avoid possible emissions into the atmosphere.

Carbon capture and storage, when deployed effectively, can contribute significantly to achieving climate goals by reducing emissions from industries and power generation (Kim et al., 2021). However, its successful implementation requires a combination of technological advancements, supportive policies, financial incentives, and effective engagement with stakeholders. Ongoing research and continuous improvement in CCS technologies are essential to realizing its full potential in the low-carbon energy transition.

Some key elements of carbon capture and storage according to Baroudi et al. (2021) are presented in Figure 1.7.

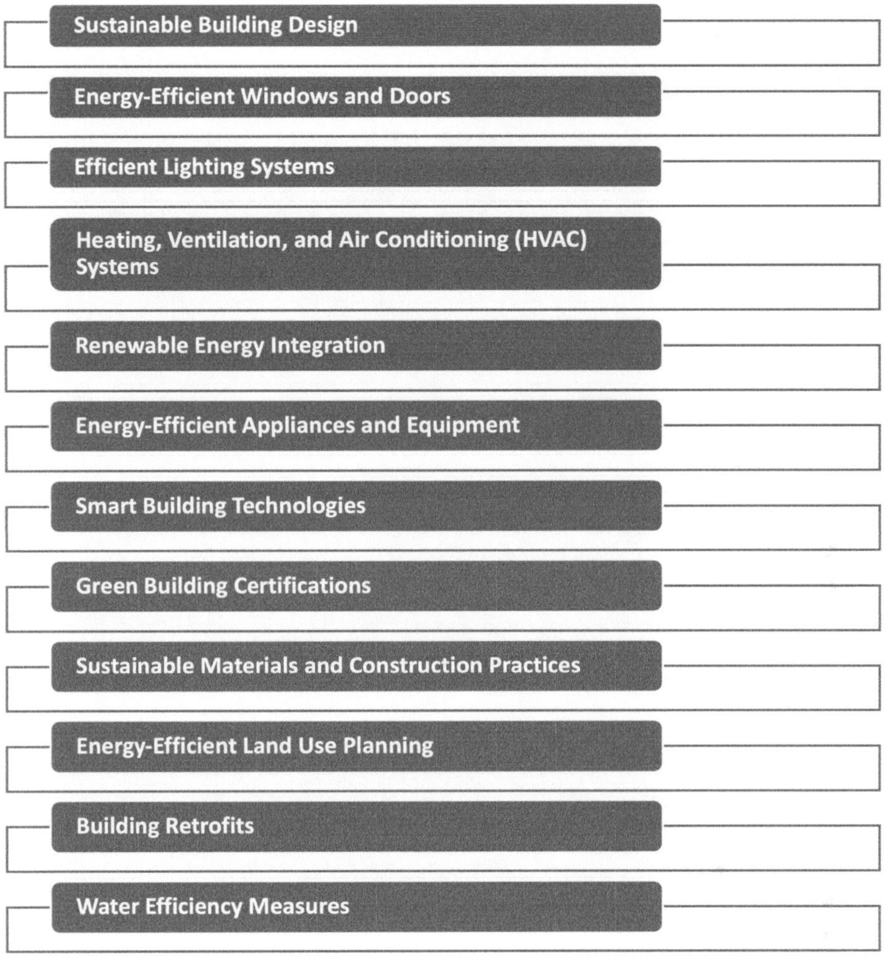

FIGURE 1.6 Key aspects of building and construction practices in the context of energy efficiency.

Source: created by authors according to Istil et al. (2023), Madushika et al. (2023).

Next, the elements of the low-carbon energy transition:

- **Smart grids and energy storage.** The other very important element in the energy transition to low-carbon technology is the smart grid, which gives a high technological and interconnected frame in managing the electricity grids with the ability to optimize them. Their incorporation supports the transition to a more sustainable and resilient energy system by accommodating renewable energy and efficient energy use and enabling consumers to actively participate in the changing energy landscape (Dileep, 2020). Energy storage is a part of the low-carbon energy transition and provides versatile and flexible solutions for the issues of both present and future that are thrown up by renewable energy integration and evolving dynamics in modern energy systems (Saraji & Streimikiene, 2023).

The key features and components of the smart grid are shown in Figure 1.8.

Smart grids include a smart metering system that allows two-way communication between consumers and utilities. These meters provide real-time information on energy consumption, enabling proper billing, demand response programs, and better energy management. Smart grids smoothly

FIGURE 1.7 Elements of carbon capture and storage.

Source: created by authors.

coordinate distributed energy resources such as solar panels, wind turbines, energy storage systems, and electric vehicles (Panda et al., 2022). This integration allows the effective use of renewable energy and increases the flexibility of the grid. Automation technologies make it possible to carry out remote monitoring and control of diverse components within the grid. The automated switches, sensors, and control apparatus further improve grid reliability, lessen the duration of outages, and optimize power distribution. Automation technologies enable the remote monitoring and control of various components within the grid. Automated switches, sensors, and control devices enhance grid reliability, reduce outage durations, and optimize energy distribution. Smart grids leverage advanced data analytics to process large volumes of data generated by sensors and smart meters. Predictive modeling helps anticipate grid behavior, optimize energy flows, and identify potential issues before they escalate (Babayomi et al., 2023). Given the increased digital connectivity, smart grids implement robust

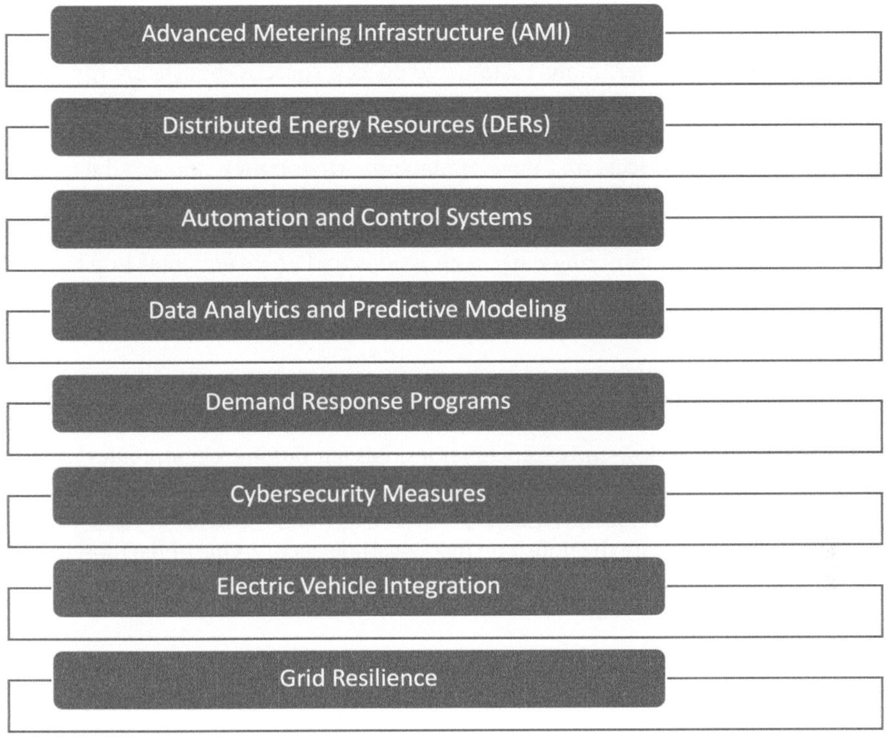

FIGURE 1.8 Key features and components of smart grid.

Source: created by authors.

cybersecurity measures to protect against cyber threats and ensure the integrity and reliability of the grid. Smart grids accommodate the integration of electric vehicles (EVs) by providing charging infrastructure and implementing smart charging strategies. This ensures optimal utilization of electricity resources and supports the growth of the electric transportation sector.

Energy storage involves capturing and retaining energy for future use, enabling a separation between energy generation and consumption. Within the framework of the low-carbon energy transition, energy storage technologies are essential for managing the variability of renewable energy sources like solar and wind. These technologies store surplus energy during periods of high availability and release it when demand surpasses the immediate supply (Sánchez et al., 2022). The main objective is to improve grid stability, reliability, and flexibility, thereby facilitating the incorporation of clean and renewable energy into the overall energy system.

- **Policy and regulatory measures.** Policy and regulatory measures are crucial in directing the course of the low-carbon energy transition. Governments and regulatory agencies create frameworks, standards, and incentives to encourage the adoption of clean energy technologies, lower greenhouse gas emissions, and support sustainable practices in the energy sector (Johnstone et al., 2021). These measures are vital for establishing a supportive environment that promotes innovation, investment, and the integration of low-carbon technologies.
- **Research and innovation.** Research and innovation are essential components of the low-carbon energy transition, propelling the creation and implementation of advanced technologies, sustainable practices, and innovative solutions to tackle climate change and environmental sustainability challenges. Within the energy sector, research and innovation are pivotal for the advancement of renewable energy technologies, enhancing energy efficiency, and discovering new methods to decrease greenhouse gas emissions.

The low-carbon energy transition entails a comprehensive overhaul of the global energy system to decrease dependence on fossil fuels, combat climate change, and enhance sustainability. This transition integrates cleaner, more sustainable energy sources, boosts energy efficiency, and adopts innovative technologies and practices across various sectors. Essential components include promoting renewable energy sources, advancing energy storage and smart grid technologies, implementing energy-efficient practices, and developing supportive policy and regulatory frameworks. Additionally, research and innovation are crucial in driving this transformation, resulting in a low-carbon energy system that is resilient, environmentally responsible, and capable of meeting increasing energy demands while minimizing carbon emissions and environmental impact. The ultimate goal of the low-carbon energy transition is to achieve a more sustainable and resilient energy future, balancing economic development with environmental stewardship (Peng et al., 2023).

1.2 POLICY DOCUMENTS OF LOW-CARBON ENERGY TRANSITION

Policy documents related to the low-carbon energy transition can vary by region, country, and jurisdiction as different governments and organizations develop their strategies and plans to address climate change and promote sustainable energy. However, examples can be given of general documents and initiatives that are commonly associated with the transition to low-carbon energy. One of the most important documents is probably the **National Climate Action Plans**. Some examples from European countries:

1.2.1 Germany

For example, Germany's Climate Action Plan 2050 (Federal Ministry for the Environment, Nature Conservation, Building and Nuclear Safety, 2016) outlines a way towards cutting at least 55% of greenhouse gas emissions by the country before 2030 and at least 85% by 2050. This plan is aimed at an energy transition (Energiewende) to be bound for greater efficiency in using energy and increased development of renewable sources of energy, while phasing out coal. The measures are undertaken at the sectoral level in the plan to deliver on the country's climate commitment under the Paris Agreement, including energy, built environment, transport, and industry and business. For both agriculture and forestry, specific emissions reduction targets are set by 2030, with guidance provided towards that end. It also allows for the mechanisms that will follow up on the progress and make the needed adjustments in the timetable of the five-yearly review of the NDCs envisaged by the Paris Agreement. It targets attaining carbon neutrality by 2050, a reduction in total emissions by 55% by 2030 with respect to the 1990 level.

The following concrete measures are outlined by the plan to achieve these targets:

1. Setting up a government commission on how to meet the challenge of fighting climate change while ensuring growth and development.
2. Increasing energy performance requirements for new buildings and buildings undergoing major renovations.
3. Developing an Integrated Comprehensive Strategy to Reduce Transport Sector Emissions.
4. Establishing a research and development program for reducing greenhouse gases from industrial processes.
5. Completely implementing all applicable regulations on fertilizer use.
6. Increasing carbon sequestration through afforestation.
7. Strengthening economic incentives, such as taxes, to induce polluters to reduce emissions.

1.2.2 France

The Integrated National Energy and Climate Plan for France (2020) is a highly detailed, cross-sectoral national plan that sets out the emissions target of France for non-ETS and non-LULUCF sectors (ESR target) pursuant to Regulation (EU) 2018/842, which aims at a reduction of 37% of these emissions by 2030 against 2005 levels. France's regional greenhouse gas emissions reduction targets are in line with its EU and international commitments. These targets include:

1. Achieving carbon neutrality by 2050, zero net emissions of a country, in line with the government's climate plan published in July 2017 and established in law.
2. Reducing greenhouse gas emissions by 40% by 2030 compared to 1990 levels, following the adoption of the Law on Energy Transition for Green Growth in 2015.
3. Complying with carbon budgets determined by the Secretary of State for the short and medium term. Such carbon budgets are allowed to specify emissions caps during successive five-year periods, with the exception of the first period, which is of four years from 2015 to 2018.

This plan is organized into two basic guiding national energy and climate programming and governance documents:

- The multi-year energy program defines the priority directions for the activities of public bodies in the energy sector over the next decade, breaking this down into two five-year periods. It regards all types of energy and many aspects of energy policy, e.g., demand-side management, promotion of renewable sources of energy, security of energy supply, limitation of the cost of energy, and ensuring balanced development of networks.
- The national low-carbon strategy serves as France's roadmap for leading climate change mitigation efforts. It offers the long-term transition of each sector of the economy towards a low-carbon economy and refers to carbon budgets in terms of national level greenhouse gas emission reduction targets for the short and mid term, with an ultimate goal of carbon neutrality by 2050.

1.2.3 Sweden

Sweden's Climate Policy Framework (Ministry of the Environment and Energy, 2017) aims for the country to achieve net-zero emissions by 2045. The plan includes measures to increase renewable energy, promote sustainable transport, and enhance energy efficiency in buildings. The purpose of implementing this framework is to establish a clear and cohesive climate policy that provides long-term signals to the market and various stakeholders. It was endorsed by Parliament with widespread support from political parties and is designed to remain robust even through political changes. This climate policy framework represents a significant reform in Sweden's approach to climate issues and serves as a crucial element in the country's commitment to align with the goals of the Paris Agreement.

1.2.4 United Kingdom

The UK's Climate Change Act (Climate Change Committee, 2008) establishes legally binding carbon budgets, and the country has set a target to achieve net-zero emissions by 2050. The UK's Ten-Point Plan for a Green Industrial Revolution outlines key areas, including offshore wind, hydrogen, and zero-emission vehicles. The Climate Change Act 2008 is up to date with all changes known to be in force on or before 4 February 2024 (Climate Change Committee, 2024).

1.2.5 Norway

Norway's Climate Agreement focuses on reducing emissions in accordance with the Paris Agreement. Norway's 2021–2030 climate action plan (Norwegian Ministry of Climate and Environment, 2021) is part of its comprehensive framework for climate policies and commitments. This plan demonstrates Norway's commitment to global climate change efforts and aligns with international agreements like the Paris Agreement. Norway aims to significantly cut greenhouse

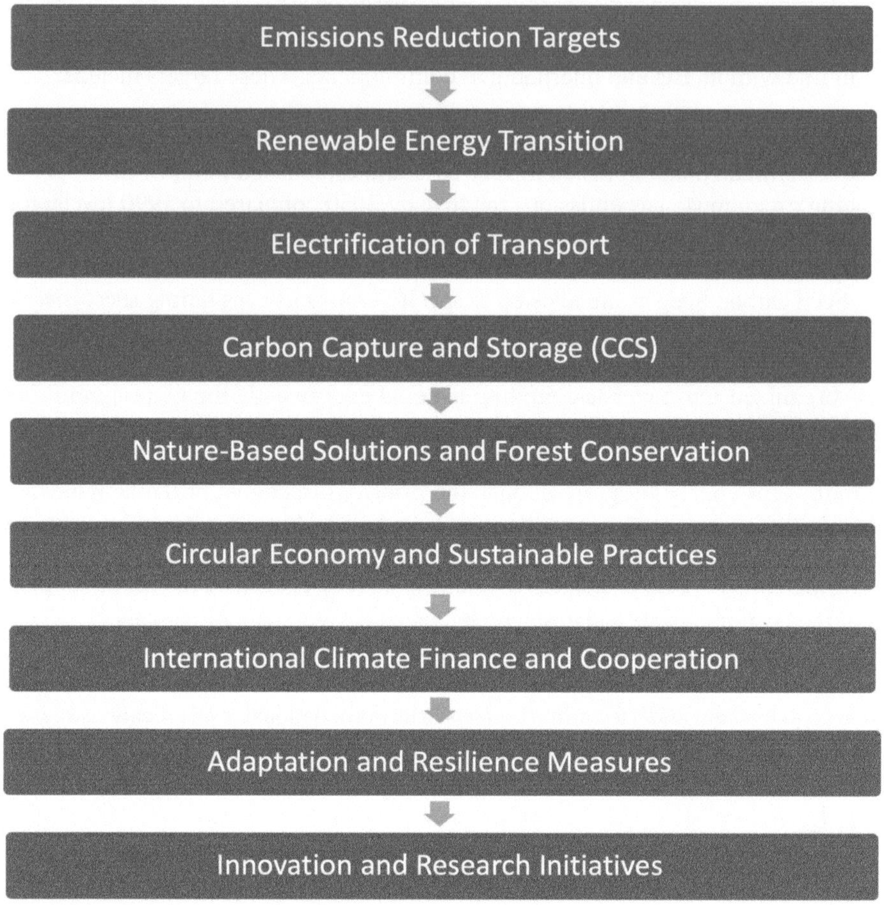

FIGURE 1.9 Key components of Norway's Climate Action Plan (2021–2030).

Source: created by authors according to Norwegian Ministry of Climate and Environment (2021).

gas emissions, promote sustainable practices, and transition to a low-carbon, climate-resilient economy. Key aspects of Norway's Climate Action Plan (2021–2030) are shown in Figure 1.9.

The climate action plan emphasizes the shift to renewable energy sources. Norway has historically relied on hydropower for electricity. The plan likely includes steps to further increase renewable energy capacity, such as wind and solar power, to lessen reliance on fossil fuels in the energy sector (Norwegian Ministry of Climate and Environment, 2021).

Norway has led the way in promoting electric vehicles (EVs) (OECD, 2022), and the climate action plan likely includes measures to speed up the electrification of the transportation sector. This might involve incentives, EV charging infrastructure development, and policies to phase out internal combustion engine vehicles.

1.2.6 Denmark

Denmark has solidified its commitment to ambitious climate action by enacting the 2020 Climate Act (The Danish Ministry of Climate, Energy and Utilities, 2020b). This legislation establishes a goal to decrease Denmark's greenhouse gas emissions by 70% in 2030, relative to 1990 levels, and it aims for climate neutrality by 2050 at the latest. Additionally, the act mandates the establishment of a new national climate target every five years with a 10-year outlook. Importantly, these new targets must not be less ambitious than the previously agreed-upon target, aligning with the principle of preventing regression outlined in the Paris Agreement. The Climate Programme 2020 (The Danish Ministry of Climate, Energy and Utilities, 2020a) marks a significant milestone as the inaugural program developed following the enactment of the Climate Act in June 2020. It serves as the first comprehensive initiative detailing how the Minister for Climate, Energy, and Utilities intends to concretely realize the likelihood of meeting the targets established in the Climate Act.

Key components of Denmark's 2020 Climate Act are shown in Figure 1.10.

One of the central objectives of the Climate Act was to achieve carbon neutrality by 2050. This ambitious target required a holistic and transformative approach to decarbonize various sectors of the economy, including energy sector, transport, and business. The Climate Act likely included specific plans and measures for reducing emissions across key sectors, such as energy, transportation, agriculture,

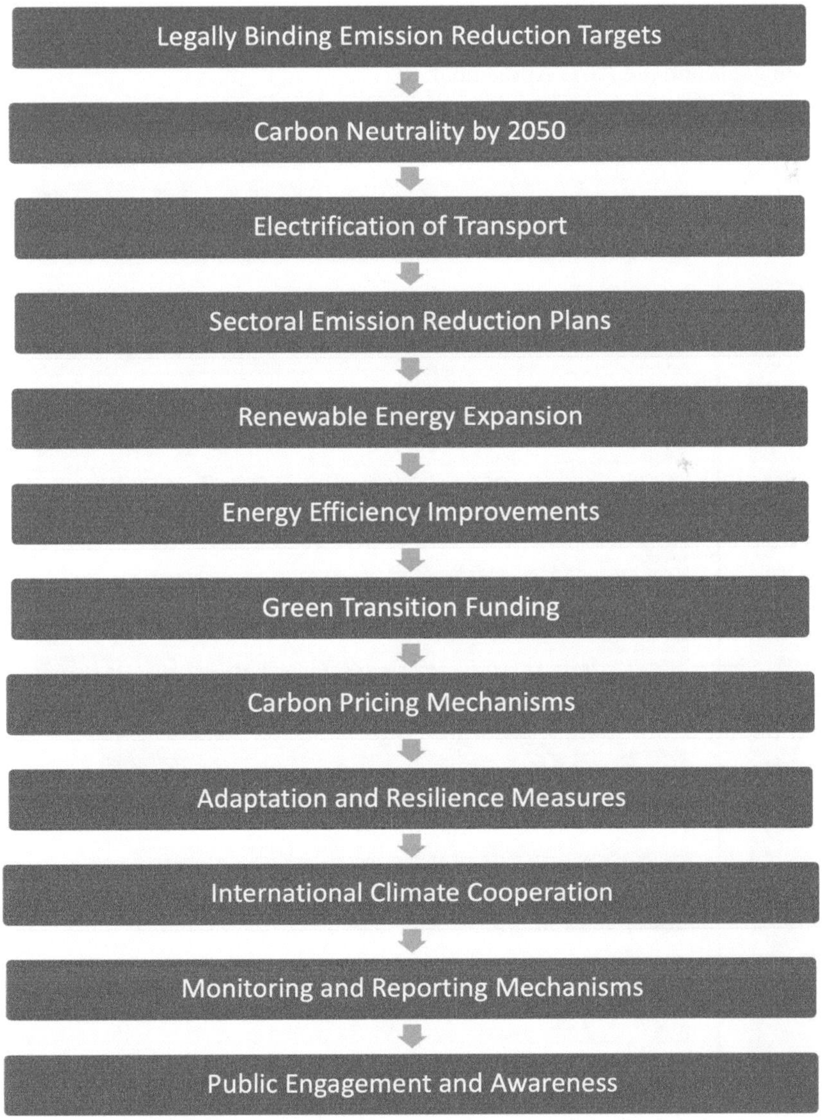

FIGURE 1.10 Key components of Denmark's 2020 Climate Act.

Source: created by authors according to The Danish Ministry of Climate, Energy and Utilities (2020a, 2020b).

and industry. Sectoral targets and strategies were likely outlined to ensure a comprehensive and balanced approach to emissions reduction. Denmark's Climate Act included provisions related to carbon pricing or market-based mechanisms to incentivize emissions reduction. This involves the establishment of a carbon market or other economic instruments to internalize the cost of carbon emissions.

1.2.7 Netherlands

The Netherlands has been actively working to reduce its greenhouse gas emissions and transition to a low-carbon economy through its Climate Agreement (National Climate Agreement, 2019). The National Climate Agreement builds upon the foundation of the Energy Accord; however, instead of solely addressing energy policy, it encompasses all greenhouse gas emissions. In 2023, the emission reduction targets outlined in the National Climate Agreement were further strengthened through the Coalition Agreement 2021–2025.

The Netherlands has committed to reducing its greenhouse gas emissions by 49% by 2030 compared to 1990 levels (National Climate Agreement, 2019). This target aligns with the European Union's climate goals and the Paris Agreement.

Key aspects of the National Climate Agreement shown in Figure 1.11.

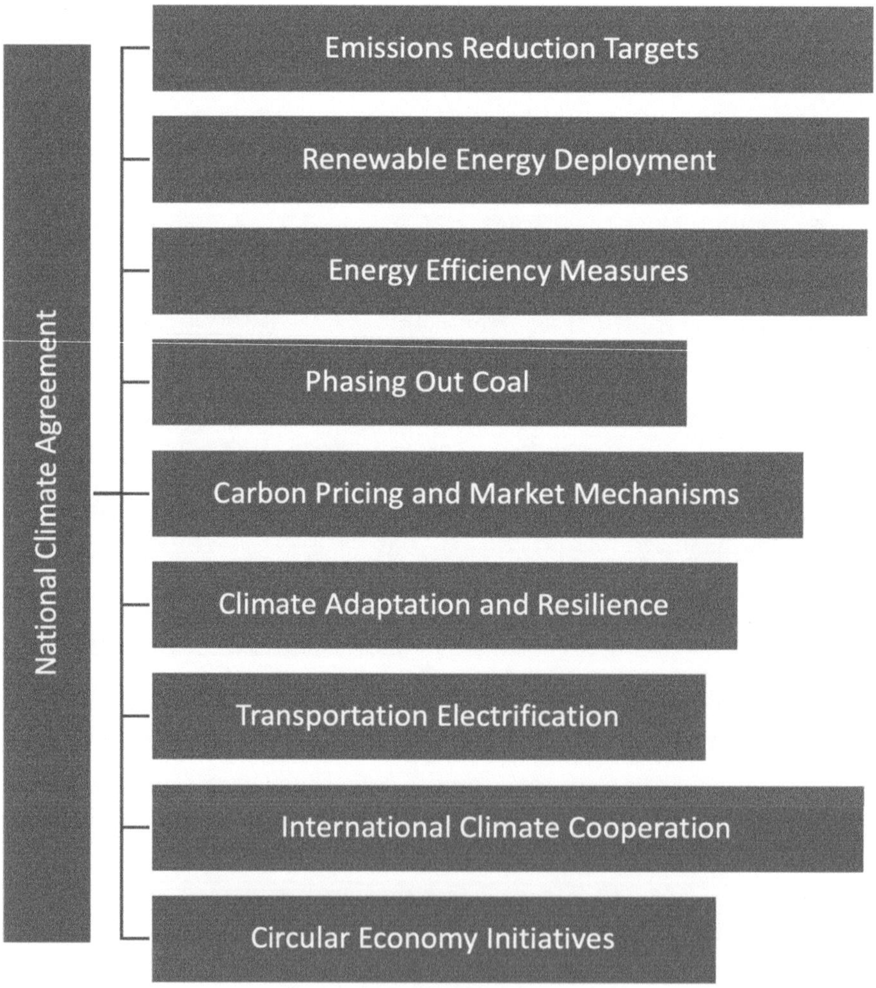

FIGURE 1.11 Key aspects of the National Climate Agreement.

Source: created by authors according to the National Climate Agreement (2019).

The Climate Agreement focuses on increasing the share of renewable energy sources like wind and solar power in the energy mix to meet climate goals. The Netherlands is adopting various energy efficiency measures across multiple sectors, including buildings, industry, and transportation, aiming to reduce energy consumption and lower greenhouse gas emissions. As part of the agreement, there are plans to phase out coal-fired power plants by 2030 (National Climate Agreement, 2019), aiding the shift to cleaner energy sources, which will help reduce emissions and improve air quality.

Alongside mitigation efforts, the Netherlands is prioritizing climate adaptation and resilience measures to tackle the impacts of climate change, such as sea-level rise, flooding, and extreme weather events. The Climate Agreement also incorporates circular economy principles, focusing on sustainable resource use, waste reduction, and recycling, which contribute to emission reductions and promote environmental sustainability.

The Netherlands' climate action plan involves engaging stakeholders, including businesses, civil society organizations, and local communities, to ensure widespread support and participation in climate initiatives. The Climate Agreement represents a comprehensive strategy to combat climate change, reduce greenhouse gas emissions, and transition to a sustainable and resilient future. Through collaborative efforts and innovative solutions, the Netherlands is striving to achieve its climate targets and contribute to global climate action.

1.2.8 Spain

Spain has been actively tackling climate change and transitioning to a low-carbon economy through its Integrated National Energy and Climate Plan (PNIEC) and related policies. The initial 2020 PNIEC (Integrated National Energy and Climate Plan 2021–2030, 2020) aimed for a 23% reduction in greenhouse gas emissions by 2030 compared to 1990 levels. However, the 2023 update has significantly raised this target to a 32% reduction. Additionally, it now aims for renewables to make up 48% of final energy consumption, with an impressive 81% in the electricity sector, and it seeks to improve energy efficiency by 44%.

As a result, the updated PNIEC expands its planned policies and measures from 78 in the 2020 version to 107. It introduces new specific actions for sectors such as railways, aviation, and navigation, rural development, better integration of renewables with the environment and territory, cybersecurity, local electricity markets, gender perspective, capacity markets, and life cycle analysis of buildings, among others. These measures are designed to strengthen the objectives and achieve greater social, economic, and environmental benefits.

All these examples are directly related with climate change mitigation through the low-carbon transition. In addition to those aforementioned, there are more documents regulating the transition of individual countries to a low-carbon economy. Some of the essential ones are shown in Figure 1.12, without country breakdown. This is because individual countries use one or more policy documents, combine them with each other, use them to complement each other, or use them separately.

Governments frequently publish roadmaps that detail plans to boost the proportion of renewable energy in the overall energy mix. These documents usually outline targets for renewable energy capacity, incentives for renewable energy projects, and strategies for integrating renewables into the grid. Energy efficiency plans describe measures and policies designed to enhance energy efficiency across various sectors, including industry, transportation, and buildings. These plans often feature targets, standards, and incentives to promote energy-efficient practices.

Some regions implement carbon pricing mechanisms as part of their low-carbon energy transition, such as carbon taxes or cap-and-trade systems. Legislation or policy documents associated with carbon pricing outline the rules and regulations governing these mechanisms. Governments may also establish programs to incentivize and support investments in clean energy projects. These programs often include financial incentives, subsidies, grants, and other mechanisms to encourage private and public investment in renewable energy and low-carbon technologies.

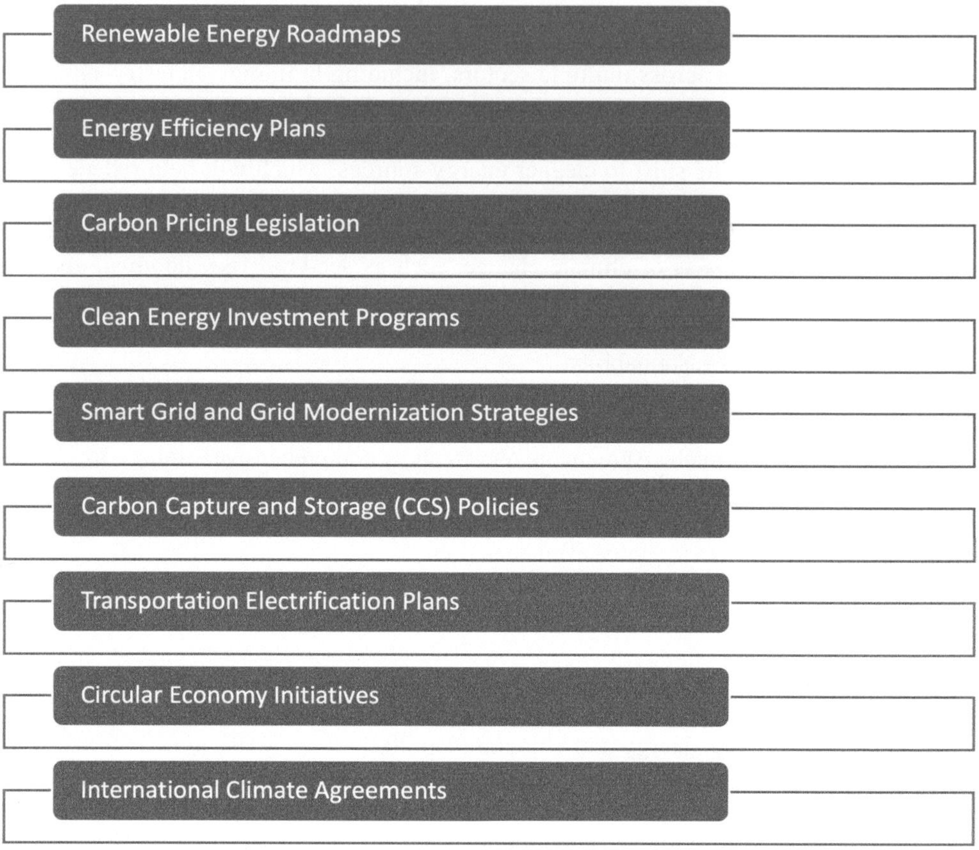

FIGURE 1.12 Documents regulating the transition to a low-carbon economy.

Source: created by authors.

To enhance the efficiency and reliability of energy infrastructure, governments may release documents outlining plans for smart grid deployment and overall grid modernization. These strategies often focus on incorporating digital technologies, energy storage, and demand response mechanisms.

Policies related to carbon capture and storage outline strategies for capturing and storing carbon emissions from industrial processes and power generation. These documents may include regulations, incentives, and plans for the deployment of CCS technologies. Governments may release plans specifically targeting the electrification of transportation. This includes policies related to electric vehicles (EVs), charging infrastructure, and other measures to reduce emissions from the transportation sector. International agreements, such as the Paris Agreement, represent global efforts to combat climate change. Countries may release documents detailing their commitments and actions within the framework of these agreements.

1.3 TARGETS AND INDICATORS OF LOW-CARBON TRANSITION

Targets and indicators of low-carbon transition are specific goals and metrics used to measure progress and success in reducing carbon emissions and transitioning to a more sustainable and climate-friendly economy. These targets and indicators provide a framework for monitoring and evaluating efforts to mitigate climate change and achieve carbon neutrality.

Targets commonly used in the context of low-carbon transition are shown in Figure 1.13.

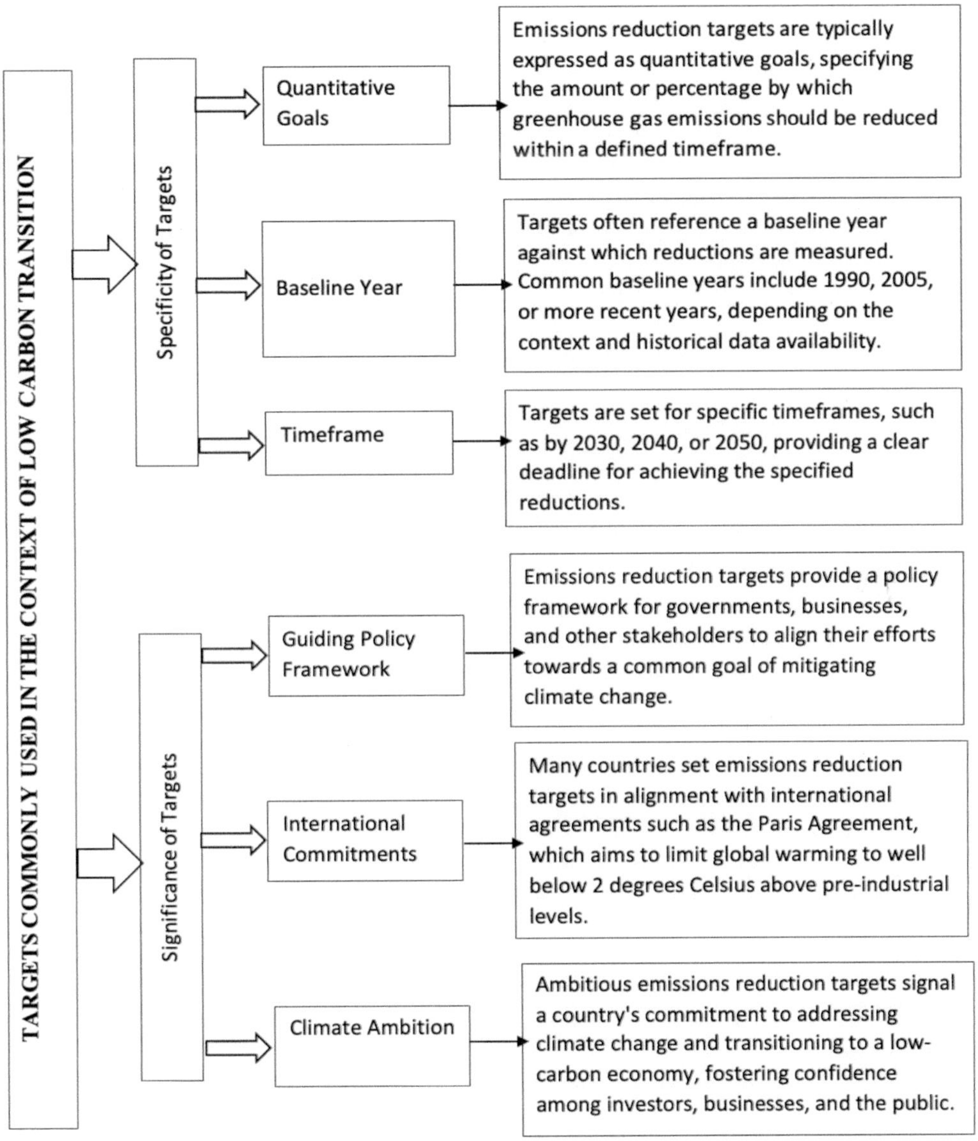

FIGURE 1.13 Targets commonly used in the context of low-carbon transition.

Source: created by authors according to Hainsch et al. (2022), Sovacool et al. (2020).

Targets often specify the percentage by which greenhouse gas emissions should be reduced relative to a baseline year. For example, a target might aim for a 30% reduction in emissions by 2030 compared to 2005 levels. Some targets set absolute emission reduction goals, aiming to achieve a specific level of emissions by a certain year, irrespective of baseline emissions. For instance, a target might aim to achieve emissions of no more than X metric tons of CO_2 equivalent by 2050. Targets are often referenced against a baseline year, typically chosen to provide a historical comparison and to assess progress over time. Common baseline years include 1990, 2005, or more recent years depending on data availability and relevance to historical emissions trends. In some cases, targets may use an adjusted baseline to account for changes in emissions due to factors such as population growth, economic development, or changes in land use. This adjusted baseline provides a more accurate reflection of emissions trends and the effort required to achieve reductions.

Targets are set for specific timeframes, ranging from short-term goals (e.g., 2025, 2030) to long-term objectives (e.g., 2040, 2050). Short-term targets provide milestones for immediate action, while long-term targets offer a vision for sustained emission reductions and climate resilience. In addition to long-term goals, many climate plans include interim targets at regular intervals (e.g., every five or 10 years) to track progress and adjust strategies as needed. These interim targets help ensure accountability and maintain momentum towards long-term objectives.

Emissions reduction targets may be set for specific sectors of the economy, such as energy, transportation, industry, buildings, agriculture, and waste. Sectoral targets recognize the unique characteristics and emissions profiles of each sector and tailor mitigation strategies accordingly. While sectoral targets focus on specific areas, they are often integrated into broader climate plans to ensure coherence and synergy across sectors. Integrated approaches recognize the interconnectedness of emissions sources and the need for holistic solutions.

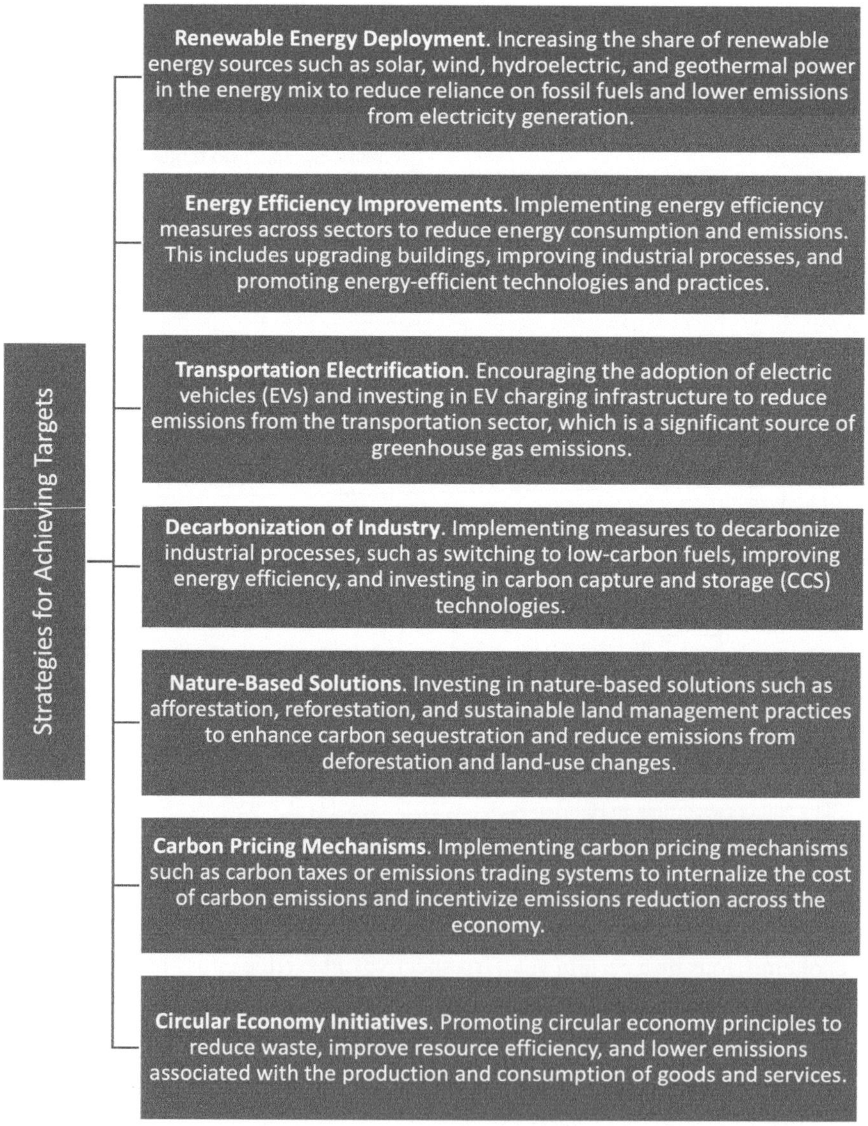

FIGURE 1.14 Strategies for achieving targets.

Source: created by authors.

Some climate agreements and policies recognize the principle of common but differentiated responsibilities, acknowledging that historical emissions and capacities vary among countries. As such, targets may be differentiated based on countries' historical emissions, development levels, and capabilities. Setting equitable targets requires balancing the need for global emissions reductions with considerations of fairness and justice. Equitable burden-sharing principles aim to ensure that all countries contribute to climate action according to their respective capabilities and responsibilities.

Climate agreements often include provisions for periodic reviews of targets and commitments to assess progress, review scientific evidence, and update policies as needed. These reviews provide opportunities to strengthen targets, adjust strategies, and incorporate new knowledge and technologies. Many climate agreements incorporate ratcheting mechanisms, whereby countries are encouraged to progressively strengthen their targets over time. This iterative process ensures that targets remain ambitious and aligned with the latest scientific understanding and evolving climate risks.

The specificity of emissions reduction targets encompasses quantitative goals, baseline years, timeframes, sectoral considerations, equity principles, and mechanisms for review and ratcheting. By defining targets with clarity and precision, policymakers can provide a roadmap for effective climate action and drive meaningful progress towards a low-carbon and climate-resilient future.

Targets of the low-carbon transition are specific goals set by governments, organizations, and stakeholders to reduce greenhouse gas emissions and combat climate change. These targets provide a framework for guiding policies, actions, and investments towards achieving a more sustainable and climate-friendly economy. Reaching these goals requires coordinated efforts from both developed and developing countries, each with their own set of challenges and opportunities.

For developed countries, the emphasis may be on transitioning from existing fossil fuel infrastructure to renewable energy sources, enhancing energy efficiency, and investing in new technologies. These nations typically have more resources for research and development, and their policies can establish standards that influence global markets.

Developing countries, on the other hand, may focus on leapfrogging to cleaner technologies without going through prolonged phases of heavy fossil fuel dependency. They might emphasize sustainable development practices, capacity building, and receiving international support in the form of funding and technology transfer.

Overall, achieving the low-carbon transition targets involves:

- Setting clear, achievable goals for emission reductions.
- Implementing policies and regulations to promote renewable energy and energy efficiency.
- Developing and adopting innovative technologies and sustainable practices.
- Ensuring international cooperation and support, particularly for developing countries.
- Engaging all sectors of society, including businesses, communities, and individuals, to contribute to the transition.

This global effort aims to create a resilient, low-carbon future, balancing economic development with environmental sustainability.

Some key strategies for developed countries to achieve these goals are shown in Figure 1.15. Some key strategies for developing countries to achieve these goals are shown in Figure 1.16.

Developing countries face several common challenges in achieving the goals of low-carbon transition, stemming from economic, social, and environmental factors (Cantarero, 2020; Kabeyi & Olanrewaju, 2022).

1.3.1 Limited Financial Resources

Many developing countries lack the financial resources needed to invest in clean energy infrastructure, energy efficiency improvements, and climate adaptation measures. Limited access to climate finance and investment capital constrains their ability to implement low-carbon projects and technologies.

FIGURE 1.15 Strategies for developed countries to achieve targets of low-carbon transition.

Source: created by authors according to Fragkos et al. (2021).

1. Technological capacity and infrastructure constraints.

Developing countries often lack the technological capacity and infrastructure necessary to deploy clean energy technologies at scale. Limited access to advanced technologies, expertise, and research and development resources hinders their ability to transition to low-carbon energy systems.

2. Energy poverty and access challenges.

Millions of people in developing countries still lack access to modern and reliable energy services, relying on traditional biomass and fossil fuels for cooking, heating, and lighting. Addressing energy poverty while transitioning to clean energy sources requires significant investment and policy interventions.

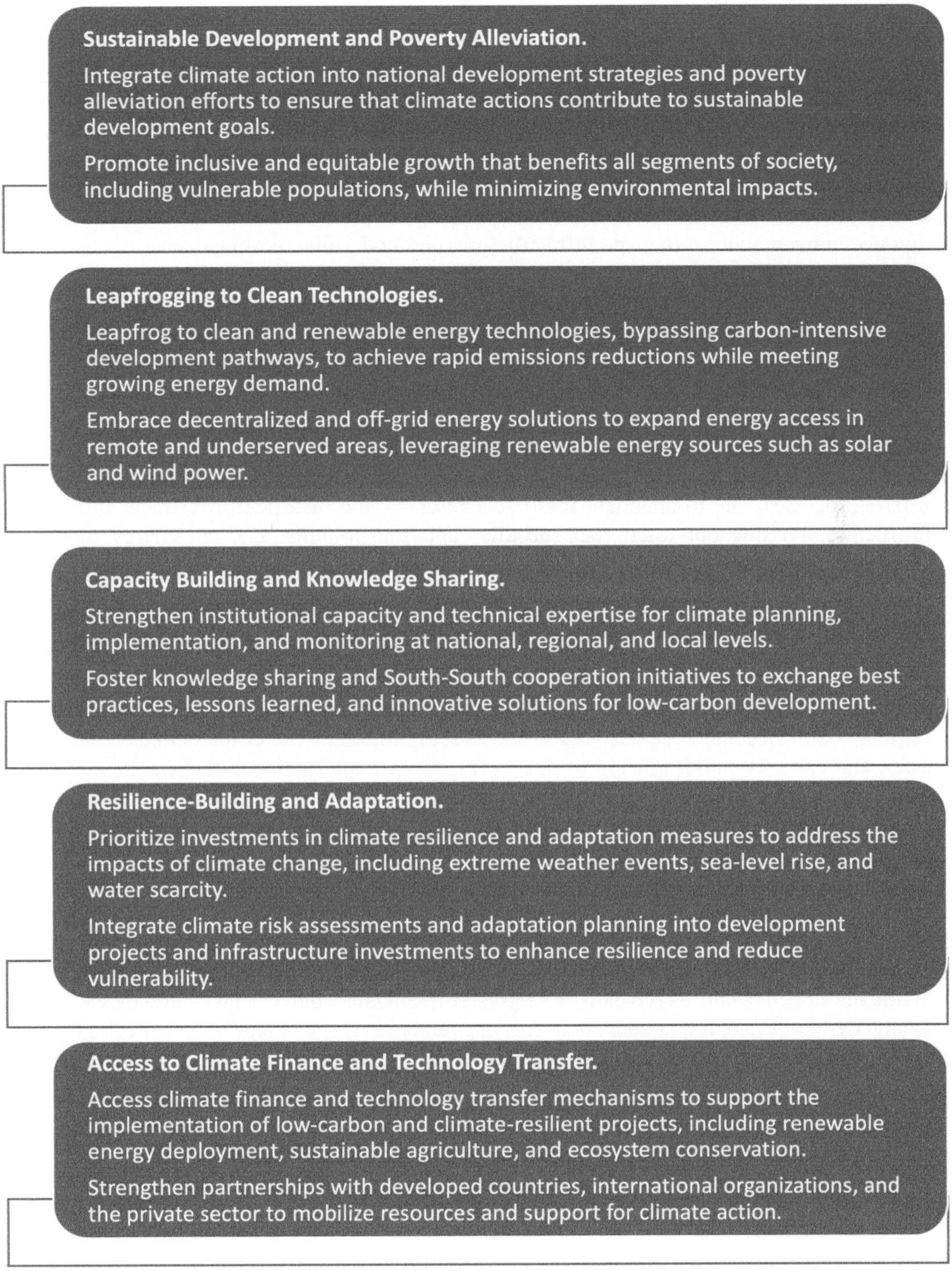

FIGURE 1.16 Strategies for developing countries to achieve targets of low-carbon transition.

Source: created by authors according to Sovacool et al. (2020).

3. Economic development priorities.

Developing countries may prioritize economic growth and poverty alleviation over environmental concerns, leading to reliance on carbon-intensive industries and energy sources for development. Balancing economic development objectives with climate objectives poses a significant challenge for policymakers.

4. Vulnerability to climate impacts.

Developing countries are often more vulnerable to the impacts of climate change, such as extreme weather events, sea-level rise, and changing precipitation patterns. Climate-related disasters can disrupt economic activities, damage infrastructure, and undermine development gains, exacerbating poverty and inequality.

5. Institutional and governance constraints.

Weak institutional capacity, governance challenges, and policy barriers can impede the effective implementation of climate policies and strategies in developing countries. Inadequate regulatory frameworks, corruption, and lack of coordination among government agencies hinder progress towards low-carbon transition goals.

6. Dependence on fossil fuels and energy subsidies.

Many developing countries rely heavily on fossil fuels for energy generation, transportation, and industrial processes, due to abundant fossil fuel reserves and historical investment in fossil fuel infrastructure. Phasing out fossil fuel subsidies and transitioning to clean energy alternatives requires political will and careful planning.

7. Limited access to clean technology and knowledge.

Access to clean energy technologies, expertise, and knowledge transfer is often limited in developing countries, hindering their ability to adopt and deploy low-carbon solutions. Technology transfer mechanisms and capacity-building initiatives are needed to bridge this gap.

8. Fragmented international cooperation and support.

Developing countries may face challenges in accessing international climate finance, technology transfer, and capacity-building support due to fragmented international cooperation, competing priorities, and donor preferences. Strengthening international partnerships and aligning support with national priorities is essential for the effective implementation of low-carbon transition initiatives.

Addressing these challenges requires coordinated efforts from governments, international organizations, the private sector, civil society, and local communities to mobilize resources, build capacity, and create enabling environments for sustainable and inclusive low-carbon development in developing countries.

While developed countries typically have greater financial resources and technological capabilities compared to developing nations, they still encounter several challenges in achieving the goals of low-carbon transition (Cantarero, 2020; Bridge et al., 2013; Sovacool et al., 2019):

1. Transitioning from legacy infrastructure.

Developed countries often have existing infrastructure and industries that are heavily reliant on fossil fuels. Transitioning away from these legacy systems to low-carbon alternatives can be complex and costly, requiring significant investment and planning.

2. Resistance to change.

Certain sectors of the economy, such as traditional energy industries and heavy manufacturing, may resist transitioning to low-carbon technologies due to concerns about economic impacts, job losses, and regulatory uncertainty. Overcoming resistance to change and fostering innovation and adaptation are key challenges.

3. Policy and regulatory uncertainty.

Changing political landscapes and policy shifts can create uncertainty for businesses and investors, impacting long-term planning and investment in low-carbon technologies and projects. Clear and stable policy frameworks are essential to provide incentives for private sector involvement in low-carbon transition efforts.

4. Addressing energy-intensive industries.

Energy-intensive industries such as steel, cement, and chemicals face unique challenges in reducing emissions while remaining competitive in global markets. Developing cost-effective technologies for decarbonizing these industries without compromising competitiveness is a significant challenge.

5. Ensuring equity and social justice.

Low-carbon transition efforts must consider the social impacts and distributional effects on vulnerable communities, workers in carbon-intensive industries, and marginalized populations. Ensuring equity, social justice, and a just transition for affected communities is a critical challenge for developed countries.

6. Balancing economic growth with emissions reduction.

Developed countries must balance the need for economic growth and competitiveness with the imperative to reduce emissions and mitigate climate change. Implementing policies that promote sustainable economic growth while achieving emissions reduction targets requires careful planning and coordination.

7. Infrastructure and land use planning.

Developing sustainable infrastructure and land use planning strategies that support low-carbon development, promote public transportation, encourage compact urban development, and protect natural ecosystems is essential but can be challenging due to competing interests and stakeholder preferences.

8. International cooperation and leadership.

Developed countries play a crucial role in international climate cooperation and must demonstrate leadership in emissions reduction efforts. Building consensus among nations, mobilizing climate finance, and fostering global partnerships for technology transfer and capacity-building are ongoing challenges.

9. Addressing climate resilience and adaptation.

While mitigation efforts are essential, developed countries also face challenges in adapting to the impacts of climate change, such as sea-level rise, extreme weather events, and changing precipitation patterns. Investing in climate resilience measures and infrastructure to protect communities and ecosystems is a priority.

10. Transitioning to circular economy models.

Shifting towards circular economy models that prioritize resource efficiency, waste reduction, and sustainable consumption and production practices is a key challenge for developed countries. Transforming consumption patterns and business models requires coordinated action across sectors and stakeholders.

Addressing these challenges requires a comprehensive and integrated approach that considers the economic, social, and environmental dimensions of sustainability. Collaboration among governments, businesses, civil society, and other stakeholders is essential to overcoming barriers and accelerating progress towards a sustainable and low-carbon future.

Indicators of low-carbon transition are metrics used to assess progress towards reducing greenhouse gas emissions, increasing energy efficiency, and transitioning to a more sustainable and climate-friendly economy. These indicators provide insights into the effectiveness of policies, initiatives, and investments aimed at mitigating climate change and achieving carbon neutrality (Mohsin et al., 2019; Mohsin et al., 2021; Streimikiene et al., 2021):

1. Greenhouse gas emissions:
   - Total emissions: tracking the total emissions of greenhouse gases, including carbon dioxide (CO_2), methane (CH_4), nitrous oxide (N_2O), and fluorinated gases (HFCs, SF_6, etc.).
   - Emissions per capita: calculating emissions on a per capita basis to account for population growth and changes in emissions intensity.
   - Emissions by sector: analyzing emissions trends by economic sectors such as energy, transportation, industry, buildings, agriculture, and waste.
2. Renewable energy deployment:
   - Renewable energy capacity: monitoring the installed capacity of renewable energy sources such as solar, wind, hydroelectric, geothermal, and biomass power.

- Renewable energy share: assessing the proportion of total energy consumption derived from renewable sources, indicating progress towards reducing reliance on fossil fuels.

3. Energy efficiency:
   - Energy intensity: measuring energy consumption per unit of GDP or economic output, indicating improvements in energy efficiency and decoupling of economic growth from energy consumption.
   - Energy consumption by sector: analyzing energy consumption trends across different sectors and identifying opportunities for efficiency improvements.
4. Carbon pricing and market mechanisms:
   - Carbon price: tracking the price of carbon emissions in carbon pricing mechanisms such as carbon taxes or emissions trading systems.
   - Emissions covered: assessing the proportion of total emissions covered by carbon pricing mechanisms and the effectiveness of these mechanisms in reducing emissions.
5. Electric vehicle adoption:
   - EV sales: monitoring the sales and adoption rates of electric vehicles (EVs) as a percentage of total vehicle sales, indicating progress towards reducing emissions from the transportation sector.
   - EV charging infrastructure: assessing the deployment of EV charging infrastructure and accessibility to support widespread EV adoption.
6. Carbon capture and storage (CCS) deployment:
   - CCS capacity: tracking the installed capacity of carbon capture and storage technologies for capturing and sequestering CO_2 emissions from industrial processes and power generation.
   - Number of CCS projects: monitoring the number of CCS projects implemented and their contribution to emissions reduction efforts.
7. Circular economy performance:
   - Recycling rates: assessing the percentage of waste materials that are recycled or reused, indicating progress towards a circular economy and reducing resource consumption.
   - Waste diversion: tracking the amount of waste diverted from landfills through recycling, composting, and other waste management practices.
8. Climate resilience and adaptation:
   - Climate risk assessments: conducting assessments to identify vulnerable areas and communities at risk from climate change impacts such as sea-level rise, extreme weather events, and water scarcity.
   - Adaptation measures: monitoring the implementation of adaptation measures such as infrastructure upgrades, land-use planning, and ecosystem restoration to enhance resilience to climate change.
9. Public awareness and engagement:
   - Public perception surveys: assessing public awareness, attitudes, and behaviors related to climate change and low-carbon transition efforts.
   - Participation in climate actions: tracking participation in climate-related initiatives, events, and advocacy campaigns to gauge public engagement and support for climate policies.
10. Policy implementation and governance:
    - Policy coherence: evaluating the alignment and effectiveness of climate policies across different levels of government and sectors of the economy.
    - Regulatory compliance: monitoring compliance with climate regulations and commitments, including reporting requirements and emissions targets.

These indicators provide a comprehensive framework for assessing progress towards low-carbon transition goals, informing decision-making, and guiding the allocation of resources and investments to accelerate the transition to a sustainable and climate-resilient future.

While indicators of low-carbon transition provide valuable insights into the progress and trends in reducing greenhouse gas emissions and transitioning to a more sustainable economy, they may not always fully reflect the real situation in countries (Mohsin et al., 2019). The *availability and quality of data* may vary across countries, impacting the accuracy and comparability of indicators. Some countries may have robust monitoring and reporting systems in place, while others may lack comprehensive data collection mechanisms. *Differences in methodologies* used to calculate indicators can affect their comparability over time and across countries. Standardized methodologies and reporting guidelines help enhance consistency and transparency in data reporting but may still result in variations in interpretation. Indicators may not capture *all relevant aspects* of low-carbon transition, leading to gaps in understanding the full scope of progress. Certain sectors or activities may be underrepresented or excluded from indicator frameworks, limiting their ability to provide a comprehensive assessment. There may be a *lag time* between the implementation of policies and initiatives and the realization of their impacts on indicators. It can take time for changes in behavior, technology adoption, and infrastructure development to manifest in measurable outcomes, leading to delays in reflecting progress accurately. *The complexity of systems*: low-carbon transition involves complex socio-economic, technological, and environmental systems, making it challenging to capture all interdependencies and feedback loops in indicator frameworks. Simplified metrics may overlook nuances and interactions within these systems, affecting the accuracy of assessments. *Policy implementation challenges*: even if indicators suggest progress in certain areas, challenges in policy implementation, enforcement, and compliance may undermine the effectiveness of efforts. Weak governance, institutional barriers, and political factors can hinder the translation of policies into tangible outcomes. *Contextual factors*: socio-economic, geographic, and cultural factors unique to each country can influence the interpretation and significance of indicators. What may be considered progress in one context may not necessarily translate to the same level of impact or relevance in another context. Indicators are *subject to interpretation* and may be influenced by subjective factors such as stakeholder perceptions, biases, and agendas. Different stakeholders may have divergent views on the significance and relevance of indicators, leading to varying assessments of progress.

Despite these challenges, indicators of low-carbon transition remain valuable tools for tracking trends, identifying areas for improvement, and informing policy decisions. By acknowledging limitations and continuously refining indicator frameworks, countries can enhance their capacity to accurately assess progress and effectively address the challenges of climate change mitigation and adaptation.

1.4 DRIVERS OF AND BARRIERS TO LOW-CARBON ENERGY TRANSITION

The factors motivating the transition to low-carbon energy are those that drive the move towards cleaner, more sustainable energy sources and practices. These factors are crucial in influencing policies, investments, and technological innovations that aim to reduce greenhouse gas emissions, combat climate change, and enhance energy security (Hainsch et al., 2022; Polzin & Sanders, 2020; Sovacool & Griffiths, 2020; Tian et al., 2022).

1. **Climate change concerns**. When discussing the drivers of the low-carbon energy transition, the primary focus is often on climate change and its effects. The increasing awareness of the negative impacts of climate change—such as rising temperatures, extreme weather events, sea-level rise, and ecosystem disruptions—creates a pressing need to reduce greenhouse gas emissions and shift to low-carbon energy sources. Exploring the first driver, climate change concerns, involves examining the various impacts of climate change and how they stimulate efforts towards adopting low-carbon energy solutions. The climate change concerns explanation is shown in Figure 1.17.

Addressing climate change concerns necessitates transformative changes in energy systems, land use, transportation, and consumption habits to decrease greenhouse gas emissions and bolster resilience to climate impacts. The transition to low-carbon energy is crucial for mitigating climate change and protecting the planet for both current and future generations.

Scientific Consensus

- The overwhelming scientific consensus on climate change, as documented by reports from the Intergovernmental Panel on Climate Change (IPCC) and other scientific bodies, highlights the reality of human-induced global warming and its adverse impacts on ecosystems, weather patterns, and communities worldwide.

Extreme Weather Events

- Climate change intensifies the frequency and severity of extreme weather events such as hurricanes, cyclones, floods, and storms, resulting in devastating humanitarian crises, economic losses, infrastructure damage, and displacement of populations.

Sea-Level Rise

- Melting polar ice caps and thermal expansion of seawater contribute to sea-level rise, posing risks to coastal communities, infrastructure, and ecosystems. Rising sea levels increase the vulnerability of low-lying areas to flooding, erosion, and saltwater intrusion, threatening livelihoods and property.

Rising Temperatures

- Increasing global temperatures due to rising concentrations of greenhouse gases in the atmosphere lead to the following consequences: including heatwaves, droughts, wildfires, and changes in precipitation patterns. These impacts threaten human health, food security, water resources, and biodiversity.

Disruptions to Ecosystems

- Climate change disrupts ecosystems and biodiversity, leading to habitat loss, species extinction, altered migration patterns, and ecosystem degradation. These changes have cascading effects on ecosystem services, agriculture, fisheries, and human well-being.

Global Food Security

- Changes in temperature, precipitation, and weather patterns affect agricultural productivity, crop yields, and food distribution systems, jeopardizing global food security and exacerbating hunger, malnutrition, and food insecurity, particularly in vulnerable regions.

Water Scarcity and Quality

- Climate change exacerbates water scarcity and quality issues, impacting freshwater availability, sanitation, and hygiene. Droughts, erratic rainfall, and increased evaporation rates strain water resources, leading to conflicts over water access and sanitation-related health risks.

Health Impacts

- Climate change affects human health through heat-related illnesses, air pollution, vector-borne diseases, and food and waterborne diseases. Vulnerable populations, including children, the elderly, and those with pre-existing health conditions, are particularly at risk.

Social and Economic Disparities

- Climate change exacerbates existing social and economic disparities, disproportionately affecting marginalized communities, indigenous peoples, and low-income populations who lack resources and adaptive capacity to cope with climate impacts.

FIGURE 1.17 The climate change concerns explanation.

Source: created by authors.

2. **Policy and regulatory frameworks**. Government policies, regulations, and targets are essential in promoting the shift to low-carbon energy. Initiatives like carbon pricing, renewable energy mandates, emissions standards, and subsidies for clean technologies provide incentives for investing in renewable energy and improving energy efficiency. Delving into the second driver, policy and regulatory frameworks, involves exploring how government actions influence and accelerate the transition to low-carbon energy (Figure 1.18).

Emission Reduction Targets

- Governments set emission reduction targets to limit greenhouse gas emissions, typically through legislation or international agreements such as the Paris Agreement. These targets provide clear objectives for climate action and guide the development of policies and strategies to achieve them.

Carbon Pricing Mechanisms

- Carbon pricing, such as carbon taxes or emissions trading systems, internalizes the cost of carbon emissions and incentivizes emission reductions. Governments establish carbon pricing mechanisms to encourage businesses and individuals to adopt cleaner technologies and practices.

Sea-Level Rise

- Melting polar ice caps and thermal expansion of seawater contribute to sea-level rise, posing risks to coastal communities, infrastructure, and ecosystems. Rising sea levels increase the vulnerability of low-lying areas to flooding, erosion, and saltwater intrusion, threatening livelihoods and property.

Renewable Energy Standards and Mandates

- Governments implement renewable energy standards and mandates requiring a certain percentage of electricity generation to come from renewable sources. These policies stimulate investment in renewable energy projects, such as solar, wind, hydroelectric, and geothermal power.

Energy Efficiency Regulations

- Governments enact energy efficiency regulations for buildings, appliances, vehicles, and industrial processes to reduce energy consumption and emissions. Standards for energy efficiency promote the adoption of energy-saving technologies and practices across sectors.

Subsidies and Incentives

- Governments provide subsidies, tax incentives, grants, and other financial mechanisms to support the deployment of low carbon technologies and projects. These incentives reduce the financial barriers to investing in renewable energy, energy efficiency, and other clean technologies.

Research and Development Funding

- Governments allocate funding for research, development, and innovation in clean energy technologies and solutions. Public investment in research institutions, universities, and private-sector partnerships accelerates technological advancements and drives down costs.

Infrastructure Investment

- Governments invest in infrastructure for renewable energy generation, transmission, and distribution, as well as charging infrastructure for electric vehicles. Funding for grid modernization, energy storage, and smart grid technologies supports the integration of renewable energy into the electricity system.

Climate Resilience and Adaptation Policies

- Governments develop policies and strategies to enhance climate resilience and adaptation, such as coastal protection measures, water management plans, and disaster preparedness initiatives. These policies help communities and ecosystems adapt to climate change impacts.

FIGURE 1.18 The role of government actions in driving the transition to low-carbon energy.

Source: created by authors according to Zhang and Andrews-Speed (2020), Johnstone et al. (2021), Polzin and Sanders (2020).

Policy and regulatory frameworks lay the groundwork for coordinated efforts and investments in the transition to low-carbon energy. They drive the systemic changes necessary to combat climate change and create a sustainable future. Effective implementation and enforcement of these policies are crucial for meeting emission reduction targets and promoting the shift to a low-carbon economy.

3. **Technological innovation and cost reduction.** Progresses in renewable energy technologies, energy storage, smart grid systems, and energy efficiency solutions have led to cost reductions and improved performance, making clean energy alternatives increasingly competitive with fossil fuels (Bogdanov et al., 2021).
4. **Energy security and resilience**. Concerns about energy security, geopolitical risks, and dependence on imported fossil fuels drive efforts to diversify energy sources and enhance resilience through investments in domestic renewable energy resources and decentralized energy systems. The main aspects of energy security and resilience as drivers of low-carbon transition are shown in Figure 1.19.

Diversification of Energy Sources

- Energy security concerns prompt efforts to diversify energy sources, reducing reliance on fossil fuels and volatile international energy markets. Transitioning to a mix of renewable energy sources such as solar, wind, hydroelectric, and geothermal enhances energy security by tapping into domestic and decentralized energy resources.

Reduction of Import Dependency

- Dependence on imported fossil fuels exposes countries to geopolitical risks, price fluctuations, and supply disruptions. Investing in domestic renewable energy sources and energy efficiency measures reduces import dependency and enhances energy independence.

Resilience to Supply Disruptions

- Vulnerabilities in traditional energy infrastructure, such as centralized power plants and long-distance transmission lines, increase the risk of supply disruptions due to natural disasters, cyberattacks, or physical damage. Distributed renewable energy systems, microgrids, and energy storage technologies improve resilience by decentralizing energy production and distribution.

Mitigation of Energy Price Volatility

- Fluctuations in fossil fuel prices can impact energy affordability and economic stability, particularly for energy-intensive industries and households. Stable and predictable costs associated with renewable energy technologies reduce exposure to price volatility and provide long-term cost savings.

Climate Resilience and Adaptation

- Climate change poses risks to energy infrastructure and operations, including increased frequency and severity of extreme weather events, sea-level rise, and temperature extremes. Investing in climate-resilient energy infrastructure, such as flood-resistant power plants and elevated transmission lines, enhances resilience to climate impacts.

Energy Access and Equity

- Energy security is closely linked to equitable access to affordable and reliable energy services for all segments of society, including rural and underserved communities. Off-grid renewable energy solutions, decentralized energy systems, and energy efficiency measures improve energy access and address energy poverty.

Reduction of Environmental Externalities

- Dependence on fossil fuels for energy production contributes to air and water pollution, ecosystem degradation, and public health impacts. Transitioning to cleaner and renewable energy sources reduces environmental externalities associated with fossil fuel extraction, combustion, and waste disposal.

Long-Term Sustainability

- Sustainable energy systems prioritize the use of renewable resources that are abundant, locally available, and environmentally benign. Investing in renewable energy technologies and energy efficiency measures ensures the long-term sustainability of energy supply while minimizing environmental degradation and resource depletion.

FIGURE 1.19 The main aspects of energy security and resilience as drivers of low-carbon transition.

Source: created by authors according to Tian et al. (2022), Kyriakopoulos et al. (2022).

5. **Economic opportunities and job creation.** The transition to low-carbon energy presents economic opportunities for industries such as renewable energy, energy efficiency, electric vehicles, and sustainable infrastructure, leading to job creation, innovation, and economic growth.
6. **Public health benefits.** Phasing out fossil fuels and reducing air pollution from energy production and transportation can lead to significant public health benefits, including improved air quality, reduced respiratory illnesses, and lower healthcare costs. Expanding on the driver of public health benefits, it's crucial to recognize how transitioning to low-carbon energy sources and reducing air pollution can lead to significant improvements in public health outcomes. Public health benefits due to the transition to low-carbon energy are shown in Figure 1.20.

Reduction in Air Pollution

- Shifting away from fossil fuels, which are major sources of air pollutants such as particulate matter (PM), nitrogen oxides (NOx), sulfur dioxide (SO2), and volatile organic compounds (VOCs), leads to improvements in air quality. Cleaner energy sources such as renewable energy and electric vehicles produce fewer harmful emissions, reducing the burden of air pollution on public health.

Lower Respiratory and Cardiovascular Diseases

- Exposure to air pollution is associated with a wide range of adverse health effects, including respiratory diseases such as asthma, chronic obstructive pulmonary disease (COPD), and lung cancer, as well as cardiovascular diseases such as heart attacks and strokes. By reducing air pollution levels, transitioning to low carbon energy sources helps prevent and mitigate these health risks.

Decrease in Premature Mortality

- Air pollution is a significant contributor to premature mortality, particularly among vulnerable populations such as children, the elderly, and individuals with pre-existing health conditions. By reducing exposure to harmful pollutants, low carbon energy transition can decrease mortality rates and improve overall public health outcomes.

Improved Indoor Air Quality

- In addition to outdoor air pollution, indoor air quality can also impact health, particularly in households reliant on traditional biomass fuels for cooking and heating. Transitioning to cleaner cooking technologies such as electric or clean-burning stoves reduces indoor air pollution exposure, benefiting the health of household members, especially women and children.

Health Equity and Environmental Justice

- Communities disproportionately affected by air pollution, such as low-income neighborhoods and communities of color, stand to benefit the most from reductions in pollution-related health risks. Low carbon energy transition promotes health equity and environmental justice by addressing disparities in exposure to harmful pollutants and improving health outcomes for marginalized populations.

Lower Healthcare Costs

- Preventing and mitigating the health impacts of air pollution through low carbon energy transition can lead to significant cost savings in healthcare expenditures. Reduced hospitalizations, emergency room visits, and medical treatments for pollution-related illnesses result in lower healthcare costs and economic benefits for society as a whole.

Enhanced Quality of Life

- Cleaner air and improved public health contribute to a higher quality of life for individuals and communities. Reduced pollution levels create safer and healthier environments for outdoor activities, work, and recreation, fostering well-being and productivity.

Positive Health Co-Benefit

- Low carbon energy transition often entails lifestyle changes and urban planning strategies that promote physical activity, active transportation, and green spaces. These co-benefits, such as increased walking and cycling, support public health by reducing sedentary behavior and promoting physical fitness.

Long-Term Health Benefits

- Investing in low carbon energy infrastructure and sustainable urban development yields long-term health benefits by preventing chronic diseases, improving life expectancy, and enhancing overall population health. These benefits extend beyond immediate air quality improvements to create healthier and more resilient communities for future generations.

FIGURE 1.20 Public health benefits due to the transition to low-carbon energy.

Source: created by authors according to Gallagher and Holloway (2020), Gao et al. (2018), Zhang et al. (2021a).

7. **Corporate sustainability and social responsibility**. Increasing pressure from consumers, investors, and stakeholders for companies to reduce their carbon footprint and adopt sustainable practices drives corporate commitments to renewable energy procurement, carbon neutrality, and environmental stewardship. Some aspects of corporate sustainability and social responsibility as drivers of the low-carbon transition are shown in Figure 1.21.

FIGURE 1.21 Some aspects of corporate sustainability and social responsibility as drivers of the low-carbon transition.

Source: created by authors according to Kumar et al. (2021), Jiang et al. (2022), Chen (2023).

The sustainability initiative within the corporates is primarily aimed at reducing greenhouse gases, conserving natural resources, and promoting renewable energy use. Prioritizing environmental sustainability in businesses leads to the mitigation of climate change and protection of ecosystems. Many companies are committing to source a significant fraction of their energy from renewables such as solar, wind, and hydroelectric power. Investments in renewable energy sources result in emission cuts of carbon, increased energy security, and resilience, as well as foster market growth for clean energy. Businesses have embraced varied energy efficiency measures that lead to reduced energy use and operating expenses and improved resource efficiency. Investments are being made in energy-efficient technologies, upgraded equipment, and process optimization. The principles of a circular economy are waste minimization, maximization of resource utility, and the promotion of reusing, recycling, and remanufacturing products.

The circular business models will actually reduce environmental impact and efficiently use resources; they stimulate innovation and build resilience. Companies have also started to scrutinize their supply chains for ethical sourcing, fair labor practices, and environmental responsibility throughout the production process. Sustainable supply chain management deals with reducing risk, enhancing transparency, and propelling social and environmental standards through suppliers. Corporate social responsibility looks towards the attainment of social equity, diversity, and inclusion with positive implications for workers, communities, and other stakeholders. They also invest in education, healthcare, work skills, and community involvement programs to help in social problems and realize inclusive growth. Business sustainability and corporate responsibility extend back to the observation of ethical standards and integrity in business dealings. The companies then abide by their principles of fair trade, human rights, labor rights, and anti-corruption in exercising responsible business that adequately meets the expectation of all stakeholders. Corporate sustainability drives innovation through enabling collaboration, knowledge sharing, and partnership across sectors. Businesses collaboratively engage with governments, NGOs, academia, and other stakeholders to develop sustainable solutions that accelerate technology adoption and solve complex sustainability challenges.

8. **Consumer awareness and demand.** Increasing public awareness of environmental issues and climate change concerns is shaping consumer preferences and behaviors. This shift drives demand for cleaner and more sustainable energy options, such as rooftop solar panels, electric vehicles, and energy-efficient appliances.

Expanding on the driver of consumer awareness and demand, it's essential to explore how changing consumer preferences and behaviors influence the transition to low-carbon energy. Consumer awareness and demand factors for the transition to low-carbon energy consumption are shown in Figure 1.22 and Figure 1.23.

Consumer awareness and demand drive market transformation towards sustainable products, services, and technologies, shaping industry practices and influencing policy decisions. By harnessing consumer preferences and behaviors, businesses, governments, and civil society organizations can accelerate the transition to a low-carbon economy and create positive environmental and social impacts.

9. **International collaboration and cooperation.** Global agreements such as the Paris Agreement and international initiatives like Mission Innovation spur collaborative and cooperative action in the realm of clean energy research, development, deployment, and capacity-building efforts. In the context of drivers, it is important to understand how such global partnerships and collective action may facilitate this transition towards a low-carbon energy future.

Growing Environmental Consciousness

- Increasing awareness of environmental issues, including climate change, pollution, and resource depletion, prompts consumers to seek products and services with lower environmental impacts. As consumers become more environmentally conscious, they demand goods and services that align with their values, including those produced with renewable energy and sustainable practices.

Preference for Clean Energy Solutions

- Consumers are increasingly interested in clean energy solutions such as solar panels, wind turbines, electric vehicles, and energy-efficient appliances. The desire to reduce carbon footprints and support renewable energy drives consumer demand for these technologies, leading to market growth and innovation.

Concerns About Air Quality and Health

- Air pollution from fossil fuel combustion poses significant health risks to communities, prompting consumers to prioritize clean air and public health. Concerns about respiratory illnesses, allergies, and other health impacts associated with air pollution motivate consumers to support low carbon energy transition efforts.

Cost Savings and Energy Efficiency

- Energy-efficient products and technologies offer cost savings over their lifecycle through reduced energy consumption and lower utility bills. Consumers recognize the financial benefits of energy efficiency and seek energy-efficient appliances, lighting, and heating systems to lower household expenses.

Demand for Sustainable Transportation

- Concerns about traffic congestion, air pollution, and greenhouse gas emissions drive consumer interest in sustainable transportation options such as public transit, cycling, walking, and electric vehicles. Government incentives, improved infrastructure, and awareness campaigns further encourage the adoption of low carbon transportation solutions.

Preference for Eco-Friendly Products

- Consumers prefer products with eco-friendly attributes, such as recyclability, biodegradability, and minimal packaging waste. Sustainable packaging, organic materials, and fair trade practices appeal to environmentally conscious consumers seeking to minimize their ecological footprint.

FIGURE 1.22 Consumer awareness and demand factors for the transition to low-carbon energy (1).

Source: created by authors according to Dubois et al. (2019), Sovacool and Griffiths (2020), Gao et al. (2020), Pettifor and Wilson (2020), Chapman and Okushima (2019).

International cooperation will be able to share best practices, research results, and technological innovations in renewable energy, energy efficiency, and measures for climate mitigation (Gielen et al., 2019). Knowledge exchange and technology transfer are crucially facilitating to any nation in accelerating the deployment of clean energy solutions and overcoming the technological barriers. In most cases, the lack of know-how, resources, and fundamental infrastructure in developing countries is a barrier to the advancement of their energy systems towards low carbon. International cooperation provides capacity-building support, technical assistance, and training programs to enhance skills and

Influence of Social and Environmental Movements

- Social and environmental movements, such as Fridays for Future, Extinction Rebellion, and the climate strike movement, raise awareness of climate change and galvanize public support for climate action. Consumer activism and advocacy campaigns pressure businesses and policymakers to adopt sustainable practices and reduce emissions.

Access to Information and Education

- Access to information and education empowers consumers to make informed choices about their energy consumption, transportation habits, and purchasing decisions. Awareness campaigns, energy efficiency labels, and environmental certifications help educate consumers about the environmental impacts of their choices and the benefits of low carbon alternatives.

Preference for Green Buildings and Homes

- Homebuyers and renters increasingly prioritize energy-efficient and sustainable features when choosing housing options. Green building certifications, such as LEED (Leadership in Energy and Environmental Design), ENERGY STAR ratings, and net-zero energy homes, appeal to consumers seeking comfortable, healthy, and environmentally friendly living spaces.

Peer Influence and Social Norms

- Social norms and peer influence play a significant role in shaping consumer behavior towards sustainability. Positive social reinforcement, peer pressure, and cultural attitudes towards environmental responsibility encourage individuals to adopt low-carbon lifestyles and influence others within their social networks.

Demand for Renewable Energy Options

- Consumers express interest in renewable energy options such as community solar programs, green power purchasing, and renewable energy tariffs offered by utilities. Demand for clean electricity sources reflects consumer preferences for supporting renewable energy development and reducing reliance on fossil fuels.

Preference for Responsible Brands

- Consumers prefer to support brands and companies that demonstrate corporate social responsibility, sustainability commitments, and ethical business practices. Brands that prioritize environmental stewardship, social equity, and transparency in their operations and supply chains attract environmentally conscious consumers and build brand loyalty.

FIGURE 1.23 Consumer awareness and demand factors for the transition to low-carbon energy (2).

Source: created by authors according to Dubois et al. (2019), Sovacool and Griffiths (2020), Gao et al. (2020), Pettifor and Wilson (2020), Chapman and Okushima (2019).

capabilities in energy planning, policy development, and project implementation. *International partnerships* mobilize financial resources, investments, and funding mechanisms to support low-carbon energy projects in developing countries (Sarkodie et al., 2020). Climate finance mechanisms such as the Green Climate Fund, bilateral aid, and multilateral development assistance provide grants, loans, and concessional financing to support clean energy initiatives and climate resilience efforts.

Harmonizing policies, regulations, and standards across countries promotes consistency and coherence in climate action and low-carbon energy transition. International agreements such as the Paris Agreement establish common goals, frameworks, and guidelines for emissions reduction, renewable energy deployment, and climate resilience planning. *Collaborative research and development initiatives* bring together scientists, engineers, and policymakers from different countries to address shared challenges and develop innovative solutions. International research networks, consortia, and partnerships foster interdisciplinary collaboration and advance knowledge in areas such as clean energy technologies, carbon capture, and sustainable development.

International cooperation drives climate diplomacy and negotiations to secure ambitious climate commitments, agreements, and outcomes. Conferences of the Parties (COP) to the United Nations Framework Convention on Climate Change (UNFCCC) provide platforms for dialogue, negotiation, and consensus-building among countries on climate-related issues.

Regional and bilateral agreements facilitate cooperation on clean energy, emissions reduction, and climate resilience efforts among neighboring countries and trading partners. Agreements such as energy partnerships, trade agreements, and regional cooperation frameworks promote shared goals and joint action on climate and energy challenges. International collaboration involves engaging the private sector through partnerships, alliances, and initiatives that promote corporate sustainability and investment in clean energy. Public-private partnerships leverage private sector expertise, innovation, and resources to scale up clean energy projects, improve energy access, and drive economic development.

10. **Social and cultural shifts.** Changing societal attitudes, cultural norms, and values towards sustainability, conservation, and environmental stewardship contribute to a broader societal consensus and support for low-carbon energy transition efforts.

The impact factors for changes in societal values, norms, attitudes, and behaviors for the transition to a low-carbon energy future are presented in Figure 1.24 and Figure 1.25.

The drivers interplay with and reinforce each other in the creation of momentum towards transition to more sustainable, resilient, and low-carbon energy futures. Policies, investments, and collaborations can most effectively leverage these drivers to upshift the pace of transition in terms of energy transition, which will contribute to reducing impacts related to climate change. A multifaceted interplay of factors drives the transition to low-carbon energy. Economic incentives, like cost savings or job creation, stimulate investment in renewable energy and energy efficiency measures. Environmental imperatives driven by climate change and air pollution concerns propel the adoption of cleaner sources of energy and sustainable practices. Policy frameworks ensure such actions through a series of regulations and mechanisms with goals on reduction percentages of renewable energy.

Energy security considerations are more likely to diversify energy sources so that there is less dependence on fossil fuels and greater resilience to any possible future supply disruptions. Public health benefits, in terms of reduced air pollution and improved quality of life, underline the importance of a transition to cleaner energy systems. Corporate sustainability efforts and social responsibility commitments foster new innovation and investment into low-carbon technologies. International collaboration in knowledge sharing, technology transfer, and finance support is the main driver for climate mitigation effort towards global sustainability. It is leadership from individuals to communities and from businesses to governments that inspire collective action to drive progress towards a more sustainable and resilient energy future. Social and cultural changes drive changes in attitude, values, and behaviors driving consumer preferences towards support for sustainable practice. Interdisciplinary advances in science and technology provide transformational solutions to complex problems of sustainability.

FIGURE 1.24 The impact factors for changes in societal values, norms, attitudes, and behaviors for the transition to a low-carbon energy future (1).

Source: created by authors according to Sovacool and Griffiths (2020), Wang et al. (2021), Hainsch et al. (2022), Wu et al. (2022), Wijaya and Kokchang (2023), Bazzani (2023), Radtke et al. (2022).

FIGURE 1.25 The impact factors for changes in societal values, norms, attitudes, and behaviors for the transition to a low-carbon energy future (2).

Source: created by authors according to Sovacool and Griffiths (2020), Wang et al. (2021), Hainsch et al. (2022), Wu et al. (2022), Wijaya and Kokchang (2023), Bazzani (2023), Radtke et al. (2022).

1.5 POLICY AREAS OF LOW-CARBON ENERGY TRANSITION

The policy areas of the low-carbon energy transition encompass a variety of measures and strategies designed to facilitate the shift towards cleaner, more sustainable energy systems.

1.5.1 Renewable Energy Promotion

Policies promoting the development and deployment of renewable energy sources, such as solar, wind, hydroelectric, and geothermal power, are central to the low-carbon energy transition. These policies often include measures like feed-in tariffs, renewable energy targets, tax incentives, and subsidies to encourage investment and market growth in renewable energy.

The renewable energy promotion policy area includes several measures and strategies to accelerate the development, deployment, and integration of renewable energy sources into the energy mix. The main measures are presented in Figure 1.26.

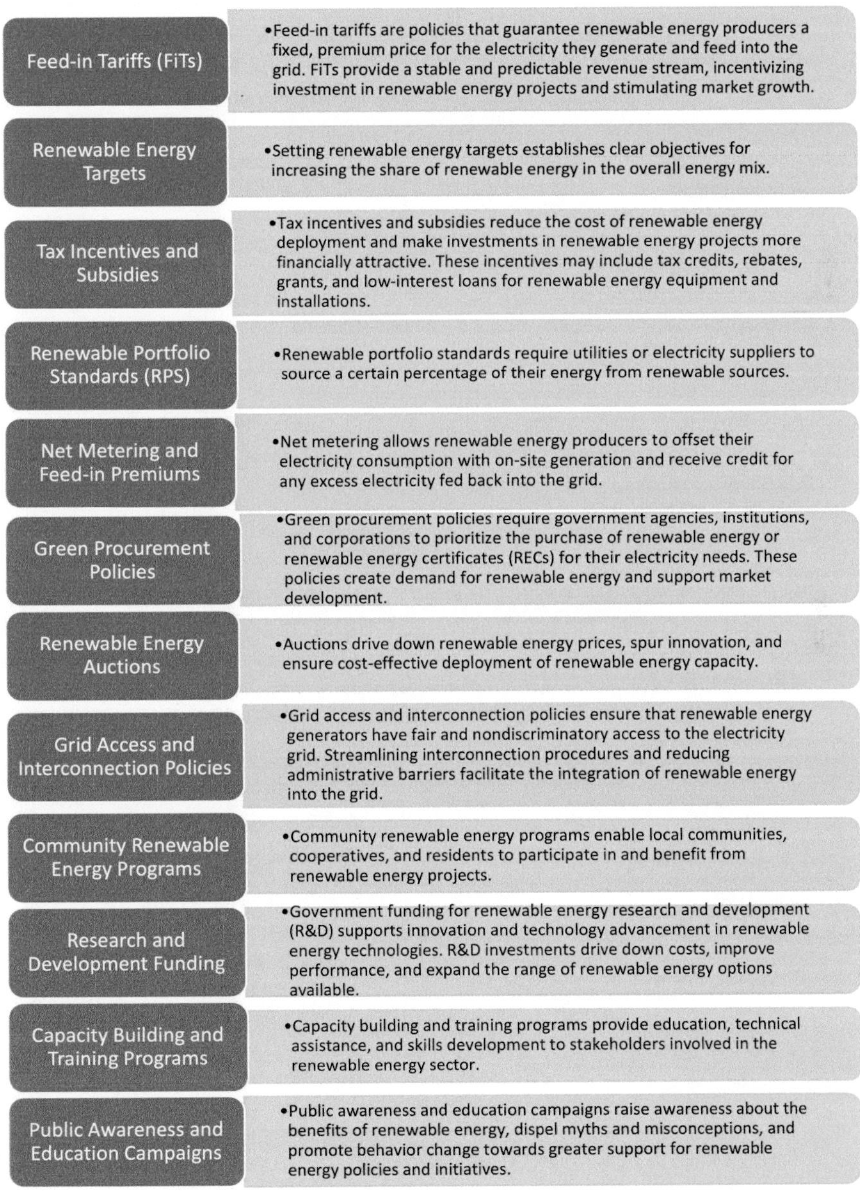

FIGURE 1.26 Measures promoting the development of renewable energy sources.

Source: created by authors according to Majid (2020), Xu et al. (2019), Adams and Acheampong (2019), Cantarero (2020), Yan et al. (2024).

Policies promoting renewable energy are crucial in accelerating the transition to a low-carbon energy future. They drive investments in renewable energy infrastructure, reduce greenhouse gas emissions, and enhance energy security and resilience. By creating favorable market conditions, these policies encourage innovation and the widespread adoption of renewable energy technologies, ultimately contributing to the achievement of Sustainable Development Goals and the reduction of greenhouse gas emissions.

1.5.2 Energy Efficiency Standards and Regulations

Energy efficiency policies are practices that aim to reduce energy consumption in buildings, transportation, and industry as well as by appliances through the development of standards, regulations, and other incentives. Examples include building codes, efficiency labeling of appliances, fuel efficiency standards for

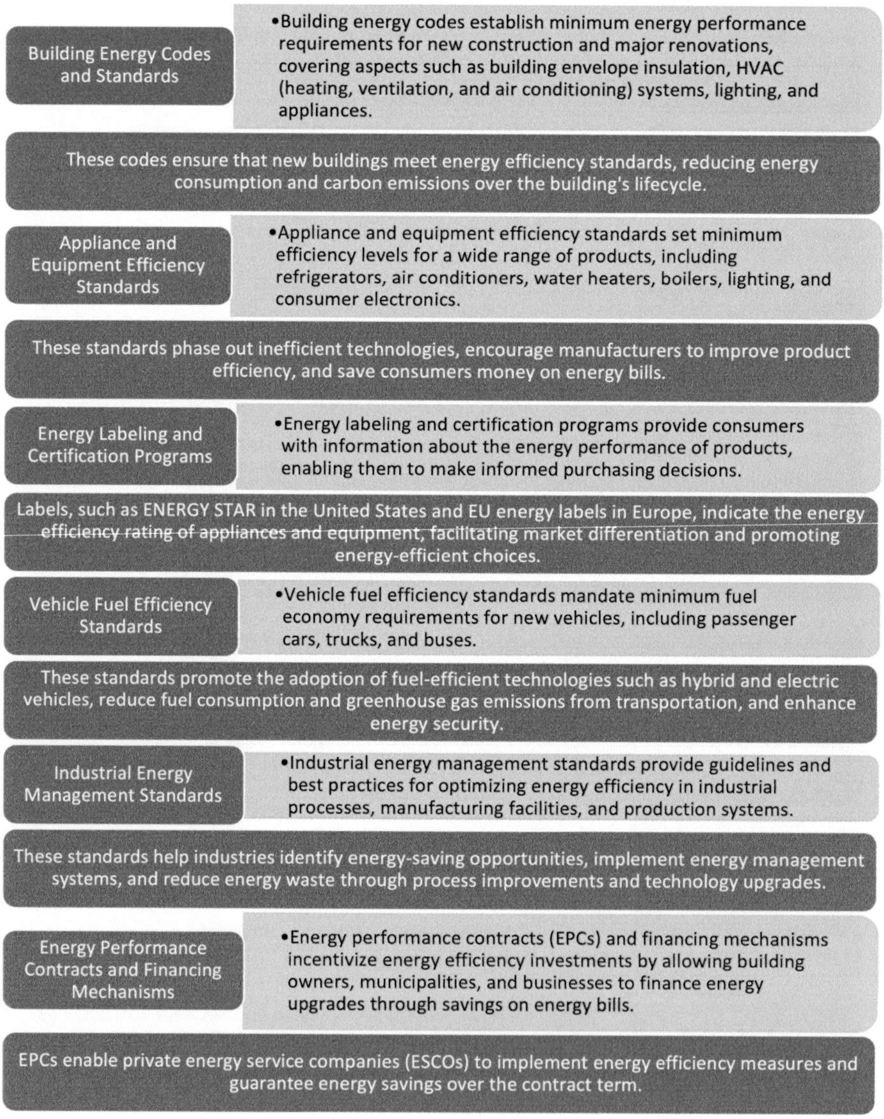

FIGURE 1.27 The policy area of Energy Efficiency Standards and Regulations (1).

Source: created by authors according to Bashir et al. (2022), Economidou et al. (2020), Bertoldi and Mosconi (2020), Brockway et al. (2021), Guo and Yuan (2020), Paramati et al. (2022), Zhao et al. (2022).

vehicles, and programs on industrial energy management to adopt energy-saving technologies and practices. In this respect, energy efficiency standards and regulations include many different types of measures and rules intended to reduce energy consumption, improve energy use, and stimulate energy-efficient technology across the board. The results of the detailed analysis are shown in Figure 1.27 and Figure 1.28.

Energy efficiency standards and regulations are essential policy tools for reducing energy consumption, lowering greenhouse gas emissions, and promoting sustainable development. By establishing clear benchmarks, providing market signals, and incentivizing investments in energy efficiency, these policies drive innovation, create jobs, and deliver economic, environmental, and social benefits for society as a whole.

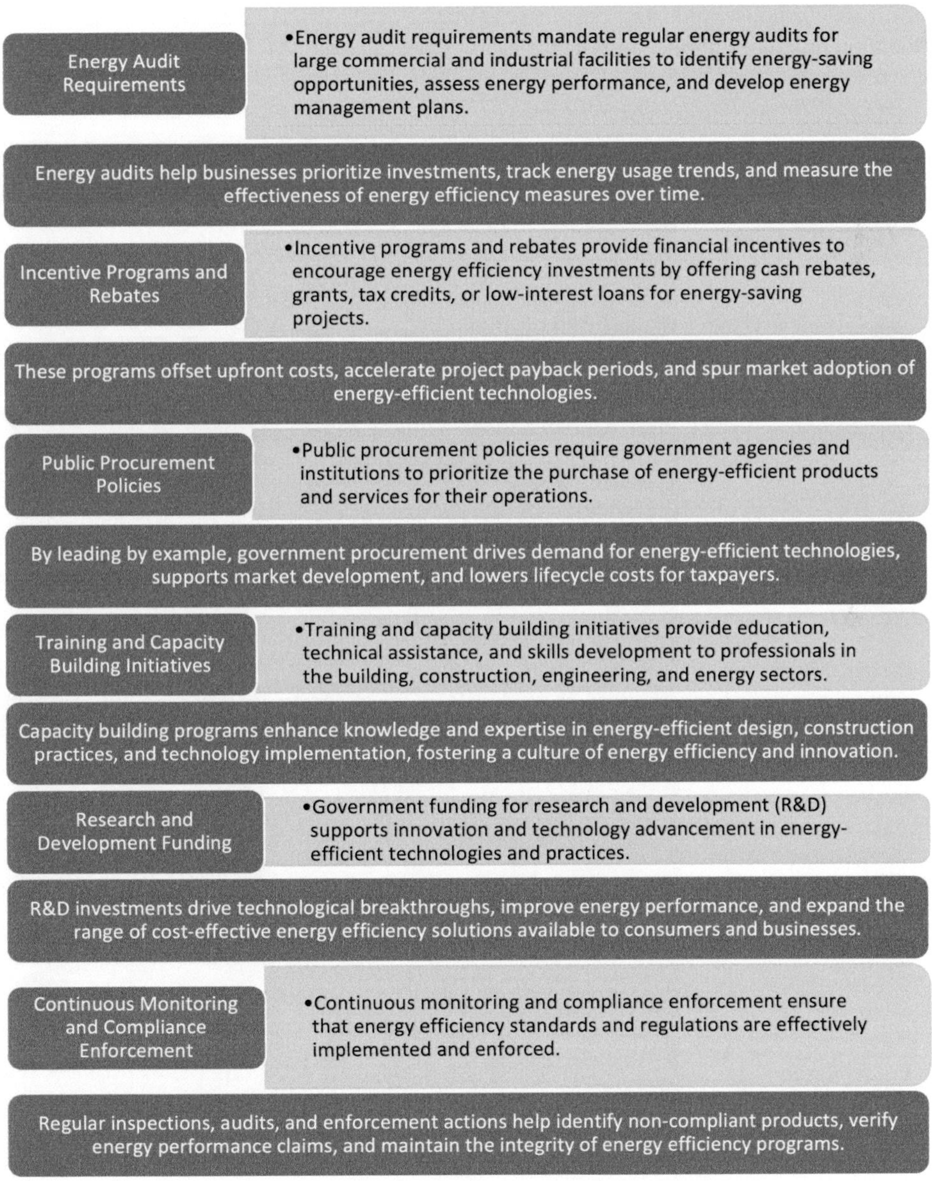

FIGURE 1.28 The policy area of Energy Efficiency Standards and Regulations (2).

Source: created by authors according to Bashir et al. (2022), Economidou et al. (2020), Bertoldi and Mosconi (2020), Brockway et al. (2021), Guo and Yuan (2020), Paramati et al. (2022), Zhao et al. (2022).

1.5.3 Carbon Pricing Mechanisms

Carbon pricing policies internalize the cost of carbon emissions through their system, which gives economic incentives for reduction. By placing a price on carbon emissions, businesses and consumers are pushed into cleaner technologies and behaviors. This area widens within the policy perspective of carbon pricing mechanisms by implementing market-based instruments for cost internalization concerning carbon emissions, giving economic incentives for reduction, and propelling economies towards being low-carbon. Carbon Pricing Mechanisms are shown in Figure 1.29.

The allocation of carbon pricing revenue can vary depending on policy design and objectives. Revenue may be used to fund climate mitigation and adaptation programs, invest in clean energy infrastructure, compensate vulnerable communities impacted by the transition, or provide dividends or rebates to households and businesses (Mayer et al., 2021). Transparent and equitable revenue allocation enhances public support for carbon pricing policies.

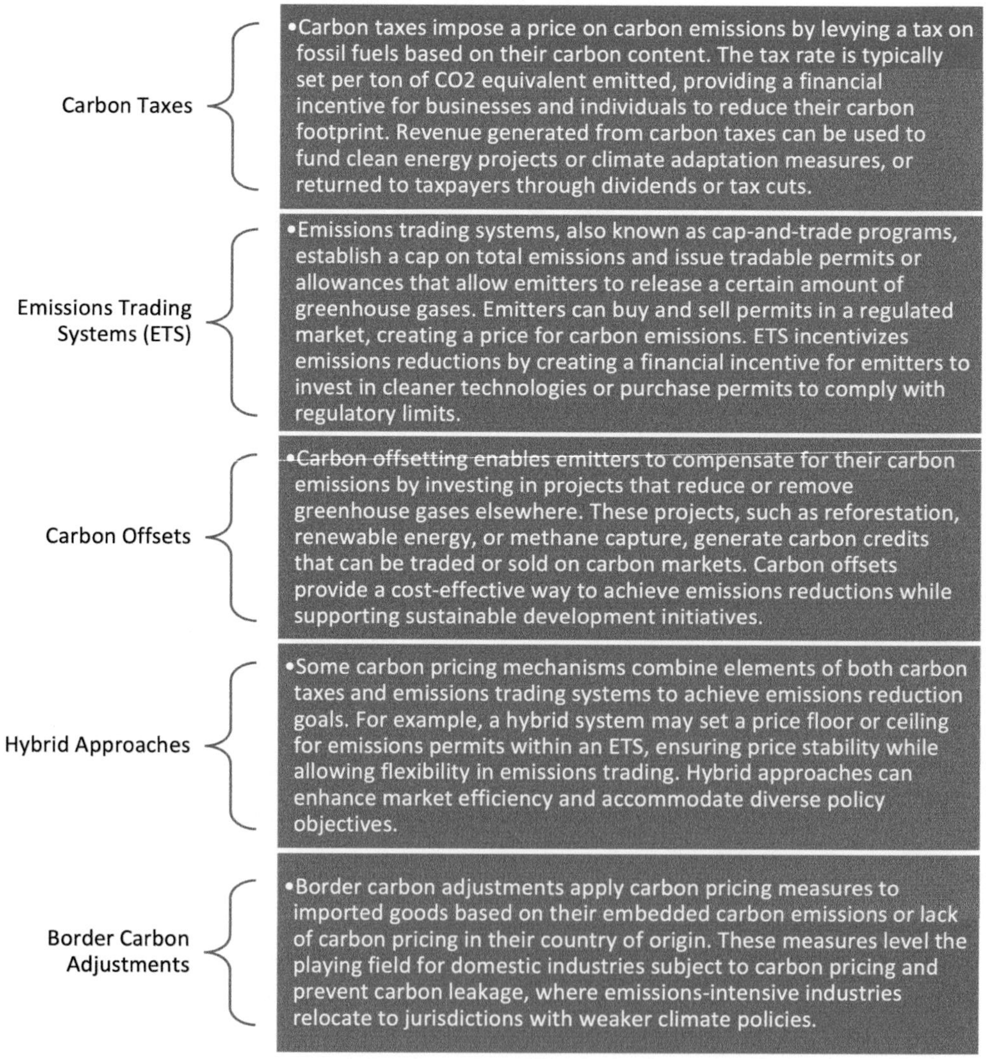

FIGURE 1.29 Carbon pricing mechanisms.

Source: created by authors according to Aiyegbusi et al. (2022), Green (2021), Gugler et al. (2021), Fageda and Teixidó (2022).

Market stability mechanisms, such as price floors, ceilings, and reserve mechanisms, help stabilize carbon markets and prevent extreme price volatility. Price floors set a minimum price for carbon permits to ensure that emissions reductions occur even when market prices are low. Price ceilings establish an upper limit on permit prices to protect industries from excessive compliance costs (Lyons and Viswanath-Natraj, 2023). Reserve mechanisms allow regulators to intervene in the market by buying or selling permits to maintain price stability.

Linking carbon markets allows for the trading of emissions permits across jurisdictions, creating a larger and more liquid market for carbon allowances. Linked markets increase flexibility, reduce compliance costs, and promote cost-effective emissions reductions by allowing emitters to access a broader pool of mitigation options. International cooperation and harmonization of carbon pricing policies facilitate market linkage and alignment of climate goals (Green, 2021; Shi et al., 2022).

Sectoral carbon pricing approaches target specific industries or sectors with tailored emissions reduction measures. Sectoral approaches may include performance standards, carbon intensity benchmarks, or sector-specific carbon taxes to address unique challenges and opportunities for emissions reduction (Hepburn et al., 2020; Feindt et al., 2021). Sectoral policies complement broader carbon pricing mechanisms and promote sectoral decarbonization pathways.

Carbon pricing mechanisms are powerful policy tools for reducing greenhouse gas emissions, fostering innovation, and driving the transition to a low-carbon economy. By internalizing the social cost of carbon emissions and creating economic incentives for emissions reduction, carbon pricing encourages sustainable development, enhances climate resilience, and promotes global cooperation in addressing the climate crisis.

1.5.4 Emission Reduction Targets

Setting ambitious emission reduction targets at the national, regional, and international levels provides a clear signal for climate action and guides policy development. These targets may be legally binding or voluntary commitments to achieve specific greenhouse gas reduction goals over time.

The importance of emission reduction targets as a low-carbon energy transition policy area cannot be overstated. These targets serve as essential policy instruments for driving climate action, shaping regulatory frameworks, and mobilizing efforts to mitigate greenhouse gas (GHG) emissions. The importance of emission reduction targets in the context of low-carbon energy transition is shown in Figure 1.30.

Emission reduction targets are indispensable policy tools for driving low-carbon energy transition, advancing climate action, and securing a sustainable and resilient future for all. By setting clear objectives, fostering collaboration, and mobilizing resources, emission reduction targets empower countries to take ambitious and decisive action to combat climate change and build a more sustainable and prosperous world.

1.5.5 Regulatory Framework for Clean Technology Deployment

Regulatory frameworks facilitate the deployment of clean technologies such as electric vehicles, energy storage, smart grids, and carbon capture and storage (CCS). Policies may include research and development funding, technology demonstration programs, and regulatory support for clean technology deployment. The term "Regulatory Framework for Clean Technology Deployment" refers to a set of laws, regulations, policies, and standards designed to facilitate the deployment, adoption, and integration of clean technologies into various sectors of the economy. This regulatory framework aims to create an enabling environment that supports the widespread adoption of clean technologies, such as renewable energy, energy efficiency, carbon capture and storage (CCS), electric vehicles, and sustainable infrastructure.

Guiding Climate Action

- Emission reduction targets provide a clear and measurable framework for guiding climate action at the national, regional, and international levels. By setting specific objectives for reducing GHG emissions, countries can align their policies, strategies, and investments with long-term climate goals and international commitments, such as the Paris Agreement.

Catalyzing Policy Development

- Emission reduction targets stimulate the development of comprehensive climate policies and regulatory measures to achieve emissions reductions across all sectors of the economy. Targets drive innovation, investment, and deployment of low-carbon technologies and practices, accelerating the transition to a more sustainable and resilient energy system.

Mobilizing Stakeholder Engagement

- Emission reduction targets mobilize stakeholder engagement and cooperation among governments, businesses, civil society organizations, and the public. Targets create a shared sense of urgency and responsibility for addressing climate change, fostering collaboration, knowledge sharing, and collective action towards achieving common climate objectives.

Providing Long-Term Certainty

- Emission reduction targets provide long-term certainty and predictability for investors, businesses, and decision-makers, enabling strategic planning and investment in low-carbon infrastructure and technologies. Clear and ambitious targets signal market demand for clean energy solutions, driving innovation, cost reductions, and market expansion in renewable energy and energy efficiency sectors.

Driving Technological Innovation

- Emission reduction targets stimulate technological innovation and research and development (R&D) in clean energy technologies, carbon capture and storage (CCS), energy storage, and other low-carbon solutions. Targets create market incentives for the development and deployment of innovative technologies that can help achieve emissions reductions cost-effectively and at scale.

Fostering International Cooperation

- Emission reduction targets promote international cooperation and collaboration by providing a common framework for countries to work towards shared climate objectives. Targets facilitate dialogue, negotiation, and coordination among nations, driving ambition and ambition ratcheting in global climate negotiations and fostering trust and solidarity among nations.

Enhancing Accountability and Transparency

- Emission reduction targets enhance accountability and transparency by requiring countries to track, report, and verify their progress towards achieving emissions reductions. Regular reporting and assessment of emission inventories, mitigation actions, and progress towards targets enable peer review, scrutiny, and assessment of countries' performance in meeting their climate commitments.

Promoting Adaptation and Resilience

- Emission reduction targets complement adaptation and resilience efforts by reducing the severity and impacts of climate change, thereby enhancing the resilience of communities, ecosystems, and economies to climate-related hazards. Achieving emission reduction targets contributes to climate stabilization, reducing the need for costly adaptation measures and safeguarding against future climate risks.

FIGURE 1.30 The importance of emission reduction targets in the context of low-carbon energy transition.

Source: created by authors.

Key components according to Lowitzsch et al. (2020), Hepburn et al. (2021), and Zhang et al. (2021b) of a regulatory framework for clean technology deployment are shown in Figure 1.31.

Overall, a robust regulatory framework for clean technology deployment offers clarity, certainty, and incentives for stakeholders to invest in and adopt clean energy solutions. This drives the transition to a low-carbon and sustainable future. Such regulatory frameworks are essential in shaping market dynamics, stimulating innovation, and achieving climate and energy policy objectives.

FIGURE 1.31 Key components of a regulatory framework for clean technology deployment.

Source: created by authors.

1.5.6 Infrastructure Investment and Modernization

In addition, policies promoting investments and modernization support the transition of infrastructure to low-carbon energy systems. These entail financial structures for the support of renewable energy infrastructures, grid modernization projects, electric vehicle supply equipment installations, and development and application of energy storage facilities to support renewables integration while enhancing the system's reliability and resilience.

Investment in modern infrastructure is essential for deploying low-carbon energy technologies such as solar, wind, hydroelectric, and geothermal energy. Upgrading transmission and distribution networks, building new grid infrastructure, and modernizing energy storage systems are critical for integrating renewable energy sources into the power system. These improvements reduce our dependence on fossil fuels and help lower greenhouse gas emissions. Enhanced infrastructure also facilitates the adoption of energy-efficient technologies and practices across various sectors, including buildings, transportation, industry, and agriculture.

Investing in energy-efficient buildings, smart meters, electric vehicle charging infrastructure, and improved industrial processes reduces energy consumption, lowers operational costs, and mitigates environmental impacts. These investments also bolster energy system resilience and reliability against climate hazards like extreme weather events, wildfires, and sea-level rise. Strengthening infrastructure, hardening critical assets, and diversifying energy sources decrease vulnerabilities and enhance system resilience, ensuring a stable energy supply during emergencies.

Infrastructure modernization promotes environmental sustainability by reducing pollution, conserving natural resources, and mitigating environmental impacts. Investments in sustainable transportation systems, green buildings, renewable energy infrastructure, and ecosystem restoration projects contribute to climate mitigation, biodiversity conservation, and ecosystem resilience. Infrastructure investment can promote equity and social justice by ensuring equitable access to clean energy, transportation, and other essential services. Targeted investments in underserved communities, low-income neighborhoods, and marginalized populations help address energy poverty, reduce disparities in access to clean energy technologies, and promote social inclusion and economic empowerment.

1.5.7 Energy Transition Planning and Governance

Effective energy transition planning and governance frameworks are essential for coordinating actions across government agencies, industry stakeholders, and civil society. Policies may establish national or regional energy transition plans, institutional mechanisms, and coordination bodies to facilitate collaboration and ensure alignment with climate and energy goals.

1.5.8 Just Transition Policies

Just transition policies aim to ensure that the benefits and costs of the energy transition are distributed equitably across society, particularly for workers and communities affected by the shift away from fossil fuels. These policies may include workforce training programs, economic diversification initiatives, and social safety nets to support impacted workers and regions.

Just Transition Policies refers to a set of strategies, measures, and interventions designed to make the transition to a low-carbon economy fair, inclusive, and equitable for all stakeholders, especially those in sectors and regions heavily reliant on fossil fuels and other high-carbon industries. These policies recognize that decarbonization efforts can lead to job losses, economic dislocation, and social disruption. They aim to mitigate these impacts while promoting sustainable development and social justice (Wang & Lo, 2021).

Key elements of just transition policies are shown in Figure 1.32.

Just transition policies are grounded in principles of social justice, equity, and human rights, recognizing that considerations of fairness, inclusion, and respect for the dignity and well-being of all individuals and communities must guide the transition to a low-carbon economy. By prioritizing the needs of workers, communities, and vulnerable populations, just transition policies seek to ensure that no one is left behind in the shift towards a sustainable and equitable future.

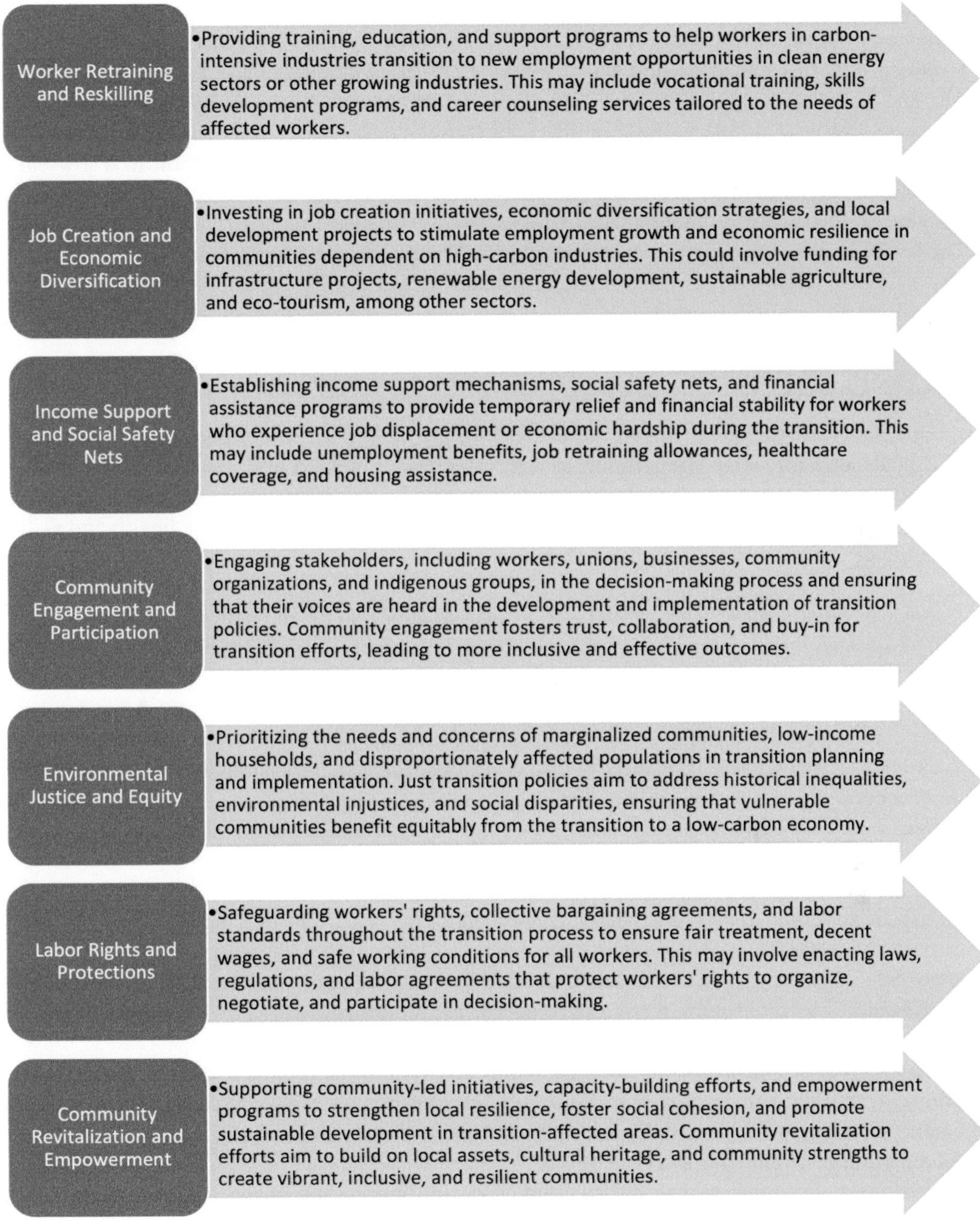

FIGURE 1.32 Key elements of just transition policies.

Source: created by authors according to Hepburn et al. (2021), Zhang et al. (2021).

Just transition policies are essential for addressing the social and economic inequalities inherent in the transition to a low-carbon economy. By prioritizing the needs of workers, communities, and marginalized populations, these policies ensure that the benefits of the transition are equitably distributed and that no one is left behind. Upholding labor rights, ensuring fair wages, and providing adequate protections for workers are fundamental principles of just transition policies. By safeguarding workers' rights to decent work and safe working conditions, these policies help mitigate the potential negative impacts of economic restructuring and job displacement.

In conclusion, just transition policies represent a holistic approach to addressing the social, economic, and environmental dimensions of the low-carbon energy transition. By centering principles of equity, inclusion, and justice, these policies help build a more sustainable, resilient, and equitable future for all. As we navigate the challenges and opportunities of the transition to a low-carbon economy, it is imperative to prioritize the well-being and dignity of workers, communities, and future generations, ensuring that our efforts leave no one behind.

These policy areas provide a comprehensive framework for guiding the transition to low-carbon energy systems, addressing climate change, promoting sustainable development, and building a more resilient and equitable future.

1.6 CONCLUSIONS

1. One general conclusion drawn from the concept of low-carbon energy transition is the urgent need for close cooperation to address climate change and achieve sustainability goals. Transitioning to a low-carbon energy system is not optional but essential to mitigate climate change impacts, reduce greenhouse gas emissions, and protect the health and well-being of current and future generations.

This conclusion underscores the critical importance of adopting clean energy technologies, implementing supportive policy frameworks, mobilizing investment in sustainable infrastructure, and fostering international cooperation to accelerate the transition. Additionally, it emphasizes the need for equitable and inclusive approaches that prioritize social equity, labor rights, and environmental justice to ensure that the benefits of the transition are shared by all members of society.

In essence, a low-carbon transition is transformative for a more sustainable, resilient, and fair future. Accepting the transition and committing to ambitious climate action can deal with the urgent challenges posed by climate change and unlock new opportunities for economic growth, job creation, and better quality of life worldwide.

2. Comprehensive and integrated policy frameworks play a part in driving progress towards climate mitigation and sustainability objectives. These policy documents emphasize the importance of setting clear emission reduction targets, implementing supportive regulatory mechanisms, and mobilizing resources to accelerate the transition to a low-carbon economy.

There seems to be a need for coordinated action across multiple sectors and levels of government to achieve ambitious climate goals. It highlights the importance of policy coherence, stakeholder engagement, and adaptive governance structures to address complex challenges and ensure effective implementation of climate policies.

Furthermore, these policy documents emphasize the importance of international collaboration and cooperation in addressing global climate challenges. They underscore the need for countries to work together to share knowledge, best practices, and technology transfer to support low-carbon development pathways worldwide.

3. It is the significance of setting ambitious, measurable, and time-bound goals to drive progress towards sustainability and climate objectives. These targets serve as guiding principles and benchmarks for policymakers, businesses, and stakeholders, providing a clear roadmap for action and accountability. Additionally, the development and tracking of indicators play a crucial role in monitoring progress, assessing performance, and identifying areas for improvement in achieving low-carbon transition goals. Indicators provide valuable insights into trends, patterns, and outcomes related to energy consumption, emissions

reduction, renewable energy deployment, energy efficiency, and other key metrics. The establishment of targets and indicators facilitates transparency, accountability, and stakeholder engagement, fostering a culture of data-driven decision-making and evidence-based policymaking. By regularly monitoring progress and reporting on outcomes, countries and organizations can demonstrate their commitment to climate action, build trust with stakeholders, and mobilize support for further efforts to accelerate the low-carbon transition.

4. One significant conclusion drawn from examining the drivers of and barriers to the low-carbon energy transition is the pivotal role of multi-faceted strategies that address both enabling factors and challenges in driving progress towards a sustainable and resilient energy future. This conclusion underscores the importance of leveraging key drivers such as climate change concerns, policy and regulatory frameworks, energy security, public health benefits, corporate sustainability, consumer awareness, international collaboration, and social and cultural shifts to accelerate the transition to a low-carbon economy.
5. One notable conclusion drawn from examining policy areas of the low-carbon energy transition is the recognition of the interconnectedness and interdependence among various policy domains in driving effective climate action and sustainability outcomes.

Such policy must be viewed as integrated and systemic in the development, implementation, and evaluation of it, as it touches on a number of elements in an energy transition: renewable promotion, energy efficiency standards and regulations, carbon pricing mechanisms, emission reduction targets, infrastructure investment, and research and innovation support. Coherence is important in making proper policies, coordination, and alignment across different areas of policies, so that synergies are maximized, trade-offs are minimized, and complementary outcomes are realized. Policymakers can ensure an enabling environment for the transition to a low-carbon economy that is going to support sustainable growth, social equity, and environmental protection by designing coordinated policies regarding energy, transport, industry, agriculture, and finances.

REFERENCES

Adams, S., & Acheampong, A. O. (2019). Reducing carbon emissions: The role of renewable energy and democracy. *Journal of Cleaner Production*. www.sciencedirect.com/science/article/pii/S0959652619331154

Aiyegbusi, O., Yalamova, R., & others. (2022). Meeting CO2 targets with carbon pricing through taxation and trading. *Earth Sciences and Environment*. https://wseas.com/journals/eshc/2022/a24eshc-011(2022).pdf

Ali, M. M., Al-Kodmany, K. M., & Armstrong, P. J. (2023). Energy efficiency of tall buildings: A global snapshot of innovative design. *Energies*, *16*(4): 2063. https://doi.org/10.3390/en16042063

Andersson, E., Arfwidsson, O., Bergstrand, V., & Thollander, P. (2017). A study of the comparability of energy audit program evaluations. *Journal of Cleaner Production*, *142*: 2133–2139. https://doi.org/10.1016/J.JCLEPRO.2016.11.070

Ashrafian, T. (2023). Enhancing school buildings energy efficiency under climate change: A comprehensive analysis of energy, cost, and comfort factors. *Journal of Building Engineering*, *80*, https://doi.org/10.1016/j.jobe.2023.107969

Babayomi, O., Zhang, Z., Dragicevic, T., & Hu, J. (2023). Smart grid evolution: Predictive control of distributed energy resources—A review. *International Journal of Electrical Power & Energy Systems*, *143*, 108149. https://www.sciencedirect.com/science/article/pii/S0142061522008080

Baidya, S., & Nandi, C. (2020). Green energy generation using renewable energy technologies. In: Bhoi, A., Sherpa, K., Kalam, A., & Chae, G. S. (Eds.), *Advances in Greener Energy Technologies (Green Energy and Technology)*. Springer, Singapore. https://doi.org/10.1007/978-981-15-4246-6_16

Baroudi, H. A., Awoyomi, A., Patchigolla, K., Jonnalagadda, K., & Anthony, E. J. (2021). A review of large-scale CO2 shipping and marine emissions management for carbon capture, utilisation and storage. *Applied Energy*, *287*(1). https://doi.org/10.1016/j.apenergy.2021.116510

Bashir, M. F., Ma, B., Bashir, M. A., & Radulescu, M. (2022). Investigating the role of environmental taxes and regulations for renewable energy consumption: Evidence from developed economies. *Economic Research*. https://hrcak.srce.hr/file/435668

Bayendang, N. P., Kahn, M. T., & Balyan, V. (2023). Combined cold, heat and power (CCHP) systems and fuel cells for CCHP applications: A topological review. *Clean Energy*, *7*(2): 436–491. https://doi.org/10.1093/ce/zkac079

Bazzani, G. (2023). Climate solidarity: A framework and research agenda for low-carbon behavior. *Sociological Forum*, *38*(1): 95–110. https://doi.org/10.1111/socf.12885

Bertoldi, P., & Mosconi, R. (2020). Do energy efficiency policies save energy? A new approach based on energy policy indicators (in the EU Member States). *Energy Policy*. www.sciencedirect.com/science/article/pii/S030142152030077X

Bogdanov, D., Ram, M., Aghahosseini, A., Gulagi, A., & others. (2021). Low-cost renewable electricity as the key driver of the global energy transition towards sustainability. *Energy*, *239*: 121752. https://doi.org/10.1016/j.energy.2021.121752

Bridge, G., Bouzarovski, S., Bradshaw, M., & Eyre, N. (2013). Geographies of energy transition: Space, place and the low-carbon economy. *Energy Policy*, *53*: 331–340.

Brockway, P. E., Sorrell, S., Semieniuk, G., & Heun, M. K. (2021). Energy efficiency and economy-wide rebound effects: A review of the evidence and its implications. *Renewable and Sustainable Energy Reviews*. www.sciencedirect.com/science/article/pii/S1364032121000769

Cakir, M., Çankaya, I., Garip, I., & Colak, I. (2022). Advantages of Using Renewable Energy Sources in Smart Grids. *10th International Conference on Smart Grid (icSmartGrid)*, 436–439. https://doi.org/10.1109/icSmartGrid55722.2022.9848612

Cantarero, M. M. V. (2020). Of renewable energy, energy democracy, and sustainable development: A roadmap to accelerate the energy transition in developing countries. *Energy Research & Social Science*, *68*: 101607. www.sciencedirect.com/science/article/pii/S2214629620302917

Carpejani, P., Jesus, É. T., Catapan, B. L., Costa, S. E., Lima, E. P., Tortato, U., Machando, C. G., & Richter, B. K. (2019). Affordable and clean energy: A study on the advantages and disadvantages of the main modalities. In: Leal Filho, W., Borges de Brito, P., & Frankenberger, F. (Eds.), *International Business, Trade and Institutional Sustainability. World Sustainability Series*. Springer, Cham. https://doi.org/10.1007/978-3-030-26759-9_35

Chapman, A., & Okushima, S. (2019). Engendering an inclusive low-carbon energy transition in Japan: Considering the perspectives and awareness of the energy poor. *Energy Policy*, *137*: 111017. www.sciencedirect.com/science/article/pii/S0301421519306044

Chen, T., Yin, X., & Wang, G. (2021). Securing communications between smart grids and real users; providing a methodology based on user authentication. *Energy Reports*, *7*: 2273–2283. www.sciencedirect.com/science/article/pii/S2352484721007277

Chen, W. (2023). Digital economy development, corporate social responsibility and low-carbon innovation. *Corporate Social Responsibility and Environmental Management*. https://onlinelibrary.wiley.com/doi/abs/10.1002/csr.2443

Climate Change Committee. (2008). *The UK Climate Change Act*. www.legislation.gov.uk/ukpga/2008/27/contents/enacted

Climate Change Committee. (2024). *The UK Climate Change Act*. www.legislation.gov.uk/ukpga/2008/27/contents

The Danish Ministry of Climate, Energy and Utilities. (2020a). *Climate Programme 2020*. https://unfccc.int/sites/default/files/resource/ClimateProgramme2020-Denmarks-LTS-under-the%20ParisAgreement_December2020_.pdf

The Danish Ministry of Climate, Energy and Utilities. (2020b). *Climate Act*. https://en.kefm.dk/Media/1/B/Climate%20Act_Denmark%20-%20WEBTILG%C3%86NGELIG-A.pdf

Dileep, G. (2020). A survey on smart grid technologies and applications. *Renewable Energy*, *146*: 2589–2625. https://doi.org/10.1016/j.renene.2019.08.092

Dubois, G., Sovacool, B., Aall, C., Nilsson, M., & others. (2019). It starts at home? Climate policies targeting household consumption and behavioral decisions are key to low-carbon futures. *Energy Research & Social Science*, *52*: 144–158.

Economidou, M., Todeschi, V., Bertoldi, P., & D'Agostino, D. (2020). Review of 50 years of EU energy efficiency policies for buildings. *Energy and Buildings*. https://www.sciencedirect.com/science/article/pii/S0378778820317229

Fageda, X., & Teixidó, J. J. (2022). Pricing carbon in the aviation sector: Evidence from the European emissions trading system. *Journal of Environmental Economics and Management*. www.sciencedirect.com/science/article/pii/S0095069621001352

Federal Ministry for the Environment, Nature Conservation, Building and Nuclear Safety (BMUB). (2016). *Climate Action Plan 2050 Principles and Goals of the German Government's Climate Policy*. Druck- und Verlagshaus Zarbock GmbH & Co. KG. https://cdn.climatepolicyradar.org/navigator/DEU/2016/climate-action-plan-2050_78fb149034cf590cc4f93a369e6ed8c3.pdf

Feindt, S., Kornek, U., Labeaga, J. M., Sterner, T., & Ward, H. (2021). Understanding regressivity: Challenges and opportunities of European carbon pricing. *Energy Economics*. www.sciencedirect.com/science/article/pii/S0140988321004266

Fragkos, P., van Soest, H. L., Schaeffer, R., Reedman, L., & others. (2021). Energy system transitions and low-carbon pathways in Australia, Brazil, Canada, China, EU-28, India, Indonesia, Japan, Republic of Korea, Russia and the United States. *Energy*, *228*: 120503. https://doi.org/10.1016/j.energy.2020.120503

Gallagher, C. L., & Holloway, T. (2020). Integrating air quality and public health benefits in US decarbonization strategies. *Frontiers in Public Health*, *8*: 563358. https://doi.org/10.3389/fpubh.2020.563358

Galvão, J. R., Moreira, L., Gaspar, G., Vindeirinho, S., & Leitão, S. (2017). Energy system retrofit in a public services building. *Management of Environmental Quality*, *28*(3): 302–314. https://doi.org/10.1108/MEQ-02-2014-0028

Gao, J., Kovats, S., Vardoulakis, S., Wilkinson, P., Woodward, A., & Li, J. (2018). Public health co-benefits of greenhouse gas emissions reduction: A systematic review. *Science of the Total Environment*, *627*: 388–402. https://doi.org/10.1016/j.scitotenv.2018.01.144

Gao, L., Hiruta, Y., & Ashina, S. (2020). Promoting renewable energy through willingness to pay for transition to a low carbon society in Japan. *Renewable Energy*, *159*: 698–708. www.sciencedirect.com/science/article/pii/S0960148120312921

Gielen, D., Boshell, F., Saygin, D., Bazilian, M. D., & others. (2019). The role of renewable energy in the global energy transformation. *Energy Strategy Reviews*, *24*: 38–50. www.sciencedirect.com/science/article/pii/S2211467X19300082

Green, J. F. (2021). Does carbon pricing reduce emissions? A review of ex-post analyses. *Environmental Research Letters*. https://iopscience.iop.org/article/10.1088/1748-9326/abdae9

Green, J. F., & Gambhir, A. (2020). Transitional assistance policies for just, equitable and smooth low-carbon transitions: Who, what and how? *Climate Policy*, *20*(8): 946–962. https://doi.org/10.1080/14693062.2019.1657379

Gugler, K., Haxhimusa, A., & Liebensteiner, M. (2021). Effectiveness of climate policies: Carbon pricing vs. subsidizing renewables. *Journal of Environmental Economics and Management*. www.sciencedirect.com/science/article/pii/S0095069620301285

Guo, R., & Yuan, Y. (2020). Different types of environmental regulations and heterogeneous influence on energy efficiency in the industrial sector: Evidence from Chinese provincial data. *Energy Policy*. www.sciencedirect.com/science/article/pii/S0301421520304705

Hainsch, K., Löffler, K., Burandt, T., Auer, H., & others. (2022). Energy transition scenarios: What policies, societal attitudes, and technology developments will realize the EU Green Deal? *Energy*, *239*: 121752. https://doi.org/10.1016/j.energy.2020.120503

Hepburn, C., Qi, Y., Stern, N., Ward, B., & Xie, C. (2021). Towards carbon neutrality and China's 14th five-year plan: Clean energy transition, sustainable urban development, and investment priorities. *Environmental Science and Policy*. https://doi.org/10.1016/j.ese.2021.100130

Hepburn, C., Stern, N., & Stiglitz, J. E. (2020). "Carbon pricing" special issue in the European economic review. *European Economic Review*. www.ncbi.nlm.nih.gov/pmc/articles/PMC7180378/

Integrated National Energy and Climate Plan 2021–2030. (2020). https://energy.ec.europa.eu/system/files/2020-06/es_final_necp_main_en_0.pdf

Integrated National Energy and Climate Plan for France. (2020). https://energy.ec.europa.eu/system/files/2022-08/fr_final_necp_main_en.pdf

International Renewable Energy Agency. (2017). *Renewable Energy Benefits: Understanding the Socio-Economics*. www.irena.org/-/media/Files/IRENA/Agency/Publication/2017/Nov/IRENA_Understanding_Socio_Economics_2017.pdf?la=en&hash=C430B7EF772BA0E631190A75F7243B992211F102

Istil, S. A., Górecki, J., & Diemer, A. (2023). Study on certification criteria of building energy and environmental performance in the context of achieving climate neutrality. *Sustainability*, *15*(3): 2770. https://doi.org/10.3390/su15032770

Jiang, Y., Hu, Y., Asante, D., & Ampaw, E. M. (2022). The effects of executives' low-carbon cognition on corporate low-carbon performance: A study of managerial discretion in China. *Journal of Cleaner Production*, *356*: 131829. www.sciencedirect.com/science/article/pii/S0959652622016225

Johnstone, P., Rogge, K. S., Kivimaa, P., & Fratini, C. F. (2021). Exploring the re-emergence of industrial policy: Perceptions regarding low-carbon energy transitions in Germany, the United Kingdom, and Denmark. *Energy Research & Social Science, 70*: 101663. www.sciencedirect.com/science/article/pii/S2214629620304643

Kabeyi, M. J. B., & Olanrewaju, O. A. (2022). Sustainable energy transition for renewable and low carbon grid electricity generation and supply. *Frontiers in Energy Research, 9*: 743114. https://www.frontiersin.org/articles/10.3389/fenrg.2021.743114/full

Kim, A., Kwon, S. H., Cui, Y., Gao, Y., Jeong, S., Kim, K. S., Lee, S., & Pyo, S. G. (2021). A study on carbon capture and storage and its feasibility in energy-environment technologies. *Journal of Nanoscience and Nanotechnology, 21*(3): 2039–2045.

Kumar, A., Connell, J., & Bhattacharyya, A. (2021). Co-opetition for corporate social responsibility and sustainability: Drivers and success factors. *Sustainability Accounting, Management and Policy Journal, 12*(2): 291–315. www.emerald.com/insight/content/doi/10.1108/SAMPJ-03-2020-0063/full/html

Kumar, S. S., & Rathore, K. (2023). Renewable energy for sustainable development goal of clean and affordable energy. *International Journal of Materials Manufacturing and Sustainable Technologies*. https://www.semanticscholar.org/paper/Renewable-Energy-for-Sustainable-Development-Goal-Kumar-Rathore/ca39b5bf7c0328eb38ab445efddb5dc964c7eef4

Kyriakopoulos, G. L., Streimikiene, D., & Baležentis, T. (2022). Addressing challenges of low-carbon energy transition. *Energies, 15*(15): 5718. https://doi.org/10.3390/en15155718

Lackner, M. (2021). Energy efficiency: Comparison of different systems and technologies. In: Lackner, M., Sajjadi, B., & Chen, W. Y. (Eds.), *Handbook of Climate Change Mitigation and Adaptation*. Springer, New York. https://doi.org/10.1007/978-1-4614-6431-0_24-2

Liu, J., Bao, L., Ye, B., & Wang, J. (2023). Low-carbon progressive design for sustainable development of buildings. *Indoor and Built Environment, 32*: 1638–1656. https://doi.org/10.1177/1420326X231158020

Lowitzsch, J., Hoicka, C. E., & van Tulder, F. J. (2020). Renewable energy communities under the 2019 European clean energy package—governance model for the energy clusters of the future? *Renewable and Sustainable Energy Reviews*. https://doi.org/10.1016/j.rser.2019.109489

Lyons, R. K., & Viswanath-Natraj, G. (2023). What keeps stablecoins stable? *Journal of International Money and Finance*. www.sciencedirect.com/science/article/pii/S0261560622001802

Madushika, U. G., Ramachandra, T., Karunasena, G., & Udakara, P. A. (2023). Energy retrofitting technologies of buildings: A review-based assessment. *Energies, 16*(13): 4924. https://doi.org/10.3390/en16134924

Majid, M. A. (2020). Renewable energy for sustainable development in India: Current status, future prospects, challenges, employment, and investment opportunities. *Energy, Sustainability and Society*. https://link.springer.com/article/10.1186/s13705-019-0232-1

Maradin, D. (2021). Advantages and disadvantages of renewable energy sources utilization. *International Journal of Energy Economics and Policy, 11*(3), 176–183. https://econjournals.com/index.php/ijeep/article/view/11027

Maware, C., & Parsley, D. M. (2022). The challenges of lean transformation and implementation in the manufacturing sector. *Sustainability, 14*(10): 6287. https://doi.org/10.3390/su14106287

Mayer, J., Dugan, A., Bachner, G., & Steininger, K. W. (2021). Is caheprbon pricing regressive? Insights from a recursive-dynamic CGE analysis with heterogeneous households for Austria. *Energy Economics*. www.sciencedirect.com/science/article/pii/S0140988321005181

Ministry of the Environment and Energy. (2017). *Sweden Climate Policy Framework*. Government Offices of Sweden. https://climate-laws.org/documents/the-swedish-climate-policy-framework_1f39?q=Sweden+Climate+Policy+Framework+&l=france&c=Policies&id=the-swedish-climate-policy-framework_747b

Mohsin, M., Rasheed, A. K., Sun, H., Zhang, J., & Iram, R. (2019). Developing low carbon economies: An aggregated composite index based on carbon emissions. *Sustainable Energy Technologies and Assessments, 33*: 11–22. https://doi.org/10.1016/j.seta.2019.07.006

Mohsin, M., Taghizadeh-Hesary, F., & Panthamit, N. (2021). Developing low carbon finance index: Evidence from developed and developing economies. *Finance Research Letters, 38*: 101168. https://doi.org/10.1016/j.frl.2020.101168

Najjarpour, M., & Tousi, B. (2024). Probabilistic reactive power flow optimization of distribution system in presence of distributed units uncertainty using combination of improved taguchi method and dandelion algorithm. *International Journal of Engineering, 37*(1): 37–47. https://doi.org/10.5829/ije.2024.37.01a.04

Nath, U. K., & Sen, R. (2021). A comparative review on renewable energy application, difficulties and future prospect. *2021 Innovations in Energy Management and Renewable Resources (52042)*: 1–5. https://doi.org/10.1109/IEMRE52042.2021.9386520

National Climate Agreement. (2019). https://climate-laws.org/documents/national-climate-agreement_1b62?q=netherland+climate+agreement&id=national-climate-agreement_aecf

Norwegian Ministry of Climate and Environment. (2021). *Norway's Climate Action Plan for 2021–2030. Meld. St. 13 (2020–2021) Report to the Storting (White Paper).* www.regjeringen.no/contentassets/a78ecf5ad-2344fa5ae4a394412ef8975/en-gb/pdfs/stm202020210013000engpdfs.pdf

OECD. (2022). *OECD Environmental Performance Reviews: Norway 2022, OECD Environmental Performance Reviews.* OECD Publishing, Paris. https://doi.org/10.1787/59e71c13-en.

Palm, J., & Bryngelsson, E. (2023). Energy efficiency at building sites: barriers and drivers. *Energy Efficiency, 16*(2): Article 7. https://doi.org/10.1007/s12053-023-10088-7

Panda, S., Mohanty, S., Rout, P. K., Sahu, B. K., & Parida, S. M. (2022). An insight into the integration of distributed energy resources and energy storage systems with smart distribution networks using demand-side management. *Applied Sciences, 12*(17): 8914. www.mdpi.com/2076-3417/12/17/8914

Paramati, S. R., Shahzad, U., & Doğan, B. (2022). The role of environmental technology for energy demand and energy efficiency: Evidence from OECD countries. *Renewable and Sustainable Energy Reviews.* www.sciencedirect.com/science/article/pii/S136403212101008X

Peng, K., Feng, K., Chen, B., et al. (2023). The global power sector's low-carbon transition may enhance sustainable development goal achievement. *Nature Communications, 14*: 3144. https://doi.org/10.1038/s41467-023-38987-4

Perrone, D., Castiglione, T., Morrone, P., Pantano, F., & Bova, S. (2023). Energetic, economic and environmental performance analysis of a micro-combined cooling, heating and power (CCHP) system based on biomass gasification. *Energies, 16*(19): 6911. https://doi.org/10.3390/en16196911

Pettifor, H., & Wilson, C. (2020). Low carbon innovations for mobility, food, homes and energy: A synthesis of consumer attributes. *Renewable and Sustainable Energy Reviews, 119*: 109629. www.sciencedirect.com/science/article/pii/S1364032120302458

Polzin, F., & Sanders, M. (2020). How to finance the transition to low-carbon energy in Europe? *Energy Policy, 147*: 111863. www.sciencedirect.com/science/article/pii/S0301421520305802

Popescu, M. (2022). Energy efficiency in electric transportation systems. *Energies, 15*(21): 8177. https://doi.org/10.3390/en15218177

Radtke, J., Yildiz, Ö., & Roth, L. (2022). Does energy community membership change sustainable attitudes and behavioral patterns? Empirical evidence from community wind energy in Germany. *Energies, 15*(3): 822. www.mdpi.com/1996-1073/15/3/822

Ruparathna, R., Hewage, K. N., & Sadiq, R. (2016). Improving the energy efficiency of the existing building stock: A critical review of commercial and institutional buildings. *Renewable & Sustainable Energy Reviews, 53*: 1032–1045.

Sánchez, A., Zhang, Q., Martín, M., & Vega, P. (2022). Towards a new renewable power system using energy storage: An economic and social analysis. *Energy Conversion and Management, 263*: 115792. www.sciencedirect.com/science/article/pii/S0196890421012322

Saraji, M. K., & Streimikiene, D. (2023). Challenges to the low carbon energy transition: A systematic literature review and research agenda. *Energy Strategy Reviews, 49*: 100117. www.sciencedirect.com/science/article/pii/S2211467X2300113X

Sarkodie, S. A., Adams, S., & Leirvik, T. (2020). Foreign direct investment and renewable energy in climate change mitigation: Does governance matter? *Journal of Cleaner Production, 263*: 121346. www.sciencedirect.com/science/article/pii/S0959652620313093

Shi, B., Li, N., Gao, Q., & Li, G. (2022). Market incentives, carbon quota allocation and carbon emission reduction: Evidence from China's carbon trading pilot policy. *Journal of Environmental Management.* www.sciencedirect.com/science/article/pii/S0301479722012233

Soares, F. J., Madureira, A. G., Pages, A. P., Barbosa, A., Coelho, A., Cassola, F., Ribeiro, F., Viana, J., Andrade, J. P., Dorokhova, M. F., Morais, N., Wyrsch, N., & Sørensen, T. (2021). FEEdBACk: An ICT-based platform to increase energy efficiency through buildings' consumer engagement. *Energies, 14*(6): 1524. https://doi.org/10.3390/en14061524

Song, J., Dessie, B. B., & Gao, L. (2023). Analysis and comparison of daylighting technologies: Light pipe, optical fiber, and heliostat. *Sustainability, 15*(14): 11044. https://doi.org/10.3390/su151411044

Sovacool, B. K., & Griffiths, S. (2020). The cultural barriers to a low-carbon future: A review of six mobility and energy transitions across 28 countries. *Renewable and Sustainable Energy Reviews, 119*, 109568. https://www.sciencedirect.com/science/article/pii/S1364032119307774

Sovacool, B. K., Lipson, M. M., & Chard, R. (2019). Temporality, vulnerability, and energy justice in household low carbon innovations. *Energy Policy, 128*: 495–504. https://doi.org/10.1016/j.enpol.2019.01.010

Sovacool, B. K., Turnheim, B., Martiskainen, M., Brown, D., & Kivimaa, P. (2020). Guides or gatekeepers? Incumbent-oriented transition intermediaries in a low-carbon era. *Energy Research & Social Science*, *66*: 101490. www.sciencedirect.com/science/article/pii/S2214629620300670

Streimikiene, D., Kyriakopoulos, G. L., & Lekavicius, V. (2021). Energy poverty and low carbon just energy transition: Comparative study in Lithuania and Greece. *Social Indicators Research*, *158*: 89–113. https://doi.org/10.1007/s11205-021-02685-9

Sukhawatthanakun, K., Roopsing, T., & Silpcharu, T. (2023). Industrial procurement management efficiency guidelines: Perform excellence through organisational change strategies. *International Journal of Professional Business Review*, *8*(4): e01770. https://doi.org/10.26668/businessreview/2023.v8i4.1770

Tian, J., Yu, L., Xue, R., Zhuang, S., & Shan, Y. (2022). Global low-carbon energy transition in the post-COVID-19 era. *Applied Energy*, *306*: 117968. https://doi.org/10.1016/j.apenergy.2021.117968

Wang, T., Shen, B., Springer, C. H., & Hou, J. (2021). What prevents us from taking low-carbon actions? A comprehensive review of influencing factors affecting low-carbon behaviors. *Energy Research & Social Science*, *71*: 101826. www.sciencedirect.com/science/article/pii/S2214629620304199

Wang, X., & Lo, K. (2021). Just transition: A conceptual review. *Energy Research & Social Science*. https://doi.org/10.1016/j.erss.2021.102291

Wang, Y., Wang, J. Q., Ma, S., Jiao, X., Wang, H., & Pu, W. (2023). Research on procurement management of engineering projects in state-owned machinery manufacturing enterprises. *Highlights in Science, Engineering and Technology*, *62*: 133–136. https://doi.org/10.54097/hset.v62i.10435

Wijaya, D. I., & Kokchang, P. (2023). Factors influencing generation Z's pro-environmental behavior towards Indonesia's energy transition. *Sustainability*, *15*(18): 13485. www.mdpi.com/2071-1050/15/18/13485

Wu, X., Xu, Y., Lou, Y., & Chen, Y. (2018). Low carbon transition in a distributed energy system regulated by localized energy markets. *Energy Policy*, *122*: 474–485. https://doi.org/10.1016/j.enpol.2018.08.008

Wu, Y., Martens, P., & Krafft, T. (2022). Public awareness, lifestyle and low-carbon city transformation in China: A systematic literature review. *Sustainability*, *14*(16): 10121. www.mdpi.com/2071-1050/14/16/10121

Xu, X., Wei, Z., Ji, Q., Wang, C., & Gao, G. (2019). Global renewable energy development: Influencing factors, trend predictions and countermeasures. *Resources Policy*. https://www.sciencedirect.com/science/article/pii/S0301420719303174

Xu, Z. Y., Mao, H. C., Liu, D. S., & Wang, R. (2018). Waste heat recovery of power plant with large scale serial absorption heat pumps. *Energy*, *165*(part B): 1097–1105. https://doi.org/10.1016/J.ENERGY.2018.10.052

Yan, X., Abdalla, A. A., Zhu, G., Uslu, Y. D., & others. (2024). Does natural resources matter? Nexus among renewable energy policies, technological innovation, environmental protection, and economic growth. *Energy Strategy Reviews*. www.sciencedirect.com/science/article/pii/S2211467X23002225

Zhang, J., Ouyang, Y., Ballesteros-Pérez, P., & Li, H. (2021a). Understanding the impact of environmental regulations on green technology innovation efficiency in the construction industry. *Sustainable Cities and Society*. https://doi.org/10.1016/j.scs.2020.102647

Zhang, S., An, K., Li, J., Weng, Y., & Zhang, S. (2021b). Incorporating health co-benefits into technology pathways to achieve China's 2060 carbon neutrality goal: a modelling study. *The Lancet Planetary Health*, *5*(11): e793–e801. https://doi.org/10.1016/S2542-5196(21)00252-7

Zhang, S., & Andrews-Speed, P. (2020). State versus market in China's low-carbon energy transition: An institutional perspective. *Energy Research & Social Science*, *61*: 101344. www.sciencedirect.com/science/article/pii/S2214629620300803

Zhao, X., Mahendru, M., Ma, X., Rao, A., & Shang, Y. (2022). Impacts of environmental regulations on green economic growth in China: New guidelines regarding renewable energy and energy efficiency. *Renewable Energy*. www.sciencedirect.com/science/article/pii/S0960148122000866

2 Energy Poverty

Dalia Streimikiene

2.1 ENERGY POVERTY CONCEPTS

2.1.1 Energy Inequalities and Justice

The concepts of energy deprivation, energy poverty, fuel poverty and linked issues of energy affordability, vulnerability, equity, inequality and justice have been widely discussed on international and country levels. Nonetheless, the forms of energy deprivation or drivers of it differ significantly among countries.

The large body of literature shows that energy poverty is closely linked to material deprivation and poor or marginalized groups of society are especially vulnerable to energy or fuel poverty, as health problems are specifically impactful for elderly, chronically sick or disabled people (Buzar, 2007; Liddell et al., 2012; Moore, 2012; Brunner et al., 2013; Bouzarovski and Petrova, 2015; Pye et al., 2015; Rademaekers et al., 2016; Thomson et al., 2019). Energy inequalities and consequently distributive injustice appear because of household's income inequality, differences in access to energy infrastructure and the energy efficiency situation of buildings, including installations and electric appliances used at homes (Jenkins et al., 2016; Dubois, 2012; Nussbaumer et al., 2012; Day et al., 2016; Bouzarovski and Tirado Herrero, 2015; Maxim et al., 2016). Scholars agree that in order to combat this distributional injustice, vulnerable groups of the society need to be targeted with social support and other state aid (Boardman, 2010; Day and Hitchings, 2011; Bouzarovski et al., 2012; Sovacool et al., 2014; Deller, 2016).

On a macro-scale energy poverty is analyzed based on energy services deprivation indicators and based on energy inequality data as energy inequalities are linked first of all with the fact that specific groups of society are being affected by energy services deprivation in different way and with different consequences (Wright, 2004; Liddell et al., 2012; Day, 2010; Hils, 2011; Tirado Herrero and Ürge-Vorsatz, 2012; Viggers et al., 2013; Snell et al., 2014; Thomson and Snell, 2013; Bouzarovski, 2014; Simcock et al., 2016). Therefore, it possible to look on energy poverty from three different angles: energy affordability, inequality in energy service deprivation among different groups of society and the energy efficiency situation. This allows one to define what are the major types of energy poverty prevailing in different countries.

Some resources, like energy infrastructure, are unevenly distributed in the territory and distributional justice is required for fair treatment of all people. Therefore, the location of energy infrastructure is the main reason of distributional energy injustices (Jenkins et al., 2016).

Non-recognition of the need of specific social groups in terms of energy affordability can be treated as recognition injustice. This lack of recognition of energy justice has been obvious in the United Kingdom policies dealing with fuel poverty in the country, as institutions didn't take into account the particular needs of various groups of society, like the elderly, sick or other vulnerable populations requiring higher-than-average room temperatures because of health reasons (Boardman, 1991).

There is also a problem of unequal representation of societal groups in various institutions like international, national and local administrative bodies and other non-governmental institutions having an impact on decision-making. There are also gender, age, social status etc. inequalities in government and other institutions as some social groups are underrepresented in decision-making bodies relevant to the energy sector. Therefore, according to energy procedural justice this type of

DOI: 10.1201/9781003517481-2

energy inequality and injustice needs to be taken into account and policies and measures proposed to reduce energy inequalities (Bickerstaff et al., 2013).

So, in current debate on energy inequalities, energy justice can be treated as a crosscutting issue in the social science research agenda. Therefore, energy justice principles should be applied in energy policies, energy supply systems, energy activism etc. A conceptual analysis of energy equity, inequality and energy justice issues is necessary. The three major types of energy justice can be defined as: distributional, recognition and procedural (Bickerstaff et al., 2013; Jenkins et al., 2014, 2016).

Therefore, by analyzing the energy justice concept, the main focus should be allocated to the following questions: where energy inequalities or energy injustices emerge, which society members are ignored and what measures should be applied for revealing and reducing energy injustice.

Starting from distributional energy justice, it is necessary to acknowledge the physically uneven allocation of resources and linked benefits and costs as well unequal distribution of associated responsibilities for these resources. It is necessary to ensure equity in resource distribution among all members of society notwithstanding income, race, age, gender etc. Research dealing with distributional energy justice requires scholars to study where energy injustice occurs around the world. The important role is allocated to the location of energy generation facilities such as power plants as many concerns about justice arise among communities in the nearby environment. Concurrently, energy poverty research has raised the distributional justice issues linked to rising energy prices and associated energy burdens (Bickerstaff et al., 2013).

Energy recognition justice is linked to the requirement that all individuals must be fairly represented in decision-making. For all social groups the complete and equal political rights should be allocated including the energy sector. Problems linked to the lack of recognition of specific societal groups are evident from cultural and political domination of some groups and misrepresentation of other societal groups. There is problem not just of recognition but also misrecognition. Recognition-based energy justice requires scholars to analyze the situation of society members who are ignored or misrepresented in political and other processes. Scholars showed in their research that there is struggle of ageing or disabled population groups for equal representation of their needs in the resource distribution processes within society (Jenkins et al., 2016; Sovacool et al., 2014).

Energy procedural justice deals with the problems of access to decision-making procedures that rule the distribution of resources analyzed earlier. It shows the necessity of equitable procedures by engaging all stakeholders in a non-discriminatory way. Procedural justice in energy is reinforced by providing equal access to legal systems and decision-making processes. Therefore, energy procedural justice requires scholars to study the equity in decision-making processes and engagement of various stakeholders. Warren and McFadyen (2010) in their study revealed the impact of community ownership in wind farms on decision-making processes including acceptance of wind farms by local communities. Energy cooperatives are new form of community owned renewable energy generation facilities showing a good example of how to deal with Not in My Backyard (NIMBY) syndrome and reduce social barriers for renewable energy project acceptance. Therefore, the consolidation of three energy justice mechanisms can be achieved by ensuring better engagement via local knowledge mobilization, better information dissemination and disclosure, and better representation of all stakeholders in various institutions (Jenkins et al., 2014).

The main role of the energy justice concept is to reveal three main forms of energy inequalities and to provide energy-just policy recommendations through the whole energy system. Energy policies and measures often target just specific issues of the energy sector to the detriment of the overall effectiveness of the energy sector. Therefore, the holistic "systems" thinking is necessary for dealing with energy inequalities and implementing an energy justice approach in policies.

Hall (2013) has distinguished that it is necessary to "widen the scope of energy justice . . . to dislodge debates from where they currently stand . . . towards a more nuanced understanding of energy"; therefore, the energy justice concept offers numerous visions in this respect and allows one to find policies and measures to reduce energy inequalities and injustice.

2.1.2 Energy Vulnerability and Resilience

Based on literature the main measure of energy vulnerability is the expected poverty, where energy vulnerability is addressed by applying a measure of probability that the household will fall into poverty (Middlemiss and Gillard, 2015). This definition is useful for policy analysis for assessing energy vulnerability in order to compare poverty alleviation measures (Bouzarovski and Tirado Herrero, 2015; Dall'O et al., 2015).

The main literature body dealing with poverty is dealing with individuals that were below the poverty line at some point in time. Poverty research has focused stereotypically on the drivers of poverty, the measures for escape or decline of poverty, and more recently on specific poverty escape trajectories, such as escape over time, compared to only temporary ones (Redmond and Hutton, 2000).

Poverty is usually measured ex-post. The literature on the dynamics of poverty pays less attention to households close to the poverty line, many of which may be especially vulnerable ex-ante to short-term rather than persistent escapes from poverty. Assessing the poverty vulnerability could be used for establishing effective preventive measures to ensure that households do not fall into poverty and are resilient to the possible shocks or various stressors. The term "vulnerability" shows that vulnerability can be described in different settings like vulnerability to poverty, an indication of poverty, or part of the multi-dimensional type of poverty, and it is typically being underlined by risk of falling into poverty (Pritchett et al., 2000).

The main approach followed in this book is "vulnerability to poverty." It indicates vulnerability as being a reason for poverty, not a symptom of poverty or a certain component of poverty. However, this definition of vulnerability is also valuable for measuring vulnerability and can used to make comparative assessment of poverty measures (Pritchett et al., 2000). Additionally, though in literature we can find that vulnerability is defined in various ways, there is a main approach applied in literature, meaning that vulnerability is more than ordinary exposure to risk; it also deals with deprivation (Chaudhuri et al., 2002).

Research about vulnerability usually deals with modelling of risks that can influence consumption outlines of individuals and result in a loss of their well-being (Dercon and Krishnan, 2000; Chaudhuri et al., 2002).

In terms of vulnerability to poverty, Heitzmann et al. (2002) present a vulnerability analysis using a "risk chain" that breaks down vulnerabilities into several categories like risk, response and outcomes. Other scholars prefer welfare promoting factors, by addressing the links among welfare, risks and vulnerability. Dercon and Krishnan (2000) developed a system consisting of human capital, remittances, savings, education, health, etc. necessary to deliver prosperity, income, and opportunities for individuals. The risks associated with these internal welfare factors can make households vulnerable and are also incorporated in this system. For example, disasters can damage wealth or reduce output, the uncertainties of public services linked to health and education systems in the country can affect these capacities, and poor contract performance can affect returns such as earnings.

Thus, the factors contributing to energy shortage increase at homes of individuals can be attributed to the concept of energy vulnerability, which can be basically defined as the tendency of a person to become unable to ensure the physically and socially necessary level of energy consumption. The energy vulnerability concept highlights the importance of energy resilience (Bouzarovski, 2014). Energy insecurity (Paugam, 1995) is also important in this sense as it emphasizes that drivers of energy vulnerability may belong to circumstances within the household or outside. In Table 2.1 external and internal drivers of energy vulnerability are described.

TABLE 2.1
Drivers of Energy Vulnerability

Drivers	Description	Type
Low access to energy services	Deprived infrastructure of energy supply and low convenience of access to energy carriers suitable for satisfying household's energy needs	External drivers
Low energy affordability	High cost of energy and low incomes of households, high taxes on energy sources and insufficient social support measures.	External and internal
Low energy supply flexibility	Inability to move to select energy service provider that is appropriate to meet household's needs.	Internal
Low energy efficiency	High energy losses during energy conversion processes in residential sector	Internal
Unsatisfied energy needs	Discrepancy between energy needs of households and available energy supply due to various social, economic, health etc. reasons	Internal
Insufficient knowledge about energy saving	Insufficient knowledge about energy saving practices at home and available support programs	Internal and external

Source: created by authors based on Bouzarovski et al., (2012); Thomson et al. (2019).

One can notice from Table 2.1 that the greatest energy vulnerability problems are internal and depend on the household. Therefore, based on these drivers it is possible to identify specific groups of households that may face energy poverty in the future. Also, it allows one to identify the main social, economic, political and infrastructural factors that have contributed to the emergence of energy vulnerable households in the present time. This is important in developed countries for households living in temporary housing arrangements. This is usually young people who are renting their houses and other residents of informal settlements (Bouzarovski et al., 2012). In developing countries, more important issues of energy vulnerability are linked to the lack of technical availability of energy carriers and financial shortages not allowing one to meet the socially necessary energy needs of a household.

Energy vulnerability has been analyzed by scholars at various levels ranging from nations to the individual, but energy vulnerability is evident at the individual level as even in the case of countries that are not considered as energy vulnerable, households may still be energy vulnerable due to various internal factors described in Table 2.1. Definitely, the energy vulnerability of near energy poor households can lead them to fall into energy poverty in the case of various levels of risk, especially when they do not have the capacity or prohibition to manage consumption smoothly enough. In addition, with higher energy vulnerabilities, households close to the poverty line may choose safer but less profitable investments compared to households at energy poverty risk. Therefore, such types of investments would further undermine their ability to build energy resilience and eventually lift themselves out of poverty for a long time period. Therefore, it is crucial to study energy vulnerability of near energy poor households' energy poverty dynamics and to define the best pathway to energy poverty alleviation. Hence, energy resilience can be defined as "the ability of individuals, households, countries, and systems to mitigate, adapt, and recover from various energy supply linked and other shocks with the aim of reduction of chronic energy vulnerability." In analysis energy poverty dynamics, energy resilience can be seen as a set of specific skills that allows individuals to escape energy poverty on a permanent basis, even facing exposure of various shocks and stressors. The next section of the book analyzes energy poverty and its drivers by taking into account the definition of energy resilience as a set of capacities that can help households avoid energy vulnerability which can be seen as future energy poverty in the long run.

2.1.3 Energy Poverty Concepts Differences in EU Countries

The European Commission has adopted the term energy poverty or energy poverty, and the Energy Poverty Observatory (European Commission, 2021; Eurostat, 2021) has proposed four key indicators for measuring it in the European Union: the portion of people unable to keep their homes warm; the share of the population with arrears on utility bills; the share of high energy expenditure in total revenue (indicator 2M) and the indicator of low absolute energy expenditure (M/2). The low absolute energy expenditure indicator 2M reflects the percentage of households whose energy consumption in total income is twice the national median. The M/2 indicator shows the portion of households whose absolute energy costs are twice the national median. There are other supplementary indicators of energy poverty developed by the Energy Poverty Observatory, like energy prices for households; the share of population with leak, damp or rot in their dwelling; the share of energy expenditure in income by income quintile etc.

A theory of capabilities was also developed to underpin the system for measuring energy poverty. The theory of capabilities has been discussed in Day (2010) and other papers, but without providing an indicator framework for a systematic energy poverty review and lacking a clear theoretical model.

The capabilities approach is based on two related concepts: functionings and capabilities, linked to individuals (Day et al., 2016). Functionings are described as "beings and doings." They addressed such issues like being in good health and activities like working to earn money. Therefore, the individual capabilities are the actual opportunities for people to carry out the given functionings, if one chooses to do this. Because theorists such as A. Sen (1993) put a high importance on freedom to choose, he argued that capabilities of individuals should be the main object of concern instead of functionings. This is due to the fact that an emphasis on functionings in reality is dictation and requirement to live in a particular way. Promoting capabilities would allow one to maximize opportunities of individuals, but leaves room for them to decide freely what kind of life they would like to live. According to A. Sen, the development programs should be targeting the growth of the individuals' capabilities and the success of these programs should be assessed taking into account the achieved increase in capabilities of people.

The principal contribution to the development of theory of capabilities in order to address energy poverty issues has been made by Sovacool et al. (2014). Sovacool et al. (2014) postulated essential findings necessary for understanding energy affordability in terms of theory of capabilities based on equity. In research performed by Dall'O et al. (2015) energy use was conceptualized also from the point of view of theory of capabilities, based on the work of Amartya Sen (1993) and Martha Nussbaum (Nussbaumer et al., 2012). The link between energy consumption and human welfare has been established and energy poverty has been conceptualized on the basis of these links. Perceptions of energy poverty and vulnerability in relation to human potential allow for a wide range of policy and instrumental frameworks to achieve greater synergies between different policies. This conceptual approach allows one to analyze the problems of energy poverty from the perspective of climate change and the future of a carbon-free society. There were also other concepts of energy poverty like Boardman (1991) and Hills' (2011) energy concepts developed in the UK based on the relevant energy poverty issues for this country described earlier. It is necessary to stress that countries have applied different measures for combating energy poverty in their policy agendas based on their situation and therefore following the concept of energy poverty.

There is an opinion among scholars that the two terms (fuel and energy poverty) have a similar meaning in developed countries (Boardman, 2010; Bouzarovski et al., 2012) and the difference between these concepts was not distinguished by scholars for many years. However, some scholars like Li et al. (2014) started to argue that fuel and energy poverty have different meanings as they reveal different problems. For example, scholars (Li et al., 2014) insist that fuel poverty is mainly a problem of rich developed countries having a cool climate like the UK, Ireland and New Zealand. At the same time, according to Li et al. (2014) energy poverty arises in all climates but typically in poor developing countries, and both fuel and energy poverty are experienced together in a cold climate, as people can experience difficulties in getting access to electricity or up-to-date cooking services and also have problems linked to indoor heating. Therefore, based on such distinctions, energy affordability, thermal comfort and low income high cost are indicators of fuel poverty and access to electricity and modern cooking facilities are indicators of energy poverty.

However, this distinction shows several fundamental problems. It is necessary to find the answer to what kind of definition of energy deprivation should be applied for families living in relatively wealthy countries with a warm climate. Healy (2004) and Tirado-Herrero et al. (2014) discovered that there is high number of families lacking satisfactory energy services at home in such countries like Portugal, Greece, Spain, Italy and Malta having a warm climate. The vulnerable households in these countries also experience problems linked to heating in winter and cooling in summer. Energy poverty in these countries can be credited to a mixture of insufficient income, limited infrastructure, energy efficiency problems etc. Another example can be found in Central Asia and South American countries with middle-income per capita and also facing problems of inadequate energy services at home due to limited energy accessibility and affordability. Therefore, it is necessary to stress that both fuel and energy poverty can be treated as synonyms and such an approach is being followed in this book. Both these definitions of energy deprivation indicate several important energy circumstances available in the residential sector that do not allow for equal participation in the lifestyles and customs characteristic to that specific society related to limited energy accessibility, affordability and low energy use efficiency (Buzar, 2007).

Table 2.2 presents the main concepts of energy poverty.

Taken together, all key explanations of energy poverty can be defined as the disproportionate share of household energy costs in total income and the inability to use enough energy services to meet necessary needs due to unreasonable prices or other constraints, such as poorly insulated buildings, lack of energy regulation options or renovation for various reasons, or a ban on switching to another energy supplier or insufficient development of the necessary energy infrastructure etc. Therefore, it can be observed that energy poverty is a multifaceted and broad topic, encompassing economic, social, behavioral, political, and other country-specific conditions (BPI, 2014; Thomson and Snell, 2013).

Countries have applied different definitions of energy poverty. B. Boardman (2010) highlighted that "fuel poverty depends on the definition; but the definition depends on who you want to focus on and this involves political judgment." This is particularly important if policy-makers seek to develop their policies and measures on specific definitions. For example, in the UK, which is acknowledged for its pioneer energy poverty definitions and studies, factual policies and measures are not associated with the scientific definitions of energy poverty. Therefore, there is a clear gap between theory and policy. More complex analysis and research is needed on the country level to identify the relevant target groups for energy poverty alleviation policies in specific country. Those groups who are suffering from energy poverty most need to be clearly identified.

For example, in the United Kingdom, the definition of fuel poverty including its measurements has been evolving through the years. Recent definition allows one to compare the targeting households' groups by using two different methods. Until 2013, the official energy poverty indicator in UK was the 2M indicator and then the Low-Income High-Cost indicator was introduced in 2013. Selection of the Low-Income High-Cost indicator had an impact on the size of the energy poor households group in the UK. The number of energy poor people has declined from 3.2 to 2.28 million. people since the new official energy poverty measure was introduced in the UK. Moreover, according to the previous energy poverty definition, which is the 2M measure, single people were dominating in the energy-poor group (53%) and households with children made just 11%. With the introduction of a new energy poverty framework, more households with children (27%) and fewer single households (26%) were formally acknowledged as energy poor. The energy-poor people group was identified throughout the official measurement and it was just partly overlapping with the group of people classified as energy-poor based on their self-definition. Therefore, just one in seven

TABLE 2.2
Energy Poverty Concepts

Sources	Energy poverty concepts
Boardman (1991)	Fuel or energy poverty show the inability of households to purchase heating and other energy services necessary at home if they need to devote more than 10% of their income to cover their energy bills.
Foster et al. (2000)	Energy poverty can be assessed as the proportion of households living below the poverty line established in the country on the basis of total household expenditure.
Hills (2011)	Energy poverty is connected to low incomes and high energy consumption, supposing that households live in energy inefficient buildings and they also experience income poverty; therefore they have to pay high energy bills in order to keep their homes warm in line with recommended heating standards.
Barnes (2007)	Energy poverty is assessed on the basis of the energy poverty line, which is defined as the minimum final energy consumption. Therefore, energy poverty is comprehended as the energy consumption limit needed to preserve their good health. Higher energy consumption outside this limit ensures a higher level of economic prosperity.
Thomson at al. (2019)	Energy poverty reveals a condition in which households use a disproportionate share of their income to meet their energy requirements, and this is being achieved at the expense of reducing food consumption. In such situations households are forced into energy expenditures by accepting reduced levels of thermal comfort, or they are even forced to reduce food and energy consumption.
Day et al. (2016)	Energy poverty is the inability of households to meet the needs arising directly or indirectly due to the absence of access to the affordable, reliable and secure energy services needed to meet their energy needs.
De Quero and Lapostolet (2009)	Energy poverty is a state in which families face a number of essential difficulties in meeting their basic energy needs, due to financial constraints and/or poor conditions of their houses
Bouzarovski and Tirado Herrero (2015)	Energy poverty is a state in which families are "trapped" in poorly insulated energy inefficient buildings with outdated, inflexible district heating and the ban on disconnecting from district heating and switching to other energy suppliers.
Bouzarovski and Cauvain (2016)	Energy poverty is a situation where families do not have access to the necessary heat in their homes due to low income or cannot provide the necessary thermal comfort due to the low energy efficiency of their homes.
Thomson and Snell (2013)	Energy poverty is the inability of households to properly heat their houses due to damp and moldy buildings and arrears on utility bills.
Rademaekers et al. (2016)	Energy poverty is a state in which families have to choose whether to heat or eat.
Bouzarovski (2014)	Energy poverty is a situation where families are unable to get access to necessary socially desirable energy services at their homes.
Pye et al. (2015)	Energy poverty is a situation where households are unable to heat their homes sufficiently or do not have access to other essential energy services at an affordable price in their houses.
OECD/IEA (2010)	Energy poverty means that households in energy poverty are not able to access basic energy services like electricity, heating, refrigeration, cooking etc.
European Commission and Social Committee (2011)	Energy poverty occurs when households find themselves in a situation where it is not possible for them to provide adequate heating for their homes at affordable prices and to have access to other energy services like electricity, lighting, internet or other facilities at affordable prices.
EUROSTAT (2021)	Indicators for measuring energy poverty are: the share of households unable to maintain adequate heating in their homes; the share of households unable to pay utility bills; the proportion of households experiencing problems with leaking roofs, damp walls etc.
European Commission (2021)	Energy poverty occurs when households experience the absence of satisfactory energy services in their homes

Source: created by authors.

energy poor people responded during the survey that they could not keep their houses adequately warm.

In Spain, having a warm climate and specific social and demographic composition, the analysis of energy poverty shows that different energy poverty indicators were applied in the country, which provided for a different trend in energy poverty. For example, the share of households spending more than 10% of their income for energy needs has increased from 2006 to 2010; however, the share of households who reported in surveys that they can't keep their homes adequately warm has decreased in the same period. Therefore, the use of different energy poverty definitions and corresponding energy poverty indicators for monitoring of factual situations leads to different target groups and different policies to support them. This makes the fight against energy poverty difficult. So, the policy-makers need to trade off among the different potential beneficiaries in shaping energy poverty alleviation policies or they need to develop complex policy packages by covering all possible groups of population that are at risk of energy poverty.

Greece represents another southern country of Europe having energy poverty problems due to the recent austerity policies. The COVID-19 pandemic has an even more severe impact on energy vulnerability of Greeks and the current low-carbon energy transition will become even more challenging for Greece. Though there are several studies exploring energy poverty in Greece (Dagoumas and Kitsios, 2014; Alexandri and Androutsopoulos, 2020) these studies do not provide a clear delineation of energy poverty and the low-carbon energy transition nexus. However, based on analysis of these studies, it possible to say that main energy poverty indicators applied in the country were showing quite negative trends. The hidden energy poverty indicators (M/2) display negative trends during the 2010–2018 period. Just 2M indicator dynamics showed positive trend in Greece during 2010–2015. Though the share of households with arrears on energy bills have been declining in Greece since 2016, Greece distinguishes itself with a very high share of arrears on energy bills comparing with other EU countries; therefore more efforts are necessary for Greece to catch up to other EU member states according to this indicator of energy poverty. Policies aiming at energy poverty alleviation in Greece are mainly short-term policies like use of Social Residential Tariffs to protect vulnerable groups of the population and provide disconnection protection. Yet, Greece distinguishes itself with limited long-term energy poverty alleviation measures especially in the field of financing and funding of improvements in the energy situation, especially energy renovation of residential buildings. The main dominating policies and measures are linked to information dissemination and awareness raising.

Hungary is post-socialist country in terms of the links among energy, welfare and housing conditions. In the last decade about 30% of people in Hungary were experiencing energy poverty. The definition of energy poverty is based on an assessment of proportion of households having a high energy burden or the high ratio between energy expenditures and income. The share of energy consumption expenditures in the household's income has been steadily increasing since 2000 (Tirado Herrero, 2013). Energy poverty in Hungary is essentially a post-socialist phenomenon depending mostly on the central planning legacy of a socialist country. The structural legacies of the past are primarily evident from the country's poorly insulated and energy inefficient residential buildings, many of which were constructed during the socialist period. For example, the detached houses built before the 1990s are the most common residential buildings in rural areas and are the most popular type of dwelling making up more than 50% of Hungary's total living space (Tirado Herrero, 2013). Because of the peculiarities of these residential buildings, they are also the most energy inefficient type of building with space heating demands ranging from 300 to 500 kWh/m^2 (Tirado Herrero, 2013).

Therefore, in Hungary, energy inefficient residential buildings supplied with district heating are the main reason for the unconventional typology of energy poverty due to the high energy costs, the lack of individual heating regulation and inability to switch to other heating sources (Tirado Herrero and Ūrge-Vorsatz, 2012). Due to high natural gas import dependency and extensive replacement of coal and oil boilers with more efficient gas boilers during 1990–1998, the subsidization of domestic energy prices was the main policy to deal with energy affordability in Hungary.

Households face energy vulnerability through a number of "coping strategies" (Brunner et al., 2013) and the main response of families to energy poverty is reflected in the big number of the population with arrears on utility bills. During 2005–2013 the share of population with arrears on utility bills increased from 15 to 25% due to continuous growth of energy prices. Furthermore, in addition to the increase in energy poverty, the vulnerable households deliberately chose to defer paying high energy bills to avoid running out of money during the winter, therefore passing on the financial consequences of their energy poverty for utilities. This energy poverty coping strategy put the households at risk of indebtedness, which will lead to the disconnection from suppliers. However, the disconnection of individual households from district heating is not technically possible in such types of residential buildings. Therefore, indebtedness and non-payment of households creates the energy vulnerability loop by putting utilities in a difficult financial situation with possibilities to invest in district heating system maintenance or upgrade. In addition, energy poverty results in substitution of modern energy carriers such as natural gas with solid polluting fuels for residential heating. This process is called "energy degradation" (Bouzarovski et al., 2012) and is a negative trend that occurred recently in several post-socialist countries in Central and Eastern Europe including Hungary as well (Fankhauser and Tepic, 2007). In 2011 more than 36% of Hungarians were dependent on solid fuels in their homes compared to 14% in 2005, though 96% of settlements in Hungary are connected to natural gas pipelines. More than half of all households in the bottom decile are using solid fuels for heating their homes. The lower-income households are consuming solid fuels due to energy poverty and this is indicator of increasing energy inequality of households in Hungary. Switching to firewood is more common in rural areas, where it is more accessible. Location interacts with other issues like household's age as older people being more deprived though they have greater needs for heating due to the health reasons (Tirado Herrero, 2013).

The situation of energy vulnerability in Hungary policy-makers to take actions towards solving these issues. However, these measures have brought short-term benefits to low-income households by allowing for a reduction in energy burdens because they have preferentially supported urban consumers of natural gas and district heating, while failing to provide relief to households in rural areas relying on bottled natural gas or firewood used for heating. Hungary has similar results according to data collected from the National Survey on Living Conditions and Household Budget. When comparing the 2M and the share of people unable to keep they houses adequately warm, the different results were obtained. The 2M method shows that less energy pours people (10%) in the country and low-income high-cost indicator indicates 21% of the population as energy pour. The 2M indicator showed that single households are dominating and low-income high-cost indicator showed that households with children are dominating in energy poverty. Both measurements provided that energy poverty is more spread among people living in detached houses.

Bulgaria, a neighboring country next to Hungary, does not yet have an official definition of energy poverty in national legal acts, however the similar problems of energy poverty are obvious in Bulgaria. The new National Recovery and Resilience Plan foresees that the legal definition of energy poverty will be adopted in 2022. There are few most important reasons of energy poverty in Bulgaria, related to high energy prices, low incomes and energy inefficient residential buildings in Bulgaria. There is opinion that definition of energy poverty should be built on objective indicators rather than on self-assessment. The main focus should be putted household's disposable income level, after covering the energy costs necessary for providing adequate warm in houses (Tzanev, 2019). There are several possible definitions of energy poverty due to the several causes of it like low household's income, high residential energy prices, and energy inefficient pour quality housing stock for residents.

Five energy poverty definitions and corresponding subjective, objective, and absolute indicators were developed in Bulgaria (Tzanev, 2019):

1. Households that spend more than 10% of their income on energy can be treated as energy pour.
2. Households which cannot afford adequate heating for they houses can be treated as energy pour. This concerns access to and benefits from heating.

3. Households that cannot pay energy bills on time and have accumulated arrears of utility bills are treated as energy pour.
4. If households have the disposable income below the average monthly cost of consumer goods after paying for the adequate heating of their homes, they are considered as energy pour.
5. If households do not have access to modern energy services they are treated as energy pour.

Comparing these five approaches to energy poverty, one can notice that the first indicator is less appropriate to Bulgaria's situation as Bulgarians spend more than 10% of their income on energy in average (Eurostat, 2021). The third indicator is also not relevant as it problematic to evaluate it in Bulgaria due to a lack of data. The fourth indicator is useful for qualitative and quantitative assessment of energy poverty and allows to differentiate energy poverty situation during winter and summer seasons. The report (Tzanev, 2019) recommends to divide all population into following groups: energy independent, at energy risk, i.e., households which are just above or below the level of providing adequate heating for their homes, and energy vulnerable households, i.e., households do not able to afford adequate heating for their homes.

Lithuania is a former Soviet state in northern Europe, experiencing a similar situation as Hungary and Bulgaria from the Soviet legacy, but situated in a colder climate zone. Though country does not have the official definition of energy poverty, the Lithuanian state seems to recognize the problem and make extraordinary efforts to address it. The National Energy and Climate Plan identifies energy poverty as a state where it is difficult or impossible for households to have adequate home heating or other essential energy services such as lighting, hot water supply etc. The plan foresees a number policies and measures to struggle against energy poverty, like financial support measures to cover utilities bills, building renovation programs, and information dissemination and education of energy customers. In addition to existing measures, a number of other policies and measures are foreseen in this plan, including greater protection of energy vulnerable consumers, the quickening of energy-efficient building renovation programs for households.

In practice, the term of socially vulnerable households is applied for residents entitled for social support provided to cover heating and hot water supply costs. Up to a certain income level, the entitled support covers all household expenditures linked to heating and hot water supply. Households which income is above the threshold but are still considered pour and still are entitled to subsidy in order to pay no more than 10% of the difference between their income and the threshold, which is a certain poverty line. Prior to COVID-19, the eligibility criteria were stricter, however during the pandemic, the criteria were relaxed to include more low-income households in support scheme. In addition, there is socially appropriate support for energy efficiency improvements of residential buildings, like JESSICA program. In the framework of JESICA program a subsidy for energy-efficient modernization of multi-apartment buildings is provided for low-income households to cover 100% of all energy renovation costs. This support is being provided just for only a few percent of the most vulnerable population, which is certainly less than the total population facing energy poverty in Lithuania (Streimikiene et al., 2021).

The long-term energy renovation strategy in Lithuania was adopted in 2020. The new strategy emphasizes the importance of renovation in the new National Recovery and Resilience Facility (RRF) plan. The strategic goal is to achieve situation than building stock do not use fossil fuels anymore by 2050. As residential buildings are among main energy consumers in Lithuania, energy renovation of residential buildings allows to achieve the double dividends, like energy poverty and GHG emission reduction.

Therefore, the energy deprivation problems in households are commonly described by the term of energy or fuel poverty. This conception is applied to apprehend the problems of insufficient energy accessibility or affordability experienced by vulnerable population, also encompassing infrastructural, economic, social, education and health concerns. Though there are disagreement among scholars on the same meaning of energy and fuel poverty concepts, in this book energy poverty and fuel poverty are treated as synonyms.

The various conceptual limitations connected to energy poverty has formed the necessity for investigating the association between energy poverty and other related concepts like energy equity, inequality, justice, affordability, vulnerability etc. by taking into account the wider socio-spatial area that underline the shortage of adequate energy services in households. Consequently, some concepts of energy poverty require broader theoretical statements that support all these paradigms. Taking into account all terminological messiness around the definitions of energy poverty, in this book just initial arguments on the fundamental associations among energy accessibility, affordability, equity, justice and state policy are provided. In summary it is necessary to stress that that all forms of household's energy deprivation have the same outcome: a shortage of adequate energy services at home and related discomfort and troubles.

Countries have applied different definitions of energy poverty. And it is possible to state, that energy poverty depends on the definition; but the definition depends on focus and this involves political judgment. Nevertheless, energy poverty definitions, drivers of energy poverty and target groups are very important for policy-makers for developing policies and measures to overcome energy poverty problems. Though there are many studies on energy poverty and various concepts of energy poverty exists, there is a clear gap between theory and policy and more complex analysis and research of country situation is necessary to define the relevant target groups for energy poverty alleviation policies in specific country.

By summarizing all key explanations of energy poverty, it can be defined as the disproportionate share of household energy costs in total income and the incapability to use enough energy services to meet necessary needs due to unreasonable prices or other constraints, such as poorly insulated buildings, lack of energy regulation options or renovation for various reasons, or a ban on switching to another energy supplier, or insufficient development of the necessary energy infrastructure etc. Therefore, energy poverty is a multifaceted topic, encompassing economic, social, behavioral, political, and other country-specific conditions.

Though there are various policies and measures created to address specific energy poverty issues, energy efficiency improvement can be considered as the major policy to deal with energy poverty problem in long-term as it assists families with energy savings by such measures as upgrades of heating and cooling systems, energy insulation of buildings, efficient electric appliances and lighting including energy saving behavior patterns. Energy efficiency improvements allows to reduce energy bills and save money for other purposes, has positive impact on living conditions, comfort, safety and health. Energy efficiency programs should target low-income vulnerable families as they are experiencing high energy low-income households.

Best practice in improving energy efficiency and energy affordability for low-income households may include setting state-level cost and savings targets. Utilities can improve their efficiency programs by including the use of various sources of funding, the adaptation of health and safety measures, cooperation with communities and the prioritization of austerity measures. Advanced financing options for apartment building owners and rural households can also be applied through public service obligation (PSO) measures. On-bill financing energy renovation programs that allow customers to save energy and recover their costs of renovation from reduced energy bills due to energy saving achieved. For multi-apartment buildings, important way to address individual incentives is to align efficiency incentives with building refinancing and renovation deadlines.

2.2 ENERGY POVERTY INDICATORS

There is no one, commonly established definition of energy poverty. For example, nutritional poverty is described by a minimum daily food intake, based on calories. However, it is not possible to measure universally what is absolute minimum level of necessary daily energy consumption. There are also measures of income poverty-established by World Bank. There is rate of $1.90 per day, established on the basis of purchasing power parity of world poorest countries (Ferreira et al., 2015). Nevertheless, there is no such common universal energy poverty measure. Many energy poverty measures were developed by scholars to help policy-makers to combat energy poverty by shaping relevant policies and measure all over the world and addressing the main issues of energy poverty in specific countries or regions (Nussbaumer et al., 2012; Legendre and Ricci, 2015; Culver, 2017; Tirado Herrero, 2013; Zhao et al., 2021; Seuret-Jimenez et al., 2020; Sokolowski et al., 2020; Spiliotis et al., 2020; Villalobos et al., 2021).

The analysis of various energy poverty concepts allowed to highlight that the lack of energy services can be understood as the broadest energy poverty definition, however it cannot be measured directly. The main problem is that there is no commonly defined basket of necessary energy services for households (Culver, 2017). So, in order to distinguish form energy, pour and not energy pour people, it is necessary to establish basic energy services basket and to define the lowest amount of necessary energy services from this basket. For example, cooking and lighting are the most common basic energy needs (Zhang et al., 2019). Nonetheless, it is not clear how to address energy needs linked to refrigeration, heating or cooling, mobility etc. So, the minimum necessary level of energy consumption cannot be defined and measured objectively; therefore, there are many alternative ways proposed to address energy poverty metrics. The main measurements are constructed on a surveillance of energy accessibility, the input of energy, the energy use outcomes evaluation and the quality of energy delivered (Wirawan and Gultom, 2021).

It is necessary to stress that the lack of energy services is not the real measure of energy poverty as it is necessary to take into account can the energy services be replaced by each other, also it is not clear which energy services are essential. Moreover, definition of energy poverty level is subjective (Siksnelyte-Butkiene, 2021).

As measures to reduce energy poverty were developed for different purposes, they can be classified according following major themes: 1) energy access 2) energy inputs, 3) outcomes of energy use; 4) the quality of energy delivered (Culver, 2017).

Therefore, observing of energy accessibility situation is often used as energy poverty metrics due to the problems of measuring use of necessary energy services. Energy access can be used as metrics of individuals access to electricity and modern cooking energy carriers. The electricity access is evaluated based on indicator of the electrification rate. This indicator is calculated as the share of the population that have been connected to the power supply network. The access to modern cooking is being evaluated by the use of liquified petroleum gas, electricity, biogas and modern biomass stoves for cooking. The number of population which do not have access to modern cooking is expressed as the share of population that apply traditional fuels for cooking: wood, charcoal, waste, crop residues etc. (IEA, 2011).

The use of indicators of accessibility for energy poverty metrics has many limitations (Siksnelyte-Butkiene, 2021). The simplicity of this indicators is major limit as by assessing access to electricity or modern cooking services, as it deals just with electricity and cooking issues as other energy needs remain uncovered by such indicators. In addition, the energy access indicators cover just household energy consumption. The lack of electricity for industries, public buildings, street lighting is not addressed by this energy poverty metrics. Besides that, energy accessibility does not take into account mobility, heating, cooling or other energy needs in households (Siksnelyte-Butkiene et al., 2021). Energy access indicators also do not show the quality of energy consumed. There is situation then households have connected to electricity however due to blackout they often do not have electricity in their homes. There are measurers of the quality of energy services provided, including the availability, reliability, adequacy, affordability, convenience, and safety (Siksnelyte-Butkiene et al., 2021; Culver, 2017).

Availability and reliability of electricity is measured by the number of hours of electricity supply during a day and whether those hours are then the electricity is mostly needed. Adequacy of electricity supply is measured by the fluctuations of voltage. Electricity affordability is usually measured by the price of electricity or the ratio between the price and real income of households.

Convenience and safety issues are linked to the collection and use of energy carriers for meeting households' energy needs (Mendoza et al., 2019).

Finally, assessment of energy accessibility by indicator of the share of the population using traditional fuels does not take into account consumer behavior factors. This complicates understanding of the main reasons of energy poverty. Families using traditional fuels for cooking is bad proxy for the families not having access to modern cooking services, as many people have access to modern cooking services but still continuing to use biomass. This is practice is named as fuel stacking (Masera et al., 2000).

Also, indicators on access to modern cooking services does not deliver information on the reasons of the energy poverty. In addition, energy poverty is usually higher in rural areas comparing to the income poverty, showing that the main issue is the distribution of modern stoves, not low income of population or high prices of energy. The main problem in rural areas is energy access, but, in urban areas the income poverty is major problem revealing that not access but energy affordability is the major cause of energy poverty. Application of energy access indicators as the measures of energy poverty effects the energy sector development programs as the main targets are: expanding access to electricity and distribution of modern cookstoves. But there are other measures, like solar home systems which can improve the living conditions and standards a lot but these issues are not addressed in energy access indicators (Kizilcec et al., 2021). Therefore, for developing relevant policies and measures necessary to overcome energy poverty, and promote economic and social development of pour countries, energy access indicators are not recommended by several scholars (Pagliaro and Meneguzzo, 2020).

Therefore, though energy access indicators are a simple metrics of energy poverty and is convenient in terms of data collection. It is binary indicator, "having/not having connection to the grid" and "having/not having access to modern cooking services." United Nations Development Program (UNDP), International Energy Agency (IEA), and the World Bank Group etc. have databases on electricity access and traditional cooking fuels usage collected mainly by national reports of countries (UNDP and WHO, 2009; IEA, 2011, 2016).

In conclusion, energy access information is simple and easy to collect but it does not capture important aspects of energy poverty like it does not consider energy used outside the household, does not measure other types of household energy consumption; does not take into account real usage for energy services and fact that people continue to use traditional fuels for cooking even having access to electricity as they do not have financial resources to pay for these modern energy services.

Because of serious limitations of using energy access indicators as energy poverty metrics, several scholars proposed alternative definitions of energy poverty by taking into account energy inputs, measured by amount of energy consumed or as disposable income spent on energy consumption by households. The main assumption is energy consumption reveals access to energy services. Energy input metrics varies based how energy consumption is measured and how the energy poverty line is set. It is necessary to stress that energy inputs in household can be easily measured, however the setting of threshold that shows the energy poor consumption level is very difficult due to geographical, cultural and other reasons (Bouzarovski and Tirado Herrero, 2015; Fizaine and Kahouli, 2019).

For energy consumption measurement of energy inputs primary, final and useful energy metrics were applied. By applying primary energy consumption indicators per capita national energy consumption is being measured. In this case energy poverty threshold is fixed by evaluating the average energy consumption per capita in poor countries. These countries were identified as poor based on low national income or the Human Development Index (Goldemberg and Johansson, 1995). So, this approach outlines energy poverty as energy consumption per capita in poor countries. However, this indicator does not reveal the distribution of energy consumption in the economy and does not provide the number of energy poor population in a specific country. It is possible to use final energy consumption metrics as well. Foster et al. (2000) described the energy poverty threshold as the average energy consumption of the low income households, i.e. families with incomes below 10% of the national poverty line. This indicator shows the energy poverty as the energy consumption volume of poor households. Other scholars, like Barnes et al. (2011) defined the energy poverty line as the energy consumption of people at the income level at which energy use is not increasing with the rise of their income. Therefore, below this threshold, the energy use of households is at the level needed just for ensuring the minimal living standards.

The study by (Goldemberg and Johansson, 1995) has evaluated the minimal energy requirements on the basis of valuation of energy necessary to supply a normative basket of fundamental energy services. This metrics shows which and how much energy services is needed by taking into account the efficiency of energy technologies used for cooking and lighting. This method suppose that the threshold of energy poverty is changing over the time by taking into account the increase of energy efficiency of appliances.

Energy Access-Consumption Matrix was developed by Pachauri and Spreng (2004). This matrix takes into account the actual availability of different energy carriers as the rows and the amount of specific energy carrier actually used as the columns. The population with each quantity of energy of energy used is counted in each cell by showing the useful energy distribution in the country (Pachauri and Spreng, 2004). The energy poverty threshold was set as a frontier across the Energy Access-Consumption Matrix.

Scholars defined energy poverty metrics also based on income. Popular metrics of energy poverty is the amount of money spent on energy by households by defining energy poverty as the limit expressed by fixed percentage of income spent on energy (Foster et al., 2000). In addition, Reddy (2003) proposed to include in this energy poverty metrics, the expenditures on energy appliances necessary to ensure energy consumption of households as the share of total income spent on energy consumption. Hammond et al. (2007) was analyzing the absolute energy expenses for defining energy poverty line.

It is necessary to stress that methods based on expenditure inputs to energy consumption are even more problematic than just energy consumption input metrics. This is due the fact that there is big difference among countries, among urban and rural areas, among income deciles etc. All these metrics depend on the cost and energy efficiency of the used electric appliances. In addition, study by (Dutt and Ravindranath, 1993) showed that the low-income households usually pay more per unit of energy than the rich households.

By summing up, measures of total primary or final energy use is useful for energy poverty metrics, but due to the changes of poverty threshold over the time and due to increase in energy efficiency of electric appliances, this indicator is very subjective and does not provide necessary information for policy-makers because of different country conditions. Though energy consumption or the income spent on energy services can be applied to measure energy poverty as data can be easily collected however there is no agreement among scholars on definition of the energy poverty threshold. Metrics of energy inputs is helpful for comparative assessment of countries; however, theses metrics are to a lesser extent useful for developing policies and measures to combat energy poverty (Adom et al., 2021; Streimikiene et al., 2021).

There is another approach to measure energy poverty based on energy poverty outcomes assessment (Santillan et al., 2020). The main assumption used by following this methodology is that without fundamental energy services households will face various negative consequences. There are several methods of assessing outcomes of energy poverty. First method is based on the individual energy use negative health impacts of the energy poor households, like the morbidity rate due to respiratory infections. Environmental impacts can be measured as the rate of deforestation due to high energy poverty in country. These all metrics have many limits as they merge the dependent and independent variables. In addition, the energy poverty outcomes are measured by opportunity costs. Mirza and Szirmai (2010) analyzed opportunity cost like time lost when collecting fuel instead of spending it at school or in production of goods (Recalde et al., 2019). The opportunity costs depend on the remoteness to fuel resources which is being gathered.

By summing up, energy poverty outcomes can be assessed by negative health and environmental outcomes, by opportunity costs and absence of alternatives for households. It is obvious, that energy poverty outcomes can apprehend the essence of energy poverty, nevertheless, the data is not easily collected. According to Sen (1993) energy poverty outcomes measured in terms of available choice. As theory of capabilities argues the energy poverty is outlined by the lack of opportunities to live as individual would prefer to live and make his/her free choices. Therefore, Sen's capability approach is also applied for defining energy poverty. Following this approach, the Asian Development Bank has expressed energy poverty as the lack of alternatives for adequate, affordable, reliable, high-quality, environmentally safe and healthy energy access (Masud et al., 2007). If household has a choice on energy services, it means that h./she has already escaped from energy poverty.

Energy poverty outcomes is difficult to measure and interpreter for policy-makers. Especially problematic is assessment of opportunity costs, requiring high research competences.

The quality of energy delivered is also popular approach in assessing energy poverty (Recalde et al., 2019; Spiliotis et al., 2020; Sokolowski et al., 2020). The Multi-Tier Framework (MTF) was created by World Bank for assessing the attributes of energy supply. This is useful for assessing of necessary energy services to support necessary quality of life and economic and social development of the country. The main assumption by applying this energy poverty assessment method is that energy services require to ensure definite quality of energy supply which can be characterized by several attributes. In this case energy supply has to satisfy several requirements. The Multi-Tier Framework is the main energy poverty measure built on the quality of energy services approach. The Multi-Tier Framework evaluates the energy services quality in households, energy use for productive purposes, and energy supply to communities. The energy supply index for households consists of electricity, cooking, and space heating. There are several attributes of quality for electricity, heating and cooking established. For electricity these are the main quality attributes: capacity, duration, quality, reliability, affordability, legality, and health and safety. In addition, for every mentioned attribute limit is established based on five tiers of access. An aggregated index as mathematical average describing energy access in each mentioned domains is developed by providing the overall energy access index (Culver, 2017).

The MTF is useful tool for tracking progress achieved in energy access and contributes to energy poverty reduction. It captures various important aspects of energy service. MTF can be used to track progress within countries and make comparative assessment of countries progress. The MTF includes all issues of SDG 7: achieving universal access to reliable, affordable and modern energy services. (UNGA, 2015). For measuring progress towards SDG 7, a limit is defined below which energy poverty is defined. Practical Action (2014) requires achieving Tier 3 for energy poverty reduction.

The extensive data requirements make MTF framework difficult to apply. The disaggregated data very useful decision-making but overall index is difficult to use for monitoring progress and informed policy-making. In addition, cooking, electricity, and heating are weighted equally in index. Groh et al. (2016) provided critique of the MTF and suggested several important improvements on the background of empirical study conducted in Bangladesh. They showed that the final tier assignment is very sensitive to subjective limit set for tier assignments, therefore the information provided by MTF should be treated with precaution.

By summing up, the multi-tier framework is the most comprehensive metrics of energy poverty as it deals with energy services in the most comprehensive way. It provides information on energy access and energy use, takes into account household energy consumption, also energy for productive use, and energy usage for communities. This indicator shows incremental energy poverty and shows the requirement for action in specific area to combat energy poverty. However, this indicator is complicated, requires a lot of data and efforts to develop it. Also, methodological weaknesses need to be taken into account.

As analysis of various approaches used for assessment of energy poverty shows that due to the multifaceted nature of energy poverty various approaches can be applied for measuring energy poverty, the main indicators being used to measure the level of energy poverty in EU will be analyzed in further. The system of indicators applied reflects the different aspects of energy poverty available EU Members States comprising 27 different countries. According to the recommendations of the European Commission (European Commission, 2020), the energy poverty indicator can be divided into four groups:

- Indicators comparing energy costs with revenues;
- Indicators based on self-assessment;
- Indicators based on direct measurements;
- Indirect indicators.

The first group includes indicators that calculate the ratio of energy expenditure to household income. Energy poverty is considered to be experienced by households above a certain threshold. Various methods can be used to set the threshold: a pre-determined share of expenditure (the indicator of a 10% share of energy expenditure in revenue is used very often), the national median share of energy expenditure in revenue multiplied by two, and so on.

For self-assessment indicators, households are directly asked questions about energy poverty, such as whether the household can afford to heat their homes adequately. The answer to this kind of question is subjective, but that subjectivity reflects the respondents' perceptions of concepts that are often not specified in the definitions of energy poverty.

Indicators based on direct measurements are the opposite of indicators based on self-assessment. Measurements of physical indicators show an objective situation (e.g., temperature level) and therefore subjectivity due to different perceptions of the current situation should be avoided. On the other hand, setting a "threshold" level to identify energy-deprived populations still retains a subjective component due to different perceptions of basic needs. For some people, temperatures below hygiene standards are sufficient, while others believe that higher temperatures meet basic needs. The situation is similar with other energy services: some people need bright lighting, others have enough dusk, and so on.

Indirect indicators measure energy poverty by assessing the relevant factors (difficulties in paying bills, housing quality indicators, number of disconnections, etc.). Such indirect access reveals different nuances of energy poverty and the wider context, but only partially reflects the level of energy poverty. For example, the difficulties declared by the population to pay the bills depend significantly on the cultural aspects (the perceived need of the population to pay the bills every month) and the conditions for the payment of the bills and the possible sanctions; the number of disconnections depends directly on regulation and so on. As a result, the key benefits of indirect indicators are that they reflect some of the less pronounced nuances of energy poverty.

2.3 DRIVERS OF ENERGY POVERTY

Based on literature energy poverty term is very broad and encompasses energy accessibility and availability of energy infrastructure, health and equity, affordability, efficiency and other issues (Bouzarovski, 2014; Hall, 2013). It is also linked to energy vulnerability and resilience, energy security, energy inequality and justice, socio-technical energy sector transition issues.

Energy poverty is a major problem throughout the world including Europe. Households living at risk of poverty or poverty are often forced to limit the heating of their homes and suffer discomfort to reduce energy costs and save cash on other necessary needs, like food, clothes etc. Or individuals living at risk of poverty cannot afford to meet other needs due to comparatively high energy prices, especially for district heating. There is important term—energy burden of household, showing the share of household income used for paying energy costs. This is so called indicator of energy affordability. Based on literature, households having 6% of the share of income spent to pay energy bills or with energy burden higher tor equal to 6% are experiencing a high energy burden and consequently have a low energy affordability. The main drivers of energy burdens increase are: physical home conditions, capacities of households to invest in energy efficiency improvements at homes, and the convenience of energy efficiency programs and availability of various energy saving support incentives in the country.

According to many authors (Tirado Herrero and Urge-Vorsatz, 2012; Bouzarovski and Tirado Herrero, 2015) in Eastern European countries, there are pitfalls of inefficient heat consumption, as low-income poor or vulnerable households, usually residing in old buildings and energy inefficient buildings and do not having enough income to renovate or move to better housing. Nor do they have the possibility to control their heat consumption levels or switch to another heat supply source, as it is legally forbidden to disconnect from the district heating system and switch to other heat supply sources. It must also be emphasized that rising energy prices do not affect the level of consumption because of low energy price elasticity, and due to fact that energy is a necessary commodity for households.

There is no universally defined and agreed term for the definition of energy poverty, but in the scientific literature, household's energy poverty usually means that households spend too much of their real income to cover their energy bills or that households are unable to satisfy their basic energy supply requirements. For example, classical indicator of energy poverty is inability of household's to ensure necessary heating comfort at their homes due to insufficient income and or too high prices of energy services (Bouzarovski and Petrova, 2015). Although the existence of problems of energy poverty, energy poverty or energy affordability is widely known around the world, there is still much debate among scientists as to the definition of energy or fuel poverty.

Several methods are also used to define the level of energy poverty, which can be distinguished on the basis of material energy needs or cost of energy. A fairly widespread approach in assessing energy poverty is evaluation of energy burden, as mentioned above, i.e. assessment of energy costs as a share of total household expenditure (Barnes, 2007). However, in less developed countries energy poverty is also being measured in terms of the type of energy consumed by households or in terms of the country-specific expenditure poverty line (Foster et al., 2000).

The main methods for defining energy poverty can be structured as follows:

- The amount of physical energy needed to meet all the necessary household's energy needs like lighting, heating, cooking, etc.
- a definite percentage of energy expenditure in overall household's expenditures;
- the volume of physical energy consumed or the type of energy source or energy carrier used by households living at the income poverty line;
- a margin of income under which energy consumption and/or energy costs are not expected to change, indicating that the lowest possible level of energy consumption is being achieved.

These four main methods for defining energy poverty have their advantages and disadvantages (Foster et al., 2000).

The energy poverty definition under the first method is built on the tactics used to define the minimum necessary food intake to ensure a healthy life. In adapting this approach to energy, it is important to outline a minimal quantity of energy, which is crucial for meeting the vital energy consumption needs of households, such as cooking, lighting and heating. However, due to differences in local conditions, it is very difficult to describe the exact minimum level of energy consumption needed to meet the unavoidable daily energy needs. Due to different cultures, habits, climatic or geographic conditions, infrastructure of energy supply and natural resource availability, the different countries have diverse cooking and/or heating practices. As a result, households in diverse parts of the world have different essential energy needs.

The second methodology for determining energy poverty is based on the indication that the assessment of energy poverty can be based on the share of household income paid for meeting their essential energy needs. In addition, low-income households typically spend more of their income on energy needs than high-income households. Empirical studies have shown that households devote between 5% and 20% of their total income to energy needs (Boardman, 1991). According to the UK approach, where energy costs exceed 10%. household income, which affects the overall well-being of households. 1988 In the United Kingdom, this 10% was twice the median of household energy expenditure, which is why this threshold was chosen. Hence, if family or individual have to spend 10% or more of income to meet their fundamental energy needs, they face a shortage of other basic goods and the necessities needed to maintain a healthy life. The highest disadvantage of this approach is related to the fact, that methods built on physical consumption of energy cannot be universally useful to assess energy poverty in all countries. This is because of certain differences in local conditions. Therefore, a country-specific threshold built on twice the median share of energy expenditures or indicator 2M used by the European Energy Scarcity Observatory is usually fixed. This avoids restrictions by applying the same 10% level in dissimilar countries. Nevertheless, an approach based on a double median share of energy costs does not reflect the impact of rising energy prices on energy poverty: if prices rise from different energy sources at the same rate, the median energy costs increase and the levels of energy poverty remain stable. This distorts estimates of the scale of energy poverty.

The third way to measure energy poverty is to assess the level of energy consumption of households that is below a certain level of expenditure or income poverty in a given country. These cost-related poverty line settings are well defined in many countries. Usually, household energy consumption surveys are employed to determine the average level of energy consumption in a country which is considered to be below the energy poverty line. However, this approach also has weaknesses, as the established poverty line is not connected to the country's energy policies, making it difficult to control and to develop necessary energy and climate change mitigation policies and measures. The assessment of energy poverty in this way is similar to the assessment of the overall level of poverty in countries.

The fourth method is to assess the links between energy demand and income of households. It is anticipated that once a certain income level is reached, families or individuals turn to use energy increasingly on continuous basis. In the case of poor families, this means that their energy consumption does not growth with the rise of their income. This method is complex and necessitates collection of many data and analysis of information mainly collected by comprehensive surveys. The major strength of this approach is that it measures energy poverty in terms of actually consumed energy by taking into account differences of local conditions, diverse policies applied in these countries and the level of energy prices available in the country (Foster et al., 2000).

There have been many studies (Boardman, 2010, 1991; Bouzarovski, 2014; Bouzarovski et al., 2012; Bouzarovski and Tirado Herrero, 2013) that examine the main causes of energy deprivation: low income, inefficient houses with poor living conditions, non-existence of energy saving appliances, deprived energy infrastructure, high prices of energy services and fuels etc. In addition, high levels of energy deprivation are linked to high poverty risks, a high proportion of older people in the country, inability to meet essential food needs, poor public health protection systems etc.

Various studies have shown that high levels of energy deprivation are directly related to poor health outcomes. Of course, poor public health is the result of a variety of factors, including unhealthy lifestyles and other harmful choices, poor medical services in the country, poor quality of the environment, especially the high air pollution. Nonetheless, people in energy poverty tend to reside in colder houses and cannot afford other essential things for living like food, and do not receive the necessary medical services (Isherwood and Hancock, 1979; Liddell et al., 2012). Usually, then the national energy poverty rate is below 5%, the at-risk-of-poverty rate is below 25% (Simcock et al., 2016). Low-income households are finding it increasingly difficult to pay their energy bills, as rising energy prices make it even more difficult to distribute limited incomes, making it impossible to raise funds to renovate low-quality homes and install energy-efficient appliances, wasting expensive and increasingly expensive energy, and a closed cycle of energy deprivation is taking place.

Simcock et al. (2016) study discussed the concept of fuel poverty based on UK example. Yearly statistics of UK show a significant rise in number of deaths in the UK in winter season (Office for National Statistics, 2014). Importantly, this rise of death during winter is much higher than in other states having much cooler climates. It has therefore been revealed buy scholars that this problem is linked to low energy efficiency of residential houses in UK, providing that for low-income, sick, old and disables people the heating of their homes become very expensive and unaffordable. These statistics on ‘excess’ deaths in the UK have been described as a ‘national scandal’ (Osbaldeston, 1984), showing that the lives of older and disabled people are less valued by politicians than by other groups of the people. The concerns about the heating affordability have been always high on the public policy agenda in UK. This is because of high concerns about public health adversely affected by cold housing. Brenda Boardman (1991) was the first researcher to publicize the problem and define the concept of fuel poverty in the UK. According to Boardman (2010), “fuel poverty” is the inability of a low-income families to obtain the necessary level of heating comfort if households have to devote more than 10% of their income to meet basic energy supply requirements. The 10% threshold was defined based on results of a study showing that fuel poverty occurs when the burden of energy consumption is twice the national median (Isherwood and Hancock, 1979). This energy poverty indicator was corrected in 2013, and a new “low-income, high-energy” condition was initiated to measure fuel or energy poverty. The main sense of this new concept is that energy-poor families reside in energy inefficient houses and face income poverty because of very high energy and fuel bills necessary to heat their homes in order to ensure a healthy indoor temperature (Hills, 2011).

As a result, UK policy is now focusing on improving energy affordability by providing direct payments to help older and low-income households to rise their heating affordability during winter, with an increasing focus on increasing energy efficiency. The main negative effects of fuel or energy shortages identified in UK are: poor health and the incapability to meet other important needs because of high energy bills. Other studies have also examined energy deprivation issues (Redmond and Hutton, 2000), but just after the famous publication by John Hills (2011) the sufficient attention has been paid to this topic in UK and other countries. J. Hills (2011) introduced the term “low cost high income” in her study. According to J. Hills (2011), the results of fuel scarcity are not only related to unsatisfactory heating and affect people’s health (Hills, 2011), but it also hinder the fulfilment of other necessary living requirements because of household budget constraints or circumstances where households have to choose between “heating or eating.” Thus, low-income people are forced to choose whether to pay their energy bills or pay for they food. Therefore, according to A. Sen’s definition (Sen, 1993), fuel shortages affect people’s ability to perform important daily functions and have a negative impact on their free choice, i.e. choosing to be healthy, educated, rich etc. R. Moore (2012) in his study further explored energy poverty, by applying such term as “insolvency for energy services” or “cold house” in his analysis of energy poverty issues. Other scholars (Lampietti and Meyer, 2002) studied energy or fuel poverty in UK. It is necessary emphasize that UK policy and research on fuel and energy poverty have influenced or even initiated various studies of energy poverty around the world.

Researchers in New Zealand were also dealing with the problems of energy efficiency in residential houses and the low heating affordability for low-income and elderly people by relating these problems to an increase of mortality and morbidity during the cold period (Viggers et al., 2013).

A study by Bouzarovsky and Petrova (2015) attempted to create a global term for defining energy poverty, focusing on the notion that low-income families facing energy poverty are somehow excluded from important activities which is necessary to maintain a healthy life and participate in societal life.

H. Thomson et al. (2019) emphasized the importance for a mutual definition of energy poverty in the European Union to establish necessary policies for energy poverty alleviation. The problem of energy poverty urgently needs to be addressed through a range of policies. A common energy poverty for EU would allow energy problem to be recognized and to increase energy poverty political visibility in EU (European Economic and Social Committee, 2011). This would also help to resolve the problem of terminology disagreement and provide an explanation of this problem and identify the necessary solutions. The link between energy poverty and other policies should also be identified by strengthening policy synergies. Nonetheless, other scholars like D. Deller (2016), disagreed with this opinion and proved that the common definition of energy poverty in the EU is problematic, as EU member states (MS) have encountered a different aspects of energy affordability and have country-specific energy poverty drivers and consequences. D. Deller (2016) and U. Dubois (2012) argued that different indicators and concepts could be used to address energy poverty in order to address its different challenges in different states. Generally, energy poverty in EU MS has been examined using data from an EU survey on household heating capacity, damp and mold houses, indebtedness, energy bill arrears etc. (Healy, 2004).

Other researchers in post-Soviet Europe have mainly investigated the affordability of district heating, predicting the links between poor quality of residential buildings and energy poverty in those countries (Thomson and Snell, 2013; Buzar, 2007). According to Bouzarovski (2014), energy poverty in Central and Eastern European nations is often connected with failed restructuring and privatization in the energy and other spheres, from soviet past inherited low-quality, inefficient multi-flat buildings, restricted access to energy infrastructure, and a ban on switching to another heat supplier, high energy prices and small incomes, aging and shrinking communities etc. As a result, Eastern European countries with a large number of low-quality apartment buildings are generally at higher risk of energy vulnerability and energy poverty. Furthermore, people residing in energy inefficient houses spend the higher shares of their income on heating because of low energy use inefficiency of their houses. Typically, these families are already vulnerable because of their overall low income.

Hence, countries having high rates of energy poverty have high risk of poverty rates. It is clear that high rates of poverty in countries are connected to high rates of unemployment, aging, poor health situation, and low availability of public support. Therefore, high heating costs contribute significantly to the increase of energy poverty risks for low-income families.

Energy poverty is also a fairly common problem in Lithuania, touching various groups of society. Various studies on energy poverty have shown that high energy poverty rates of former socialist countries in Central and Easter Europe was linked to the rising energy prices because of restructuring of the energy sector and the liberalization of energy prices since the collapse of the Soviet Union. Researchers have provided evidence that in the six years since 2006, energy prices rose by 22% in Czech Republic, by 33% in Poland and by 69% in Hungary. Thus, energy poverty in former socialist states is caused not only by abruptly rising energy prices, but also by poor buildings conditions and limited social support by the governments having limited state budgets (Bouzarovski and Tirado Herrero, 2015).

K. Droutsa et al. (2016) studied the problems of energy poverty in post-crisis Greece and identified the key causes of energy poverty, that are mainly attributed to high energy prices and low income of population due to 2008 economic crisis. In Ireland, which has analogous climatic conditions and similar residential buildings stock as in the United Kingdom, energy poverty is described as the inability to provide sufficient home heating because of income constraints or the inability to attain adequate levels of heating comfort at homes because of energy efficient houses. In Ireland, the key result of energy poverty is also linked to the negative health consequences of energy inefficient, cold houses (Healy, 2004).

In France, for policy reasons the “energy insecurity” term is applied in order to describe families that are experiencing problems in accessing energy due to a lack of needed fiscal resources or poor conditions of residential buildings. Although the term energy insecurity places less emphasis on the problem of heating, it still allows for better consideration of the problems of pour housing stock and heating efficiency in France (Dubois, 2012). It should be emphasized that most research on energy poverty in the European countries has focused on residential heating. Consequently, according scholars, energy poverty is primarily connected to the problem of an increase in mortality rate during winter months and the signal that these premature deaths occur in older and more vulnerable households (El Ansari and El_Silimy, 2008). While older and sick people are more exposed to the cooler climate, this segment of population requires additional heating services compared to other types of households. As elder people use to spend utmost of their time at home compared to the working age population, therefore, heating discomfort problems have a greater negative impact on their quality of life and health. Knowing the differentiated energy needs of older and disabled households or people with chronic illnesses, as well as households with young children, particular attentiveness should be provided for energy poverty alleviation in these groups of society due to ensure recognitional energy justice. These segments of the population are also politically disregarded, as they find it difficult to influence governmental decisions and to force politicians to acknowledge their needs and make the decisions necessary for their well-being, therefore the problem of procedural energy injustice is evident in this case.

Some studies have also highlighted the problem of disability or age overestimation in addressing energy deprivation (Day and Hitchings, 2011). Nevertheless, a greater focus on certain social groups can lead to improved consideration of the problem of energy poverty. Failure to recognize the differences between social groups and their inherent problems of energy poverty can lead to many mistakes in the development of policies and measures to address energy poverty, to the failure to identify specific needs and to underestimating energy vulnerabilities.

Researchers have also presented another approach to measuring energy poverty based on the availability of energy services (Modi et al., 2005). This approach does not make any proposals on the amount of energy needed to meet energy services. P. Nussbaumer et al. (2012) developed an integrated multi-dimensional energy poverty indicator that integrates the fuels needed for modern cooking, access to electricity for lighting, refrigerator, television/radio, telephone and more indicators. Another integrated measure of energy poverty, the Total Energy Access indicator, has been introduced by other institutions to assess the affordability of household energy according to minimum standards established for such energy services like lighting, food cooling, cooking, space heating or cooling (Practical Action, 2014). This indicator also supposes that households should not use more than 30 minutes per day for gathering the energy carriers needed to meet essential energy needs. This service-oriented energy poverty assessment method is one of the utmost far-reaching ways to identify and measure energy poverty.

Several studies have examined the main drivers of energy poverty in Europe. Traditionally, researchers studying energy poverty have faced such problems as a mix of rising energy prices and declining incomes as well as poor energy insulation of houses. So, energy-inefficient housing and the use of outdated household electric appliances and poor-efficiency heating systems provide the basis for high energy bills, while creating the conditions for energy poverty.

Many researchers confirmed that low energy efficiency of residential buildings is a major cause of energy poverty, leading to discomfort from thermal energy consumption, which in turn leads to high humidity and formation of mold in houses. In general, the main results of such energy deprivation are poor health outcomes and various manifestations of social exclusion (Boardman, 2010). Thus, the main factors in energy poverty are low real incomes of households, high prices of energy supply, energy inefficient buildings and appliances including energy inefficient behavior. Therefore, it is essential to emphasize that energy poverty is caused by material deprivation that goes outside the low-income-related poverty. Also, in developed countries having low levels of energy poverty, poor families face energy poverty (Maxim et al., 2016). For example, in Finland, where energy consumption per capita is high and the climate is cool, the share of lowest income decile families living in low-quality houses and having debt for energy bills is twice higher than the national average. According to A. Bouzarovski (2014), high levels of energy poverty in Eastern European countries are associated with insufficient investment in energy efficiency and insufficient levels of social support.

As mentioned earlier, countries are experiencing different problems linked to energy poverty. In the United Kingdom and Ireland, the coincidence of low-income people living in energy-inefficient housing has led to great levels of energy poverty, although the climate in these countries is relatively mild (Liddell et al., 2012). In Italy, having a warm clime, energy poverty is associated with low levels of energy affordability because of high energy prices. In Austria and Germany, energy poverty is mainly associated with energy-inefficient residential buildings, resulting in the rise of energy bills as the consequence of rising energy prices (Brunner et al., 2013).

Researchers have identified additional demographic factors affecting energy poverty in the country. Vulnerable families, having children, retired, and elderly persons, the disabled and chronically sick people, are at greater risk of energy poverty or are more energy vulnerable population groups (Snell et al., 2014; Wright, 2004). Research has also shown that post-communist countries, like Lithuania, Poland, Hungary etc., are "trapped" in low-energy efficiency houses equipped with no flexible district heating systems and they do not have the ability to control heating or to switch to energy suppliers. In addition, the limited renovation opportunities due to both low income and various organizational barriers and insufficient state support for housing renovation have a negative impact on the increase of energy vulnerability and energy poverty (Tirado Herrero and Urge-Vorsatz, 2012; Bouzarovski and Tirado Herrero, 2015). Inadequate energy services led to wider inequalities in housing conditions, low heating comfort and increased issues of distributional energy justice. A. Maxim et al. (2016) found that tenants not possessing regulated rents and mortgage owners face significantly bigger problems of energy deprivation primarily due to financial constraints and the inability to improve the energy efficiency of their houses in comparison with homeowners or the tenants whose rents are regulated by the heating bill. In addition, research has shown that households residing in detached houses are more prone to energy poverty than households residing in multi-apartment buildings due to the negative impact of many external walls on thermal comfort in the detached house. In addition, it has been found that for the households in rural areas it is more difficult to heat their homes properly and they are more likely to experience energy deprivation (Thomson and Snell, 2013). The main reasons for this are limited energy supply and high levels of poverty in rural areas due to limited infrastructure for energy supply in remote and distant areas. Lack of sufficient state support is also a significant determinant of energy poverty in rural areas (Bouzarovski and Tirado Herrero, 2015).

The key drivers of energy poverty are grouped and described in Table 2.3.

Therefore, the main drivers of energy poverty are physical, behavioral, economic, social and political. Comparing these types with drivers of energy vulnerability given in Table 2.1 one can notice that they are overlapping; however this different grouping proposed in Table 2.3 allows one to highlight the policy dimension of energy poverty.

It is also important to analyze energy poverty dynamics, which allows for monitoring households' poverty status over several years. The track of energy poverty dynamics allows us to define sustained poverty escapes, referring to families that escape from energy poverty for several years. Transitory poverty escapes allow one to track households that used to be in energy poverty; however, they succeeded in escaping from energy poverty, but later they fell back into energy poverty. Impoverishment as a term describes the process of a household falling into energy poverty. And finally, the chronic energy poverty defines a long-term energy poverty situation of households persisting during many years, which is often being transmitted to other generations.

As analysis of energy poverty and its drivers are researched through the years, they have shown the emergence of energy poverty and related energy inequality and energy justice, energy vulnerability and energy resilience concepts. In the following sub-section of this book the various energy poverty concepts are systematized by discussing country differences in energy poverty understanding.

TABLE 2.3
The Physical, Economic, Behavioral and Policy Drivers of Energy Poverty

Drivers	Description
Physical	The limited physical infrastructure of energy supply does not provide alternative energy supply options for households
	Limited choice of energy suppliers or service as well as restrictions for switching to other energy supplies
	Old and inefficient residential buildings
	Outdated heating systems, dirty fuels and high fuel costs
	Poor energy insulation of residential buildings, leaking roofs and walls, inefficient HVAC systems etc.
	Energy inefficient outdated lighting systems and electric appliances like refrigerators, washing machines etc. and lighting sources
	Extreme weather conditions that require extensive heating and cooling at home
Economic	Long-lasting economic problems and hardship
	Increasing energy resources prices
	Incapability to cover up-front costs of energy efficiency investments in buildings and heating or cooling systems, new efficient electric appliances etc.
	Inability to get credit for financing energy efficiency upgrades in homes
Behavioral	Lack of information about energy efficiency programs and bill payment assistance
	Lack of understanding about possible energy saving measures and energy saving behavior including the positive impacts and other benefits
	Non-compliance of energy services with special energy needs of the vulnerable population like the elderly, the disabled, families with children or chronically sick individuals entitled to increased energy consumption
	Behavioral barriers of energy savings are linked with limited awareness among households about the potential gains from improved efficiency
Policy	Inadequate state support, high energy prices and low incomes are the major drivers of energy poverty
	Lack of policies and measures for assistance in paying bills or other energy efficiency programs
	Non-favorable energy utility rate design practices, like high fixed user charges, limiting the household's ability to react to increased energy bills by implementing energy efficiency measures

Source: created by authors based on Bouzarovski (2014); Bouzarovski et al. (2012); BPI (2014); Bouzarovski and Tirado Herrero (2015).

2.4 POLICIES AND MEASURES TO FIGHT ENERGY POVERTY

The main strategies to reduce energy poverty are reducing energy consumption by increasing energy efficiency and saving energy, using sustainable and affordable urban mobility solutions, freezing or reducing public transport fares, especially for people with low income, and restoring public transport services, investing in social housing and energy retrofitting of residential buildings, and investing in the creation of green jobs.

One can notice that energy poverty alleviation is closely linked to climate change mitigation policies, such as energy use efficiency improvement and promotion of use of renewable energy sources.

The following climate change mitigation policies and measures are able to fight energy poverty (Streimikiene et al., 2021):

- Support to energy insulation and installation of renewable residential buildings.
- Energy audits.
- Information dissemination.

Addressing energy poverty through the financing and funding of household energy improvements is a constructive approach. By directly facilitating the enhancement of building insulation, cooling systems, energy storage, heating systems, household appliances, and the use of renewable energy technologies, it is possible to make a significant impact on energy saving, GHG emission reduction and energy poverty alleviation. It's important to explore various methods for financing climate change energy poverty mitigation policies and measures, such as utilizing tax incentives, grants, and loans for specific improvements. Additionally, funding from public sources on the global, European, national, regional, or local levels, private funding, public-private partnerships and levies, including electricity, natural gas or heating oil levies, can all play a role. Nevertheless, we must be mindful of the regressive impacts that electricity levies used to fund solar panels may have on energy poor households. These households might end up shouldering a larger share of their income for these levies without being able to access the benefits.

Energy audits are a highly effective measure for mitigating climate change, as they play a crucial role in ensuring energy and cost savings in households with limited access to energy resources. During these audits, professionals visit energy-vulnerable households to provide personalized advice on how to improve their specific energy consumption situations. These efforts are particularly successful because they are often conducted in collaboration with other important social organizations, such as social workers or health professionals, ensuring that the most vulnerable households receive the support they need.

Information dissemination and awareness raising are measures that indirectly help energy-poor households improve their situation by providing advice, information, or education on how to save energy and adopt energy-saving behaviors.

There are three additional policies and measures aimed at addressing energy poverty. However, these measures do not effectively address the underlying structural issues contributing to energy vulnerability. While they may provide short-term relief by reducing the financial burden of energy costs on households, they do not offer sustainable, long-term solutions to the problem. Furthermore, these measures not only fail to address the root causes of energy poverty but also have detrimental effects on efforts to mitigate climate change. Therefore, it is important to develop a more comprehensive approach that addresses the systemic issues and considers the long-term implications for climate change in order to effectively combat energy poverty.

Short-term energy poverty alleviation measures:

1. Disconnection protection offers safeguards against disconnection for households, particularly in colder months during winter. It is primarily aimed at supporting physically vulnerable households, including individuals with disabilities and senior citizens.
2. Social support provides financial aid to households to cover various expenses such as housing, living costs, and energy bills.
3. Financial or fiscal assistance aimed at reducing energy bills can be delivered through two primary methods. The first method involves utilizing social tariffs, which work to lower the energy bills that households are required to pay. The second method involves providing direct financial assistance to help households cover their energy bills. Additionally, fiscal assistance in the form of reduced taxes, such as VAT, is primarily applicable to district heating and electricity.

2.5 ENERGY POVERTY CASE STUDY

2.5.1 EU Energy Poverty Indicators

The study by the European Commission (2016) tested and validated a large number of energy poverty metrics. In the end the small set of four metrics was selected for measuring energy poverty in EU. In total 178 indicators were analyzed and just 42 of them were nominated based on existing statistics and data available at the EU level. Twelve supplementary indicators were gathered from Eurostat and an additional seven indicators from SILC. The remaining 12 indicators were collected from country level statistics, various projects and literature.

From 178 analyzed indicators 58 of them were connected to income or expenditures and 51 indicators addressed the physical infrastructure of energy supply. Just 10 energy demand indicators and 15 indicators linked to demographic conditions were collected on the EU level. There were 139 single metric indicators and 39 constructed indicators, from which 22% were dealing with income and/or expenditures. From 178 of analyzed energy poverty relevant indicators 10 were consensual-based; 42 were expenditure-based and 11 were outcome-based. The remaining 14 indicators were constructed by applying different approaches, like a mixture of several approaches. The remaining 101 energy poverty metrics can be considered just as supporting indicators.

Energy poverty metrics were assessed based on their efficacy and suitability for a methodical comparative assessment of energy poverty in EU MS. Several criteria were applied for selection of energy poverty indicators. A traffic light reporting system was applied for the scoring of energy poverty indicators. The score from one to three was applied, where three was set as the highest score for the selecting indicator. The indicators with score of 2.0 or below were excluded from further analysis. The scores for such criteria like comparability, quality and robustness, data availability, and complexity of construction method were used to define the best energy poverty indicators on the EU level. Finally, qualitative criteria like the strengths and weaknesses were addressed in the indicator selection (European Commission, 2016).

The diversity of analyzed energy poverty indicators shows the multi-faceted problem of energy poverty in different areas. Especially puzzling was capturing the main energy poverty problems in such a broad and diverse region like the EU (Halkos and Gkampoura, 2021).

The following two main approaches were found dominating scientific literature dealing with energy poverty measures (Karasek and Pojar, 2018): expenditure- and consensual-based approaches. Expenditure-based measures address energy poverty on the basis of various data on the household's energy expenditures and compare it with income of households in order to reveal problems of energy affordability. Consensual-based approaches try to define households that are declaring their difficulties in meeting necessary energy services or address the so-called perceived deprivation method. According to Thomson (2013) there are other energy poverty metrics, i.e., temperature-based metrics. This method is based on internal house temperature assessment. Nevertheless, due to difficulties linked to data collection this approach is very difficult to apply. Tirado Herrero and Urge-Vorsatz (2012) showed that in major Central and Eastern European where district heating is prevailing, the households can't control their heat consumption and internal temperatures as well.

The outcome-based approach was initially also excluded from energy poverty indicators list for EU, as these indicators are created by taking into account the outcomes of energy poverty, like disconnections from energy supply, arrears of utility bills and cold-weather-related deaths. The use of such indicators is problematic because of the unavailability of comparable national statistics, controlled access to utilities data and problems of causality as many outcomes can be results of other factors and energy poverty is one but not the only factor providing for such outcomes. However, arrears of utility bills finally were recommended as indicators for assessing energy poverty, as Eurostat and SILC (2014) provide good-quality data on the EU level on arrears of utility bills.

Therefore, the diversity of approaches for energy poverty assessment in EU, shows different approaches assumed to gather energy poverty information in EU, having specific strengths and weaknesses. Therefore, the combination of various approaches would be the best approach in assessing energy poverty. The expenditure-based approach provides information about the real energy expenditure level but it cannot provide whether customers reduce energy expenditures due to budget constraints or because of other reasons. The consensual-based approach allows one to show the wider issues of energy poverty, but it is based on subjective information, making it difficult to compare energy poverty across EU MS (European Commission, 2016).

The initial final set of EU energy poverty indicators that was found to be the most adequate for measuring, comparing and tracking energy poverty and its drivers and outcomes across the EU was defined:

The 2M indicators categorize as energy poor a household having the proportion of income dedicated to energy services twice above the national median level. Therefore, this indicator shows families that have an abnormally high proportion of their income spent for paying their energy bills.

The Low Income High Cost (LIHC) indicator categorizes as energy poor a household having income after paying housing bills below the poverty threshold and households having the proportion of income dedicated to energy services above the national median level. This energy poverty indicator also shows the families that spent a high proportion of income to pay energy bills, but its main focus is on households having low absolute income after paying their energy bills.

The Hidden Energy Poverty Indicator (HEP) or M/2 indicator categorizes an energy poor household as having equivalized energy expenses below half of equivalized energy spending's median. This indicator shows households whose energy spending is uncommonly low according to national standards. This metric suggests that energy poor households are forced to forgo a basic level of consumption.

The inability to keep homes warm—or the warmth indicator—categorizes energy poor households as those declaring difficulties in heating their homes during the cold season. This indicator directly reflects the problems of energy poverty. Energy poverty is a situation in which households cannot afford to heat their home adequately or are unable to meet other necessary energy needs at an affordable cost. Such a situation can be assessed in different ways, and one single metric is not enough. So, families that cannot satisfy their energy needs at an affordable cost are compromising their other necessary expenditures and this is being addressed by 2M and LIHC indicators. Or energy poor households are forced to compromise their energy expenditure and this is captured by HEP and warmth.

The main indicators that can capture spatially disaggregated energy poverty problems in the EU are shown in Table 2.4.

Therefore, based on an analysis of a wide range of indicators for capturing the most important energy poverty issues in the EU, the most relevant energy poverty indicators given in Table 2.4 are recommended for EU MS (European Commission, 2020). The main problems for defining the reliable set of energy poverty indicators are linked to defining the key components of income and energy expenditure. The selected energy poverty indicators and supporting indicators are the most suitable, seeking to save time and cost in collecting data for these indicators. These indicators of different issues of energy poverty and existing surveys can be used for collecting necessary data.

TABLE 2.4
The Main Energy Poverty Indicators Relevant to the EU

Approach	Indicators	Data
Expenditure-based approach by applying actual expenditures	2M or the share of households having the percentage of energy spending compared to equivalized disposable income twice above the national median level	HBS: equivalized income; taxes; energy expenditures
	LIHC or the share of households with the share of energy expenditures that are above the national median and with their residual income below the national poverty threshold set in the country	
	HEP or M/2 indicator or the share of households with absolute energy expenditure that are below half the national median level	
Consensual-based approach	The inability of households to keep their houses adequately warm	SILC
Outcomes-based approach	Arrears on utility bill or the share of households having arrears on their utility bills	SILC

Source: created by authors based on European Commission (2016, 2020).

These indicators are based on three different approaches and are able capture the broad and multifaceted energy poverty problem on the EU level. Expenditure-based measures of energy poverty can capture the extent of energy poverty or the number of energy poor households and the depth of energy poverty. The energy poverty gap is very important from a decision-making pint of view as it indicates the amount of financial support necessary for households to escape from energy poverty.

It is necessary to stress that the selection of energy poverty indicators is based first of all on theoretical uncomplicatedness, easiness of application and available data at EU MS level. More sophisticated indicators can be applied, like required energy spending, which requires information on energy efficiency of residential buildings; however, for this indicator it is necessary to improve the data available from existing surveys in the EU. There are also differences linked to frequency, timing, and structure of the EU Household Budget Surveys (HBS) data. It is essential to include a variable in the SILC survey addressing total energy spending. It is also very important to adapt the thresholds of energy poverty as energy indicators are very sensitive to choice of line. Small changes in set line provide for large modifications in the energy poverty assessments. The use of lines like actual medians allow one to capture distribution of energy spending across the population and can be used for two expenditure-based metrics: twice the median share (2M) and the half the median absolute expenditure (HEM) indicators.

Finally, the following energy poverty indicators were recommended for EU MS (European Commission, 2020):

1. Indicators focusing on the affordability of energy services
   - Share of population at risk of poverty (below 60% of national median equivalized disposable income) not able to keep their home adequately warm, based on the question 'Can your household afford to keep its home adequately warm?' (Eurostat, SILC [ilc_mdes01])
   - Share of total population not able to keep their home adequately warm, based on the question 'Can your household afford to keep its home adequately warm?' (Eurostat, SILC [ilc_mdes01])
   - Arrears on utility bills: share of population at risk of poverty (below 60% of national median equivalized disposable income) having arrears on utility bills (Eurostat, SILC, [ilc_mdes07])
   - Arrears on utility bills: share of population having arrears on utility bills (Eurostat, SILC, [ilc_mdes07])
   - Expenditure on electricity, gas and other fuels as a proportion of total household expenditure
   - Proportion of households whose share of energy expenditure in income is more than twice the national median share (source Eurostat, Household Budget Surveys, 2015)
   - Share of households whose absolute energy expenditure is below half the national median. (Eurostat, Household Budget Surveys, 2015)
2. Complementary indicators
   - Electricity prices for household consumers—average consumption band (Eurostat, [nrg_pc_204])
   - Gas prices for household consumers—average consumption band (Eurostat, [nrg_pc_202])
   - Gas prices for household consumers, lowest consumption band (Eurostat, [nrg_pc_202])
   - Share of population at risk of poverty (below 60% of national median equivalized disposable income) with leak, damp or rot in their dwelling (Eurostat, SILC [ilc_mdho01])
   - Share of population with leak, damp or rot in their dwelling—total population (Eurostat SILC, [TESSI292])
   - Final energy consumption per square meter in the residential sector, climate-corrected (Odyssee-MURE project database).

The European Energy Scarcity Observatory recommends four key indicators of energy poverty:

- Inability to keep home adequately warm.
- High share of energy expenditure in income (2M)—the share of energy expenditure is higher than the national median share of energy expenditure multiplied by two.
- Arrears on utility bills.
- Hidden energy poverty (HEP)—absolute energy costs are less than half the median cost.

It should be noted that the choice of these indicators was largely determined by the information collected in the standardized household surveys (SILC) and the Household Budget Survey (HBS), although the surveys themselves did not focus on energy poverty. Thus, it can be argued that the indicators were chosen to link the available information to the problem of energy poverty, rather than to assess what data should be collected in order to objectively assess the level of energy poverty. Nevertheless, these indicators are also recommended in the European Commission's document SWD (2020) 960/F1.

The practical experience of the European Union in assessing energy poverty is described in Table 2.5.

As Table 2.5 shows, although many countries focus on the indicators recommended by the European Energy Scarcity Observatory, the National Energy and Climate Action Plans mention a much wider range of energy scarcity assessment options, indicating the potential for better adaptation or policy-making to the national context indicator. In France, for example, a specific variation of the "low-income and high-energy cost" indicator is used, in which households with an income not exceeding three deciles of income are considered to be in energy poverty if their energy expenditure exceeds 8% income. Country-specific indicators limit comparability with other countries and need to be very well thought out to reflect the concept of energy poverty. Although the Latvian National Energy and Climate Action Plan referred to the indicators recommended by the European Energy Scarcity Observatory, in practice the determination of the number of energy deprived households is exclusively related to the social support system: energy deprived households are those that receive material support for use or receive housing assistance. It is clear that such simplification is beneficial from an administrative point of view, but it moves away from the specific nature of this phenomenon in terms of energy supply, which risks losing the relevance of policies.

TABLE 2.5

EU Countries Use Energy Poverty Indicators (According to the National Energy and Climate Action Plans).

Country	An energy poverty indicator is used
Austria	SILC indicators are mentioned
Belgium	Specific indicators (number of warning letters, number of terminated contracts etc.)
France	Household energy costs below three deciles of income exceed 8% income; feeling cold indicator from the survey
Germany	Comparison of minimum cost of living with income
Italy	Ad-hoc indicator, value close to relative poverty
Denmark, Estonia, Latvia, Lithuania, Malta, Spain	Indicators recommended by the European Laboratory for Energy poverty
Bulgaria, Cyprus, Greece, Hungary, Luxembourg, the Netherlands, Poland, Slovenia, Sweden	The National Energy and Climate Action Plans do not mention the energy poverty indicator

Source: created by authors based on European Commission, SWD (2020) 960 final. Brussels, 14.10.2020.

When developing the concept of assessment of energy poverty and policy measures in the field of households adapted to Lithuania, it is expedient to use the most common indicators as a basis but to assess their application by taking into account the national context, complementarity of indicators, showing energy deprivation, experience in other countries and practical aspects.

2.5.2 Assessment of Energy Poverty Indicators in Lithuania

The starting point was the assessment of the most practical energy poverty indicators recommended by the European Energy Scarcity Observatory: the inability to provide adequate heat for housing, high share of energy costs in relation to income, inability to pay for utilities bills on time and hidden energy poverty.

The practical application of the hidden energy poverty indicator is complicated by the fact that absolute energy costs depend on many factors and abnormally low energy costs can be attributed not only to energy poverty but also to high energy efficiency, small space and other factors that are not necessarily reflected in energy costs. For example, in energy efficient housing, the absolute heat consumption is much lower than in old non-renovated housing. Assessing the energy poverty in light of housing characteristics would probably be the most logical way to remove uncertainties, but the detailed information on housing needed for this is not currently available at the EU level and Lithuania as well. On the other hand, even with all the information on housing, it would be necessary to take into account the fact that energy demand is determined not only by technical characteristics but also by behavioral characteristics and the lifestyle of households. Behavioral changes focused on energy adequacy also play an important role, with households deliberately seeking to reduce their energy consumption in order to minimize their impact on the environment, which could also be mistakenly identified as energy poverty due to low energy costs.

The indicator of inability to pay on time for utility bills provides valuable and prompt information on the difficulties experienced by the households, but it is not possible to consider it as the main indicator that a) it is targeted at households that receive energy as utilities but does not adequately reflect consumers once a year purchasing fuel; (b) the inability to pay for utilities on time depends heavily on the behavior of the individual household (e.g., according to all other indicators, energy poverty is more common in households having retired people with delays in paying utility bills being the least common in this group) and cultural characteristics.

In view of the uncertainties discussed, it is more appropriate to consider the indicators of hidden energy deprivation and inability to pay for utilities as complementary to the overall energy deprivation picture in the country but not as the main indicators of energy poverty in the country. The remaining two indicators (the share of the population unable to heat their homes enough and the share of households that spend a large part of their income on energy) are already included in the National Progress Plan and mentioned in the Lithuanian National Energy and Climate Action Plan. The problems of practical application of these indicators in Lithuania were analyzed in detail and specific directions for improvement were defined further.

2.5.2.1 Proportion of Households That Spend a Large Proportion of Their Income on Energy

At present, the 2M indicator is calculated by evaluating the proportion of households with a share of energy expenditures in their income that is twice as high as the national median. As in other EU countries, data for calculating this indicator are taken from a household budget survey (HBS) conducted every four years or less frequently. Thus, the latest available data for the calculation of this indicator are from 2016. A survey of household budgets completely complicates the monitoring of the energy poverty situation. Moreover, a more detailed analysis has shown that, according to the methodology of the household budget survey, utility costs are not recalculated to reflect annual energy costs.

In order to use this indicator to monitor energy poverty, it is appropriate to set a fixed level (threshold) in line with national specificities, which, unlike the median share of expenditure, should not change from year to year. Although this level can in principle be freely chosen according to policy objectives, it is useful to take into account the median share of household energy expenditure in recent years for the assessment of national specificities, as shown in Figure 2.1. There is not much data to calculate these indicators. Until now, it has been possible to use only data for 2016. The data on the household budget survey and 2M indicator calculated based on these surveys and presented in the National Progress Plan are not suitable for the correct calculation of energy poverty as the national median share of energy expenditure is not possible to evaluate correctly due to the nature of the household budget survey and non-recalculation of utility costs.

In this context, alternative sources were used—energy costs disaggregated from the Statistics on Income and Living Conditions (SILC) surveys based on more detailed data provided for year 2019. According to the survey data, UAB Vilmorus 2020 households' survey data was also used. It should be noted that the indicators presented in Figure 1.1 were obtained by different methods, and different surveys conducted by Statistics of Lithuania and Vilmorus UAB differ. They differ in terms of representativeness and response trends, and rough calculations were used to disaggregate housing costs. As a result, the data are not fully comparable but provide a fairly good overview of the situation.

Figure 2.1 shows a declining trend in the share of energy expenditure relative to income from 8.7% in 2016 to just over 7% in 2019. The median share of energy expenditure in household income was even higher at 10% in the post-crisis period due to lower household incomes and higher energy prices, but these periods no longer reflect the situation in recent years. On the other hand, although data from different years and sources are not directly comparable, comparing the survey results with the total energy consumption statistics shows that Vilmorus UAB respondents tended to provide lower energy costs, while in the SILC many respondents indicated increased energy costs.

As mentioned earlier, the level of the share of energy expenditure to income used for monitoring (the threshold of the share of energy expenditures in income indicates energy poverty) can be chosen taking into account various factors, as well as comprehensibility and ease of interpretation.

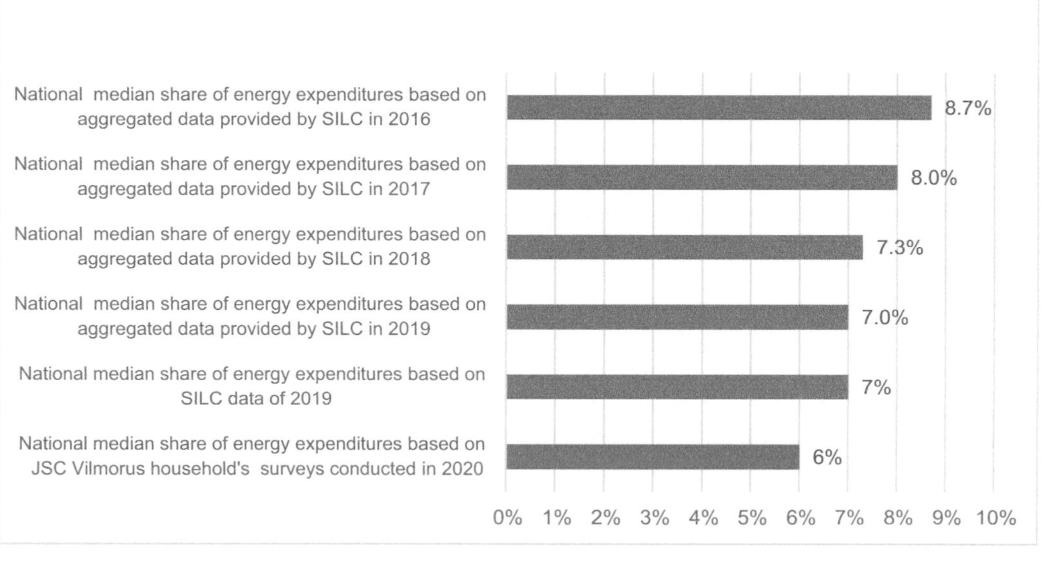

FIGURE 2.1 National median of the share of energy expenditure in 2016–2020.

Source: created by authors.

Based on data from recent years, it is appropriate to calculate the share of households with more than 15% of their energy needs to monitor the level of energy poverty. The following thresholds can be applied as an additional indicator:

More than 10%. This threshold of income spent to cover energy expenditure is widely used in the scientific literature and can therefore be used for various comparisons. In addition, it indicates the risk of reaching higher income share thresholds.

More than 20%. This threshold of income to energy expenditure in the national context can be considered as extremely painful for households.

The indicator of the population that spends a large share of income on energy costs reflects the objective side of energy poverty, but subjective factors are important in the assessment of energy poverty and have a significant impact on decision-making.

The similar shortcoming is characteristic for M/2 or hidden energy poverty indicators, which do show the behavioral energy consumption patterns and energy efficiency performance of residential buildings. Therefore, this indicator also needs to be revised by addressing these important factors having an impact on low share of energy expenditures in total income of households.

2.5.2.2 Individuals Living in Households Who Cannot Afford Adequate Warmth in Their Homes Due to Lack of Funds

The share of households that cannot afford adequate heat in their homes (maintaining a suitable temperature) is currently estimated from the SILC surveys. This indicator is probably receiving the most public attention, as for many years Lithuania has been in the worst position among EU MS except Bulgaria (see Figure 2.2).

The tendency that the inability to heat homes is higher in more southern European countries raises some doubts about the reliability of this indicator, and in the northern countries, except

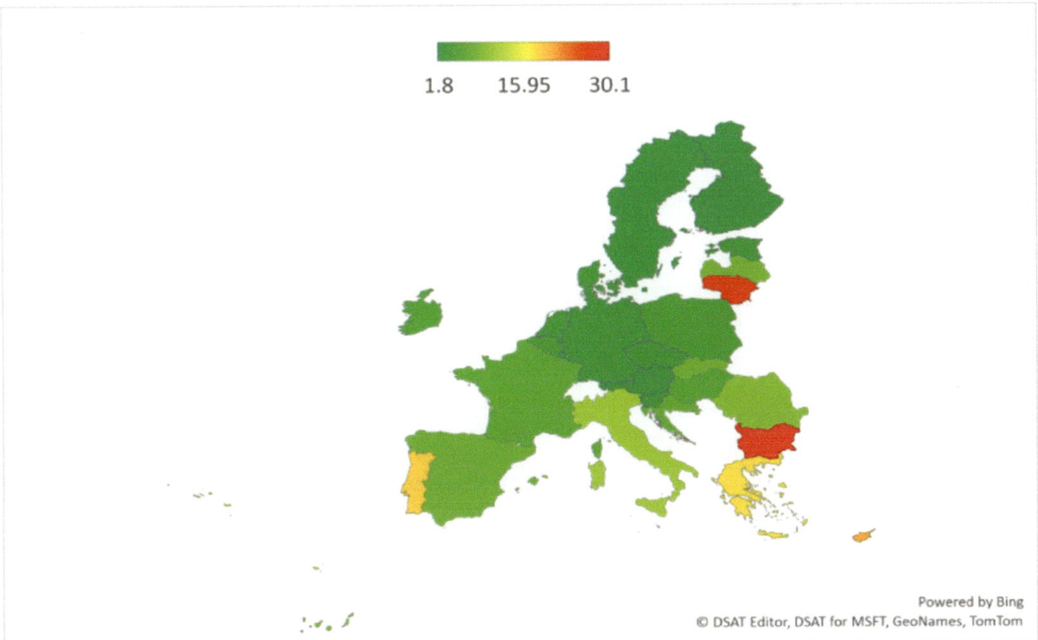

FIGURE 2.2 Proportion of people in the European Union who cannot afford adequate heat in their homes.

Source: created by authors based on EUROSTAT (2022).

Lithuania, the possibilities to ensure adequate heat levels in dwellings are assessed better. Although the indicator itself focuses on maintaining a suitable housing temperature and theoretically includes both heating in winter and cooling in summer, only Cyprus and Malta have the highest number of heating days, with Northern and Central European countries having the highest total number of days. Differences in income across countries also do not adequately explain the level and dynamics of this indicator.

The significance of the indicator of the population who cannot heat their homes sufficiently in Lithuania has decreased only slightly over time and in 2020 it reached 23.1% (see Figure 2.3), although the dynamics of most other related indicators (energy prices, number of recipients of heating compensation and compensation paid to them, share of the population having arrears in utility bills etc.) showed much more positive trends in Lithuania. In Figure 2.3 the dynamics of the indicator in question in some EU countries are also presented.

In Estonia, this indicator has been consistently low (1.1–4.2%) and has not exceeded the EU average, which has been steadily declining from 10.8% in 2008 down to 6.9% in 2019. In Latvia in 2005–2007 this indicator was close to the situation in Lithuania, but in recent years it has approached the European Union average. Despite a significant downward trend (starting at around 3 percentage points annually starting in 2017), in Bulgaria 27.5% residents are still not able to provide adequate heating for their homes. In recent years, Lithuania has also seen moderate trends in the share of the population unable to heat enough, but a more significant decrease was recorded only in 2020, when this share decreased by 3.6 percentage points.

Given the dynamics of this indicator, it was assumed that this situation was affected by methodological shortcomings. The questionnaires of the last 10 years were analyzed; the three periods used in the Lithuanian SILC data were identified in Figure 2.4. Given the similarity of countries and the proximity of energy markets, Figure 2.4 also shows the dynamics of this indicator in Latvia.

In the first period, the population was asked about the possibility of sufficient heating in the context of other indicators of the financial situation (in addition to the household's ability to spend a week away from home or to eat meat, fish or a vegetarian meal); in the second period the question was moved to the context of housing conditions. In addition to questions about poor housing

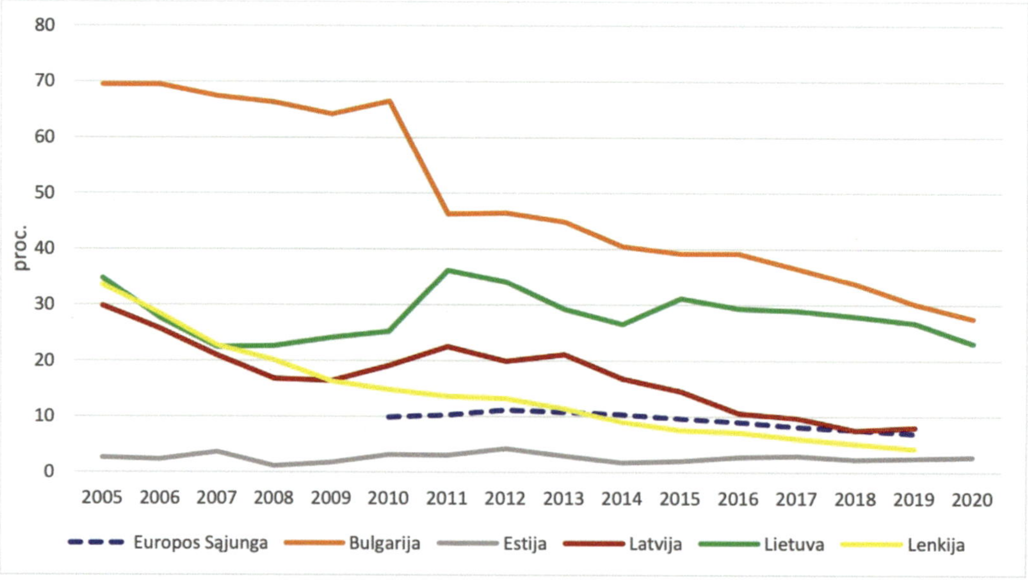

FIGURE 2.3 Change in the share of persons unable to heat their houses sufficiently.

Source: created by authors based on EUROSTAT (2022).

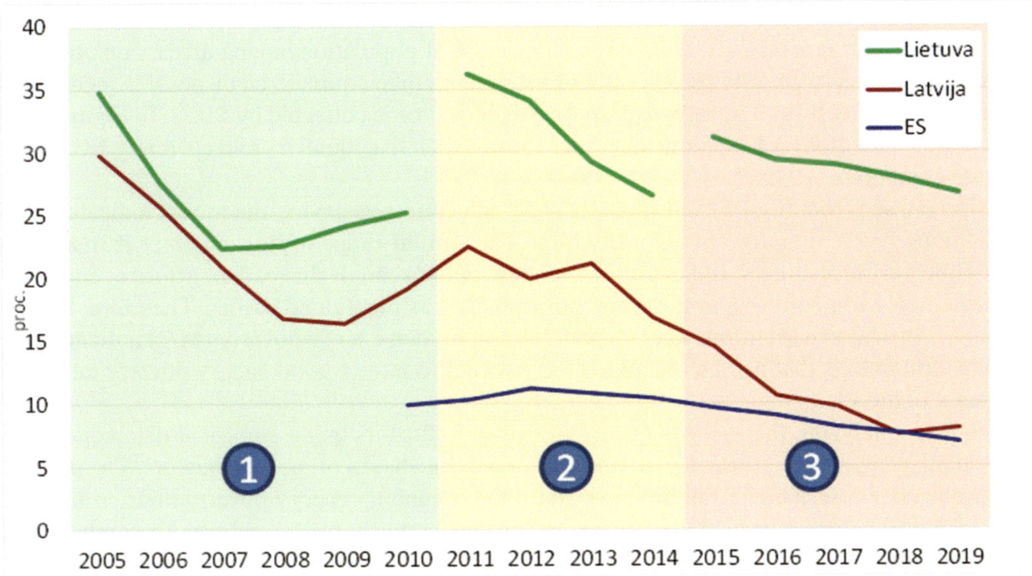

FIGURE 2.4 The share of population unable to sufficiently heat their homes and the period of the questionnaire.

Source: created by authors.

problems (dripping roofs etc.) and lack of daylight in the third period, the question about possible use of electric heaters when it is cold was included. Such a remark has already been mentioned in the methodological material for interviewers, but just since 2015, it was included in the questionnaires. As can be seen from Figure 2.4, the jumps in the proportion of the population that could not adequately heat their home were observed during the periods of the questionnaire change. Of course, there were other factors that simultaneously affected the dynamics of this indicator, but it is necessary to highlight that wording of the question had a significant impact on the results of surveys.

A representative survey of population in Lithuania conducted by UAB Vilmorus in 2020 also confirms this assumption. When asked with a remark about heating from district heating networks, 34.5% of respondents indicated that they could not adequately heat their house. Meanwhile, when they were asked whether the income received by a household allows for sufficient heating of a dwelling, those who are unable to heat their dwelling indicated only 12%. Population. Another 17.7% of respondents replied that they could heat their home sufficiently if they limited the satisfaction of other needs. Depending on the wording of the question, not only the overall proportion of respondents but also the structure of answers is changing among the groups of population. Naturally, asking for an additional note about homes heated by district heating results in higher levels of energy poverty in urban areas and in apartment buildings. This can be linked to the inability of people to regulate heating and therefore the internal temperature of their homes if they are living in such types of buildings.

Examining the questionnaires used in other European Union countries, there is a tendency to present the question used in most countries to calculate this indicator in the context of household finances (corresponding to the first period in Lithuania), thus reflecting both energy and financial aspects and better conveying the concept of energy poverty. In Ireland, meanwhile, the question is whether the household maintains adequate heat in the dwelling—if not, the question follows whether this is due to financial (inability) or other reasons. Such a survey strategy makes it possible to refine the responses to the energy poverty situation and to identify households with energy poverty rather than technical or other heating constraints.

2.5.2.3 Arrears on Utility Bills

The indicator of arrears on utility bills shows the share of population having arrears on utility bills or population at risk of poverty (below 60% of national median equivalized disposable income) having arrears on utility bills. The information for this indicator is collected by SILC. In Figure 2.5 the trends of this indicator for Lithuania and other former socialist countries and currently EU member states are compared.

As one can see from Figure 2.5 Lithuania is among countries having the lowest indicators of the share of households with arrears on utility bills. The similar range of this indicator is in all Baltic States. However, according to indicators of the share of households that cannot afford adequate heat in their homes, Lithuania was in the worst position, as was mentioned earlier. Therefore, based on the analysis provided earlier, the warmth and 2M and hidden energy poverty (M/2) indicators have several shortcomings that need to be addressed in order to have a good energy poverty monitoring instrument in the EU.

Therefore, analyzing the main EU energy poverty indicators using empirical data collected for the Lithuanian case study allowed us to find additional drawbacks of these indicators. The proposals for improvement described earlier are also useful for advancing energy poverty metrics and monitoring in the EU. This is especially important for developing targeted policies to combat energy poverty in the EU.

The energy poverty indicators should serve to monitor the success of policies implemented to alleviate energy poverty. As mentioned before, climate change mitigation policies such as improving energy efficiency and promoting renewable energy usage have a direct long-term impact on energy poverty alleviation in contrast to short-term policies targeting low-income populations through social benefits, protection from disconnection of electricity, social tariffs for electricity, etc.

In Lithuania, a comprehensive set of policies is in place to address energy poverty. These policies include measures such as disconnection protection and financial assistance to reduce energy bills through reduced VAT. District heating is made accessible to all residents, while households with

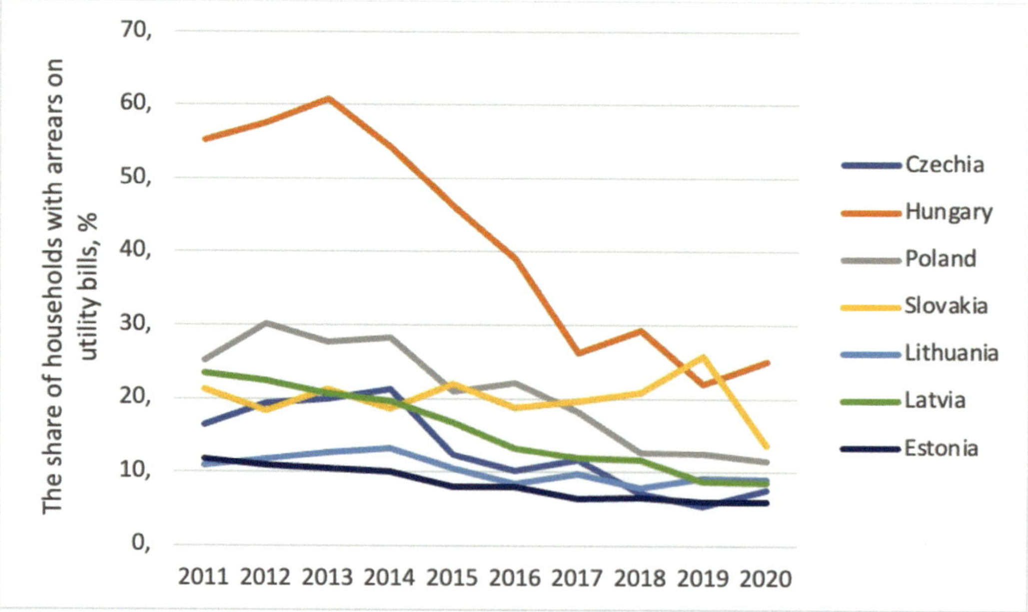

FIGURE 2.5 The share of households with arrears on utility bills, %.

Source: created by authors based on EUROSTAT (2022).

low incomes receive compensation for heating, cold, and hot water costs. Furthermore, a social support system offers benefits to individuals whose income falls below the national threshold, ensuring that a safety net is in place to help those in need.

Lithuania has implemented a range of long-term policies to address energy poverty. These measures include providing financial support through the Modernisation Programme to promote the use of renewable energy sources in residential buildings. Soft loans with a fixed 3% interest rate are also available to encourage energy efficiency and the adoption of renewable energy sources in residential buildings under the Multi-apartment Buildings Renovation Programme. Furthermore, the country promotes renewables through a Feed-in Tariff for Photovoltaics in residential houses and encourages the participation of prosumers of renewable energy sources in line with Lithuanian laws and regulations.

2.5.2.4 Conclusions

There are important tradeoffs among the simplicity, comprehensiveness, and accuracy of energy poverty indicators. The simplicity criteria for indicators selection are very important for data collection and clear communication for policy-makers. Comprehensiveness criteria of energy poverty indicators selection requires one to take into account fundamental energy services in the household. The accuracy criteria need to take into account the usability of energy to deliver necessary energy services. There are also other important criteria like comparability, quality and robustness, data availability and complexity of construction method etc.

The energy poverty indicators can be analyzed and systematized based on various approaches, like access of population to electricity and modern cooking fuels; energy consumption inputs; the outcomes of energy use and the quality of energy delivered. Energy poverty indicators can be also grouped according to approach applied: expenditure-based approach; consensual-based approach, outcomes-based approach and temperature-based approach.

Quantifying energy poverty based on energy consumption or the expenditure-based approach is the most popular approach around the world as it is able to capture all types of energy use in a single metric, but it depends on several changing assumptions. In the EU three main energy poverty indicators are expenditure-based. However, the problem of setting the threshold for energy poverty is a very sensitive issue for these indicators. Without a clear theoretical reference for an energy poverty line, all the methods of setting an energy poverty threshold for expenditure-based indicators are subject to the arbitrary assumptions and subjective decisions of scholars or policy-makers.

Just one consensual-based indicator—inability to keep the house warm—was proposed by Energy Poverty Observatory for EU MS; however, this indicator is very subjective and very sensitive to formulation of question and even for the place in which part of questionnaire the question is allocated. Therefore, this indicator should be treated with precaution.

Energy poverty indicators based on the outcomes of energy consumption like arrears on utility bills is attractive, objective as based on information from utilities and a simple indicator. The indicator of arrears of utility bills is recommended by the European Commission as well and this is the also the only outcomes-based indicator proposed to measure energy poverty at the EU level.

Most important for applying energy poverty indicators is to make comparisons on the energy poverty situation in the countries on the basis of several indices at one time. Nevertheless, combining several indicators into a single index produces an energy poverty indicator that can be neither meaningful nor easy to interpret.

It is necessary to stress the importance of policy implications of energy poverty indicators design. Different understandings about who the energy poor are and why they are poor will lead to different policy decisions and finally choices of different targeted groups.

More empirical work is needed to demonstrate the relevance of energy poverty indicators recommended by European Commission to monitor the energy poverty level at EU MS due to methodological drawbacks of expenditure-based and consensual-based indicators.

Therefore, in the absence of one best energy poverty indicator, policy-makers must critically evaluate existing energy poverty indicators and use them in accordance with particular policy aims. If the main purpose is communication, for selection of energy poverty indicator, the simplicity and transparency is the key issue. For the cross-country comparative energy poverty assessment and setting energy poverty reduction targets at the EU level, a more comprehensive measure of energy poverty is necessary. Also, the key issue is to address how well the changes of energy poverty indicators will really influence reductions in energy poverty of a specific country.

Traditional short-term energy poverty alleviation policies are the protection of vulnerable households against disconnection, particularly in colder months during winter, financial aid to households to cover various expenses such as housing, living costs, and energy bills, and financial or fiscal assistance aimed at reducing energy bills like utilizing social tariffs, etc.

The long-term energy poverty alleviation measures are closely linked to climate change mitigation policies and measures and cover state support to energy insulation and installation of renewable energy sources in residential buildings, support of energy audits in households, information dissemination and awareness raising about energy efficiency improvement and energy saving measures, energy renovation and renewable energy support programs in households etc.

REFERENCES

Adom, P. K.; Amuakwa-Mensah, R.; Agradi, M. P.; Nsabimana, A. 2021. Energy poverty, development outcomes, and transition to green energy. *Renewable Energy*, 178, 1337–1352.

Alexandri, E.; Androutsopoulos, A. 2020. Multicriteria evaluation of ecolabels for the energy upgrade of dwellings in Greece. *International Journal of Sustainable Energy*, 39, 67–87. https://doi.org/10.1080/14786451.2019.1644337.

Barnes, D. F. 2007. The challenge of rural electrification. In D. F. Barnes (Ed.), *The Challenge of Rural Electrification: Strategies for Developing Countries* (pp. 1–17). Washington, DC: RFF Press.

Barnes, D. F.; Khandker, S. R.; Samad, H. A. 2011. Energy poverty in rural Bangladesh. *Energy Policy*, 39(2), 894–904.

Bickerstaff, K.; Walker, G.; Bulkeley, H. 2013. Introduction: Making sense of energy justice. In K. Bickerstaff, G. Walker, H. Bulkeley (Eds.), *Energy Justice in a Changing Climate*. London: Zed Books, pp. 1–13.

Boardman, B. 1991. *Fuel Poverty: From Cold Homes to Affordable Warmth*. London: Bellhaven.

Boardman, B. 2010. *Fixing Fuel Poverty: Challenges and Solutions*. London: Earthscan.

Bouzarovski, S. 2014. Energy poverty in the European Union: Landscapes of vulnerability. *Wiley Interdisciplinary Reviews: Energy and Environment*, 3, 276–289.

Bouzarovski, S.; Cauvain, J. 2016. Spaces of exception: Governing fuel poverty in England's multiple occupancy housing sector. *Space and Polity*, 20, 310–329.

Bouzarovski, S.; Petrova, S. 2015. A global perspective on domestic energy deprivation: Overcoming the energy poverty-fuel poverty binary. *Energy Research & Social Science*, 10, 31–40.

Bouzarovski, S.; Petrova, S.; Sarlamanov, R. 2012. Energy poverty policies in the EU: A critical perspective. *Energy Policy*, 49, 76–82.

Bouzarovski, S.; Tirado Herrero, S. 2015. The energy divide: Integrating energy transitions, regional inequalities and poverty trends in the European Union. *European Urban and Regional Studies*. https://doi.org/10.1177/0969776415596449.

BPI. 2014. *Buildings Performance Institute Europe. Alleviating Fuel Poverty in the EU*. Investing in Home Renovation, A Sustainable and Inclusive Solution, Brussels, Belgium.

Brunner, K.-M.; Mandl, S.; Christanell, A. 2013. From local to national: Tackling fuel poverty in Austria. In *ECEEE Summer Study Proceedings*. Belambra Presqu'Tle de Giens, France.

Buzar, S. 2007. When homes become prisons: The relational spaces of post-socialist energy poverty. *Environment and Planning A*, 39, 1908–1925.

Chaudhuri, S.; Jalan, J.; Suryahadi, A. 2002. *Assessing Household Vulnerability to Poverty: A Methodology and Estimates for Indonesia*. Columbia University Department of Economics Discussion Paper No. 0102-52. New York: Columbia University.

Culver, L. C. 2017. *Energy Poverty: What You Measure Matters, The Stanford Natural Gas Initiative, Pre-symposium White Paper for: Reducing Energy Poverty with Natural Gas: Changing Political, Business, and Technology Paradigms, May 9 & 10, 2017*. Stanford University. https://ngi.stanford.edu/sites/g/files/sbiybj14406/f/NGI_Metrics_LitReview%282-17%29.pdf

Dagoumas, A.; Kitsios, F. 2014. Assessing the impact of the economic crisis on energy poverty in Greece. *Sustainable Cities and Society*, 13, 267–278. https://doi.org/10.1016/j.scs.2014.02.004.

Dall'O, G.; Sarto, L.; Sanna, N.; Tonetti, V.; Ventura, M. 2015. On the use of an energy certification database to create indicators for energy planning purposes: Application in northern Italy. *Energy Policy*, 85, 207–217.

Day, R. 2010. Environmental justice and older age: Consideration of a qualitative neighbourhood-based study. *Environment and Planning A*, 42, 2658–2673.

Day, R.; Hitchings, R. 2011. 'Only old ladies would do that': Age stigma and older people's strategies for dealing with winter cold. *Health and Place*, 17, 885–894.

Day, R.; Walker, G.; Simcock, N. 2016. Conceptualising energy use and energy poverty using a capabilities framework. *Energy Policy*, 93, 255–264.

De Quero, A.; Lapostolet, B. 2009. *Groupe de travail: Précarité énergétique*. Paris. www.hclpd.gouv.fr/IMG/pdf/Rapport_Pelletier_precarite_energetique_cle0ddb6a.pdf

Deller, D. 2016. *Energy Affordability in the EU: The Risks of Metric Driven Policies (SSRN Scholarly Paper No. ID 2850876)*. Rochester, NY: Social Science Research Network. https://papers.ssrn.com/abstract=2850876.

Dercon, S.; Krishnan, P. 2000. Vulnerability, seasonality, and poverty in Ethiopia. *Journal of Development Studies*, 36(6), 25–50.

Droutsa, K.; Kontoyiannidis, S.; Dascalaki, E. G.; Balaras, C. A. 2016. Mapping the energy performance of hellenic residential buildings from EPC (energy performance certificate) data. *Energy*, 98, 284–295.

Dubois, U. 2012. From targeting to implementation: The role of identification of fuel poor households. *Energy Policy*, 49, 107–115.

Dutt, G. S.; Ravindranath, N. H. 1993. *Bioenergy: Direct Applications in Cooking*. Washington, DC: Island Press, pp. 1–1177.

El Ansari, W.; El-Silimy, S. 2008. Are fuel poverty reduction schemes associated with decreased excess winter mortality in elders? A case study from London. *UK Chronic Illness*, 4, 289–294.

European Commission. 2016. *Under the Pilot Project 'Energy Poverty—Assessment of the Impact of the Crisis and Review of Existing and Possible New Measures in the Member States, Framework Contract ENER/A4/516-2014*. Final, Trinomics B.V., Rotterdam, Netherlands. https://ec.europa.eu/energy/sites/ener/files/documents/Selecting%20Indicators%20to%20Measure%20Energy%20Poverty.pdf (accessed on 10 January 2022).

European Commission. 2020. *Energy Poverty Observatory*. https://www.energypoverty.eu/indicators-data. (accessed on 10 March 2020).

European Commission. 2021. *EU Energy Poverty Observatory*. https://www.energypoverty.eu/about/about-observatory.

European Economic and Social Committee. 2011. *Opinion of the European Economic and Social Committee on 'Energy Poverty in the Context of Liberalisation and the Economic Crisis' (Exploratory Opinion) (2011/C 44/09)*. https://eur-lex.europa.eu/LexUriServ/LexUriServ.do?uri=OJ:C:2011:044:0053:0056:EN:PDF.

EUROSTAT. 2015. *Household Budget Survey 2015 Wave EU Quality Report*. https://ec.europa.eu/eurostat/documents/54431/1966394/HBS_EU_QualityReport_2015.pdf.

EUROSTAT. 2021. *European Union Statistics on Income and Living Conditions*. https://ec.europa.eu/eurostat/web/microdata/european-union-statistics-on-income-and-living-conditions.

EUROSTAT. 2022. *Database. Population and Living Conditions*. https://ec.europa.eu/eurostat/data/database (accessed on 10 January 2022).

Fankhauser, S.; Tepic, S. 2007. Can poor consumers pay for energy and water? An affordability analysis for transition countries. *Energy Policy*, 35, 1038–1049.

Ferreira, F. H. G.; Chen, S.; Dabalen, A.; Dikhanov, Y.; Hamadeh, N.; Jolliffe, D.; Narayan, A.; Prydz, E. B.; Revenga, A.; Sangraula, P.; Serajuddin, U.; Yoshida, N. 2015. A global count of the extreme poor in 2012: Data issues, methodology and initial results. *Tech Rep*. World Bank Group, pp. 1–68.

Fizaine, R.; Kahouli, S. 2019. On the power of indicators: How the choice of fuel poverty indicator affects the identification of the target population. *Applied Economics*, 51, 1081–1110.

Foster, V.; Tre, J.-P.; Wodon, Q. 2000. Energy prices, energy efficiency and fuel poverty. *Tech Rep*, September. World Bank Group, pp. 1–7.

Goldemberg, J.; Johansson, T. B. 1995. Energy as an instrument for socio-economic development. *Tech Rep.* New York: United Nations Development Programme, pp. 1–80.

Groh, S.; Pachauri, S.; Narasimha, R. 2016. What are we measuring? An empirical analysis of household electricity access metrics in rural Bangladesh. *Energy for Sustainable Development*, 30, 21–31.

Halkos, G. E.; Gkampoura, E.-C. 2021. Coping with energy poverty: Measurements, drivers, impacts, and solutions. *Energies*, 14(10), 2807. https://doi.org/10.3390/en14102807.

Hall, S. 2013. Energy justice and ethical consumption. *Local Environment*, 18(4), 422–437.

Hammond, A.; Kramer, V. J.; Katz, R. S.; Tran, J. T.; Walker, C. 2007. The next billion: Market size and business strategy at the base of the pyramid. *Tech Rep.* Washington, DC: World Resources Institute and International Finance Corporation, pp. 1–149.

Healy, J. 2004. *Housing, Fuel Poverty and Health: A Pan-European Analysis*. New York: Routledge.

Heitzmann, K.; Canagarajah, R. S.; Siegel, P. B. 2002. Guidelines for assessing the sources of risk and vulnerability. *Social Protection Discussion Paper 0218*. Washington, DC: World Bank

Hills, J. 2011. Fuel poverty: The problem and its measurement. *Interim Report of the Fuel Poverty Review. CASE Report 69*. London: Department for Energy and Climate Change.

IEA. 2011. Energy for All: Financing access for the poor. *Tech Rep.* Paris: International Energy Agency, pp. 1–52.

IEA. 2016. WEO energy access database. *Tech Rep.* International Energy Agency.

Isherwood, B. C.; Hancock, R. M. 1979. *Household Expenditure on Fuel: Distributional Aspects*. London: Economic Advisor's Office.

Jenkins, K.; McCauley, D.; Heffron, R.; Stephan, H.; Rehner, R. 2016. Energy justice: A conceptual review. *Energy Research & Social Science*, 11, 174–182. https://doi.org/1016/j.erss.2015.10.004.

Karasek, J.; Pojar, J. 2018. Programme to reduce energy poverty in the Czech Republic. *Energy Policy*, 115, 131–137.

Kizilcec, V.; Parikh, R.; Bisaga, I. 2021. Examining the journey of a pay-as-you-go solar home system customer: A case study of Rwanda. *Energies*, 14, 330.

Lampietti, J.; Meyer, A. 2002. *When Heat Is a Luxury: Helping the Urban Poor of Europe and Central Asia Cope with the Cold*. Washington, DC: World Bank.

Legendre, B.; Ricci, O. 2015. Measuring fuel poverty in France: Which households are the most fuel vulnerable? *Energy Economics*, 49, 620–628.

Li, K.; Lloyd, B.; Liang, X.-J.; Wei, Y.-M. 2014. Energy poor or fuel poor: What are the differences? *Energy Policy*, 68, 476–481.

Liddell, C.; Morris, C.; McKenzie, S. J. P.; Rae, G. 2012. Measuring and monitoring fuel poverty in the UK: National and regional perspectives. *Energy Policy*, 49, 27–32.

Masera, O. R.; Saatkamp, B. D.; Kammen, D. M. 2000. From linear fuel switching to multiple cooking strategies: A critique and alternative to the energy ladder modeF5. *World Development*, 28(12), 2083–2103.

Masud, J.; Sharan, D.; Lohani, B. N. 2007. Energy for all: Addressing the energy, environment and poverty nexus in Asia. *Tech Rep.* Asian Development Bank, pp. 1–123.

Maxim, A.; Mihai, C.; Apostoaie, C.-M.; Popescu, C.; Istrate, C.; Bostan, I. 2016. Implications and measurement of energy poverty across the European Union. *Sustainability*, 8, 483.

Mendoza, C. B.; Cayonte, D. D. D.; Leabres, M. S.; Manaligod, L. R. A. 2019. Understanding multidimensional energy poverty in the Philippines. *Energy Policy*, 233,110886.

Middlemiss, L.; Gillard, R. 2015. Fuel poverty from the bottom-up: Characterising household energy vulnerability through the lived experience of the fuel poor. *Energy Research & Social Science*, 6(Supplement C), 146–154.

Mirza, B.; Szirmai, A. 2010. Towards a new measurement of energy poverty: A cross-community analysis of rural Pakistan. *Tech Rep.* United Nations University, pp. 1–41.

Modi, V.; McDade, S.; Lallement, D.; Saghir, J. 2005. *Energy and the Millennium Development Goals*. Energy Sector Management Assistance Programme. New York: United Nations Development Programme.

Moore, R. 2012. Definitions of fuel poverty: Implications for policy. *Energy Policy*, 49, 19–26.

Nussbaumer, R.; Bazilian, M.; Modi, V. 2012. Measuring energy poverty: Focusing on what matters. *Renewable and Sustainable Energy Reviews*, 16, 231–243.

OECD/IEA. 2010. *Energy Poverty. How to Make Modern Energy Access Universal*? World Energy Outlook for UN General Assembly on Millennium Goals. https://iea.blob.core.windows.net/assets/fdbdd604-de2c-4977-8a3f-20f93e68e738/HowtoMakeModernEnergyAccessUniversal.pdf

Office for National Statistics. 2014. *Excess Winter Mortality in England and Wales*. http://www.ons.gov.uk/ons/rel/subnational-health2/excess-winter-mortality- in-england-and-wales/index.html (accessed on 22 December 2014).

Osbaldeston, J. 1984. Fuel poverty in UK cities. *Cities*, 1, 366–373.

Pachauri, S.; Spreng, D. 2004. Energy use and energy access in relation to poverty. *Economic and Political Weekly*, 39(3), 271–278.

Pagliaro, M.; Meneguzzo, F. 2020. Distributed generation from renewable energy sources: Ending energy poverty across the world. *Energy Technology*, 8, 2000126.

Paugam, S. 1995. The spiral of precariousness: A multidimensional approach to the process of social disqualification in France. In G. Room (Ed.), *Beyond the Threshold: The Measurement and Analysis of Social Exclusion*. Bristol: The Policy Press, pp. 49–79.

Practical Action. 2014. Poor peopled energy outlook 2014. *Tech Rep*. Practical Action, pp. 1–74. https://infohub.practicalaction.org/bitstream/handle/11283/556929/Poor%20people's%20energy%20outlook%202014.pdf;jsessionid=4A5E9B817CE6777528EB4FBD6321F227?sequence=1.

Pritchett, L.; Suryahadi, A.; Sumarto, S. 2000. Quantifying vulnerability to poverty: A proposed measure, applied to Indonesia. *Policy Research Working Paper no. 2437*. Washington, DC: The World Bank.

Pye, S.; Dobbins, A.; Baffert, C.; Brajkovíc, J.; Grgurev, I.; Miglio, R.; Deane, P. 2015. Energy poverty and vulnerable consumers in the energy sector across the EU. In *Analysis of Policies and Measures London*. Brussels, Belgium: European Commission.

Rademaekers, K.; Yearwood, J.; Ferreira, A.; Pye, S.; Hamilton, I.; Agnolucci, P.; Grover, D.; Karásek, J.; Anisimova, N. 2016. *Selecting Indicators to Measure Energy Poverty*. Trinomics, Rotterdam, The Netherlands.

Recalde, M.; Peralta, A.; Oliveras, L.; Tirado-Herrero, S.; Borrell, C.; Palenda, L.; Gotsens, M.; Artazcoz, L.; Mari-DeirOlmo, M. 2019. Structural energy poverty vulnerability and excess winter mortality in the European Union: Exploring the association between structural determinants and health. *Energy Policy*, 233, 110869.

Reddy, B. S. 2003. Overcoming the energy efficiency gap in India's household sector. *Energy Policy*, 31, 1117–1127.

Redmond, G.; Hutton, S. 2000. Poverty in transition economies: An introduction to the issues. In S. Hutton (Ed.), *Poverty in Transition*. London: Routledge, pp. 1–13.

Santillan, O. S.; Cedano, K. G.; Martinez, M. 2020. Analysis of energy poverty in 7 Latin American countries 仏 energy poverty index. *Energies*, 13, 1608.

Sen, A. 1993. Capability and well-being. In A. Sen, M. Nussbaum (Eds.), *The Quality of Life*. Oxford: Oxford University Press, pp. 30–53.

Seuret-Jimenez, D.; Robles-Bonilla, T.; Cedano, K. G. 2020. Measurement of energy access using fuzzy logic. *Energies*, 13, 3266.

Siksnelyte-Butkiene, Indre. 2021. A systematic literature review of indices for energy poverty assessment: A household perspective. *Sustainability*, 13(19), 10900. https://doi.org/10.3390/su131910900.

Siksnelyte-Butkiene, L.; Štreimikiene, D.; Lekavičius, V.; Balezentis, T. 2021. Energy poverty indicators: A systematic literature review and comprehensive analysis of integrity. *Sustainable Cities and Societies*, 67, 102756.

Simcock, N.; Walker, G.; Day, R. 2016. Fuel poverty in the UK: Beyond heating? *People, Place and Policy*, 10, 25–41.

Snell, C.; Bevan, M.; Thomson, H. 2014. *Fuel Poverty and Disabled People: The Impact of Policy Change*. Kendal: EAGA Charitable Trust.

Sokolowski, J.; Lewandowski, R; Kielczewska, A.; Bouzarovski, S. 2020. A multidimensional index to measure energy poverty: The Polish case. *Energy Sources, Part B: Economics, Planning, and Policy*, 15, 92–112.

Sovacool, B. K.; Sidortsov, R. V.; Jones, B. R. 2014. Deciphering energy justice and injustice. In B. K. Sovacool, R. V. Sidortsov, B. R. Jones (Eds.), *Energy Security, Equality and Justice*. Abingdon and New York: Routledge.

Spiliotis, E.; Arsenopoulos, A.; Kanellou, E.; Psarras, J.; Kontogiorgos, P. 2020. A multi-sourced data based framework for assisting utilities identify energy poor households: A case-study in Greece. *Energy Sources Part B-Economics Planning Policy*, 15, 49–71.

Streimikiene, D.; Kyriakopoulos, G.; Lekavičius, V.; Siksnelyte-Butkiene, I. 2021. Energy poverty and low carbon just energy transition: Comparative study in Lithuania and Greece. *Social Indicators Research*, 2021, 1–53.

Thomson, H.; Simcock, N.; Bouzarovski, S.; Petrova, S. 2019. Energy poverty and indoor cooling: An overlooked issue in Europe. *Energy and Buildings*, 196, 21–29. https://doi.org/10.1016/j.enbuild.2019.05.014.

Thomson, H.; Snell, C. 2013. Quantifying the prevalence of fuel poverty across the European Union. *Energy Policy*, 52, 563–572.

Tirado Herrero, S. 2013. *Fuel Poverty Alleviation as a Co-Benefit of Climate Investments: Evidence from Hungary*. Budapest: Central European University.

Tirado Herrero, S.; Ürge-Vorsatz, D. 2012. Trapped in the heat: A post-communist type of fuel poverty. *Energy Policy*, 49, 60–68.

Tzanev, D. 2019. *Energy Poverty in the Light of Local Elections: Analytical Rationale*. http://www.eneffect.bg/images/upload/123/EnergyPovertyAnalysisfinal-zg-cover.pdf.

UNDP and WHO. 2009. The energy access situation in developing countries. *Tech Rep*. New York: United NationalDevelopment Programme and the World Health Organization, pp. 1–142.

Viggers, H.; Howden-Chapman, R.; Ingham, T.; Chapman, R.; Pene, G.; Davies, C.; Currie, A.; Pierse, N.; Wilson, H.; Zhang, J.; Baker, M.; Crane, J. 2013. Warm homes for older people: Aims and methods of a randomised community-based trial for people with COPD. *BMC Public Health*, 13, 176.

Villalobos, C.; Chávez, C.; Uribe, A. 2021. Energy poverty measures and the identification of the energy poor: A comparison between the utilitarian and capability-based approaches in Chile. *Energy Policy*, 152, 112146, https://doi.org/10.1016/j.enpol.2021.112146.

Warren, C. R.; McFadyen, M. 2010. Does community ownership affect public attitudes to wind energy? A case study from south-west Scotland. *Land Use Policy*, 27(2), 204–213.

Wirawan, H.; Gultom, Y. M. L. 2021. The effects of renewable energy-based village grid electrification on poverty reduction in remote areas: The case of Indonesia star. *Energy and Sustainable Development*, 62, 186–194.

Wright, F. 2004. Old and cold: Older people and policies failing to address fuel poverty. *Social Policy & Administration*, 38, 488–503.

Zhang, T.; Shi, X. P.; Zhang, D. Y.; Xiao, J. J. 2019. Socio-economic development and electricity access in developing economies: A long-run model averaging approach. *Energy Policy*, 232, 223–231.

Zhao, J.; Jiang, Q. Z.; Dong, X. C.; Dong, K. Y. 2021. Assessing energy poverty and its effect on CO_2 emissions: The case of China. *Energy Economics*, 97, 105191.

3 Energy Security

Indre Siksnelyte-Butkiene

3.1 PERCEPTION OF THE ENERGY SECURITY CONCEPT

While energy security has always been a critical consideration, its heightened significance has become especially apparent in the face of recent global challenges. Ensuring energy security has risen to the forefront of agendas at both national and regional levels worldwide. The necessity to enhance energy security amidst current global challenges is driven by factors such as geopolitical conflicts, as well as economic and social difficulties faced by energy consumers and the entire energy sector. These challenges are associated with climate change, the impacts and threats of global pandemics, and the scarcity of energy resources (Siksnelyte-Butkiene, 2023).

The military conflict in Ukraine, coupled with following energy supply and affordability challenges faced by European countries, clearly illustrates the critical significance of energy security. Recent challenges in the energy sector underscore the imperative for heightened attention from a country's government to ensure the energy independence of a country in order to ensure the well-being of its citizens. Simultaneously, factors such as the COVID-19 pandemic, challenges associated with global warming, the increasing energy demand, worries linked to depletion of energy sources, and issues related to energy poverty emphasize the importance of promoting renewable energy and increasing energy efficiency at both the national and household levels.

Energy security is a fundamental pillar of the energy system of a country and stands as a key element in international relations (Marhold, 2021). However, the concept's perception is uncertain and context-dependent, resulting in no one universally accepted definition and clear or standardized content. Numerous attempts have been made in the scientific literature to define the concept of energy security. Scientific studies reveal variations in the perception of energy security among different cultures and regions. It is essential to note that these differences are also influenced by factors such as gender, political orientation, resource availability, and dependency of a country on import, among others (Sovacool et al., 2012). For instance, Demski et al. (2018) identified that public concerns about energy security tend to be higher in countries with lower economic performance and human well-being indicators.

The traditional comprehension of energy security revolves around the concern of guaranteed access to energy. This traditional view encompasses several key dimensions that collectively protect the energy system of a country such as energy supply security, self-reliance, diversification of energy sources, and cost-effectiveness. Central to this understanding is the assurance of a secure and uninterrupted energy supply, where a country aims to guarantee reliable access to the required energy resources necessary for sustaining its economic and societal functions. Self-sufficiency of energy resources is the other central element meaning a country's capability to meet its energy demands from domestic energy sources. By reducing dependence on external suppliers, self-sufficiency increases a country's resilience against potential disruptions in the global energy market. Energy diversification is another cornerstone of traditional comprehension of energy security, supporting a diversified mix with a range of energy sources. This strategy aims to mitigate risks associated with overreliance on a single energy source, thereby increasing the resilience and adaptability of energy infrastructure of a country. The other critical consideration is affordability, declaring the importance of ensuring energy prices at levels that are reasonable for all energy consumers (households and industry). This aspect seeks to ensure that energy will be economically viable, supporting the economic performance and well-being of people.

DOI: 10.1201/9781003517481-3

The widely known "4As" framework—consisting availability, affordability, accessibility, and acceptability—aptly characterizes this traditional perception of energy security (APERC, 2007). In a traditional framework, the acceptability dimension is primarily understood in terms of amount of greenhouse gas (GHG) emissions or the portion of renewable energy in the energy portfolio.

Over time, the comprehension of energy security has evolved significantly, extending beyond its traditional association with the exhaustion of fossil fuel resources and energy access issues. In recent years, a broader and more interdisciplinary approach has emerged, covering a diverse range of issues. This expanded perspective includes considerations related to climate change, impact on the environment, globalization, and the overall well-being of people.

The recent energy security studies place significant emphasis on environmental considerations. For instance, Cervan et al. (2022) underscore the significance of environmental impact in shaping the energy security of a country. The authors define energy security as the sustained availability of energy while concurrently ensuring affordability and minimizing environmental impacts. This perspective underscores the growing recognition that a sustainable and secure energy future must not only ensure a stable and affordable energy supply but also prioritize minimal impact to the environment.

Furthermore, there is a notable tendency of increasing significance of political and governmental aspects in energy security assessment research. For example, Kisel et al. (2016) and Khan and Dhakal (2022) underscore the increasing significance of political and governmental dimensions in shaping the overall energy security landscape. These considerations reflect an understanding that effective energy security strategies must consider the influence of political frameworks, government policies, and international relations when shaping the energy sector.

According to the Cherp and Jewell (2014) a comprehensive understanding of the energy security concept should give the answers to three fundamental questions: "energy security for whom?" "Energy security for which values?" and "Energy security from what threats?" These questions are considered as essential as they facilitate the exploration of diverse dimensions, including both sectorial considerations such as end-users, infrastructure, and resources and a geographical perspective covering national, regional, and global scales. In their framework, Cherp and Jewell (2014) characterize energy security as the "low vulnerability of vital energy systems," wherein vulnerability is explained as "a combination of exposure to risks and resilience capacities." The provided definition underscores the importance of not only minimizing the potential risks but also supporting the resilience capacities of energy systems.

This shift in perception reflects a recognition that energy security is linked to complex interrelated factors that extend beyond only resource availability. The current perception of energy security involves a more holistic assessment, recognizing the interconnections of energy systems with broader environmental, social, and economic contexts. As highlighted by Fang et al. (2018), the capability to evaluate opportunities for increasing energy security becomes a crucial aspect of fostering optimal, clean, and low-carbon development across the entire energy system. In essence, this evolution in the understanding of energy security aligns with the imperative to address not only the urgent challenges related to energy supply but also the long-term sustainability and resilience of energy systems in the face of global concerns and challenges.

The diversities in perception and the increasing emphasis on environmental, social, and governmental-/policy-related considerations led to notable disparities among studies presenting indicator sets and indices for the measurement of energy security. These diversities underscore the complexity of the nature of conceptualizing and assessing energy security. Also, these disparities show the critical importance of ensuring that the chosen definition and the selected indicators will be not only relevant but also adaptable to the changing global concerns. Therefore, regular reassessment is necessary to adapt the dynamic changes in the global environment, advancements in technologies, and interactions in global and regional challenges associated with energy accessibility. It allows one to align the measurement frameworks with emerging priorities, reflecting changes and addressing new emerging concerns.

3.2 ENERGY SECURITY INDICATORS

Numerous tools have been developed to assess energy security across different countries, country groups, or regions. Studies that use different composite indicators or sets of indicators can be systematically classified into several categories based on their research focus, which is, the primary objectives of the investigations. A visual representation of these research directions is presented in Figure 3.1.

There are many scientific studies focusing on development of a novel composite indicator or set of indicators aimed to assess the energy security level, either on a country scale or at a global scale. The review and analysis of these recently used and developed composite indicators provide valuable perspectives on the understanding of energy security. A comprehensive examination of these composite indicators, introduced or applied in recent years, is analyzed in Table 3.1. This analysis enhances the comprehension of how energy security is defined and measured in the current studies, offering insights into global and country-specific concerns.

Other research studies concentrate on introducing new methodologies for assessing energy security. For instance, Yang et al. (2022) conducted a study that presented an approach based on the Driving Forces-Pressures-State-Impacts-Responses (DPSIR) model for assessing energy security in China. The research involved a comprehensive examination of interrelated linkages and impacts within the energy security framework. Berdysheva and Ikonnikova (2021) developed two indices to evaluate energy trade and security dynamics globally, specifically focusing on the security of energy consumption and production. Wang et al. (2021) contributed to the field by presenting an assessment tool designed to evaluate energy economic security, considering four dimensions: energy industry development, energy utilization level, energy supply capacity, and the environmental impact. Studies presenting methodological innovations contribute to the advancement of understanding and assessing various dimensions of energy security within the scientific discourse.

The other part of scientific studies has been specifically oriented towards conducting partial assessments of energy security from specific perspectives. For instance, Kosai and Unesaki (2020) introduced a composite indicator to evaluate nuclear energy supply security. The presented indicator found practical application in the assessment of nuclear energy supply security in the case study of the United States. Similarly, Fuentes et al. (2020) introduced the Power System Security

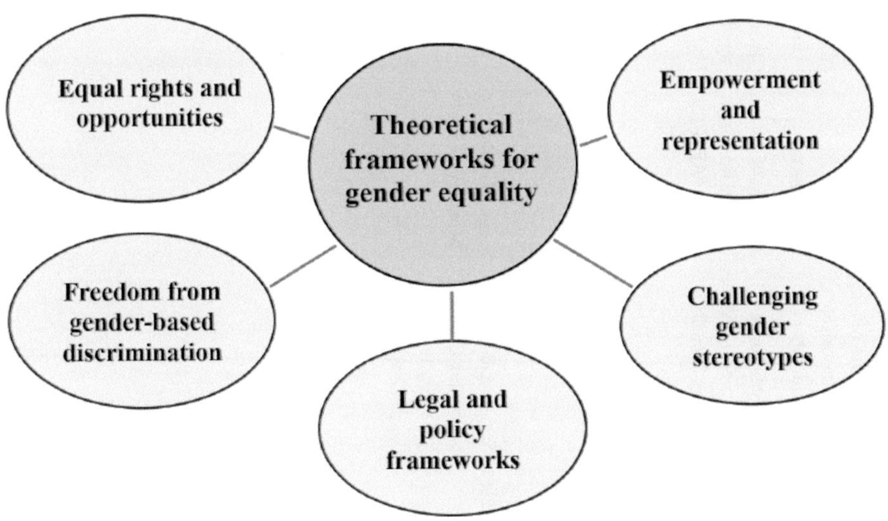

FIGURE 3.1 The main research directions of studies dealing with energy security issues.

Source: developed by the authors.

TABLE 3.1
Energy Security Indicators Sets and Indices

Approach	Name of composite indicator	Source	Case study location	Number of indicators used	Input data	Experts' involvement	Process of criteria selection	Weighting method	Contribution to the field
Traditional	Energy Security Indicators (4 A's of Energy Security)	Asia Pacific Energy Research Centre (APERC, 2007)	Global applied	5 compound indicators	Diversification of primary energy demand; net energy import dependency; non-carbon based fuel portfolio; net oil import dependency; Middle East oil import dependency	N/A	N/A	N/A	The indicators proposed are simple and the methodology can be easily adapted to other research. Therefore, the set of indicators is widely applied.
Interdisciplinary	Simple and Complex Energy Security Indicators and Metrics	Sovacool and Mukherjee (2011)	–	320 single and 52 compound indicators	Supply and production security; dependency; diversification; stability of price; access and equity; decentralization; energy affordability; research and innovation; reliability and safety; resilience and adaptive capacity, energy efficiency and energy intensity; investment and employment; land use; water; pollution; governance; trade and regional interconnectivity; competition and markets; and knowledge and access to information	Yes	Literature, Delphi method, experts interviews/ survey	N/A	The proposed indicators cover all dimensions of sustainability and it is one of the first attempts to stress the importance of social and environmental aspects for the strength of energy security of a country.

Traditional	Energy Security Risk Index	Global Energy Institute (2012)	25 OECD countries	29	Global fuels; fuel imports; energy expenditures; price and market volatility; energy use intensity; electric power sector; transportation sector; environmental indicators	N/A	Literature	N/A	The index allows one to compare the level of energy security risk across different countries.
Traditional	Energy Security Index	Franki and Viskovic (2015)	Croatia	6	Electricity price; renewable fraction; import; fossil fuel consumption; air pollution	No	Own selection	Own estimation—equal weights	The index is created and applied for a case study of Croatia.
Traditional	Singapore Energy Security Index	Ang et al. (2015)	Singapore	22	Energy intensity; affordability; dependency; fuel mix; reserves; oil consumption; technology diversity; Electricity load factor; system interruption; energy consumption; generation; transport fuel diversity; air pollution; renewable fraction; modal share of public transport	No	Literature, own selection	Own estimation—subjective weighting	An index to measure the level of energy security in Singapore was proposed and applied in practice.
Traditional	Energy Security Index	Erahman et al. (2016)	Global applied (71 countries)	14	Energy production; self-sufficiency; reserves; diversification; affordability; electrification rate; population relying on traditional energy sources; vehicle ownership; air pollution; intensity; efficiency	No	Literature	Principal component analysis, equal weighting	Indicator set to perform global analysis of energy security is proposed and applied in practice.

(*Continued*)

TABLE 3.1 *(Continued)*
Energy Security Indicators Sets and Indices

Approach	Name of composite indicator	Source	Case study location	Number of indicators used	Input data	Experts' involvement	Process of criteria selection	Weighting method	Contribution to the field
Traditional	Energy Security Indicators for Energy Resource-Poor Economies	Li et al. (2016)	Japan, Singapore, South Korea, Taiwan	9	Energy self-sufficiency; diversity; energy intensity; efficiency; air pollution; consumption; share of fossil fuels	No	Own selection	Principal component analysis	Indicators set to measure the level of energy security of resource-poor economies were presented.
Interdisciplinary	–	Augutis et al. (2017)	Lithuania	68	Installed capacity; power units' lifetime; energy consumption and generation by sources; interchangeability; possibility to choose energy supplier; import; market indicators; political risk; affordability; commitments regarding renewable fraction; commitments regarding efficiency; commitments regarding GHG emissions; public acceptance regarding nuclear energy	Yes	Own selection, experts	Own estimation—equal weights	A set of indicators was created to reflect the country situation after significant changes in the sector.

Interdisciplinary	–	Zhang et al. (2017)	China	20	Energy production; self-sufficiency; reserves; diversity of energy consumption; price stability; affordability; share of electricity; energy intensity; efficiency; power plant utilization; forest coverage; water and air emissions; health; investment in energy industry; research intensity; expenditure on energy saving and environmental protection	Yes	Literature	Experts, fuzzy analytic hierarchy process (AHP)	The indicator set for energy security measurement in China's provinces are presented and applied for assessment at a regional level. The innovative approach, which used advanced multi-criteria techniques, is applied. The study can serve as a good methodological example for further research.
Interdisciplinary	Energy Security Index	Wang and Zhou (2017)	Global applied (162 countries)	23	Energy production; consumption; import; reserves; days to obtain permanent electricity connection; efficiency; access to energy sources, affordability; renewable fraction; energy intensity; air pollution; political stability; regulatory quality; government effectiveness; control and corruption; gross domestic product (GDP)	No	Literature, own selection	Principal component analysis, own selection—subjective weight allocation	A framework to perform global analysis and identify disparities among countries is presented.

(*Continued*)

TABLE 3.1 *(Continued)*
Energy Security Indicators Sets and Indices

Approach	Name of composite indicator	Source	Case study location	Number of indicators used	Input data	Experts' involvement	Process of criteria selection	Weighting method	Contribution to the field
Interdisciplinary	Energy Supply Security Index	Li and Zhang (2018)	China	12	Energy production; reserves; production diversity; energy dependence; affordability; energy consumption; air and water pollution; renewable fraction; investments in infrastructure; energy efficiency	No	Literature	Entropy weight method based on Shannon's entropy	An innovative approach based on advanced techniques to forecast energy supply security was developed.
Traditional	–	Le and Nguyen (2019)	Global applied (74 countries)	10	Energy production; consumption; intensity; CO_2 emissions; renewable fraction	No	Literature	N/A	The study proposed a methodology to measure the impact of energy security on the economic growth.
Traditional	China energy security index	Song et al. (2019)	China	18	Reserves of resources; production; dependency; energy mix; installed capacity; air pollution; renewable fraction; electricity share in total energy consumption; coal consumption for power generation; energy intensity; affordability	No	Literature	Own estimation	A new index for China energy security measurement at the country level was proposed and applied for a case study in different years from 1990 to 2014.

Interdisciplinary	Power System Security Index	Fuentes et al. (2020)	Argentina	45	Energy recourse reserves; energy access; adequacy; efficiency; diversification; import diversification and dependency; political stability; potential of renewable energy; interconnections; electricity consumption; affordability; intensity; renewable fraction; compactivity of business; research and development (R&D) expenditures	No	Literature, own selection	Own estimation—equal weights	A novel electricity security index was introduced and applied for a case study of Argentina. The proposed methodology underscores the necessity of assessing energy security across diverse energy systems to determine areas requiring policy enhancements.
Interdisciplinary	Dynamic Energy Security Index	Wang et al. (2020)	China	17	Import dependency; energy reserves; concentration; self-sufficiency; intensity; energy mix; affordability; investments; efficiency; air quality	No	Literature	Own estimation—equal weights	The Dynamic Energy Security Index, employing functional data analysis, was introduced and implemented in a case study of China. This study makes a contribution to the discussions on energy security in China and its provinces.

(Continued)

TABLE 3.1 *(Continued)*
Energy Security Indicators Sets and Indices

Approach	Name of composite indicator	Source	Case study location	Number of indicators used	Input data	Experts' involvement	Process of criteria selection	Weighting method	Contribution to the field
Interdisciplinary	Energy Security Index for Cape Verde	Coutinho et al. (2020)	Cape Verde	20	Self-sufficiency; dependency on imports; energy diversification; geographic distribution of energy infrastructure; affordability; accessibility; innovations; capacity margins; intensity; direct employment; government investment; air quality; information accessibility	Yes	Literature, Delphi survey	Delphi survey	The Energy Security Index for Cape Verde was introduced and practically implemented. The presented index is founded on a transparent and sound methodological basis, ensuring the repeatability and adaptability of the proposed methodology to other studies in the field.
Traditional	Energy Security Index of Pakistan	Bin Abdullah et al. (2020a)	Pakistan	22	Energy generation; consumption; diversification; air quality; accessibility; efficiency; dependency on imports; renewable fraction	Yes	Literature, experts survey	Principal Component Analysis	The Energy Security Index for Pakistan was developed and implemented to assess performance in the period spanning 2000 to 2018. It should be noted that this index is based on traditional understanding of energy security.

Traditional	Weighted Energy Security Index	Rodriguez-Fernandez et al. (2020)	27 EU member states	3	Socioeconomic risk index; energy dependency on imports; concentration	No	Literature	Own estimation—equal weights	The index was introduced and applied in a case study across the EU countries. Diverging from conventional methodologies, this index is distinctive in its concentration on the security of a single energy resource.
Interdisciplinary	Energy Security Level	Augutis et al. (2020)	The Baltic States	24 compound indicators	Energy generation; consumption; installed capacity; renewable fraction; energy reserves; market competitiveness; dependency on imports; diversification; affordability; household energy expenses; political risk	Yes	Literature, own selection	Experts survey	The approach measuring the energy security level was developed and applied for the case study of the Baltic States. The presented methodology contributes to the current understanding of the energy security concept with the involvement of market-, policy-, and household expenses indicators.

(*Continued*)

TABLE 3.1 ***(Continued)***
Energy Security Indicators Sets and Indices

Approach	Name of composite indicator	Source	Case study location	Number of indicators used	Input data	Experts' involvement	Process of criteria selection	Weighting method	Contribution to the field
Interdisciplinary	Security of Nuclear Energy Supply	Kosai and Unesaki (2020)	The United States	6	Energy generation; diversity in energy imports; GDP; population; public opinion/acceptance	No	Literature, own selection	Principal component analysis, equal weights	The methodology for assessing the security of nuclear energy supply was introduced and implemented in a case study of the United States. The created indicator is unique in its focus on measuring the security of one energy resource.
Interdisciplinary	Provincial Energy Security Index	Li et al. (2020)	China	12	Energy generation; consumption; diversity; storage capacity; renewable fraction; infrastructure; intensity; air quality; sewage ratio; affordability; subsidies; investments	Yes	Own selection	Entropy method, order relation method (experts)	New index for regional level analysis was proposed and applied. The study makes a contribution to the discussions on energy security in China and its provinces.

Interdisciplinary	China's Energy Security Index	Gong et al. (2021)	China	13	Energy generation; consumption; air quality; investments; affordability; GDP	No	Literature	Entropy weight method	New index for assessing energy security in China at the regional level was introduced and applied. The study makes a contribution to the discussions on energy security in China and its provinces.
Interdisciplinary	Energy Security Index of Pakistan	Bin Abdullah et al. (2020b), Bin Abdullah et al. (2021)	Pakistan	39	Energy generation; consumption; renewable fraction; affordability; import; intensity; efficiency; level of corruption; effectiveness, of governance; oil stock; air quality; forest coverage	No	Literature	Principal component analysis	The index for evaluating energy security in Pakistan was introduced and applied to assess performance over the past three decades and to project potential performance in the future.
Interdisciplinary	Energy Security Performance Framework	Huang et al. (2021a)	34 OECD countries	20	Energy generation; consumption; renewable fraction; reliance on imports; national reserves; efficiency; air quality; forest coverage; political stability; expenditure on R&D; perception of corruption; affordability; duration to get electricity	No	Literature	Principal component analysis	The proposed set of indicators for measuring energy security was introduced and applied across 34 OECD countries. The framework proposed enables the repetition of the research and the monitoring of progress achieved.

(*Continued*)

TABLE 3.1 *(Continued)*
Energy Security Indicators Sets and Indices

Approach	Name of composite indicator	Source	Case study location	Number of indicators used	Input data	Experts' involvement	Process of criteria selection	Weighting method	Contribution to the field
Interdisciplinary	–	Sotnyk et al. (2021)	Ukraine	15	Energy generation; consumption; energy reserves; reliance on imports; depreciation of fixed assets; investments; renewable fraction; efficiency; energy deprivation; financial burden of renewable energy; capacity development; fluctuations in energy supply	No	Literature, own selection	N/A	An enhanced set of indicators to assess energy security in Ukraine was proposed, justifying the incorporation of supplementary composite indicators in consideration of global and domestic threats.
Interdisciplinary	Energy Security Index	Stavytskyy et al. (2021)	45 European and West Asian countries	29	Energy consumption; reliance on imports; energy export; depletion; efficiency; intensity; renewable fraction; air quality	No	Own selection	Own estimation—equal weights	An energy security index for the comprehensive assessment of international energy security was developed and applied for an analysis of 45 countries. The study aimed to identify and analyze changes over the course of the past two decades.

Interdisciplinary	Energy Economic Security Index	Wang et al. (2021)	China	24	Investments in energy; employment; benefit to the energy industry; energy price fluctuations; energy generation; consumption; efficiency; elasticity; reliance on imports; air quality; composition of the energy mix	No	Literature, own selection	Principal component analysis	An index considering economic aspects of energy security at the regional level were introduced and applied for the assessment of regions in China. The study makes a contribution to the discussions on energy security in China and its provinces.
Interdisciplinary	–	Huang et al. (2021b)	China	10	Energy consumption; reserves; investments; energy diversification; GDP; air quality; spending on pollution control; R&D investments; input of human resources in R&D	Yes	Own selection	Fuzzy best worst method (BWM)	A new approach, which makes a contribution to the discussions on energy security in China and its provinces, was introduced and applied in practice.

(*Continued*)

TABLE 3.1 *(Continued)*
Energy Security Indicators Sets and Indices

Approach	Name of composite indicator	Source	Case study location	Number of indicators used	Input data	Experts' involvement	Process of criteria selection	Weighting method	Contribution to the field
Interdisciplinary	–	Zhang et al. (2021)	China	28	Energy generation; diversification; efficiency; self-sufficiency; affordability; households' expenditure for energy needs; heated area per household; accessibility; reliability infrastructure intensity; efficiency; air quality; investments; expenditures on environmental protection	Yes	Literature, own selection	Fuzzy AHP, pairwise comparisons	A novel and innovative approach for assessing regional energy security was introduced and applied in a case study. The advanced methodology employed in this study can potentially serve as a guiding framework for future research in the field.

Interdisciplinary	–	Kharazishvili et al. (2021)	Ukraine	42	Energy generation; consumption; GDP; intensity; reserves; reliability; air quality; imported energy; participation in international markets; composition of the energy mix; renewable fraction; infrastructure; self-sufficiency; energy suppliers; technological dependence; legal frameworks; stability in regulations and policies; capital investments; salaries in the energy sector; affordability; quality of public policies; quality of energy services; personnel competence	No	Literature, own selection	Principal component analysis	A novel approach for the analysis and strategic management of energy security is introduced and implemented in a case study of Ukraine.
Interdisciplinary	–	Wu et al. (2021)	Global applied (125 countries)	18	Energy generation; consumption; diversification; reserves; import; self-sufficiency; GDP; accessibility; diversification of energy supply; affordability; renewable fraction; air quality; intensity; efficiency; governmental efficiency and spending; political stability	No	Literature, own selection	Principal component analysis	A set of indicators suitable for global-level analysis was selected, and an innovative approach for measuring energy security was applied to conduct a comprehensive assessment at the global level.

(Continued)

TABLE 3.1 *(Continued)*
Energy Security Indicators Sets and Indices

Approach	Name of composite indicator	Source	Case study location	Number of indicators used	Input data	Experts' involvement	Process of criteria selection	Weighting method	Contribution to the field
Interdisciplinary	Bangladesh Energy Security Framework	Khan and Dhakal (2022)	Bangladesh	23	Energy production; security of supply; dependency on imports; composition of the energy mix; price; infrastructure; subsidies; market decentralization, international geopolitics; economic and environmental acceptability; efficiency; institutional frameworks and governance; adaptability of technologies; safety aspects and reliability; resilience	Yes	Literature, Delphi method, experts interviews/ survey	N/A	The study introduces a set of indicators aimed at addressing energy security concerns in Bangladesh and highlights substantial disparities in the perception of the term among stakeholders and energy experts.
Interdisciplinary	Energy Security Performance	Ha and Thanh (2022)	Global applied (85 countries)	7	Energy consumption; intensity; air quality; intensity of CO_2; renewable fraction; foreign investments; GDP; level of industrialization; government efficiency	No	Own selection, literature	N/A	The study has empirically demonstrated a nonlinear relationship between a country's participation in global value chains and the performance of its energy security.

Interdisciplinary	Energy Security Index	Cervan et al. (2022)	9 countries of South America	15	Energy consumption; access to energy; quality; affordability; supply; composition of the energy mix; intensity; air quality; renewable fraction	No	Literature	N/A	A novel index is introduced, encompassing diverse dimensions of energy security with particular emphasis on environmental considerations. This attention to environmental issues serves as guidance and recommends the inclusion of more environmental aspects in future studies, highlighting their significance to a country's energy security.
Interdisciplinary	Pakistan's Multidimensional Energy Security Index	Bin Abdullah et al. (2022)	Pakistan	27	Energy consumption; accessibility; affordability; intensity; efficiency; reliance on import; corruption level in a country; renewable fraction; effectiveness of governance; air quality; forest coverage	Yes	Own selection, experts survey	Principal component analysis	A proposed energy security index for Pakistan has been applied to assess the situation over the past three decades, examining it as a whole and across each dimension. The methodological approach employed in this assessment is valuable for future studies and can provide guidance for future research.

(Continued)

TABLE 3.1 *(Continued)*
Energy Security Indicators Sets and Indices

Approach	Name of composite indicator	Source	Case study location	Number of indicators used	Input data	Experts' involvement	Process of criteria selection	Weighting method	Contribution to the field
Interdisciplinary	–	Tutak and Brodny (2022)	12 EU member states	17	Energy consumption; efficiency; supply; intensity; reliance on imported energy; composition of the energy mix; renewable fraction; net trade; GDP; intensity of GHG emissions; affordability; income of households	No	Literature	Entropy weight method	A set of indicators, aligning with the priorities of the EU energy policy, was formulated and implemented in a case study for 12 EU countries. The suggested set demonstrates adaptability and can be easily applied for a case study encompassing all EU member states.
Interdisciplinary	–	Yang et al. (2022)	China	20	Energy consumption; intensity; air quality; investments; economic output; industrialization and urbanization; GDP; population	No	Literature, own selection	Entropy weight method	A novel approach to analyze energy security was introduced, employing the DPSIR model to evaluate the energy security level and interrelations among its various dimensions.

Source: developed by the authors.

Index, specifically developed for assessing the level of electricity supply security. The index was employed in a case study of Argentina. Rodriguez-Fernandez et al. (2020) contributed by introducing the Weighted Energy Security Index, specifically designed to assess security of gas supply. The proposed index found application in a case study of EU countries. These targeted assessments underscore the importance of evaluating energy security across diverse energy systems and provide valuable insights into energy security within specific energy domains, offering targeted perspectives for policy refinement.

Within the scientific literature, endeavors to predict energy security level are evident. Some authors have employed known and widely recognized indices for this purpose. For instance, Ziemba (2022) predicted energy security level by using the International Energy Security Risk Index, presented by the Global Energy Institute over a decade ago (in 2012). The other authors have developed new composite indicators. For example, Li and Zhang (2018) created the Energy Supply Security Index and conducted short-term forecasts intended for China. The forecasting methodologies, whether relying on already established methodologies or new composite indicators, contribute to the ongoing efforts to predict and assess energy security dynamics.

Also, researchers have shown interest in exploring the multifaceted relationships between energy security and various other aspects. The connections between energy security and critical factors such as monetary policy, participation in global value chains, macroeconomic stability, income inequality, energy poverty, and food security have been empirically demonstrated in the scientific literature. Such kinds of investigations enhance a thorough understanding of the relationships between energy security and broader socio-economic and global performance.

Some scientific papers have delved into the examination of external factors directly influencing the level of energy security. It is scientifically proven that factors such as climate change, disruptions in resource supply, improper political actions, inappropriately organized energy markets, and limited energy infrastructure flexibility can have negative effects on a country's energy security. These findings show the importance of considering a broad spectrum of external elements when evaluating and enhancing energy security.

The comprehensive and multi-directional examination of energy security in the scientific literature, marked by the number of diverse composite indicators and analytical methodologies applied across different countries globally, shows the complex nature of the concept and variety of challenges faced in recent years. It can be stated that the concerns of recent years have fundamentally changed the conceptualization of energy security. This transformation can be clarified by analyzing the indicators incorporated into the latest energy security assessment frameworks.

Many composite indicators have been developed and used for the evaluation of energy security. The analysis of studies in the Web of Science database in recent years (2015–2022) allowed identification of 39 composite indicators and indicator sets aimed at assessing the energy security of countries or regions or conducting global-level assessment. Through a systematic literature review, these identified indicator sets and indices were broadly classified based on the approach followed. Some studies adhered to the traditional view of energy security, whereas more recent research takes an interdisciplinary approach, considering environmental, social, governmental, and geopolitical dimensions of the issue. It's important to mention that the categorization of developed and used indicators to interdisciplinary or traditional approaches is subjective and based on contextual assumptions. Table 3.1 presents the key features of composite indicators and frameworks created and utilized in recent years to assess the level of energy security.

The increasing among of research on energy security assessment has introduced diverse indices and assessment frameworks, highlighting the multi-dimensional nature of the problem. Importantly, energy security is interlinked with various other concerns, including energy poverty, efficiency of resources, and economic, environmental, social, and foreign policy. Through an analysis of scientific literature, it becomes evident that indicators are becoming more sophisticated, encompassing a broader range of dimensions, and aiming to support future decisions in energy policy.

Following an in-depth analysis of recent composite indicators, it is evident that, despite variations in the range and quantity of included indicators or reflected dimensions, the interdisciplinary approach to the problem is prevailing. Notably, there is an increasing trend of involving experts in the selection of indicators or determining their weights. This practice contributes to the development of more comprehensive composite indicators and helps mitigate subjectivity. Scientific methods, particularly principal component analysis and the entropy weighting method, are becoming more prevalent for determining weights, replacing the previously dominant subjective weighting allocation practice.

The present understanding of energy security is closely related with the paradigm of sustainable development and diverges significantly from the conventional "4As" framework. By incorporating technological, geopolitical, social, economic, and environmental dimensions, the energy security approach aligns closely with the holistic perspective of sustainable development. Therefore, it becomes imperative that this expanded approach would be applied in the practice of formulating national or regional energy policies to ensure alignment with the broader and enduring objectives of sustainable energy development.

In the traditional "4As" approach, connections with indicators that reflect environmental and social impacts were primarily centered around metrics like GHG emissions or renewable energy fraction. However, the current perception has evolved to encompass a broader set of aspects, including impact to health, job creation, quality of services, etc. It is essential to underscore that the selection of indicators for measuring energy security requires careful consideration, guided not only by indicator popularity or alignment with national or regional goals but also by an assessment of potential risks to a country's energy security level. For instance, although renewable energy indicators often address environmental and social dimensions, in many cases they overlook critical factors such as: supply chain stability of required equipment, timely equipment servicing, or energy system flexibility issues.

A diverse range of energy security dimensions and indicators have been used in recent composite indicators. Table 3.2 provides an overview of the latest and most popular dimensions identified in recent scientific studies (2020–2022).

It is important to note that the identified dimensions of energy security are interconnected. Authors categorize and name dimensions differently as well as include similar indicators in different dimensions. For instance, the availability dimension often encompasses indicators related to infrastructure and energy supply, although some authors identify them as distinct dimensions. Similarly, indicators concerning resource depletion or the development of new resources are typically part of the availability dimension, while others consider them in separate dimensions. The same applies to the acceptability dimension, where indicators related to air pollution or renewable energy may be categorized as individual dimensions. Nevertheless, recent trends indicate a growing emphasis on including indicators related to environmental impact, governance, and technological development when measuring and assessing energy security. Therefore, the changes in the prevailing understanding of energy security are evident.

The comprehensive analysis of composite indicators and their input data has provided valuable insights into the prevailing understanding of energy security. Accordingly, seven dimensions have been identified as having the biggest impact on the current perception; these dimensions are related to energy resources availability, energy affordability, environmental impact, social impact, technological development, geopolitical relations, and government effectiveness. Although accessibility and affordability dimensions still remain fundamental in assessing energy security, there has been a notable evolution in the acceptability dimension. Now, it includes considerations of environmental and social impacts, driven by rising concerns about climate change and societal well-being. Recent geopolitical tensions, military conflicts, and disputes among some countries have underscored the critical role of geopolitical relations for the country's energy security. Furthermore, recent studies have placed significant emphasis on indicators related to government effectiveness, such as measures of corruption levels. Also, a separate dimension encompasses indicators associated with the

TABLE 3.2
Variety of Dimensions in Recent Composite Energy Security Indicators

Dimensions	Source
Environment, depletion	Bin Abdullah et al. (2020b, 2020a, 2022); Coutinho et al. (2020); Wang et al. (2020); Zhang et al. (2021); Wang et al. (2021); Stavytskyy et al. (2021); Huang et al. (2021a); Gong et al. (2021); Tutak and Brodny (2022); Cervan et al. (2022)
Availability, infrastructure, new resources	Li et al. (2020); Bin Abdullah et al. (2020b, 2020a, 2022); Coutinho et al. (2020); Wu et al. (2021); Stavytskyy et al. (2021); Zhang et al. (2021); Huang et al. (2021b); Khan and Dhakal (2022); Ha and Thanh (2022)
Affordability	Bin Abdullah et al. (2020b, 2020a, 2022); Coutinho et al. (2020); Wu et al. (2021); Zhang et al. (2021); Huang et al. (2021b); Cervan et al. (2022); Khan and Dhakal (2022)
Governance	Bin Abdullah et al. (2020b, 2020a, 2022); Coutinho et al. (2020); Wu et al. (2021); Zhang et al. (2021); Huang et al. (2021b); Huang et al. (2021a)
Efficiency, developability	Bin Abdullah et al. (2020b, 2022); Coutinho et al. (2020); Wu et al. (2021); Zhang et al. (2021); Stavytskyy et al. (2021); Huang et al. (2021a); Ha and Thanh (2022)
Energy supply	Li et al. (2020); Huang et al. (2021a); Gong et al. (2021); Wang et al. (2020); Wang et al. (2021); Cervan et al. (2022)
Technology	Augutis et al. (2020); Bin Abdullah et al. (2020b, 2020a, 2022); Coutinho et al. (2020); Zhang et al. (2021); Huang et al. (2021b)
Accessibility	Wu et al. (2021); Huang et al. (2021b); Stavytskyy et al. (2021); Cervan et al. (2022); Khan and Dhakal (2022)
Acceptability, adaptability	Li et al. (2020); Wu et al. (2021); Huang et al. (2021b); Ha and Thanh (2022); Khan and Dhakal (2022)
Consumption	Wang et al. (2020); Wang et al. (2021); Stavytskyy et al. (2021); Gong et al. (2021)
Sustainability	Li et al. (2020); Bin Abdullah et al. (2020b); Coutinho et al. (2020); Ha and Thanh (2022)
Social sustainability; socio-political	Coutinho et al. (2020); Augutis et al. (2020); Zhang et al. (2021); Huang et al. (2021a); Tutak and Brodny (2022)
Regulation	Bin Abdullah et al. (2020b, 2022); Coutinho et al. (2020)
Energy, energy industry development	Huang et al. (2021b); Wang et al. (2021); Tutak and Brodny (2022)
Economic, external influences of policies and market, investment	Augutis et al. (2020); Li et al. (2020); Huang et al. (2021b); Gong et al. (2021)

Source: produced by the authors.

advancement of new technologies, research investments, and efficiency enhancements. Figure 3.2 presents the main dimensions for indicators set under the current perception of the energy security concept.

- *Availability.* Including indicators related to energy availability is crucial for assessing the energy security of a country, because they provide insights into the reliability and adequacy of energy supply within a country. By evaluating factors such as energy production capacity, reserves, and infrastructure, policymakers can assess the country's ability to satisfy present and future energy needs. Consideration of energy availability indicators allows one to identify vulnerabilities and risks related to potential supply disruptions. Countries with diverse and enough energy sources are better equipped to sustain disruptions caused by geopolitical conflicts, natural disasters, infrastructural accidents, etc. Energy availability indicators also directly affect economic stability and growth. Adequate energy supply is required for different industrial activities and stimulates economic development of

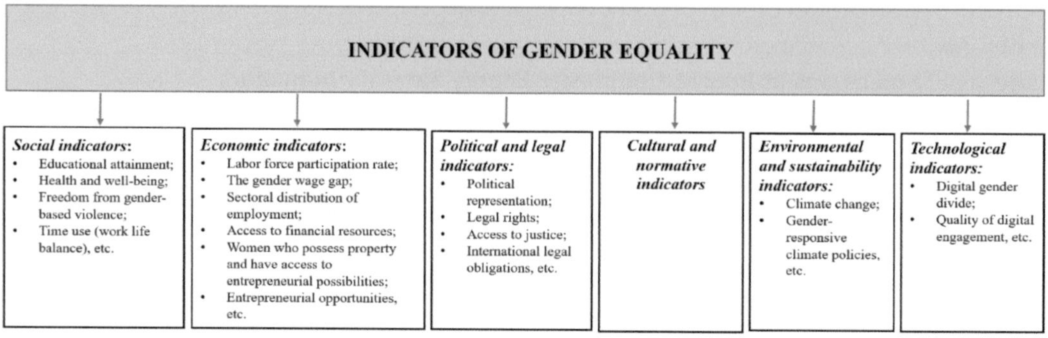

FIGURE 3.2 Dimensions for indicators set under the current perception of energy security.

Source: developed by the authors.

a country. Conversely, shortages of energy or energy disruptions can lead to economic downturns with many side effects such as job losses, decreased domestic and international competitiveness, etc. The analysis of the energy availability dimension is essential for planning and implementing low-carbon energy transition strategies. It is necessary to emphasize that energy availability is closely linked to national security of each country, as uninterrupted energy supply is essential for maintaining essential services, defense capabilities, and stability in society. Indicators related to energy availability allows one to identify potential threats in national security on which basis strategic decisions to enhance resilience of a country can be made.

- *Affordability.* Energy affordability indicators reflect several crucial aspects of a country's energy system. First of all, they reflect social equity and well-being through affordability of energy services to all segments of society (households, business, and industry). Access to affordable energy is vital to meet basic needs, improve living standards, and enhance overall well-being of society. Affordable energy contributes also to economic competitiveness, where lower energy prices enable one to operate more efficiently, attract investments, and contribute to economic growth. It is necessary to mention that assessment of energy affordability indicators is also important for managing the transition to sustainable energy sources as transitioning to renewable energy may require initially higher costs for energy consumers. Therefore, the analysis of these indicators can serve as a metric for evaluating equitable pricing mechanisms.
- *Environmental impact.* The involvement of indicators reflecting environmental impact of energy production and consumption is essential for seeking sustainable energy development. By assessment indicators such as GHG emissions, air and water pollution, land degradation, and others, the negative impact on ecosystems and natural resources can be minimized. Such indicators allow one to track progress towards low-carbon energy transition and climate change mitigation targets outlined in international agreements. By aligning energy policies with international climate change objectives, countries can enhance energy security by reducing dependence on fossil fuels and mitigating climate-related risks. Also, environmental indicators have a clear link with public health, particularly in terms of air quality and pollution-related health issues. For example, such air pollutants as particulate matter (PM2.5), sulfur dioxide (SO_2), or nitrogen oxides (NOx) allow one to evaluate the impact on people's health in terms of energy-related activities. Therefore, by incorporating environmental considerations into energy planning and infrastructure development, a country can mitigate the impact to the climate change and enhance energy security of a country. Also, it is worth it to mention that, by demonstrating a commitment

to environmental sustainability, countries can enhance their credibility, attract investment, and foster collaboration with other nations on energy and climate issues. Therefore, these indicators serve also to strengthen international reputation and collaboration.

- *Social impact.* Social indicators allow one to consider whether energy services are equally accessible to all segments of society, regardless of socioeconomic status, geographic location, or demographic characteristics. Assessing indicators such as energy poverty rates, access to modern energy services, or affordability helps identify energy access disparities and areas for improvement to ensure equality among different groups of people. As energy plays a vital role in improving the quality of life and well-being of people, social indicators provide insights into the extent to which energy services contribute to enhancing living standards, health outcomes, and overall quality of life. Monitoring health indicators enables policymakers to identify potential health hazards and implement appropriate mitigation measures. It is important to highlight, as energy security is closely related to economic development, such social indicators as energy-related employment rate or income level in the sector provide insights into the socio-economic benefits derived from energy investments and infrastructure development. Therefore, social indicators help monitor progress towards achieving social equity and supporting inclusive development through energy sector.
- *Technological development.* The level of technological innovation has a clear relationship with the level of energy security of a country. First of all, technological development indicators allow one to consider the resilience and reliability of energy infrastructure. Assessment of such indicators as grid reliability, maintenance standards, and technological innovation levels enables policymakers to identify vulnerabilities, overcome areas for improvement, and enhance the overall resilience of energy networks against disruptions and emergencies. Also, technological indicators are related to the achievements in energy efficiency and the adoption of new efficient technologies and practices, as enhancing energy efficiency reduces energy waste, decreases energy expenditures, and mitigates environmental impacts, contributing to energy security objectives. Technological indicators also show insights regarding the deployment of renewable energy sources (RES), technological advancements in smart grid technologies, or utilization of energy storage solutions. Overall, they reflect the level of innovation, research, and development activities in the energy sector, driving technological advancements in the whole sector.
- *Geopolitical relations.* Geopolitical relations directly affect a country's access to diverse energy sources and supply routes as geopolitical relations impact the stability and security of energy transit routes, including pipelines, maritime shipping lanes, and transportation corridors. Geopolitical tensions, conflicts, and regional instability pose significant risks to energy security by threatening the stability of energy-producing regions and disrupting global energy markets. Monitoring indicators such as geopolitical risk, possible conflicts, and regional security threats helps identify potential sources of instability and assess their impact on energy supply chains. Developing risk mitigation strategies to address geopolitical risks are essential to ensure energy security and minimize the impact of geopolitical disruptions on energy markets. Overall, consideration of geopolitical indicators allows one to identify vulnerabilities related to reliance on specific energy suppliers or transit countries. Therefore, timely decision-making can reduce the risk of supply disruptions due to geopolitical tensions, conflicts, or diplomatic disputes, enhancing energy security.
- *Government effectiveness.* Government effectiveness directly impacts the development and implementation of energy policies and regulations and plays a critical role in crisis response and resilience planning. Assessing indicators such as policy coherence, regulatory stability, and institutional capacity helps evaluate the government's ability to

formulate and enforce energy policies that promote security, sustainability, and resilience. Effective governance ensures the alignment of energy objectives with broader national priorities and facilitates the implementation of measures to enhance energy security. Also, government effectiveness affects the investment climate for the development of energy infrastructure and deployment of new technologies. Government effectiveness indicators also include measures of corruption and governance integrity, such as the level of corruption perception, anti-corruption policies, and institutional accountability mechanisms.

In summary it can be stated that including indicators related to energy availability in energy security assessments provides policymakers with critical insights into the reliability, resilience, and sustainability of a country's energy supply. These indicators inform strategic decision-making, support economic growth, and contribute to national security and environmental sustainability objectives. The consideration of indicators reflecting energy affordability issues in energy security assessments allows policymakers to consider social equity concerns and potential of economic development, alleviate poverty, enhance energy access and security, and efficiently support the transition to sustainable energy sources. Incorporation of indicators reflecting the impact to the environment encourages people to develop the energy sector in a more sustainable way, mitigate environmental concerns, protect public health, enhance resilience of the energy system, and strengthen international cooperation. These indicators provide valuable insights regarding the relationship between energy systems and the state of the environment and can guide policy decisions towards implementing long-term energy security and environmental objectives. Consideration of social indicators in energy security assessments allows one to promote equity, protect public health and safety, stimulate economic development, and uphold human rights and social responsibility. These indicators provide valuable insights regarding energy systems' impact to the society and help guide policy decisions towards achieving inclusive, equal, and sustainable energy development. Incorporation of technological indicators into the energy security assessment framework provides a comprehensive understanding of the technological landscape of the energy system and identification of areas for improvements in order to achieve long-term energy security objectives. Geopolitical indicators provide valuable insights for policymakers regarding the geopolitical landscape and help to identify potential risks and vulnerabilities to safeguard national interests in an increasingly complex geopolitical environment. In a such a way countries can strengthen their energy security position, mitigate risks, and ensure sustainable access to reliable and affordable energy resources. Involvement of government effectiveness indicators in energy security assessment framework gives valuable insights for policymakers and allows one to identify areas for improvement in order to strengthen energy resilience and governance capacity. By ensuring effective governance, countries can enhance regulatory stability, attract investment, mitigate corruption risks, improve response on possible crises, foster public trust on energy institutions, and contribute to a more robust and sustainable energy system of a country.

Recent challenges, including the COVID-19 pandemic, global energy crisis, geopolitical and military conflicts, geopolitical rivalries, and various economic uncertainties, have underscored the relevance of energy security on both national and regional levels. This significance is reflected in scientific studies, which have actively delved into energy security issues in recent years. The current perception of the concept of energy security is interdisciplinary, incorporating a diverse range of indicators crucial for fostering sustainable development in the energy sector. Therefore, the methodology for measuring energy security should be integrated into each country's policy documents, serving as a vital tool to support national autonomy and the well-being of people. Moreover, the selection of indicators should align with both national and global concerns, necessitating regular reassessment to accommodate evolving environmental dynamics, technological advancements, and changing priorities on global, regional, and national scales.

3.3 POLICIES AND MEASURES TO OVERCOME ENERGY SECURITY CHALLENGES

Overcoming energy security challenges requires a multifaceted approach involving various policies and measures. Some of the main policy directions can be singled out with the successful implementation of which governments can enhance energy security, reduce risks associated with supply disruptions, and promote a resilient and sustainable energy future. The main policy directions are singled out and presented in Figure 3.3.

Although this division is not exhaustive and it is possible to divide policies and applied measures according to different aspects, such a division perfectly reveals the components and basis of energy security.

- *Diversification of energy sources* in an energy mix reduces dependence on a single energy source and mitigates risks associated with disruptions of energy supply. One of the most preferable alternatives for the diversification under the current global context is the development and deployment of RES. The most significant measures in implementing resource diversification policy that can be mentioned are such measures as investments in energy storage technologies and grid infrastructure.
- The overall energy demand can be reduced by implementing *energy efficiency-oriented policy*, thereby reducing reliance on imported energy resources and enhancing energy security of a country. By implementing an energy efficiency-oriented policy, the government can introduce requirements for the construction sector, introduce and apply standards for appliance, and regulate efficiency in the industrial sector to promote the adoption of new energy-efficient technologies and practices. All these measures together with others such as energy audits, retrofitting existing infrastructure, and promoting energy conservation behaviors help optimize energy use and reduce vulnerability to supply disruptions, which directly affects the energy security of a country.

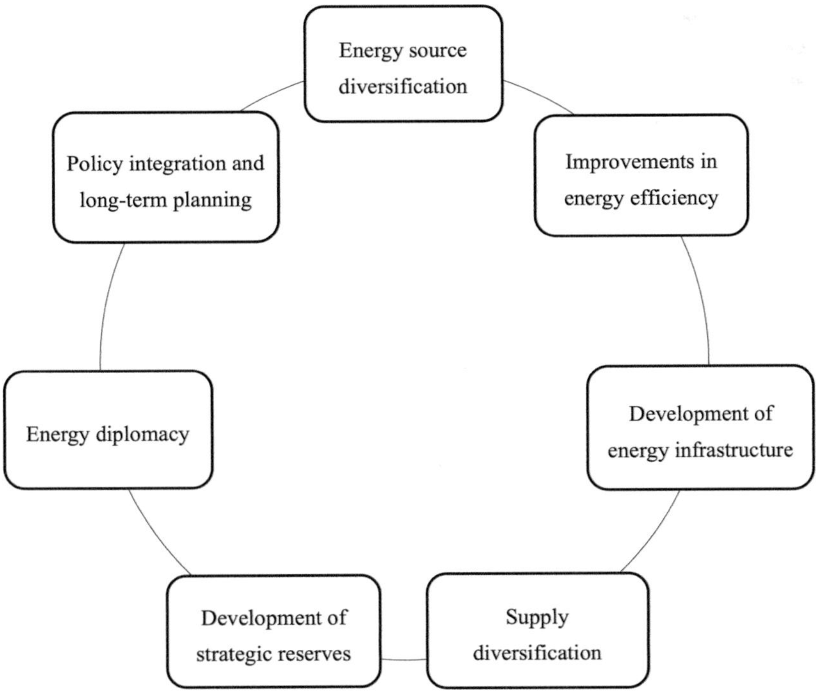

FIGURE 3.3 Policy directions to overcome energy security.

Source: created by the authors.

- Investment in *development of technologically advanced and resilient energy infrastructure* is essential for seeking to ensure reliable energy supply and minimize possible risk of disruptions. Upgrading aging energy infrastructure, growth of energy capacity, and enhancement of interconnectivity between energy grids of different countries enhances energy security and overall resilience of the energy system. Based on the policy implemented, the new investments can be directed into power generation facilities, transmission or distribution networks, pipelines, energy storage facilities, and other areas based on each country's specific context and long-term energy plans.
- *Diversification of alternative supply routes and suppliers* in energy sources contributes to lower vulnerability to disruptions in energy supply chains. In implementing a supply diversification-oriented policy, such measures as investments in the development of multiple transportation routes, promotion of energy trade agreements, cooperation with neighboring countries, and participation in regional energy markets is implemented. It allows one to minimize reliance on single transit corridors and unfavorable geopolitical situations.
- *Development of strategic reserves* and preparation of mechanisms for possible emergency response are very important for mitigating the impact of supply disruptions and ensuring the level of energy security of a country. The most applied measures used to implement this direction-oriented policy are establishment of strategic energy stockpiles, alternative supplier route development, development of contingency plans, strengthening of emergency preparedness, and response capabilities.
- *Energy diplomacy* and various cooperation actions with energy-producing and energy-transit countries enhances access to energy sources and allows one to strengthen the energy market's stability. Some main measures to implement this policy direction can be mentioned such as diplomatic relations development and negotiation due to energy trade agreements, development of new partnerships, diplomatic relations strengthening among countries, etc.
- *Integration of energy security considerations into national energy policies, strategies, and long-term planning frameworks* ensures a coordinated and proactive approach to managing the energy security of a country. Some measures used to implement this policy direction can be mentioned such as development of energy security strategies, adoption of a holistic approach to energy policy-making, and implementation of monitoring and response systems to evolve energy security challenges.

The issues of energy security are at the top of agendas of the EU. Therefore, various initiatives are implemented to strengthen national and regional energy security at the national level of member states. The following subsection presents a comparative assessment of energy security across the EU countries based on a multi-criteria assessment technique. The provided assessment serves as an illustrative example of how energy security assessments can be arranged to effectively address the complex challenges facing countries and regions worldwide. Also, the performed assessment follows recommendations and requirements for assessment in a sustainable way.

3.4 COMPARATIVE ASSESSMENT OF ENERGY SECURITY ACROSS THE EU COUNTRIES: A CASE STUDY

Numerous scientific studies address various energy security issues from analysis of different methodological aspects related to the perception and measurement of energy security to empirical studies that apply existing energy security indicator sets or indices or suggest their modifications. It can be stated that, mainly, quantitative methods are used for energy security assessments, especially in cross-country comparisons or evaluation of energy security dynamics. This subsection presents a comparative assessment of energy security across the EU countries, including a detailed representation of the methodology created to perform an assessment in accordance with requirements for sustainable decision-making.

3.4.1 Basis of the Framework

The recent energy security literature put emphasis on the current issues of security experienced due to the global crisis such as geopolitical conflicts, scarcity of energy resources, and global warming. These concerns caused a lot of problems for EU countries as well, especially due to high dependency on imported energy and high concentration in energy markets. As was mentioned before, the EU region was selected for the case study as an illustrative example to perform comparative assessment of member states. The basis of the comparative assessment is provided in Figure 3.4.

The entire assessment, from formulation of the problem to representation of the results, is based on the Bellagio Sustainability Assessment and Measurement Principles (STAMP) (Pinter et al., 2012). The Bellagio STAMP is recognized as a suitable tool for decision-making and helps one to perform research in a sustainable way as much as possible. These principles provide guidance on how to define the main vision of the assessment, pay attention to all sustainability dimensions, determine the scope of the assessment, guide the process of indicator selection, ensure transparency of the research and continuity, and provide guidance for representation of the results. Therefore,

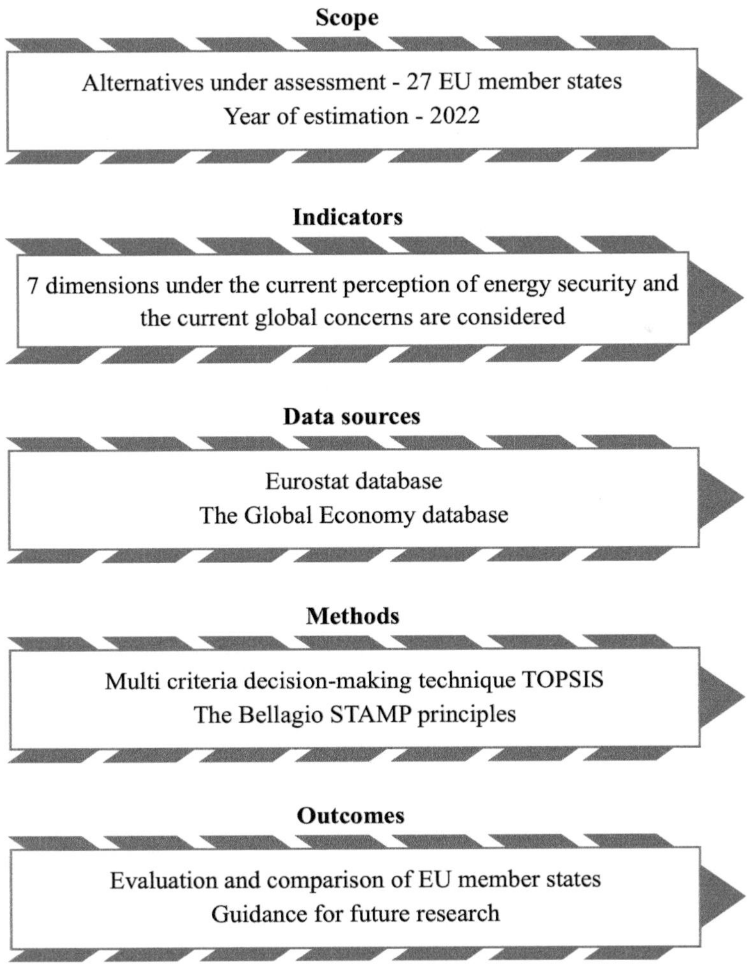

FIGURE 3.4 Basis of the comparative assessment.

Source: created by the authors.

the guidance of the Bellagio STAMP allows one to reflect sustainability issues and perform the whole assessment from the main vision to communication of the results in a sustainable way, and it is a good tool to use as a basis for the illustrative example, which will serve as guidance for future research. The Bellagio STAMP principles are presented in Figure 3.5.

As the Guiding Vision principle emphasizes that the well-being and needs of people in future generations should be considered, the goal of the assessment is in line with broader goals of sustainable energy development in the region. The provided framework helps to evaluate effort towards implementing sustainable energy development at national and regional levels.

The principle of Essential Considerations highlights the need to consider all sustainability dimensions as well as their interactions between each other, the trends and drivers, possible risk, and uncertainties. Based on this, the indicators set developed considers all dimensions of sustainability and takes into account not only critical aspects for energy security but also takes into account the current perception of the energy security concept and considers the current global concerns and uncertainties that allow one to perform a comprehensive and relevant evaluation. The involvement of relevant indicators enables one to identify the current trends, enablers, and barriers to strengthening energy security.

The principle of Adequate Scope calls attention to the selection of a proper time horizon and appropriate geographical scope. In this case study, all EU member states are included in the comparative assessment. It gives the possibility to assess the current situation and achievements made of the entire region, as well as the impact of the implemented policies at a regional and national level. Since the issues of energy security in the region have become critical in recent years, countries' performance in the study is assessed using the latest available statistics (year 2022). Also, the developed framework allows one to evaluate short- and long-term effects as well as repeat assessment in the future.

The Framework and Indicators principle emphasizes the need to create a set of indicators with standardized and comparable indicators, which reflect all the domains of the problem under analysis. Also, the principle stresses the importance of using new and reliable data sources as well as

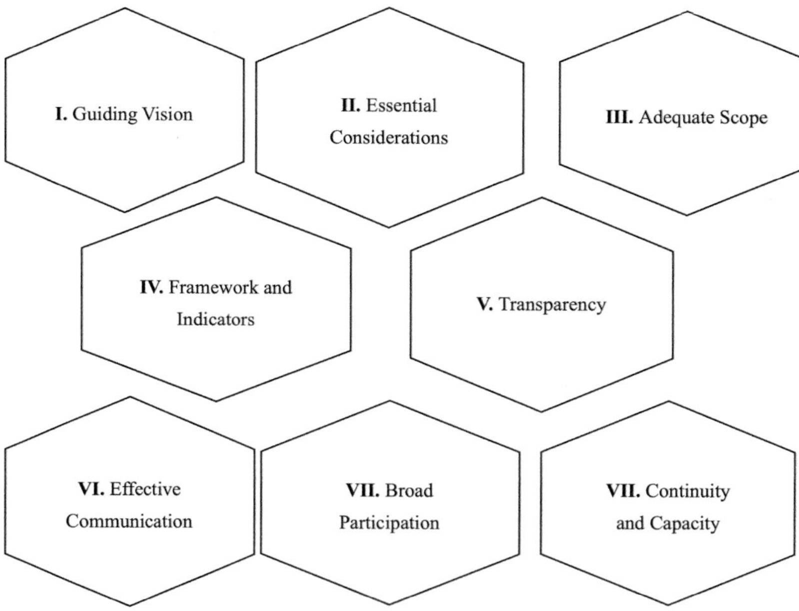

FIGURE 3.5 The Bellagio STAMP principles.

Source: created by the authors.

justified and scientifically recognized tools for the assessment. The developed indicators set reflects all the dimensions identified under the current perception of energy security and current issues facing the energy sector globally (7 dimensions singled out after detailed analysis of scientific literature). The developed framework allows one to monitor the progress in the future and to identify areas for improvement. The data used are publicly available and new (the newest statistical data are from 2022); the databases are reliable and recognized (Eurostat database and the Global Economy database). The multicriteria decision-making tool the Technique for Order of Preference by Similarity to Ideal Solution (TOPSIS) is used for the assessment and countries comparison.

The Transparency principle stresses the need to use publicly available and reliable data, to present all assumptions applied, to identify data sources and methods used, and justify indicators, weight, etc. The current assessment carefully follows all the requirements of the Transparency principle.

The Effective Communication principle is implemented through usage of visual instruments and clear and plain language in order to simplify interpretation and reach a broader audience. Also, the presentation and interpretation of the results follows objectivity requirements.

The Broad Participation principle is implemented through involvement of several indicators that are self-reported. However, in future studies this principle can be implemented more widely, for example with engagement of stakeholders or experts for the identification of indicators' weight or creation of possible scenarios, etc.

The principle of Continuity and Capacity seeks that the evaluation be repeated in the future and it would be easy to adapt the framework developed to changing conditions. In this research, the repeatability of the assessment is ensured through the selection of reliable and publicly available databases, transparency and justification of methodology, and justification of indicators' relevance and their weights. It enables future reassessment and adaptation to changing conditions, as well as tracking the progress achieved.

3.4.2 Development of Indicators Set

Sets of indicators were created based on a systematic scientific literature review, results of which were presented in subsection 2.2. The review and analysis of the recent energy security composite indicators and indices allowed one to single out 7 dimensions, which are important to consider in order to correspond current perceptions, challenges, and global concerns of energy security. Based on detailed examination of available data in databases, which are publicly available, two databases were selected for this research to develop a set of indicators for comparative assessment of energy security in the EU member states. Also, when selecting the indicators, it was important that they be representative, comprehensible and specific, sensitive, and verifiable. The data source to reflect indicators for Availability, Affordability, Environmental Impact, Social Impact, and Technological Development dimensions were selected from the Eurostat database. Indicators reflecting Geopolitical Relations and Government Effectiveness were selected from the Global Economy database. Finally, 18 indicators were chosen to correspond to the problem under analysis. The developed set of indicators and their measurement are provided in Table 3.3.

To reflect the availability dimension in measurement of energy security, three indicators had been selected, and these are: energy imports dependency, energy self-reliance, and energy supply diversity. The inclusion of these indicators reflects a comprehensive approach to measuring the availability dimension of energy security. These indicators collectively provide insights into a country's vulnerability to external shocks, the resilience of its energy system, and the flexibility of the energy supply mix. It is crucial to consider these indicators when strategies to enhance energy security are developing in order to ensure a stable, reliable, and secure energy supply in a country.

Energy imports dependency shows the share of a country's energy needs that is met through imported energy. This indicator is critical to reflect the availability dimension, because it indicates possible vulnerability to external supply disruptions. High dependency on imported energy puts a country at a disadvantage to face geopolitical risks, trade disputes, and supply disruptions. Such

TABLE 3.3
Set of Indicators. Source: created by the authors.

Dimension	Indicator	Measurement	ID	Data source	Ideal value
Availability	Energy imports dependency	Percentage	AV1	Eurostat database	Min
	Energy self-reliance	Percentage	AV2	Eurostat database	Max
	Energy supply diversity	HHI (0–1)	AV3	Eurostat database	Min
Affordability	Electricity prices for household	EUR/kWh (all bands consumption; all taxes and levies included)	AFF4	Eurostat database	Min
	Electricity prices for non-household	EUR/kWh (all bands consumption; all taxes and levies included)	AFF5	Eurostat database	Min
	People having arrears on utility bills	Percentage of total population	AFF6	Eurostat database	Min
Environmental impact	Energy efficiency	Rate (final energy consumption/primary energy consumption	EI7	Eurostat database	Max
	Share of RES in gross final energy consumption	Percentage	EI8	Eurostat database	Max
Social impact	Persons living in homes with leakages	Percentage of total population	SI9	Eurostat database	Min
	Persons unable to warm homes enough	Percentage of total population	SI10	Eurostat database	Min
	Health impacts of air pollution	Premature death rate per 100,000 people	SI11	Eurostat database	Min
Geopolitical relations	Political stability	Index (−2.5 weak; 2.5 strong)	GR12	The Global Economy database	Max
Government effectiveness	Corruption perceptions	Index (100 = no corruption)	GE13	The Global Economy database	Max
	Effective governance	Index (−2.5 weak; 2.5 strong)	GE14	The Global Economy database	Max
	Regulatory quality	Index (−2.5 weak; 2.5 strong)	GE15	The Global Economy database	Max
Technological development	Energy productivity	Euro per kilogram of oil equivalent (EUR/KGOE)	TD16	Eurostat database	Max
	Government budget allocations for R&D (energy sector)	Percentage of total government budget allocations for R&D	TD17	Eurostat database	Max
	Usage of Internet of Things (IoT)	Percentage of individuals	TD18	Eurostat database	Max

events like political instability in energy supplier or transit countries, various sanctions, or natural disasters can significantly affect the energy supply. Also, the dependency on imported energy can affect the economic stability of a country. Costs of imported energy can influence inflation rate, trade balances, or stability of currency; sharp energy prices increase can lead to economic instability. Undoubtedly, this indicator is also related to the strategic autonomy of a country. Countries with lower energy dependency can better control their energy policies and are less influenced by external pressures. This autonomy supports long-term strategic planning and resilience of the energy system.

Energy self-reliance indicates the ability of a country to meet its energy needs through domestic production. This indicator indicates the country's energy sovereignty and resilience to supply shocks. Countries with high energy self-reliance are better equipped to handle international supply shocks. Also, internal generated energy allows one to reduce reliance on potentially unstable suppliers. It is also important to highlight that the development and usage of internal energy sources, especially RES, supports sustainable development, reduces impact on the environment, and increases implementation of international climate change commitments. An aspiration to achieve higher self-reliance goes along with improvements in innovation environment and investment in local energy technologies.

Energy supply diversity measures the range of different energy sources and suppliers a country uses. This indicator is vital for the availability dimension. A diverse energy supply reduces the risk associated with reliance on a single source or supplier. Diversification of energy sources helps mitigate the impact of supply disruptions, price spikes, and resource depletion. Also, a varied energy mix provides higher flexibility in responding to changes in demand and supply conditions. Encouraging a diverse energy supply fosters innovation and competition within the energy sector. This can lead to the development of new technologies, increased efficiency, and lower energy costs. It is necessary to mention that diversity in energy supply often includes an increased share of RES in the energy mix that helps to reduce GHG emissions and address the global concerns of climate change.

The inclusion of electricity prices for household and non-household consumers and the rate of people having arrears on utility bills as indicators for the affordability dimension provides a fundamental picture of energy affordability issues. These indicators reflect the economic burden of energy costs, the prevalence of energy poverty, and the overall effectiveness of policies aimed to ensure affordable energy for energy consumers. It is important to monitor these indicators and implement targeted measures to improve energy affordability, ensuring that all citizens have access to essential, safe, and reliable energy services.

Electricity prices for household and non-household consumers are essential indicators for the affordability dimension when measuring energy security. First of all, they show cost burden on households, where high price can have a significant financial burden, especially on low-income families. Affordable electricity is critical to ensure that all citizens have access to basic energy services. Second, for non-household consumers, electricity prices impact production costs and overall competitiveness. High energy costs can lead to higher prices for goods and services, potentially reducing a country's economic competitiveness in the global market. Third, electricity price affects investment and overall economic growth. Businesses are more likely to invest in regions where energy costs are predictable and affordable. The new investments foster economic development, job creation, etc. Also, energy prices reflect energy equity issues in a country, therefore it is important to monitor them and shape energy policy to ensure fair access to energy across different sectors of society, which supports inclusive economic growth.

The share of people having arrears on utility bills indicates financial strain on households. High share indicates that a significant portion of the population is struggling to pay their utility bills and suggests that energy prices are unaffordable for many, leading to potential other financial difficulties and reduced quality of life. Also, this indicator is a measure of energy poverty, where individuals or families cannot afford essential energy services. High levels of arrears are a strong signal that energy poverty is a pressing issue, which can have severe health and social consequences. It is important to monitor this indicator in order to implement effective social and economic policies in a country. This indicator can be used as feedback on whether subsidies, support programs, or energy efficiency measures are reaching those persons who need it.

The inclusion of energy efficiency and the share of RES in gross final energy consumption as indicators for the environmental dimension is justified by their significant impact on reducing environmental harm and promoting sustainable practices. These indicators reflect critical aspects of reducing GHG emissions, conservation of natural resources, improvement air quality, and achieving Sustainable Development Goals. Monitoring and improvements in these indicators are crucial in order to design effective strategies for the insurance of an environmentally secure and sustainable energy future.

Energy efficiency is measured as the ratio of useful output from energy inputs to the total energy consumed. This indicator is crucial for the environmental dimension because it has a clear relationship to the amount of GHG emissions. Improvements in energy efficiency directly reduce the amount of energy consumed for the same level of service, leading to lower GHG emissions. This is essential in mitigating climate change and meeting international environmental targets and commitments. Also, improvements in energy efficiency reduce energy resource demand and decrease the environmental impact associated with extraction, transportation, and processing of energy sources. This includes reduced land degradation, water use, and air pollution. The optimization of usage of available energy resources promotes economic growth and contributes to a balance between economic and environmental goals. Besides that, the increase in energy efficiency can lead to significant cost savings for energy consumers, which can be reinvested in further sustainability initiatives or other critical areas.

The share of RES in gross final energy consumption is vital for the environmental dimension, because it shows reduction of emissions in the energy sector. The usage of RES significantly reduces the carbon footprint of energy consumption and is one of the most important measures for combating climate change and achieving climate change and energy targets. As renewables are inexhaustible energy sources, the increase of RES share in the final energy mix not only promotes sustainable use of natural resources but also reduces the pressure on depleting fossil reserves. Also, a higher share of RES improves energy security by diversifying the energy portfolio and lowering reliance on imported fossils. Finally, a high RES fraction provides price stability since renewable energy costs are less volatile compared to fossil fuels.

The social dimension corresponds with the selection of indicators showing implications for public health, social equity, and overall quality of life. Three indicators have been selected, which are: the share of persons living in homes with leakages, the share of persons unable to warm homes enough, and the health impacts of air pollution as indicators for the social dimension. These indicators provide a comprehensive view of how energy security affects the well-being of individuals and communities. It is important to monitor these indicators as they can guide the development of targeted interventions that promote social inclusion and improve living conditions of people. By monitoring these indicators, countries can ensure that their energy policies not only provide reliable and affordable energy but also support a healthy and equitable society.

The share of persons living in homes with leakages indicates energy inefficiency and costs issues in a household sector. Homes with a leaking roof, windows, or walls are energy-inefficient and require more energy to maintain a comfortable indoor temperature. As a consequence, it leads to higher energy consumption and higher energy bills for households and increases the risks of falling into energy poverty. Various poor housing conditions can have an impact on people's health. For example, leakages can lead to dampness and mold growth, which are associated with respiratory problems and other health issues as well as overall well-being of individuals. Therefore, it is important to measure and monitor this indicator, because it allows one to identify the vulnerability of people on which based targeted measures can be prepared and social inequality in a country reduced.

Such indicators as the share of persons unable to warm homes enough directly reflect the energy poverty issue in a country. Suffering from energy poverty can lead to some health problems; it can especially affect the elderly, children, and people with pre-existing health problems. Also, this indicator reflects social inclusion issues, because the assurance of essential energy services for all individuals is a vital aspect of social inclusion and equity as well as a basic necessity for the comfort and well-being of people. Therefore, the inclusion of this indicator allows one to evaluate the assurance of basic quality of life conditions and the level of social welfare in a country.

The health impacts of air pollution are measured by premature death rate, which shows the lost years of life due to air pollution (concretely—particulate matter (PM2.5)). PM2.5 may be inhaled into the lungs and have a harmful effect on people's health, especially on people suffering from heart and lung illnesses. Therefore, the improvements in air pollution directly affect public health and reduce the healthcare system burden. This indicator is essential in measuring the social impact of the energy system developed and should be involved and monitored when assessing the energy security of a country.

The geopolitical relations dimension is reflected by the political stability index. This index shows the stability and predictability of a country's political environment and is one of those ensuring a favorable investment climate, continuity of energy projects, and effective management of crisis. A stable political environment is a crucial factor influencing attractiveness to investors, encouraging long-term investments in energy infrastructure, and ensuring smooth progress of energy infrastructure development. A clear regulatory framework is crucial for long-term planning, where stable policies reduce uncertainty and lead to efficient energy market functioning. Political instability can lead to disruptions in both exporting and importing countries, while stability in the political environment can foster countries' cooperation, creating new energy infrastructure and cross-border energy networks. Therefore, this indicator can be also associated with reliability of energy supply.

The government effectiveness dimension is reflected by three indicators which are as follows: corruption perceptions, effective governance, and regulatory quality. Including these indicators to express the government effectiveness dimension in measuring energy security is valuable by their impact on investments, allocation of resources, smooth policy implementation, quality of energy services, strength for energy crisis management, and whole efficiency of energy markets. The selected indicators collectively provide a comprehensive understanding of how government can influence a country's energy security and effective and sustainable energy resource management. Therefore, it is important that these indicators would be constantly monitored to improve governance practices and enhance the overall energy security framework in a country.

Corruption perceptions measures the corruption perceptions of the public sector. A low level of corruption leads to confidence of investors by attracting new investments in the energy sector while a high level of corruption in a country discourages investment due to possible increased risks and costs. Corruption is also related to efficient resource allocation, which means efficiency in services and more efficient implementation of energy projects and whole energy policy in a country.

Effective governance reflects the quality of public services (including energy) where high effectiveness can be associated with reliable energy supply and efficient energy resource management. The selected government effectiveness index evaluates not only perceptions of the quality of public and civil services but also the extent of their independence from political pressure, policy formulation, and implementation quality, as well as the compliance with commitments of the policy being implemented. Also, this indicator shows the prerequisites for effective policy formation, long-term energy security planning, and fast and coordinated responses to maintain energy security in cases of possible energy system disruptions.

Regulatory quality is expressed through the regulatory quality index, which shows the government's capacity to develop and enforce policies and regulations that encourage private sector investments and efficiency as well as ensure stable functioning and growth of the energy sector. High regulatory quality reduces uncertainty for energy market participants, promotes innovations and new technology adoption, as well as ensures competitivity in the market.

Three indicators have been selected to reflect the Technological Development dimension, and these are: energy productivity, government budget allocations for R&D in the energy sector, and the usage of IoT. The selected indicators collectively provide a comprehensive view of usage of technological advancements and innovations in the energy sector. Based on that, such aspects as efficiency, reliability, or energy systems sustainability in the context of energy security can be evaluated.

Energy productivity is an essential indicator to reflect technological development; it shows how efficient energy resources are used and is vital in order to reduce energy import dependency or to enhance energy self-sufficiency. The improvements in energy productivity lead to economic growth without additional energy consumption. It allows the economy of a country to be more resilient to possible energy price fluctuations and disruptions in energy supply. It is important to note that improvements in energy productivity contribute to the mitigation on environmental impact and support implementation of international climate change commitments. Therefore, the achievements in energy productivity help to implement a transition to a low-carbon economy and long-term energy security goals.

Government budget allocations for R&D in the energy sector support the development of new technologies and innovative solutions in the sector. Such investments are crucial in seeking to improve energy efficiency, diversify energy sources, and enlarge capabilities of energy storage and distribution. Also, investments in R&D are critical for the integration of cutting-edge technologies into the national energy system and for the assurance of competitiveness in the global energy market, which also can provide economic benefits from the export of advanced energy technologies and solutions.

The usage of IoT as an indicator for measuring energy security is important due to its significant impact on efficiency, reliability, resilience, and sustainability of energy systems. IoT technology enhances real-time monitoring, predictive maintenance, grid management, and consumer empowerment, while also facilitating the integration of RES and improving the overall security and resilience of the energy infrastructure. Therefore, monitoring the usage of IoT provides valuable insights into the technological advancements that underpin a secure and sustainable energy future of a country.

Overall, 18 indicators have been selected to reflect 7 energy security dimensions under current perception in the context of global concerns and uncertainties in order to measure energy security in all EU member states. As the EU region consists of 27 countries, which differ in area, location, socioeconomic conditions, policy implementation, etc., it is important not only to select indicators that are comparable, reliable, and corresponded to the problem under analysis, but also it is important to apply an appropriate tool for assessment and ranking.

3.4.3 Assessment and Ranking Tool

Energy security assessment and countries ranking is based on the multi-criteria technique TOPSIS (Hwang and Yoon, 1981). The selected method is among the most commonly employed tools for sustainability assessment and decision-making in the energy sector, especially when aspects under consideration are conflicting with each other. The TOPSIS method adheres to the concept of the ideal solution and is noted for its relatively simple computational process, allowing for quick evaluation results compared to other MCDM techniques. This simplicity makes it easier for decision-makers to comprehend the outcomes and the importance of the indicators and their weights used in determining the final ranking. The logical and transparent computation process is advantageous for both researchers and policymakers, facilitating the repetition of studies and the monitoring of progress at national or regional levels. Additionally, this approach enhances the appeal of using the proposed methodology for similar evaluations in other countries and regions. It also supports the implementation of the Bellagio STAMP principles due to its universality, scientific recognition, transparency, and wide and easy applicability. TOPSIS calculation is based on 7 main steps, which are as following:

Step 1. A decision matrix consisting of m alternatives and n criteria is developed for the multi-criteria evaluation:

$$D = \left[x_{ij}\right] = \begin{matrix} \\ a_1 \\ a_2 \\ \cdots \\ a_m \end{matrix} \begin{bmatrix} & X_1 & X_2 & \cdots & X_n \\ x_{11} & x_{12} & \cdots & x_{1n} \\ x_{21} & x_{22} & \cdots & x_{2n} \\ \cdots & \cdots & \cdots & \cdots \\ x_{m1} & x_{m2} & \cdots & x_{mn} \end{bmatrix} \tag{3.1}$$

Step 2. The calculation of composite scores using the TOPSIS method involves measuring the distances to both the ideal best and ideal worst solutions. The procedure is as follows:

First, the normalized matrix is obtained:

$$\bar{x}_{ij} = \frac{x_{ij}}{\sqrt{\sum_{i=1}^{m} x_{ij}^2}}; \; i = \overline{1,m},; \; j = \overline{1,n}, \tag{3.2}$$

Step 3. The weighted normalized matrix is computed as follows:

$$V = \begin{bmatrix} w_1 r_{11} & w_2 r_{12} & \cdots & w_n r_{1n} \\ w_1 r_{21} & w_2 r_{22} & \cdots & w_n r_{2n} \\ \cdots & \cdots & \cdots & \cdots \\ w_1 r_{m1} & w_2 r_{m2} & \cdots & w_n r_{mn} \end{bmatrix}; \sum_{j=1}^{n} w_j = 1 \tag{3.3}$$

Step 4. The ideal solution $A+$ and the ideal worst solution $A-$ are determined:

$$A^+ = \{\max_j v_{ij} \mid i \in I), (\min_j v_{ij} \mid i \in I'), \; j = \overline{1,n},\} = \{v^+_1, v^+_2 \ldots v^+_n\}; \tag{3.4}$$

$$A^- = \{\min_j v_{ij} \mid i \in I), (\max_j v_{ij} \mid i \in I'), \; j = \overline{1,n},\} = \{v^-_1, v^-_2 \ldots v^-_n\} \tag{3.5}$$

Step 5. The relative distances of each alternative from the ideal solution and the ideal worst solution are computed:

$$S_i^+ = \sqrt{\sum_{j=1}^{n} \left(v_{ij} - v_j^+\right)^2}, \; j = \overline{1,m}, \tag{3.6}$$

$$S_i^- = \sqrt{\sum_{j=1}^{n} \left(v_{ij} - v_j^-\right)^2}, \; i = \overline{1,m}, \tag{3.7}$$

Step 6. The relative closeness of each alternative to the ideal solution is computed:

$$C_i = \frac{S_i^+}{S_i^+ + S_i^-}; \tag{3.8}$$

Step 7. The alternatives are ranked based on the computed C_i values, where higher value means higher closeness to the ideal solution.

3.4.4 Data

Overall, 18 indicators have been selected to reflect 7 energy security dimensions under current perception in the context of global concerns and uncertainties. The indicators selected for the evaluation differ among the EU member states. This subsection will briefly review them, underlying the most significant differences. Table 3.4. summarizes all the data for the assessment.

The EU is highly dependent on imported energy. Energy imports dependency (AV1) in the region in 2022 was 63%. Despite that, some countries distinguish with low dependency on imported energy; among these Estonia can be mentioned with only 6%, Sweden with 27%, and Romania with 32%. The most dependent energy systems are on islands, for example, Malta, where 99% of energy is imported and Cyprus with 92% of imported energy. Also, highly dependent countries can be singled out: Luxemburg, Netherlands, Austria, Italy, Greece, Ireland, Belgium, Lithuania, Spain, Portugal, and Slovakia.

TABLE 3.4
Data for Calculations. Source: Eurostat, the Global Economy

	AV1	AV2	AV3	AFF4	AFF5	AFF6	EI7	EI8	SI9**	SI10	SI11*	GR12	GE13	GE14***	GE15	TD16	TD17	TD18
Belgium	73.95	16.26	0.150	0.452	0.286	3.2	0.739	13.759	15.7	5.1	44	0.58	73	1.23	1.25	7.22	1.78	14.13
Bulgaria	37.13	50.59	0.171	0.115	0.215	18.8	0.524	19.095	11.0	22.5	158	0.25	43	−0.27	0.32	2.53	0.3	1.15
Czech Republic	41.79	51.33	0.127	0.157	0.214	1.9	0.645	18.195	6.8	2.9	81	0.82	56	1.09	1.39	4.71	3.55	5.13
Denmark	42.87	36.21	0.135	0.537	0.409	3.5	0.835	41.601	16.8	5.1	21	0.87	90	1.99	1.84	17.75	5.56	14.57
Germany	68.56	29.55	0.147	0.348	0.252	4.3	0.781	20.796	12.0	6.7	39	0.61	79	1.29	1.52	10.64	6.39	7.11
Estonia	6.16	62.11	0.230	0.238	0.310	4.4	0.591	38.472	10.2	3.4	7	0.72	74	1.34	1.56	4.19	0.24	16.07
Ireland	79.16	19.03	0.150	0.317	0.312	10.6	0.834	13.107	16.6	7.2	9	0.88	77	1.55	1.64	26.77	1.16	20.57
Greece	79.60	12.30	0.294	0.259	0.260	34.1	0.771	22.678	12.5	18.7	95	0.06	52	0.45	0.46	8.32	2.11	3.65
Spain	74.35	22.08	0.215	0.345	0.260	9.2	0.717	22.116	19.7	17.1	30	0.27	60	0.92	0.8	9.26	2.61	11.77
France	51.92	42.54	0.167	0.211	0.149	7.1	0.675	20.259	18.0	10.7	30	0.33	72	1.17	1.19	10.28	6.84	10.77
Croatia	60.30	29.54	0.147	0.149	0.271	14.5	0.831	29.354	9.4	7.0	96	0.61	50	0.58	0.5	6.72	0.56	5.58
Italy	79.42	18.35	0.245	0.392	0.393	5.0	0.805	19.131	19.6	8.8	79	0.41	56	0.45	0.51	11.09	3.28	4.38
Cyprus	92.02	9.09	0.209	0.336	0.374	8.1	0.734	19.429	39.1	19.2	70	0.42	52	0.73	0.77	8.78	0.05	6.59
Latvia	38.75	43.38	0.230	0.301	0.261	5.9	0.919	43.316	17.5	7.1	75	0.48	59	0.69	1.17	5.55	3.27	4.69
Lithuania	72.43	12.58	0.310	0.247	0.399	5.5	0.854	29.599	10.9	17.5	77	0.65	62	0.99	1.3	5.87	4.2	5.31
Luxembourg	91.32	8.15	0.189	0.195	0.164	4.4	0.966	14.356	15.4	2.1	12	1.06	77	1.77	1.83	14.70	0.08	10.74
Hungary	64.18	34.35	0.164	0.109	0.285	8.5	0.767	15.190	20.4	4.7	107	0.64	42	0.53	0.41	5.39	5	5.18
Malta	99.01	1.57	0.368	0.151	0.142	5.6	0.787	13.404	6.1	7.6	37	0.91	51	0.8	0.67	4.30	2.96	8.67
Netherlands	80.23	11.64	0.139	0.098	0.196	1.5	0.774	14.972	14.8	5.3	32	0.72	80	1.58	1.71	9.85	2.07	65.69
Austria	74.45	30.02	0.136	0.236	0.214	2.6	0.873	33.758	9.1	2.7	36	0.64	71	1.47	1.28	10.63	3.62	14.57
Poland	46.03	48.10	0.151	0.171	0.183	4.5	0.735	16.879	6.0	4.9	125	0.5	55	0.26	0.72	5.27	1.59	3.33
Portugal	71.27	23.69	0.184	0.235	0.166	4.7	0.805	34.677	25.2	17.5	20	0.86	62	1	0.76	8.54	2.57	6.34
Romania	32.41	57.02	0.175	0.342	0.425	17.8	0.775	24.140	10.0	15.2	103	0.49	46	0	0.36	6.03	1.14	2.09
Slovenia	53.97	31.53	0.183	0.189	0.224	6.5	0.763	25.002	20.8	2.6	56	0.71	56	1.07	0.69	7.24	4.94	9.28
Slovakia	69.63	29.23	0.155	0.189	0.292	5.9	0.696	17.501	4.9	7.1	98	0.44	53	0.38	0.85	5.40	1.03	3.94
Finland	40.88	46.97	0.122	0.205	0.140	5.7	0.774	47.886	4.5	1.4	3	0.89	87	1.76	1.78	6.38	4.21	12.35
Sweden	26.82	52.60	0.148	0.223	0.191	3.6	0.728	66.002	7.1	3.3	6	0.9	83	1.57	1.68	10.14	4.07	15.94

* the latest available data 2021 included in the assessment

** the latest available data 2020 included in the assessment

*** as selected method can be used only with positive values, GE14 indicator for Bulgaria will be changed to 0 in the calculations.

High energy dependence is influenced by low energy self-reliance. Energy self-reliance (AV2) differs significantly among countries. However, only four EU member states can fulfil more than half their energy needs, and these are: Romania (57%), Sweden (53%), Czech Republic (51%) and Bulgaria (51%). The lowest energy self-reliance is in the most energy dependent countries: Malta (2%), Luxembourg (8%), and Cyprus (9%).

Energy supply diversity (AV3) indicates the risk associated with reliance on a single energy source, i.e., a variety of energy sources in the energy mix. Low diversity signals that a single event can significantly affect energy supply in a country. It is measured by the Herfindahl Hirschman Index (HHI) (from 0 to 1, where 0 means very high diversity and 1 is a single energy source). The lowest diversity in energy supply is in Malta (37%), Lithuania (31%), and Greece (29%).

The analysis of statistical data showed that electricity prices for households (AFF4) and non-households consumers (AFF5) differ significantly among countries. The highest prices are in such countries as Denmark, Belgium, Italy, Romania, Cyprus, Spain, and Germany, while the lowest rates are in such countries as Bulgaria, Czech Republic, France, Luxembourg, Hungary, Malta, Netherlands, Poland, and Finland.

Energy prices have a significant impact on another indicator, which is, for the share of people who cannot pay for utility services (AFF6). Although this indicator is usually assigned to the social dimension, it clearly signals energy affordability issues in countries. The statistical data signal that the Greek government should be especially concerned about the ability of the population to pay for energy services. In this country, as much as 34% of the population indicates that they have debts for utility services. It is worth noting that this indicator is determined not only by energy prices but also by the socioeconomic situation in the country. However, the provision of energy services to residents is an essential condition for every energy system, which must be ensured. Also, a very high share of people unable to pay their utility bills are in Bulgaria (19%), Romania (18%), Croatia (15%), and Ireland (11%). The least amount of people facing the problem of paying utility bills has been identified in Netherlands (1.5%) and Czech Republic (1.9 %).

Increasing energy efficiency (EI7) has been one of the main priorities of EU energy policy for many years. Increasing energy efficiency seeks to solve many problems in different fields, ranging from optimization of energy generation, to reduction of the impact to the climate change, to social and economic problems such as energy poverty reduction or increase in energy availability. In this research the energy efficiency ratio is used, which provides insight into how effectively primary energy resources are used. Energy efficiency is expressed through the ratio between final energy consumption and primary energy consumption. Statistical data show that the lowest energy efficiency ratios are in Bulgaria (0.52) and Estonia (0.59), while among the best performing countries in context of efficient energy use are Luxembourg (0.97), Latvia (0.92), Austria (0.87), and Lithuania (0.85).

Among other strategic priorities of the EU for many years is the development of renewable energy. The RES in gross final energy consumption (EI8) is 23% in the region in 2022. However, some countries perform with really high results. The most impressive result distinguishes Sweden, where 66% of energy is generated from RES, the second one is Finland with almost 48%, and the third is Latvia with 43%. The lowest shares of RES in final energy consumption were in Ireland (13%), Malta (13%), Belgium (14%), Luxembourg (14%), Netherlands (15%), and Hungary (15%).

The inclusion of social aspects in the energy policy is a very important priority of the EU energy policy. The energy efficiency of housing, the sufficient fulfilment of energy needs, and the impact of energy on people's health are essential indicators reflecting the social dimension. The big problem of energy efficiency of buildings has been identified in many EU countries. However, the most inefficient houses are in Cyprus, where 39% of the population live in homes with various leakages (SI9), Portugal—25 %, Slovenia—21 %, Hungary—20 %, Spain—20 %, Italy—20%, France—18 %, and Latvia—18 %. The most efficient dwellings according to the statistics are in Finland, Slovakia, Poland, and Malta.

The biggest shares of people unable to warm their homes enough (SI10) are in Bulgaria (23%), Cyprus (19%), Greece (19%), Lithuania (18%), Portugal (18%), Spain (17%), and Romania (15%). In the rest of the countries the share of people unable to warm their homes enough is less than 10%. Less than 3% was identified in Finland, Luxembourg, Slovenia, Austria, and Czech Republic.

The energy sector is one of the main air polluters; health impacts of air pollution (SI11) directly show how important low-carbon energy transition is to the whole society. Statistical data show that the highest problem is in Bulgaria and Poland, where the premature death rate is 158 in Bulgaria and 125 in Poland. The lowest values of lost years of life due to air pollution per 100,000 people are in Finland (3), Sweden (6), Estonia (7), and Ireland (9).

The political stability index (GR12) varies across the EU countries from the lowest 0.06 in Greece to 1.06 in Luxembourg. As the index can reach values from −2.5 to 2.5, it can be stated that EU member states can be described as moderately political stable countries. However, it can also be said that the EU region consists of very different countries with large differences in terms of political stability, corruption, effective governance, and quality of regulation. The corruption perceptions index (GE13) varies from 42 to 90. The highest corruption is determined in Hungary, Bulgaria, Romania, and Croatia. The lowest level of corruption is determined in Denmark, Finland, Sweden, and Netherlands. The governance effectiveness index (GE14) ranges between −0.27 in Bulgaria and 1.99 in Denmark. It signals significant differences among countries and their ability to implement national, regional, and international policy goals. The regulatory quality index (GE15) also differs significantly among EU member states from 0.32 in Bulgaria to 1.84 in Denmark.

Energy productivity (TD16) is 9.31 EUR/KGOE in the whole EU. However, in separate countries energy productivity differs significantly. The highest energy productivity is in Ireland (26.77 EUR/KGOE), Denmark (17.75 EUR/KGOE), and Luxembourg (14.70 EUR/KGOE). The lowest productivity is in Bulgaria (2.53 EUR/KGOE), Estonia (4.19 EUR/KGOE), Malta (4.30 EUR/KGOE), and Czech Republic (4.41 EUR/KGOE).

The highest shares of government budget allocations for energy R&D (TD17) are in such countries as France (6.84%), Germany (6.39%), Denmark (5.56%), Hungary (5%), and Slovenia (4.94%). The lowest attention to energy R&D activities is in Cyprus (0.05%), Luxembourg (0.08%), Estonia (0.24%), Bulgaria (0.3%), and Croatia (0.56%).

In terms of usage of IoT (TD18), the Netherlands stands out, with more than 65% of the population using the IoT. All other countries are well behind the Netherlands, for example, in second position is Ireland with 21%, third is Estonia with 16%, and fourth is Sweden with close to 16%. The lowest use of IoT is in Bulgaria (1.15%), Romania (2.09%), Poland (3.33%), Greece (3.65%), and Slovakia (3.94%). Overall, many EU countries still have to do a lot to achieve widespread adoption of the IoT.

3.4.5 Results and Guidance for Future Research

The results of the assessment of energy security across the EU member states are presented in Table 3.5. The highest achieving five countries in terms of energy security across the EU are Netherlands, Sweden, Finland, Denmark, and Austria. These leading countries stand out for their achievements in many assessment areas, for example the leading Netherlands distinguishes with high energy supply diversity, affordable energy for its consumers, good social dimension indicators, low corruption, effective governance, high regulatory quality, and especially large usage of IoT. Sweden is characterized by quite a low level of energy import dependency, high energy self-reliance, a high fraction of RES in final energy consumption, low impact of air pollution on people's health, stable political environment, low corruption, effective governance, and high regulatory quality.

Greece, Bulgaria, Cyprus, and Romania are rated the worst in terms of energy security across the EU. These countries are characterized by low achievement in almost all areas assessed.

TABLE 3.5
Results of the Assessment and Countries' Ranking, 2022

Countries	S_i^+	S_i^-	C_i	Rank
Netherlands	0.0382	0.0720	0.6533	1
Sweden	0.0401	0.0671	0.6259	2
Finland	0.0446	0.0657	0.5960	3
Denmark	0.0445	0.0627	0.5845	4
Austria	0.0462	0.0588	0.5602	5
Estonia	0.0499	0.0605	0.5484	6
Ireland	0.0479	0.0576	0.5458	7
Luxembourg	0.0541	0.0596	0.5240	8
Germany	0.0516	0.056	0.5207	9
France	0.0506	0.0544	0.5182	10
Czech Republic	0.0548	0.0585	0.5167	11
Slovenia	0.0529	0.0518	0.4949	12
Belgium	0.0557	0.0507	0.4767	13
Latvia	0.0559	0.0487	0.4652	14
Portugal	0.0572	0.049	0.4612	15
Malta	0.0610	0.0519	0.4597	16
Poland	0.0619	0.0514	0.4534	17
Hungary	0.0606	0.0471	0.4375	18
Lithuania	0.0600	0.0462	0.4349	19
Slovakia	0.0624	0.0478	0.4336	20
Spain	0.0579	0.0423	0.4221	21
Croatia	0.0611	0.0434	0.4154	22
Italy	0.0618	0.0428	0.4089	23
Romania	0.0674	0.0385	0.3636	24
Cyprus	0.0691	0.0354	0.3390	25
Bulgaria	0.0743	0.0377	0.3368	26
Greece	0.0742	0.0275	0.2706	27

Source: produced by the authors

3.5 CONCLUSIONS

Energy security is not a static concept; it evolves in response to changes in time, location, socioeconomic conditions, geopolitical stability, overarching global issues, etc. Each of these factors can significantly affect the perception of energy security, necessitating a dynamic and adaptable approach to energy security assessment. The detailed analysis of recent scientific studies has enabled us to identify 7 critical dimensions that are essential for analyzing and evaluating the energy security of a country or region, and these are as follows: availability, affordability, environmental impact, social impact, technological development, government effectiveness, and geopolitical relations. The identified dimensions provide a comprehensive framework to address the multifaceted nature of energy security. It is imperative that all these dimensions be incorporated to accurately reflect the recent global context.

The understanding of energy security must be responsive to the changing circumstances of the global energy environment. This adaptability is crucial for developing effective policies and strategies that ensure a stable and secure energy supply in the face of challenges. The dimensions identified through the detailed scientific literature analysis provide a robust basis for such an adaptable understanding, enabling researchers, policymakers, and stakeholders to make informed decisions that enhance the resilience and sustainability of energy systems. The identified 7 dimensions offer a perspective that is essential for addressing the complex and interdependent challenges of the current energy system.

The selected indicators for the assessment and ranking of energy security across the EU are designed to reflect the recent perception of energy security as well as current global concerns related to the development of the energy sector. The provided indicators set offers a solid foundation for comprehensive analysis, which can be used for a wide range of research purposes. Researchers can use it to track the progress, identify leading countries, determine the most effective practices in shaping the energy sector, and uncover the strategies that contribute to a high level of energy security. The provided indicator set can be adapted to specific regional contexts, available data, unique characteristics of individual countries or country groups, the specific aim of the research, and other particularities of the study. The adaptability ensures that the assessment of energy security remains relevant, comprehensive, and aligned with the latest concerns and challenges in the energy sector.

In the provided illustrative example of the energy security assessment, all indicators have equal weights. However, although this is the most common practice in this type of research, it is valuable in future studies to use experts to determine the weights of the indicators. The determination of weights for indicators in a multi-criteria assessment is a very important step that can significantly influence the outcomes and insights derived from the performed analysis. The engagement of experts to determine indicators weights is a highly recommended practice for sustainable decision-making.

In conducting further energy security studies, it is crucial not only to include all necessary dimensions and carefully select the appropriate indicators but also to adhere to principles and recommendations that facilitate sustainable research practices. The Bellagio STAMP principles provide comprehensive guidance throughout the entire process of the assessment. These principles assist in defining the main vision and scope of the assessment, ensuring that all dimensions of sustainability are considered. They guide the process of indicator selection, emphasize the importance of transparency of the research and continuity issues, and provide recommendations for the effective representation of results. Therefore, the application of the Bellagio STAMP principles is a very useful tool for future energy studies, which allows one not only to scientifically justify the research methodology but also guides the entire process of the assessment, ensuring a systematic and sustainable approach.

REFERENCES

Ang, BW; Choong, WL; Ng, TS. A framework for evaluating Singapore's energy security. *Applied Energy*, 2015, 148, 314–325.

Asia Pacific Energy Research Centre (APERC). *A Quest for Energy Security in the 21st Century: Resources and Constraints*. Asia Pacific Energy Research Centre, Tokyo, 2007, 113 p. Available at: https://aperc.or.jp/file/2010/9/26/APERC_2007_A_Quest_for_Energy_Security.pdf (accessed on 12 January 2023).

Augutis, J; Krikstolaitis, R; Martisauskas, L; Peciulyte, S; Zutautaite, I. Integrated energy security assessment. *Energy*, 2017, 138, 890–901.

Augutis, J; Krikstolaitis, R; Martisauskas, L; Urboniene, S; Urbonas, R; Uspuriene, AB. Analysis of energy security level in the Baltic States based on indicator approach. *Energy*, 2020, 199, 117–427.

Berdysheva, S; Ikonnikova, S. The energy transition and shifts in fossil fuel use: The study of international energy trade and energy security dynamics. *Energies*, 2021, 14, 17, 5396.

Bin Abdullah, F; Hyder, SI; Iqbal, R. A model for strategizing energy security dimensions and indicators selection for Pakistan. *International Journal of Renewable Energy Research*, 2020a, 10, 2, 558–569.

Bin Abdullah, F; Iqbal, R; Ahmad, S; El-Affendi, MA; Kumar, P. Optimization of multidimensional energy security: An index based assessment. *Energies*, 2022, 15, 11, 3929.

Bin Abdullah, F; Iqbal, R; Hyder, SI; Jawaid, M. Energy security indicators for Pakistan: An integrated approach. *Renewable & Sustainable Energy Reviews*, 2020b, 133, 110122.

Bin Abdullah, F; Iqbal, R; Jawaid, M; Memon, I; Mughal, S; Memon, FS; Rizvi, SSA. Energy security index of Pakistan (ESIOP). *Energy Strategy Reviews*, 2021, 38, 100710.

Cervan, D; Rodriguez, CV; Inga, C. Energy security: Multidimensional analysis for South American countries. *International Journal of Renewable Energy Research*, 2022, 12, 2. https://doi.org/10.20508/ijrer.v12i2.12982.g8500

Cherp, A; Jewell, J. The concept of energy security: Beyond the four as. *Energy Policy*, 2014, 75, 415–421.

Coutinho, GL; Vianna, JN; Dias, MA. Alternatives for improving energy security in Cape Verde. *Utilities Policy*, 2020, 67, 101112.

Demski, C; Poortinga, W; Whitmarsh, L; Bohm, G; Fisher, S; Steg, L; Umit, R; Jokinen, P; Pohjolainen, P. National context is a key determinant of energy security concerns across Europe. *Nature Energy*, 2018, 3, 10.

Erahman, QF; Purwanto, WW; Sudibandriyo, M; Hidayatno, A. An assessment of Indonesia's energy security index and comparison with seventy countries. *Energy*, 2016, 111, 364–376.

Fang, D; Shi, S; Yu, Q. Evaluation of sustainable energy security and an empirical analysis of China. *Sustainability*, 2018, 10, 5, 1685.

Franki, V; Viskovic, A. Energy security, policy and technology in South East Europe: Presenting and applying an energy security index to Croatia. *Energy*, 2015, 90, 494–507.

Fuentes, S; Villafafila-Robles, R; Lerner, E. Composed index for the evaluation of the energy security of power systems: Application to the case of Argentina. *Energies*, 2020, 13, 15, 3998.

Global Energy Institute. *International Index of Energy Security Risk, Assessing Risk in a Global Energy Market*, 2012. Available at: https://www.globalenergyinstitute.org/sites/default/files/2019-11/InternationalIndex2012.pdf (accessed on 12 January 2023).

Gong, X; Wang, Y; Lin, BQ. Assessing dynamic China's energy security: Based on functional data analysis. *Energy*, 2021, 217, 119324.

Ha, L; Thanh, TT. Global value chains and energy security: Evidence from global sample. *Energy Strategy Reviews*, 2022, 42, 100870.

Huang, BJ; Zhang, L; Ma, LM; Bai, WLYS; Ren, JZ. Multi-criteria decision analysis of China's energy security from 2008 to 2017 based on Fuzzy BWM-DEA-AR model and malmquist productivity index. *Energy*, 2021b, 228, 120481.

Huang, SW; Chung, YF; Wu, TH. Analyzing the relationship between energy security performance and decoupling of economic growth from CO2 emissions for OECD countries. *Renewable & Sustainable Energy Reviews*, 2021a, 152, 111633.

Hwang, CL; Yoon, K. *Multiple Attributes Decision Making Methods and Applications*. Springer: Berlin, Hedelberg, Germany, 1981, pp. 22–51.

Khan, MR; Dhakal, S. Do experts and stakeholders perceive energy security issues differently in Bangladesh? *Energy Strategy Reviews*, 2022, 42, 100887.

Kharazishvili, Y; Kwilinski, A; Sukhodolia, O; Dzwigol, H; Bobro, D; Kotowicz, J. The systemic approach for estimating and strategizing energy security: The case of Ukraine. *Energies*, 2021, 14, 8, 2126.

Kisel, E; Hamburg, A; Harm, M; Leppiman, A; Ots, M. Concept for energy security matrix. *Energy Policy*, 2016, 95, 1–9.

Kosai, S; Unesaki, H. Quantitative evaluation of security of nuclear energy supply: United States as a case study. *Energy Strategy Reviews*, 2020, 29, 100491.

Le, TH; Nguyen, CP. Is energy security a driver for economic growth? Evidence from a global sample. *Energy Policy*, 2019, 129, 436–451.

Li, JC; Wang, L; Lin, XS; Qu, S. Analysis of China's energy security evaluation system: Based on the energy security data from 30 provinces from 2010 to 2016. *Energy*, 2020, 198, 117346.

Li, P; Zhang, JS. A new hybrid method for China's energy supply security forecasting based on ARIMA and XGBoost. *Energies*, 2018, 11, 7, 1687.

Li, YZ; Shi, XP; Yao, LX. Evaluating energy security of resource poor economies: A modified principle component analysis approach. *Energy Economics*, 2016, 58, 211–221.

Marhold, AA. Unpacking the concept of 'energy security': Lessons from recent WTO case law. *Legal Issues of Economic Integration*, 2021, 48, 2, 147–170.

Pinter, L; Hardi, P; Martinuzzi, A; Hall, J. Bellagio STAMP: Principles for sustainability assessment and measurement. *Ecological Indicators*, 2012, 17, 20–28.

Rodriguez-Fernandez, L; Carvajal, ABF; Ruiz-Gomez, LM. Evolution of European Union's energy security in gas supply during Russia-Ukraine gas crises (2006–2009). *Energy Strategy Reviews*, 2020, 30, 100518.

Siksnelyte-Butkiene, I. Defining the perception of energy security: An overview. *Economies*, 2023, 11, 174.

Song, Y; Zhang, M; Sun, RF. Using a new aggregated indicator to evaluate China's energy security. *Energy Policy*, 2019, 132, 167–174.

Sotnyk, I; Kurbatova, T; Kubatko, O; Prokopenko, O; Prause, G; Kovalenko, Y; Trypolska, G; Pysmenna, U. Energy security assessment of emerging economies under global and local challenges. *Energies*, 2021, 14, 18, 5860.

Sovacool, BK; Mukherjee, I. Conceptualizing and measuring energy security: A synthesized approach. *Energy*, 2011, 36, 8, 5343–5355.

Sovacool, BK; Valentine, SV; Bambawale, MJ; Brown, MA; Cardoso, TD; Nurbek, S; Suleimenova, G; Li, JK; Xu, Y. Exploring propositions about perceptions of energy security: An international survey. *Environmental Science & Policy*, 2012, 16, 44–64.

Stavytskyy, A; Kharlamova, G; Komendant, O; Andrzejczak, J; Nakonieczny, J. Methodology for calculating the energy security index of the state: Taking into account modern megatrends. *Energies*, 2021, 14, 12, 3621.

Tutak, M; Brodny, J. Analysis of the level of energy security in the three seas initiative countries. *Applied Energy*, 2022, 311, 118649.

Wang, DQ; Tian, SH; Fang, L; Xu, Y. A functional index model for dynamically evaluating China's energy security. *Energy Policy*, 2020, 147, 111706.

Wang, JQ; Shahbaz, M; Song, ML. Evaluating energy economic security and its influencing factors in China. *Energy*, 2021, 229, 120638.

Wang, Q; Zhou, K. A framework for evaluating global national energy security. *Applied Energy*, 2017, 188, 19–31.

Wu, TH; Chung, YF; Huang, SW. Evaluating global energy security performances using an integrated PCA/DEA-AR technique. *Sustainable Energy Technologies and Assessments*, 2021, 45, 101041.

Yang, B; Ding, LJ; Zhan, XY; Tao, XZ; Peng, F. Evaluation and analysis of energy security in China based on the DPSIR model. *Energy Reports*, 2022, 8, 607–615.

Zhang, L; Bai, WLYS; Xiao, HJ; Ren, JZ. Measuring and improving regional energy security: A methodological framework based on both quantitative and qualitative analysis. *Energy*, 2021, 227, 120534.

Zhang, L; Yu, J; Sovacool, BK; Ren, JZ. Measuring energy security performance within China: Toward an inter-provincial prospective. *Energy*, 2017, 125, 825–836.

Ziemba, P. Energy security assessment based on a new dynamic multi-criteria decision-making framework. *Energies*, 2022, 15, 24, 9356.

4 Gender Equality

Lina Volodzkiene

4.1 GENDER EQUALITY CONCEPT IN LOW-CARBON TRANSITION

Gender equality is a complex notion that entails guaranteeing equal rights, opportunities, and treatment for all genders in legal, cultural, and social spheres. It is both a fundamental human right and a precondition for attaining social justice and fostering equity across political, economic, and social domains. Facilitating the full and equal participation of both men and women in decision-making, the workforce, and community life fosters societal development and aligns with Sustainable Development Goal 5, which promotes gender equality and women's empowerment across all sectors. Gender equality encompasses not merely individual fairness but also serves as a catalyst for a functional and just society. Promoting gender equality in the workplace by rectifying income inequities, occupational segregation, and underrepresentation in leadership positions can stimulate economic growth and enhance employee happiness (Du, 2024). Research from several regions indicates that improving gender equality substantially aids in poverty alleviation and economic advancement (Rohmatilah, 2023). This further illustrates that gender equality is essential not only for individual empowerment but also for societal progress.

Moreover, gender equality is essential for promoting sustainable development. Balanced engagement in economic activities is essential for attaining environmental, social, and economic sustainability, hence enhancing societal resilience and improving health, education, and overall well-being. Consequently, attaining gender equality is crucial for cultivating a harmonious, wealthy, and sustainable planet. It is fundamental for constructing more inclusive communities, enabling equal contributions from both men and women, which results in improved governance, economic prosperity, and enhanced social cohesion.

4.1.1 Theoretical Frameworks for Gender Equality in the Low-Carbon Transition

Gender equality is essential for fostering a just and inclusive low-carbon transition. Theoretical frameworks establish a basis for comprehending the integration of equal rights and opportunities, freedom from gender-based discrimination, challenging gender stereotypes, and empowerment into climate action, along with the role of legal and policy measures (Figure 4.1).

These frameworks underscore the necessity of dismantling gender stereotypes and guaranteeing representation at all tiers of decision-making. Intersectionality emphasizes the interplay between gender and other identities, including race and class, while legal and policy frameworks establish the criteria for integrating gender equality into climate policies. Collectively, these components provide a holistic strategy for advancing gender equality in the transition to a sustainable, low-carbon future.

Equal rights and opportunities. Gender equality is founded on the notion that men and women ought to possess equal access to essential rights, including education, employment, healthcare, and political representation. It seeks to eliminate obstacles that hinder individuals from fully engaging in society due to their gender.

In education, gender equality guarantees that both males and females possess equal access to high-quality learning opportunities and resources for academic advancement. This encompasses promoting equitable treatment in educational environments and dismantling prejudices that may restrict persons depending on gender. In the realm of employment, gender equality ensures equal access to job opportunities, equitable compensation, career progression, and leadership roles.

DOI: 10.1201/9781003517481-4

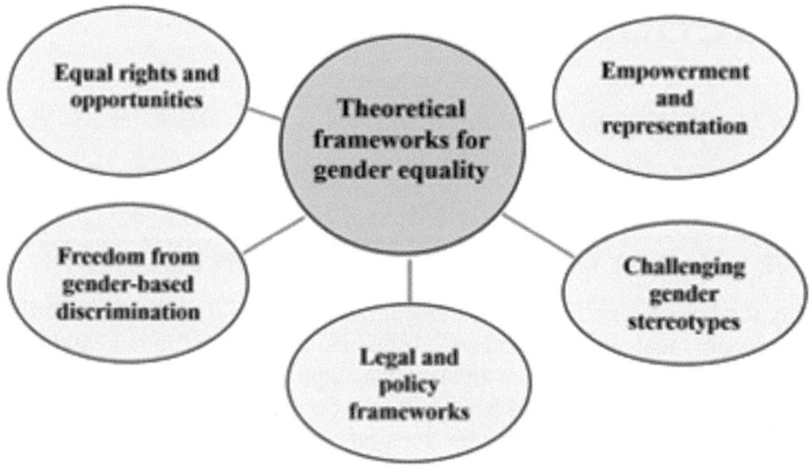

FIGURE 4.1 Theoretical foundations for integrating gender equality into climate action.

Source: created by authors.

Mitigating the gender wage disparity and decreasing the underrepresentation of women across diverse industries is essential for attaining workplace equality. Moreover, initiatives that advocate for diversity and inclusion, work-life equilibrium, and inclusive leadership are crucial for cultivating a just and equitable workplace (Du, 2024). Legal systems, especially within the European Union, have historically promoted gender equality by implementing legislation that guarantees equal remuneration and equitable treatment. In healthcare, gender equality guarantees that both men and women receive fair and courteous treatment. It tackles inequality in access to medical care, encompassing reproductive health rights, to guarantee equity in health outcomes (Protosavitska, 2023). Gender equality in politics aims to enhance women's representation and involvement in decision-making roles. Promoting female involvement in political positions is crucial for attaining gender equality in leadership and incorporating women's viewpoints in governance.

From an economic perspective, gender equality guarantees that both men and women possess equal access to financial resources, such as loans, credit, and property ownership. The economic empowerment of women can result in favorable societal effects, including poverty reduction and enhanced family welfare. Ultimately, legal safeguards are essential for guaranteeing equitable rights and opportunities for all genders. Legislation designed to prevent discrimination, harassment, and violence must be uniformly enforced for both men and women (Protosavitska, 2023). Gender equality influences all facets of society—education, work, healthcare, politics, and economy. Attaining it necessitates a thorough and culturally attuned strategy to foster equity and justice for all genders.

Freedom from gender-based discrimination is a crucial element in attaining gender equality. It involves the eradication of attitudes and behaviors that restrict individuals' chances and rights according to their gender. Gender-based discrimination is prevalent in various sectors, including employment, education, healthcare, and social participation. Discriminatory practices frequently arise from entrenched cultural norms, biases, and stereotypes that perpetuate traditional gender roles, consequently limiting opportunities and potential for women. This form of discrimination undermines fundamental human rights and constitutes a substantial obstacle to attaining gender equality. Its persistence is seen in significant domains like wage inequities, the underrepresentation of women in leadership positions, cultural conventions, and the repercussions on health and education.

An explicit illustration of this discrimination is the gender pay gap, when women frequently receive inferior remuneration compared to men for performing identical tasks. This worldwide problem contravenes the idea of equal remuneration for equivalent work and exacerbates economic disparity. In academic settings, women encounter challenges including disparities in promotion and compensation, frequently exacerbated by insufficient transparency (Morgan et al., 2018). In the business sector, biases and preconceptions persistently obstruct women's advancement into leadership positions, maintaining economic and professional inequality. The inadequate presence of women in leadership positions across diverse sectors, such as politics, industry, and academia, persists as a significant issue. Leadership is frequently linked to masculinity, and gender-specific obstacles hinder numerous women from reaching senior roles.

Cultural and societal conventions additionally perpetuate gender inequality. In numerous nations, conventional gender norms compel women to prioritize caregiving over professional pursuits, so constraining their employment participation and sustaining their subordinate status. These entrenched practices present substantial obstacles to legislative and legal reforms, hindering the attainment of major change without a comprehensive cultural makeover.

Gender-based discrimination is also widespread in healthcare and education. Women's access to healthcare is frequently hindered by prejudices within medical systems, and underprivileged groups may encounter further obstacles in getting essential services (Leite et al., 2021). Gender disparities in educational environments can result in conflicts and discrimination among students, adversely impacting their interpersonal connections and development (Namita Kumari, 2024). Initiatives to address gender-based discrimination include several techniques, such as legislating for gender equality and offering training to enhance awareness of gender biases. International human rights treaties, such the Convention on the Rights of the Child (CRC), underscore the necessity of eradicating gender discrimination and guaranteeing equal opportunity for girls and boys (Rafferty, 2013). The enduring presence of conventional gender norms and stereotypes constitutes a significant obstacle to advancement. The cultural reluctance to change, along with the constraints of legal systems, underscores the need for a holistic strategy.

Attaining genuine gender equality necessitates a synthesis of legal, educational, and cultural strategies to eradicate the obstacles that perpetuate gender-based discrimination. Proponents of gender equality strive to confront detrimental perceptions, implement equal pay laws, and advocate for equitable employment practices. Educational endeavors are essential for mitigating these biases from an early stage, cultivating a culture in which all individuals can prosper unimpeded by conventional gender norms. Ultimately, liberation from gender-based discrimination entails establishing a society in which an individual's gender does not dictate their potential, access to resources, or the esteem they are afforded. Genuine gender equality will enable both men and women to participate completely in all facets of life, unencumbered by the constraints of preconceptions and discrimination.

Empowerment and representation are crucial to promoting gender equality, guaranteeing that both women and men possess equal possibilities to engage in decision-making processes at all societal tiers. Empowerment entails equipping women with essential tools, resources, and confidence to express their rights, whereas representation guarantees their visibility and active engagement in leadership positions. Collectively, these components facilitate the dismantling of structural barriers and advance gender equality by influencing societal frameworks and policies. This response explores the complex aspects of empowerment and representation, their influence on gender equality, and the obstacles they encounter.

Empowerment includes social, political, and economic dimensions, with the objective of enhancing women's self-esteem, decision-making authority, and capacity to effect change. Education and professional development are essential elements that empower women to take control of their life and make substantial contributions to economic growth and social progress (Sen et al., 2023). Empowerment efforts can diminish poverty, hunger, and violence, while simultaneously fostering democracy and effective governance.

Global representation in leadership positions, especially in political spheres, continues to be constrained despite the implementation of women quotas and party lists (Thwala, 2023). Increased female representation in legislative bodies is favorably associated with enhanced performance on socioeconomic Sustainable Development Goals (SDGs). Nonetheless, attaining equal representation does not inherently result in gender equality inside institutions or society. Structural and cultural impediments persist in hindering women's progression into leadership roles. Gender biases, societal customs, and religious or cultural standards frequently restrict women's presence in political and leadership positions. In systems with gender quotas, women may encounter obstacles including tokenism and interpersonal conflict (Thwala, 2023). Confronting these difficulties necessitates a comprehensive strategy encompassing diversity and inclusion programs, rectifying the gender wage disparity, and promoting inclusive leadership (Du, 2024). Legal frameworks, affirmative action, and gender mainstreaming are crucial for ensuring that empowerment and representation result in enduring change. Moreover, political education, extensive access for women, and the cultivation of political and social networks are essential to improve women's representation in politics.

Although empowerment and representation are essential for attaining gender equality, considerable challenges persist. Structural and cultural impediments endure, and equal representation does not necessarily signify genuine gender equality. An integrated strategy that amalgamates legal reforms, societal transformation, and focused initiatives is essential for addressing these issues and guaranteeing that empowerment and representation yield substantial advancements in gender equality.

Challenging gender stereotypes. Attaining gender equality necessitates the deconstruction of ingrained societal conventions that allocate distinct tasks and behaviors according to gender. Such prejudices frequently constrain individuals to conventional roles, limiting their autonomy and potential. Women are generally anticipated to prioritize caregiving and domestic duties, whilst men are compelled to exemplify power, leadership, and emotional stoicism. Gendered expectations affect personal life, job interactions, media representations, and societal standards, hence perpetuating traditional gender roles.

In educational settings, instructional materials often perpetuate gender stereotypes by portraying men in professional positions and women in domestic responsibilities. Studies indicate that such depictions can reinforce notions of male dominance and constrain the recognition of women's capabilities (Huang and Liu, 2024). It is imperative to revise these representations, especially in educational materials, to equip young individuals with a more progressive comprehension of gender roles and to promote equality from an early age. The media significantly influences societal perceptions about gender. Conventional advertising frequently perpetuates detrimental stereotypes by depicting women as homemakers and caregivers (Tehseem et al., 2018). Advocating for diverse and equitable gender portrayals in media can alter public perceptions and facilitate the progression of gender equality. Gender stereotypes in the workplace adversely affect recruiting practices, frequently privileging male candidates and intensifying discrepancies, especially for women of color who encounter intersectional discrimination. Employing measures like blind recruitment and diversity training can alleviate these biases, promoting a more egalitarian workplace. Furthermore, gender inequality in the workplace is manifested through compensation disparities, inequitable growth chances, and the underrepresentation of women in leadership roles. Resolving these difficulties necessitates extensive diversity initiatives, rectifying salary disparities, and fostering inclusive leadership (Du, 2024). Cultural and societal conventions impede the attainment of gender equality, especially in areas where conventional gender roles are firmly established. In these contexts, familial and cultural factors significantly shape gender stereotypes, perpetuating workplace sexism and constraining chances for both men and women (Hoi and Xuan, 2023). Confronting these cultural norms is essential for fostering inclusive workplaces and promoting social justice.

The influence of gender stereotypes transcends employment, affecting interpersonal interactions and cultural perceptions. Benevolent sexism, which depicts women as loving and empathic, can either elevate or diminish their social position depending on the context (Fosch-Villaronga et al., 2021). Confronting these prejudices is essential for enabling individuals to seek opportunities that correspond with their abilities and ambitions, unencumbered by conventional gender norms.

Notwithstanding advancements in addressing these challenges, deeply rooted cultural beliefs and business motivations persist as substantial obstacles to attaining gender equality. An extensive plan encompassing education, media reform, and legislative initiatives is crucial for addressing prejudices and promoting equitable practices across all sectors. By promoting varied representations and confronting detrimental preconceptions, we may create an atmosphere in which individuals can make autonomous choices regarding their lives, jobs, and identities, unencumbered by antiquated gender conventions.

Intersectionality is an essential framework for comprehending how different forms of discrimination—pertaining to gender, ethnicity, class, and sexual orientation—intersect to produce distinct experiences of both oppression and advantage. This methodology is essential for formulating more inclusive and effective gender equality policies and campaigns, as it recognizes the interrelation of social identities. Intersectionality, by analyzing several dimensions of inequality, offers a profound comprehension of societal disparities and guides policies that more effectively cater to the distinct requirements of numerous groups.

Socioeconomic inequities are influenced by gender, ethnicity, class, and other social characteristics, exacerbating the obstacles individuals encounter. Policies that concentrate exclusively on gender disparity may neglect the distinct issues faced by women of color or those from lower socioeconomic strata. Intersectionality underscores the necessity of including marginalized groups in decision-making processes and formulating policies that mirror their lived experiences. Heightened knowledge of intersectionality among policymakers, educators, and the public enables stakeholders to more effectively confront social inequality. In education, intersectionality is essential for acknowledging the many identities of learners and incorporating these elements into curricula and pedagogical approaches. This necessitates a reevaluation of conventional educational frameworks to establish more equitable learning settings that address the systematic disparities encountered by students from underprivileged backgrounds (Zhang and Gao, 2024). Comprehending the intersection of race, class, and gender in educational experiences is essential for promoting diversity and inclusion in learning (Pinem, 2023). In the domain of policy, embracing an intersectional approach entails transcending a singular emphasis on gender. Policies should be formulated to address the needs of individuals most impacted by various forms of inequality, especially women from marginalized racial or ethnic groups, to accurately represent the intricate realities of people's life.

Intersectionality is essential in professional domains where underrepresentation and systemic biases obstruct the progress of minority groups. Initiatives aimed at enhancing diversity within organizations can mitigate the obstacles encountered by women, particularly those from marginalized backgrounds, in attaining leadership roles. Identifying the influence of institutional biases can improve access to culturally relevant services for marginalized populations. Moreover, the notion of intersectionality encompasses sustainability and inclusiveness initiatives. Utilizing an intersectional perspective in sustainability projects guarantees that policies more accurately represent the interests of all community members, especially those from marginalized groups, thus enhancing the breadth and efficacy of these measures.

Although intersectionality offers a thorough framework for tackling injustice, its actual use in policy and practice continues to pose difficulties. Conventional methods that concentrate on isolated facets of identity may hinder the comprehensive incorporation of intersectionality into prevalent methodology. To address these difficulties, it is essential to foster a comprehensive awareness of intersectionality among key stakeholders and formulate policies that consider the nuances of oppressed groups' experiences. This guarantees that activities for gender equality and social justice are authentically inclusive and effective, benefiting persons irrespective of their gender, ethnicity, class, or other intersecting identities.

Legal and policy frameworks. The quest for gender equality is codified in numerous international treaties, national constitutions, and statutes aimed at advancing equal rights and safeguarding against gender-based violence. An illustrative instance is the United Nations' Sustainable Development Goal 5, which embodies a worldwide pledge to eradicate gender disparities in areas including education, employment, healthcare, and governance. This objective, in conjunction with frameworks such as the Convention on the Elimination of All Forms of Discrimination against Women (CEDAW) and the Convention on the Rights of the Child (CRC), obligates state parties to combat discrimination and promote gender equality in all facets of life (Rafferty, 2013). Numerous nations have incorporated these ideas into their constitutions, guaranteeing equal legal treatment and forbidding gender-based discrimination. Some constitutions explicitly entrench gender equality, while others have implemented legislation addressing gender-based violence, including legal recourse for victims. In addition to legal measures, many efforts seek to mitigate the gender pay gap, enhance female representation in leadership positions, and guarantee equitable access to healthcare. Such initiatives are frequently bolstered by strategies such as affirmative action and gender quotas.

The efficacy of these legal and policy frameworks is predominantly contingent upon their implementation, which encounters substantial obstacles. Entrenched cultural and societal practices—such as early marriage and detrimental traditions—frequently obstruct the effective implementation of laws designed to promote gender equality (Rafferty, 2013). Moreover, insufficient political will, financial resources, and effective enforcement mechanisms constrain the efficacy of these frameworks. An important concern is the growing emphasis on the prosecution of gender-based violence. This emphasis, while significant, may divert attention from the wider structural and systemic aspects of gender inequality. Emphasizing individual accountability risks overlooking the fundamental social and economic dynamics that sustain inequality.

Notwithstanding significant advancements in establishing legal and legislative frameworks for gender equality, cultural opposition, ingrained societal norms, and enforcement difficulties persist in hindering substantial development. Attaining genuine gender equality necessitates continuous endeavors to enhance these frameworks, synchronize them with global norms, and guarantee their efficient execution. A holistic strategy must encompass both the legal and cultural dimensions of gender inequality, promoting a society in which all persons can prosper, irrespective of gender.

Gender equality is essential for facilitating a just and inclusive low-carbon transition. Integrating theoretical frameworks, like equal rights and opportunities, freedom from gender-based discrimination, and empowerment, into climate action establishes a foundation for a more equitable future. These frameworks underscore the necessity of deconstructing gender stereotypes, guaranteeing representation at all levels of decision-making, and addressing the intersectionality of gender with other social identities. Legal and policy frameworks offer the necessary institutional support to actualize gender equality, guaranteeing that all individuals can participate in and gain from the transition to a sustainable, low-carbon future. A comprehensive strategy that integrates these components will be essential for attaining enduring gender equality throughout this critical transition.

4.1.2 Specific Content and Expression of Gender Equality in Climate Action

The explicit content and articulation of gender equality in climate action emphasize how the low-carbon transition may mitigate current gender gaps while promoting inclusive growth. Climate change disproportionately affects women, especially in low-income and rural regions, owing to their responsibilities in agriculture, resource management, and energy supply. This transformation also offers distinct chances for women's empowerment, particularly in burgeoning fields like renewable energy and green technologies. Gender-sensitive energy policies, coupled with initiatives to empower women as leaders and agents of change, are crucial for fostering a more equitable and sustainable future. By tackling structural impediments and advancing gender equality, climate action can attain more efficient, resilient, and socially equitable results.

The disproportionate effects of climate change on genders, especially on women in low-income and rural areas, constitute a complex issue intricately linked to social roles, resource accessibility, and cultural norms. Women bear an unequal burden in responsibilities such as agriculture, water procurement, and energy supply, which are particularly vulnerable to environmental fluctuations including droughts, floods, and other climate-related phenomena. Climate-related vulnerabilities are exacerbated by pre-existing gender inequalities that limit women's adaptive capacities and access to resources. Rectifying these discrepancies is essential for advancing gender equality and enhancing climate resilience.

The susceptibility of women to climate change is significantly shaped by established gender norms and societal roles. These roles frequently restrict women to duties directly influenced by environmental alterations, like agriculture and domestic resource management (Adeola et al., 2024a). In certain areas, such as Africa, the disparity in access to agricultural resources—namely land, labor, and information—intensifies the effects of climate change on women, hindering their capacity to adapt and respond effectively to climate-related difficulties (Adeola et al., 2024b). In Bangladesh, women encounter heightened mental health issues stemming from augmented workloads and social concerns as male family members travel for employment due to environmental changes (Goudet et al., 2023). The restricted access of women to resources and decision-making authority exacerbates their susceptibility to climate change. Cultural practices and social conventions in numerous locations impede women's involvement in adaption methods and community decision-making processes (Assaduzzaman et al., 2023). In Burkina Faso, financial limitations and substantial domestic responsibilities influence the adaptation tactics of women farmers, thereby impacting household income and food security (Diendéré and Ouédraogo, 2023). The exclusion from decision-making is seen in coastal communities in Bangladesh, where women's absence in household and community-level decisions intensifies their vulnerability (Assaduzzaman et al., 2023).

To effectively mitigate the gendered effects of climate change, policy interventions must emphasize gender equality and social inclusion. Policies that promote women's land ownership and participation in agricultural sectors are essential for bridging the gender gap in adaption capacities (Adeola et al., 2024c). The United Nations Framework Convention on Climate Change (UNFCCC) has made progress in incorporating gender concerns into climate policies; however, significant work is still required to thoroughly address gender inequities in this area. Empowering women via education and involvement in climate change initiatives can bolster their resilience and aid in the formulation of sustainable solutions (Spitzer, 2023).

Climate change profoundly impacts mental health, with women facing unique issues stemming from their cultural roles and duties. Women frequently experience increased psychological stress due to climate-induced environmental alterations and the ensuing social uncertainties (Goudet et al., 2023). Mitigating the mental health effects of climate change necessitates gender-sensitive approaches that consider the unique needs and obstacles encountered by women in at-risk areas. Although women's susceptibility to climate change is extensively established, it is as crucial to acknowledge their potential as catalysts for change. Women possess significant traditional knowledge and abilities that can be utilized in climate adaptation and resilience-building efforts. Incorporating women into decision-making and policy formulation enables communities to devise more effective and inclusive strategies for combating climate change. This comprehensive method not only mitigates gender disparities but also enhances the overall resilience of communities to climate-related effects.

Opportunities for women's empowerment in a low-carbon economy. The shift to a low-carbon economy offers a distinct chance to advance gender equality by generating new employment and economic prospects in areas such as renewable energy, energy efficiency, and green technologies. Women, traditionally underrepresented in various sectors, will gain substantially from focused measures designed to enhance their involvement. By implementing efforts centered on training, education, and financial assistance, gender disparities can be reduced, guaranteeing that both men and women equally reap the benefits of the transition to a green economy.

The growth of employment prospects in renewable energy and green technology is a vital element of the low-carbon transition. These businesses are crucial for sustainable development and can markedly increase women's participation in traditionally male-dominated fields. The involvement of women in managing renewable energy collectives and other green initiatives clearly demonstrates the potential for socio-economic growth. Women-led renewable energy initiatives in certain areas have demonstrated a significant rise in income for recipients, highlighting the transformative capacity of gender-inclusive policy in these fields (Ingole, 2023). Training and education are essential for enabling women to engage in the green economy. Reskilling initiatives, especially in areas where women have restricted access to technology and educational resources, can unleash considerable economic potential. Gender-sensitive policies that facilitate women's involvement in green reskilling programs are essential for their active participation in the transition to a sustainable future (Adeola et al., 2024a).

Policy and financial assistance are crucial for empowering women in the low-carbon transition. Gender equality and social inclusion should be fundamental to community energy projects and other programs designed to enhance the green labor market. An intersectional approach that encompasses the technological, political, social, and institutional facets of the energy transition is essential for formulating more inclusive policies that facilitate women's access to green employment (Castán Broto, 2024). Initiatives in certain areas aimed at empowering women through sustainable economic projects have demonstrated promise; nevertheless, these efforts require enhancement to align with the overarching goals of sustainable development and to ensure sufficient employment opportunities for women.

Nevertheless, obstacles persist in completely actualizing the promise for women's empowerment within the low-carbon economy. Numerous current policies, especially in developing nations, are gender-neutral and neglect to tackle specific gender-related concerns. There is an urgent necessity for more gender-inclusive climate change policy and planning (Adeola et al., 2024b). Furthermore, the global low-carbon transition may pose challenges for developing nations, especially those that function as resource centers for industrialized economies. This dynamic may hinder these countries' advancement towards Sustainable Development Goals. The low-carbon economy presents a substantial opportunity for promoting gender equality and empowering women; nevertheless, achieving this potential necessitates intentional efforts. Through the implementation of specific policies, provision of education and financial assistance, and the rectification of gender inequities, women can assume a pivotal role in the transition to a sustainable and equitable future.

Gender-sensitive energy policies are crucial for enabling a fair transition to low-carbon energy systems, especially in developing nations where women frequently serve as the principal users of domestic energy for cooking, heating, and lighting. Women, particularly in rural regions, depend significantly on conventional energy sources like biomass and kerosene, which are detrimental to both health and the environment. These energy sources pose considerable health concerns for women and impose the responsibility of fuel collection upon them, so limiting their time and prospects for economic engagement. Adopting cleaner energy technologies such as solar, wind, and advanced cookstoves can mitigate these challenges, strengthening women's health, decreasing their workload, and improving their general well-being.

Gender-sensitive policies are essential to guarantee women's access to inexpensive, clean, and sustainable energy solutions. These policies must be formulated to meet the distinct requirements of women, equipping them with the skills, resources, and financial assistance essential for the adoption and utilization of clean energy technologies. Access to contemporary energy services is associated with the empowerment of women, enhancing their economic prospects and social standing. Access to clean fuels, improved cookstoves, and electricity has demonstrably enhanced women's labor force participation and quality of life (Chandrasekaran et al., 2023). The participation of women in decision-making processes within the energy sector is essential. The presence of women in leadership positions frequently correlates with enhanced carbon disclosure and improved efficacy of climate change mitigation efforts (Abd Majid and Jaaffar, 2023). Their involvement guarantees that energy policies are both inclusive and more attuned to the requirements of individuals most impacted by energy poverty.

Notwithstanding the significance of gender-sensitive energy policy, numerous obstacles impede their efficient execution. Women are disproportionately impacted by energy poverty, especially in rural regions where conventional gender roles restrict their access to contemporary energy solutions.

Furthermore, intersectional obstacles like socio-economic position, cultural norms, and geographic location intensify women's difficulties in obtaining energy resources and reaping the benefits of energy transitions (Castán Broto, 2024). The absence of gender-disaggregated statistics on energy use and poverty obstructs the formulation of tailored strategies, complicating the comprehensive understanding and addressing of women's special needs.

Inclusive policy design is essential to address these difficulties. Engaging women and underrepresented groups in the formulation, execution, and assessment of energy policy guarantees their pertinence and efficacy. Capacity development and education are critical elements of effective energy transitions. Women require access to energy literacy initiatives and financial resources to maximize the advantages of sustainable energy technologies (Keränen and Enwald, 2023). Moreover, implementing an intersectional framework in energy policy can effectively tackle the fundamental causes of energy disparity, ensuring that the varied experiences of women and other marginalized groups are taken into account (Castán Broto, 2024). Gender-sensitive energy policies are essential for an equitable energy transition, but they must be integrated into a comprehensive strategy that tackles the fundamental socio-economic and cultural variables affecting energy access and utilization. Structural inequities and social justice concerns must be addressed to guarantee inclusive development methods that are advantageous to all societal members, irrespective of gender. This holistic strategy will advance both gender equality and a more equitable and sustainable energy future for everyone.

Women as agents of change in climate action. Women serve as vital agents of change in climate action, providing distinct perspectives and experiences that improve the inclusivity and efficacy of climate legislation. Their involvement in decision-making about the low-carbon transition is essential for formulating comprehensive and equitable solutions. Women contribute valuable perspectives in natural resource management, community resilience, and sustainable development, which are crucial for tackling the multifaceted difficulties presented by climate change. Empowering women as leaders in these sectors may foster innovation, expedite the implementation of sustainable practices, and enhance the inclusivity of the green economy.

In numerous locations, women have exhibited leadership in climate resilience and natural resource management. Women's participation in climate mitigation activities, including conservation projects and renewable energy programs, enables them to wield decision-making authority and contribute to environmental preservation within their communities (Dhal, 2024). In agriculture-dependent areas, where women are disproportionately impacted by climate change, it is imperative to implement inclusive policies that include women's perspectives to enhance community resilience and the efficacy of climate programs (Adeola et al., 2024a).

The participation of women in leadership roles is essential for promoting sustainable development and growing the green economy. Green reskilling projects have enabled women to engage actively in the transition to a low-carbon economy, fostering gender equality and economic progress (Adeola et al., 2024c). In various regions, the incorporation of gender equality into climate action has demonstrated efficacy in producing sustainable and inclusive results, underscoring the significance of cooperative endeavors in climate change mitigation and women's advancement (Jeffs, 2023). Female entrepreneurs are integral to the green economy, especially in areas such as renewable energy and sustainable agriculture. Their leadership in the development and expansion of low-carbon initiatives is crucial for fostering innovation and the implementation of sustainable business practices. Women in entrepreneurship have exhibited a pronounced emphasis on climate action, utilizing their social networks and environmental interests to secure a competitive edge in the global market. Supporting women entrepreneurs and business leaders enables societies to leverage their potential for substantial contributions to sustainable development.

Nonetheless, despite the capacity for women to spearhead climate action, patriarchal norms and gender disparities persistently obstruct their full participation in climate change adaptation. Systemic impediments of this nature must be confronted to advance gender equality and social justice in climate projects (Adeola et al., 2024b). Moreover, numerous current programs are gender-neutral, neglecting to address the distinct requirements and obstacles encountered by women. Implementing gender-inclusive climate policies and including gender concerns in climate planning are essential measures for attaining sustainable development (Adeola et al., 2024a). Although women have demonstrated their ability to lead and innovate in climate action, substantial obstacles persist that hinder their complete involvement. Overcoming these obstacles via inclusive policies and gender-sensitive practices is essential for harnessing the potential of women as catalysts for change. By acknowledging and endorsing women's contributions to community resilience, sustainable development, and the green economy, society can formulate more inclusive and equitable strategies to address the climate problem.

Addressing structural barriers. To successfully include gender equality in the low-carbon transition, it is crucial to tackle the structural obstacles that restrict women's involvement. These obstacles encompass inequitable access to education, financial resources, and decision-making authority, with constraining social norms that limit women's responsibilities within particular sectors. Policies that guarantee equitable access to training, education, and employment opportunities within the green economy are essential for surmounting these challenges. Rectifying these deficiencies will enable women to engage comprehensively in climate action and capitalize on the prospects offered by the low-carbon transition.

Disparities in access to education and employment constitute a key obstacle for women, especially in sectors associated with the green economy. Women are markedly underrepresented in fields such as science, technology, engineering, and mathematics, which are essential for the green transition. The swift progress in digital and ecological technologies, termed the Fourth Industrial Revolution, offers both difficulties and opportunities for women. To guarantee women's inclusion in this transition, it is essential to implement specific policies that enhance female engagement in these vital sectors. Social norms and gender roles significantly impede women's engagement in the low-carbon transition. These conventions frequently result in workplace discrimination, inequitable promotion prospects, and the restriction of women to specific tasks within industry. Addressing these difficulties necessitates a comprehensive strategy, incorporating the execution of diversity and inclusion initiatives and endeavors to confront deep-seated gender biases (Du, 2024).

Gender-responsive climate funding is an essential element in tackling structural hurdles. In several developing nations, women are deprived of the financial means required to invest in renewable energy technology or establish sustainable enterprises. Gender-responsive climate finance mechanisms aim to address this disparity by equipping women with the financial resources necessary to adapt to and capitalize on climate initiatives (Adeola et al., 2024a). These measures are particularly crucial in regions where digital financial inclusion has enhanced women's access to financial services, while obstacles persist in attaining gender parity (Duvendack et al., 2023).

Robust policy frameworks are crucial for the incorporation of gender equality into climate funding. International frameworks such as the Gender Action Plan (GAP) underscore the necessity for gender-sensitive climate initiatives, directing the formulation of policies that guarantee women's equitable access to climate funding and their full participation in the low-carbon transition (Adeola et al., 2024b). These frameworks facilitate the formulation of policies that render climate finance more inclusive and equitable. Policies must not only tackle financial and educational obstacles but also embrace an intersectional approach that accounts for the intricate interactions of social, economic, and cultural issues. Mitigating gender vulnerabilities in the low-carbon transition necessitates policies that consider the varied experiences of women from diverse backgrounds and guarantee their equitable participation in climate action. Participatory governance is essential for developing inclusive and successful climate policies. Involving women and other marginalized groups in climate change governance guarantees that policies are customized to the requirements of vulnerable people and aid in enhancing resilience.

Notwithstanding considerable advancements, obstacles persist in the complete incorporation of gender equality within climate policies. The opposition to intersectional feminist methodologies and the ongoing existence of gender-neutral regulations highlight the necessity for sustained activism and policy advancement (Flavell, 2023). By tackling these structural impediments and enacting gender-responsive policies, women may become pivotal participants in the low-carbon transition, fostering a more equitable and sustainable future.

Benefits of gender equality in a low-carbon transition. Attaining gender equality in the low-carbon transition yields multiple advantages, enhancing both policy efficacy and sustainable development. Gender equality improves the inclusion of climate action by ensuring that climate policies consider the needs and contributions of all societal members. It fosters sustainable economic growth by maximizing worker potential and advancing social justice by diminishing gender inequities in access to resources and opportunities. This integration corresponds with international frameworks such as the United Nations Sustainable Development Goals (SDGs), including SDG 5 (Gender Equality) and SDG 13 (Climate Action), facilitating the attainment of overarching sustainability objectives.

Climate change affects individuals differently according to gender, with women frequently encountering heightened vulnerabilities due to socioeconomic inequality, restricted access to resources, and power disparities. Incorporating gendered aspects into climate strategies is crucial for advancing social justice and gender equality (Adeola et al., 2024a). In areas where women and children are disproportionately impacted by climate change, gender-inclusive policies are essential to prevent exacerbating existing inequality, while empowering marginalized groups and fostering sustainable development (Adeola et al., 2024c). Certain nations' strategies for climate change illustrate the significance of incorporating gender and social inequality in policy formulation. By including gender concerns in climate strategies, these nations seek to attain more effective and lasting results. Joint initiatives in climate change mitigation, rural revitalization, and women's development are essential for enhancing the efficacy of climate solutions (Jeffs, 2023).

The economic and social ramifications of gender-sensitive climate policies are substantial. Climate policy that effectively integrates gender equality can enhance environmental, economic, and social sustainability. Although there are apprehensions that climate initiatives may result in economic trade-offs, research indicates that these effects are typically minimal. Furthermore, incorporating gender equality into climate initiatives strengthens the social aspect of sustainability, promoting a more equitable approach to development. Gender mainstreaming, especially in areas such as agriculture and natural resources, is crucial for tackling structural disparities. This procedure guarantees that gender factors are adequately incorporated into policy execution and budget distribution.

Notwithstanding advancements in including gender in climate policies, considerable obstacles persist. Gender issues are frequently misperceived as solely pertaining to women, and there is an inconsistency in the implementation of gender mainstreaming across various levels of administration. Moreover, budget allocations for gender initiatives frequently prove inadequate, underscoring the necessity for enhanced competence to formulate and execute gender-responsive policies. International climate accords must embody elevated aspirations for carbon reductions while guaranteeing extensive participation and adherence. Incorporating gender concerns into these agreements can improve their efficacy and fairness. The incorporation of gender equality into climate policies poses obstacles but provides significant opportunity to improve the efficacy and equity of climate initiatives. Mitigating gender inequities and advancing social justice can substantially enhance sustainable development and synchronize climate policy with global objectives. Achieving these aims necessitates concerted efforts to surmount existing obstacles and ensure the comprehensive integration of gender considerations into climate action initiatives.

The incorporation of gender equality into climate action is crucial for mitigating the unequal effects of climate change on women and for guaranteeing that the shift to a low-carbon economy is inclusive, equitable, and efficient. Empowering women as catalysts for change, formulating gender-sensitive energy policies, and tackling structural impediments are essential measures for achieving a more resilient and sustainable future. By wholly adopting gender equality in climate policies, we may improve the effectiveness of climate programs and advance social justice, resulting in a low-carbon transition that equitably benefits all individuals. The efficacy of these initiatives depends on the persistent promotion of gender-responsive policies, guaranteeing that both men and women may participate in and gain from the global struggle against climate change.

4.2 INDICATORS AND FRAMEWORKS OF GENDER EQUALITY

Gender equality constitutes an essential human right and a vital component of social justice. It guarantees that everyone, irrespective of gender, has equal opportunities, rights, and duties across all domains of life, including education, employment, politics, and familial relations. Attaining gender equality is not merely an issue of equity but also a catalyst for sustainable development, economic advancement, and social progress. Various worldwide indicators and frameworks have been established to measure and monitor progress towards gender equality (Figure 4.2), encompassing social, economic, political, cultural and normative, environment and sustainability, and technological elements.

These instruments assist policymakers, researchers, and advocates in evaluating disparities and executing plans to mitigate gender-based inequality.

4.2.1 Social Indicators of Gender Equality

Social indicators of gender equality assess the impact of gender on access to services, opportunity, and involvement in diverse societal facets. These indicators underscore the obstacles and difficulties women encounter relative to males and offer essential insights into the societal frameworks that sustain gender inequality. Social indicators are crucial as they elucidate the profound influence of ingrained gender norms and roles on various aspects, including education, health, personal safety, and the distribution of labor within homes.

Educational attainment is a crucial social indicator, as it directly influences an individual's life chances. This statistic assesses gender inequality in literacy rates, school enrollment, and completion rates across several educational levels—primary, secondary, and tertiary. Historically, in numerous regions, girls have encountered substantial obstacles to educational access due to societal conventions that favor boys' education over that of girls. Notwithstanding advancements in recent decades, gender inequality in education endures, especially in low-income nations where poverty, cultural customs, and restricted access to educational institutions disproportionately impact girls.

In regions where females are prone to premature school dropout, these discrepancies sustain cycles of poverty and restrict women's capacity for meaningful workforce participation. In Nigeria, notable geographical discrepancies are seen, with the Northern regions, especially the North West,

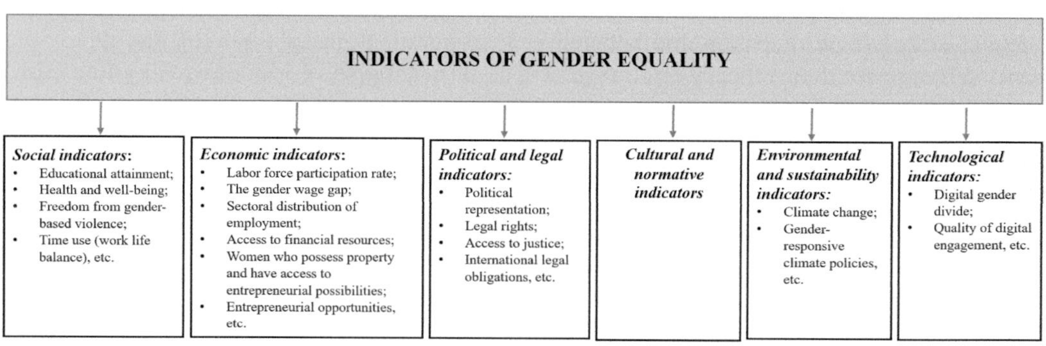

FIGURE 4.2 The complex of indicators and frameworks of gender equality.

Source: created by authors.

witnessing a decrease in educational attainment relative to the South. This is mostly attributable to the increasing population of ignorant individuals and "out-of-school" youngsters, indicative of the unequal allocation of educational resources (Suleiman et al., 2024). In Sub-Saharan Africa, gender discrepancies in rural schooling are associated with agricultural landownership, with customary practices favoring males intensifying the problem. Addressing these discrepancies necessitates comprehensive strategies that take into account sustainable development objectives (Liu, 2022).

Bridging the gender gap in education is imperative for numerous reasons. Educated women are more likely to obtain superior work prospects, contribute to economic development, and play a vital role in alleviating poverty. Moreover, education augments women's decision-making authority within households, increases their engagement in communal and political affairs, and results in improved health outcomes for women and their children. Increased educational attainment is associated with elevated income and improved developmental outcomes for children, so underscoring the enduring advantages of gender equality in education (Evans et al., 2021).

Socioeconomic position profoundly affects educational achievement, thus complicating gender differences. In Nigeria, socioeconomic status significantly influences educational inequality, with low-income families experiencing bigger disparities. Government initiatives, like subsidized educational programs aimed at low-income households, are essential for resolving these disparities (Nevo and Egenti, 2019). In Pakistan, home income and socioeconomic factors greatly influence educational achievement, with boys more likely to reach higher levels of schooling than girls (Pasha, 2024). Policies targeting female education and low-income households are crucial for promoting socioeconomic stability.

Worldwide, disparities in educational attainment based on gender varied by area. In the European Union, women typically possess higher educational attainment than men; however, variations occur among countries. Scandinavian nations demonstrate superior educational attainment, whilst Southern European nations fall short (Głodowska, 2018). In China, disparities between urban and rural areas and familial investment in education influence the duration of educational attainment, although gender does not substantially affect educational length (Zhu, 2024). In Turkey, gender and socioeconomic status continue to pose substantial barriers to school enrollment and academic performance, with females typically exhibiting poorer educational attainment than boys (Bengü Börkan and Ozan Bakış, 2016).

Rectifying gender inequities in education, although crucial, does not invariably result in commensurate increases in labor force participation. While bridging the gender gap in education facilitates enhanced career prospects, it does not ensure equal representation of women in the workforce. In numerous areas, enduring obstacles—such as gender discrimination, unpaid caregiving duties, and restricted access to leadership positions—persistently hinder women's economic involvement (Evans et al., 2021). This underscores the necessity for comprehensive programs that not only advance gender equality in school but also guarantee women equitable opportunity in the workforce.

Despite substantial advancements in mitigating gender inequities in education, numerous problems persist, especially in low-income and rural regions. Socioeconomic constraints, cultural customs, and geographical discrepancies persist in obstructing educational success for girls. Resolving these difficulties necessitates focused policies and actions that foster equitable access to school and equip females with the resources essential for academic success. Furthermore, governments must amalgamate education with comprehensive economic and social changes to guarantee that educational advancements result in enhanced gender equality in the workforce and society as a whole.

Health and well-being are vital domains for evaluating gender equality, with social variables including life expectancy, healthcare access, and reproductive health offering crucial insights into discrepancies among genders. Despite women often having a longer lifespan than males, this benefit is frequently compromised by insufficient access to healthcare services, especially in low-resource environments where maternal health services are scarce. Maternal mortality rates serve as a critical indicator of health disparities among genders, particularly in nations with elevated maternal mortality figures, where women encounter disproportionate risks during gestation and parturition due to insufficient healthcare infrastructure, a scarcity of qualified health practitioners, and obstacles to obtaining reproductive health services. In these contexts, maternal mortality signifies entrenched systemic problems that impede women's health outcomes (Bagade et al., 2022).

Another significant factor in health disparities is adolescent birth rates, which underscore concerns over gender norms and reproductive autonomy. Elevated rates of adolescent pregnancies frequently stem from insufficient information, restricted access to contraception, and societal influences that diminish young women's autonomy over their reproductive decisions. Such pregnancies can significantly hinder a girl's education and future economic opportunities, hence perpetuating cycles of gender inequity. In areas such as Latin America and the Caribbean, gender inequality and inadequate health funding adversely impact access to contemporary contraceptives, resulting in women in nations with more gender disparity being less likely to have their family planning requirements fulfilled (Moreira et al., 2023). This highlights the necessity for extensive, macro-level interventions to rectify inequality in reproductive health services.

The accessibility of reproductive health services, including contraception and prenatal care, is crucial for women's general well-being. Gender-sensitive healthcare policies that guarantee fair access to services are crucial for diminishing gender gaps in health outcomes. In sub-Saharan Africa, socioeconomic gaps intensify obstacles to healthcare access, with money, education, and media exposure significantly influencing health inequities (Alamneh et al., 2022). Mitigating these obstacles is essential for enhancing maternal mortality rates and the overall health outcomes of women of reproductive age.

Beyond maternal and reproductive health, comprehensive considerations are essential for addressing women's health throughout all life phases. Women's health evaluations frequently concentrate exclusively on reproductive health, overlooking other health requirements such as chronic illnesses and overall wellness. Incorporating both female-specific and general health issues within these assessments is essential for comprehensively addressing health disparities (Koirala et al., 2024). Gender-sensitive healthcare policies that integrate gender concerns into program evaluations and delivery systems are essential for enhancing access and the efficacy of health interventions.

Socioeconomic determinants, including income, education, and social status, profoundly affect health outcomes and life expectancy. Studies demonstrate that increased gender equality in several domains correlates with extended healthy life expectancies for all genders. Cultural and social norms, together with power dynamics and gender discrimination, persist in perpetuating systemic health inequities. Confronting these problems necessitates a cooperative strategy, prioritizing impacted groups in decision-making and fostering co-creation and shared accountability in health initiatives.

Although women often have a longer life expectancy than males, this longevity is sometimes accompanied with a greater prevalence of chronic diseases and functional impairments, which may not necessarily elevate mortality rates but considerably diminish their quality of life. This underscores the necessity for a comprehensive strategy to tackle gender disparities in health, taking into account the intricate interactions of biological, social, and cultural elements. Expanding women's health assessments and including gender considerations in healthcare policies and programs can facilitate progress towards enhanced gender equality in health outcomes.

The prevalence of **gender-based violence** is a significant indicator of gender inequality. This includes multiple forms of violence, including domestic abuse, sexual harassment, human trafficking, and detrimental traditional practices such as female genital mutilation and child marriage. Gender-based violence serves as both a catalyst and a result of gender inequality, illustrating entrenched power disparities and societal norms that either permit or endorse violence against women. Elevated instances of domestic violence or sexual harassment within a culture are unequivocal indicators of the obstacles women have in attaining social equality. Gender-based violence infringes upon women's rights and dignity, while also significantly affecting their physical and mental health, social engagement, and economic output. Women subjected to abuse frequently endure prolonged psychological trauma, encounter economic dependency, and possess an elevated chance of health complications, including sexually transmitted illnesses and unintended pregnancies.

Gender-based violence is extensively documented worldwide, with the World Health Organization indicating that around 30% of women globally have encountered some form of violence (Maria Giammarioli et al., 2024). Throughout the COVID-19 epidemic, occurrences of gender-based violence escalated markedly in numerous areas. South Asian Bangladeshi women in Toronto encountered increased marital violence attributed to patriarchal dominance, economic strain, and social seclusion (Islam and Akter, 2024). The epidemic in India intensified existing gender disparities, leading to heightened chances of domestic violence and coerced marriages for girls (Markan et al., 2022).

The origins of gender-based violence are fundamentally entrenched in society standards that confer inferior status upon women, resulting in power disparities and subjugation. In Nepal, these power dynamics are a major contributor to violence against women (Dahal et al., 2022). Economic instability and patriarchal norms exacerbate gender-based violence, as evidenced during the COVID-19 pandemic when financial troubles and the entrenchment of male domination resulted in heightened violence (Islam and Akter, 2024). The ramifications of gender-based violence are profound and extensive, impacting not just women's immediate security but also their enduring welfare. Women subjected to gender-based violence are at an increased risk of noncommunicable diseases and psychological stress (Maria Giammarioli et al., 2024). A substantial percentage of young people in Nigeria reported encountering physical and sexual assault, highlighting the pervasive nature of violence in the nation (Ndep et al., 2022).

Confronting gender-based violence is crucial for attaining gender equality and fostering an environment in which women can exist without fear and engage fully in society. Legal frameworks and social services are essential in addressing violence, as nations that adopt robust regulations experience advancements in gender equality. Innovative strategies, including the enhancement of ICT infrastructure and the provision of economic and social support, were crucial in alleviating the increase in violence during the COVID-19 pandemic (Markan et al., 2022). Community-based initiatives have been helpful in mitigating gender-based violence. Initiatives in Nigeria have aimed at altering societal narratives and empowering women via education and economic opportunities, thereby contributing to the reduction of violence (Ndep et al., 2022).

Addressing gender-based violence is essential for advancing gender equality; however, it is also imperative to examine the wider socio-economic causes that contribute to gender inequality. Insufficient access to education and economic opportunities intensifies the difficulties women encounter, especially in low- and middle-income nations where elevated levels of gender inequality correlate with heightened violence. Initiatives to counteract gender-based violence must be integrated into a broader plan that tackles the foundational causes of gender inequality. This necessitates extensive legislative reforms and societal transformations aimed at achieving gender parity, addressing not just acute violence but also the socio-economic factors that sustain inequality.

Time use, a critical social indicator, quantifies the extent to which men and women allocate their time among paid employment, menial domestic work, and caregiving obligations. In the majority of societies, women are responsible for a disproportionate amount of unpaid labor, such as domestic chores, elder care, and child-rearing. This unequal distribution of labor frequently restricts women's capacity to pursue education and personal development opportunities or engage in paid employment. Although unpaid domestic work is indispensable to the operation of households and economies, it is seldom acknowledged or compensated, which contributes to the perpetuation of economic disparities between men and women. Women's financial independence and opportunities for career advancement are diminished by their overrepresentation in unpaid labor. Additionally, the stress levels and health outcomes of women may be adversely affected by the dual burden of juggling paid employment and neglected domestic responsibilities.

Time use surveys are valuable instruments for comprehending gender disparities in labor and determining the amount of time that men and women allocate to unpaid domestic tasks in comparison to paid employment. Countries that prioritize policies that facilitate work-life balance, including affordable childcare and parental leave, exhibit reduced gender disparities in employment and improved gender equality.

These disparities were further exacerbated by the COVID-19 pandemic, which influenced the division of domestic labor within traditional gender roles. In Germany, women engaged in a greater amount of unpaid care work during periods of homeworking, while men's involvement in domestic responsibilities was restricted (Leshchenko and Chung, 2023). In the same vein, the pandemic in India resulted in an increase in unpaid care labor for women, which reduced their rest hours and increased their overall workload (Chowdhuri et al., 2023). The quality of women's jobs is significantly impacted by the unequal burden of unpaid work, which drastically restricts their participation in the labor force. In Nigeria, women's economic productivity and overall well-being are restricted by time poverty, which is the result of unpaid domestic work (Ndimele et al., 2023). During the pandemic in Kenya, young women spent a greater amount of time on domestic responsibilities and less time on paid work, which exacerbated their long-term economic vulnerability (Williams et al., 2022).

Unpaid domestic labor has substantial mental health implications. Stress and mental health disorders are more prevalent among women who are burdened with domestic responsibilities. Research conducted in Sweden and Brazil has demonstrated that women experience elevated stress levels as a result of the imbalance between personal time and domestic duties (Carneiro et al., 2023). Furthermore, the cognitive burden associated with domestic responsibilities, including planning and organizing, exacerbates stress and adversely affects the levels of life satisfaction among women (Reich-Stiebert et al., 2023).

Policy interventions and changes in societal norms are both necessary to address gender disparities in unpaid labor. The redistribution of caregiving responsibilities is contingent upon public investment in social infrastructure, including childcare and eldercare services. In Nigeria, such investments have the potential to alleviate the unequal burden of unpaid work and enhance the economic productivity of women (Ndimele et al., 2023). Encouraging men to participate more actively in domestic labor is also essential. In Germany, there was some progress during the pandemic, as men with more egalitarian attitudes increased their contribution to caregiving (Leshchenko and Chung, 2023). This suggests that altering gender role attitudes can result in more equitable divisions of labor. Policies that promote work-life balance, such as affordable childcare and parental leave, are linked to reduced gender disparities in employment and enhanced gender equality. Positive outcomes for women's participation in the labor force have been associated with policy initiatives that prioritize work-life balance.

Although research emphasizes the existence of persistent gender disparities in unpaid domestic work, it also indicates potential avenues for change. Despite the fact that the COVID-19 pandemic has exacerbated pre-existing inequality, it has initiated discussions regarding the necessity of a more equitable distribution of domestic responsibilities. It is imperative to address both structural factors—such as access to social infrastructure—and cultural norms that perpetuate traditional gender roles.

Social indices of gender equality offer critical insights into the complex problems women encounter in obtaining services, education, healthcare, and chances for personal and professional advancement. Analyzing these indicators—namely educational attainment, health and well-being, gender-based violence, and time utilization—enables a deeper comprehension of the obstacles to gender equality and the formulation of tailored policies that foster a more fair and just society. Ultimately, attaining gender equality in these social dimensions is not merely a question of equity but also a catalyst for economic growth, social cohesion, and enhanced well-being for all societal members.

4.2.2 Economic Indicators of Gender Equality

Economic indicators are essential for comprehending gender disparities in the labor market, income allocation, and resource accessibility. These metrics elucidate the influence of gender-based inequities on women's economic prospects and financial stability, hence impacting wider societal results.

The **labor force participation rate**, a key economic statistic, assesses the ratio of women engaged in paid employment relative to men. A pronounced gender disparity in labor force participation frequently indicates societal norms that impose an unequal share of unpaid domestic labor and caring responsibilities on women. The disproportionate allocation of responsibilities restricts women's capacity to engage fully in the formal economy, as numerous women are obligated to prioritize domestic obligations over paid work. In numerous nations, the idea that women should oversee domestic responsibilities and caring obligations limits their access to full-time or lucrative employment, resulting in economic dependency and diminished career progression chances. The gender gap in labor force participation is a significant indicator of economic inequality, as it sustains financial inequities between men and women by constraining women's prospects for professional advancement and economic autonomy.

Conventional gender norms persist in asserting that women must prioritize domestic responsibilities over employment, hence perpetuating the gender gap in labor force participation. The allocation of housework varies by region, with women typically dedicating more time to domestic tasks and childcare than men. This phenomenon is not exclusive to poor nations; even in more advanced areas such as the United States, women's contributions to domestic labor significantly surpass those of men (Quadlin, 2022). The economic ramifications of unpaid labor are significant yet frequently go unacknowledged. The value of unpaid domestic labor in China is substantial yet goes unrecognized in conventional economic metrics, despite women's major contributions to both the home and the wider economy (Li, 2022). In India, women's unpaid labor in family companies is frequently disregarded, resulting in an inaccurate assessment of their total economic contributions (Deshpande et al., 2021). These unrecorded contributions, while essential, perpetuate gender inequity by obscuring the actual magnitude of women's economic involvement.

The obligation of unpaid domestic labor directly affects workforce participation. In Indonesia, the underrepresentation of women in significant economic positions is intricately associated with societal norms that compel them to prioritize household duties over career advancement (Rohmatilah, 2023). In Poland, gender norms affect the allocation of household chores; nevertheless, women in less conventional homes are more inclined to equally distribute domestic responsibilities with their partners, indicating that mitigating gender wage disparities could promote equality in both domestic and professional spheres (Magda et al., 2024). These discrepancies illustrate how unpaid labor impedes women's capacity to engage fully in the formal economy, constraining their professional prospects and total financial autonomy.

Policy and structural constraints significantly influence women's labor market outcomes. Despite reduced salary and participation discrepancies, the Nordic nations continue to face challenges related to gender segregation in employment sectors (Kowalewska, 2023). In Australia, the equilibrium between familial and professional obligations is essential, especially for women in dual-income households. Gender imbalance in professional and familial responsibilities can result in financial precariousness, particularly in single-parent households, where the responsibility of managing work and caring is disproportionately shouldered by women. These structural obstacles demonstrate that, even in nations with comparatively progressive gender policies, women persistently encounter impediments that obstruct their economic equality.

While increasing women's labor force participation is crucial for advancing gender equity, it does not provide a complete answer for achieving true gender equality. In numerous contexts, women's labor force participation has diminished despite economic progress, partly due to persistent patriarchal practices and gendered conceptions of labor. Challenging these entrenched societal norms is essential for creating a more fair environment for women, both in the workplace and in the home. Comprehensive policies that facilitate work-life balance, such as paid parental leave and affordable daycare, are crucial for enabling women to progress in their careers while fulfilling domestic responsibilities. Addressing the gender discrepancy in labor force participation necessitates tackling both institutional and cultural factors that sustain inequality. Strategies to reduce the labor force participation gap must encompass policies that promote work-life balance, public investments in social infrastructure, and transformations in societal norms. Societies may attain authentic gender equality solely by addressing these intricate difficulties, allowing women to fully participate in both the workforce and familial obligations without encountering undue limitations.

The **gender wage gap** is a significant economic metric, illustrating the ongoing inequality in compensation between men and women for comparable roles. This income disparity underscores the systemic disadvantages that penalize women, even when they occupy equivalent positions to men. Notwithstanding legislative initiatives and regulatory measures, the gender wage gap remains a substantial obstacle to women's economic empowerment and equality. Occupational segregation significantly contributes to this disparity. Women frequently experience underrepresentation in lucrative industries such as technology and finance, while they are disproportionately represented in lower-paying fields such as caregiving, hospitality, and education. These areas provide not just reduced salaries but also limited prospects for career advancement, hence exacerbating the gender wage disparity. In these lower-wage sectors, women often encounter restricted promotion prospects and are disproportionately impacted by income inequality.

The gender wage gap is significantly affected by societal conventions and assumptions that sustain conventional gender roles. These norms frequently divert women from lucrative jobs and towards positions that are economically devalued. Stereotypes that designate women as caregivers or inadequately equipped for leadership roles persistently restrict their advancement in high-income professions (Salman, 2023). Societal expectations and biases establish a framework in which women encounter structural obstacles that hinder their attainment of salary equality with male colleagues. The gender wage gap significantly impacts women's economic empowerment, constraining their financial independence and capacity to amass money. This disparity constrains women's autonomy, diminishing their ability to invest in their future and jeopardizing their long-term economic stability. Furthermore, the financial disparities resulting from the pay gap exacerbate elevated incidences of melancholy and anxiety among women, especially those who earn considerably less than their male counterparts. The psychological repercussions underscore the wider societal implications of economic inequality among genders. Women continue to encounter substantial obstacles, such as the "glass ceiling," that impede their access to senior roles, even within identical businesses. The "glass ceiling" sustains wage inequality by obstructing women's progression to leadership positions that would enable them to attain competitive pay. Moreover, women frequently encounter inferior working circumstances relative to men, exacerbating the wage disparity (Antón et al., 2023). This encompasses inconsistencies in employment quality, the physical work environment, and working hours, all of which might further constrain women's economic potential.

Various governmental interventions and remedies have been proposed to mitigate the gender wage gap. Legislative initiatives like the Equal Pay Act and the Paycheck Fairness Act seek to eliminate wage discrepancies by enforcing equal remuneration for equivalent labor. The efficacy of these regulations is frequently constrained by enforcement difficulties and entrenched cultural views that oppose change. Alongside legislative initiatives, enhancing women's access to education and training, especially in lucrative sectors such as science, technology, engineering, and mathematics (STEM), is essential for reducing the wage disparity. Mitigating educational gaps, particularly in developing nations, is crucial for enabling women to contend for lucrative employment opportunities. Corporate activities significantly contribute to mitigating the pay disparity. Organizations that use clear compensation structures, adaptable work arrangements, and fair recruitment processes aid in reducing the wage disparity and fostering women's professional growth. Promoting such practices can foster an environment in which women have equitable opportunity to thrive and attain competitive salaries.

The gender wage gap is a multifaceted issue shaped by numerous structural and sociological factors; however, progress has been observed in certain domains. Evidence indicates that employment quality for women has enhanced non-monetary dimensions throughout the European Union, signifying a wider catch-up trend for women in the labor market (Antón et al., 2023). Moreover, research indicates that when accounting for variables like as occupation and work schedules, the income disparity can be markedly diminished, while it remains prevalent in numerous scenarios. Attaining genuine pay parity necessitates persistent endeavors to confront both structural impediments—such as occupational segregation and inequitable access to leadership positions—and cultural elements that perpetuate gender-based economic inequality.

The **sectoral distribution of employment** offers essential insight into gender discrepancies in the labor market, highlighting entrenched inequality that endures across many industries. Women are disproportionately represented in traditionally undervalued industries, like childcare, health-care assistance, and education, which are linked to nurturing responsibilities according to established societal standards. These sectors, although vital to societal functioning, sometimes yield lower profitability and have diminished prospects for financial autonomy and career advancement. Conversely, men predominate in lucrative sectors such as information technology, engineering, and finance (Kowalewska, 2023). This gender segregation restricts women's earning potential and perpetuates preconceptions that influence professional choices from a young age, perpetuating economic inequity.

Gender segregation in employment is a prevalent issue evident in most labor markets. In the UK, the overrepresentation of women in lower-paid sectors, such as caregiving and education, results in lower average salaries relative to male-dominated areas like finance and engineering (Leoncini et al., 2024). In 24 developed nations, including those with advanced social systems, women still encounter considerable employment segregation, indicating that structural challenges endure irrespective of a country's welfare framework (Kowalewska, 2023). In Italy, disparities in employment and salaries persist across genders, even when both men and women select identical fields of study. Although sectors with a greater percentage of women, such as education and caregiving, demonstrate advancements in narrowing gender disparities, these industries continue to provide lower absolute incomes relative to male-dominated sectors (Galos and Kulic, 2023).

Gender segregation influences not only earnings but also women's job advancement. The COVID-19 pandemic intensified these discrepancies, since women—especially those with caring duties—were disproportionately impacted due to their overrepresentation in the sectors most severely affected by the pandemic. This has resulted in enduring repercussions for women's employment rates and work hours, constraining their prospects for professional progression (Wielgoszewska et al., 2023). In the United States, despite educational progress, women in male-dominated sectors such as STEM see diminishing pay advantages, although more gender parity is noted in industries traditionally linked to women, such as education (Sharma, 2022). In Poland, women constantly have lower employment rates and elevated unemployment risks compared to men, notably affecting young women and those lacking tertiary education (Kucharski and Rutkowska, 2021).

Historical gender prejudices and societal conventions significantly influence employment trends, relegating women to subordinate roles in the workplace. These institutional impediments restrict women's progression chances and perpetuate gender disparity in the work market. Despite the growing involvement of women in the economy, conventional gender norms, along with the challenge of reconciling work and familial obligations, persist in obstructing women's career advancement and economic autonomy.

Confronting gender segregation in the workforce necessitates a comprehensive strategy that addresses the intricate interactions of societal norms, educational decisions, and systemic prejudices. Extensive legislative measures are required to establish a more equitable labor market in which all genders possess equal prospects for professional progression and financial autonomy. These policies must aim to dismantle prejudices that restrict women to lower-paying, undervalued positions, while simultaneously fostering equitable access to lucrative industries typically dominated by males. Societies can only achieve significant progress towards gender equality in the job market by tackling these systemic constraints.

Access to financial resources is essential for women's economic empowerment, profoundly influencing their financial autonomy and decision-making authority. Metrics such as women's access to credit, possession of bank accounts, and authority over property are essential in assessing their equitable engagement in the economy. Nevertheless, in specific areas, women continue to encounter legal and institutional obstacles that restrict their capacity to obtain loans or possess property. These obstacles can impede their capacity to initiate enterprises, invest in education, or accumulate assets. Restricted access to financial services can intensify gender inequality by hindering women's capacity to make autonomous economic decisions or address financial difficulties. Despite significant advancements globally, women in many regions still face obstacles that restrict their complete economic involvement.

Financial literacy is an essential element in economically empowering women. It bolsters their confidence, decision-making capabilities, and money management proficiency, which are vital for business success. Enhancing women's financial literacy can markedly improve their access to resources and networks, underscoring the necessity for governmental measures that facilitate greater accessibility to such education. Financial literacy is particularly crucial for women employed in the unorganized sector to facilitate informed decision-making and effective resource management. Investment initiatives aimed at women can augment their financial autonomy, hence fostering greater economic empowerment. Digital financial inclusion has gained significance, especially following the COVID-19 epidemic. Digital financial services enhance women's access to financial products, hence augmenting their economic roles and decision-making capacities.

Initiatives designed to enhance women's financial access are crucial for advancing economic equality. Microfinance programs, community savings and loans, and governmental initiatives are crucial in eliminating patriarchal structures and promoting women's financial autonomy. In India, financial inclusion has positively influenced women-owned firms, illustrating that policies aimed at diminishing gender inequities in financial intermediation can improve women's entrepreneurial success. The experiences of BRICS nations underscore that financial inclusion is a crucial catalyst for human development and women's empowerment, emphasizing the necessity of inclusive financial policies to promote economic growth and equality (Sahu et al., 2024). Although financial inclusion is essential for empowering women, it is imperative to tackle broader societal concerns, including gender-based discrimination and cultural biases, to guarantee women's complete economic involvement. Initiatives aimed at enhancing women's financial access must be accompanied with measures that foster gender equality and uphold women's rights. It is essential to cultivate collaborations across governmental, non-governmental, and community groups to establish a more equitable society that promotes both women's economic empowerment and their social and political participation. These many strategies are essential for attaining enduring gender equality and fostering an atmosphere in which women may flourish in all facets of life.

The proportion of **women who possess property and have access to entrepreneurial possibilities** is a vital indicator of gender equality in economic involvement. Property ownership offers financial stability and substantially increases women's negotiating power within households and communities, hence fostering their overall empowerment. Women possessing property or other assets are more capable of making decisions regarding their livelihoods, education, and family planning, so promoting more economic independence. Cultural customs, discriminatory legislation, and societal conventions frequently impede women's capacity to obtain or manage property, hence limiting their opportunities for economic advancement. In numerous nations, inheritance laws and norms preferentially benefit male heirs, resulting in restricted access to assets for women, notwithstanding their role as primary caregivers within their families. This systemic inequity sustains gender inequities and obstructs women's economic development.

Cultural and legal impediments to property ownership are widespread in numerous places. In South Africa, notwithstanding post-apartheid regulations designed to mitigate inequality, traditional roles and gender stereotypes persist in limiting women's access to land. Men frequently receive preferential treatment in land distribution owing to traditional norms that anticipate women would marry and reside with their husband's family (Masenya, 2024). In Northern Ghana, the customary land tenure system disadvantages women, as patriarchal interactions favor men in property ownership. Women in these areas frequently obtain inferior quality land and encounter substantial financial obligations, such as exorbitant royalties, which hinder their capacity to participate in productive economic endeavors and exacerbate the poverty divide (Baba Alhassan et al., 2024). The economic ramifications of asset ownership are substantial. Research from 18 developing nations demonstrates that women with asset ownership are more inclined to engage in household decision-making concerning health, expenditures, and social contacts, underscoring the empowering influence of property ownership (Amir-ud-Din et al., 2023). In India, women have a declining growth rate in land holdings, which constrains their capabilities as farmers and intensifies gender inequality in the agricultural sector. Despite a favorable trend in rising earnings for female laborers, wage inequities in agriculture continue to be a significant obstacle to women's economic empowerment (Singh et al., 2023). Socio-economic variables significantly contribute to the perpetuation of gender imbalance in property ownership. In Nigeria, religious beliefs and economic limitations hinder women's property ownership, necessitating legal reforms to enhance gender inclusion and economic empowerment (Popoola et al., 2023). In homes, wealth inequality frequently leads to women possessing lower income and asset levels compared to males. This disparity influences couples' financial management, resulting in separate money handling due to unequal wealth, hence perpetuating financial and gender inequality (Althaber et al., 2023).

Strategies to augment women's economic empowerment must concentrate on mitigating these systemic impediments. Advancing gender equality in economic involvement can substantially enhance family welfare and alleviate poverty. Empowerment strategies encompass enhancing women's access to education, property rights, and autonomous income options. Financial access is essential for advancing gender inclusion, since nations with more inclusive financial systems generally have elevated female labor participation rates. This association is more pronounced in middle-income nations with stable legal and theological frameworks that promote gender equality (Asongu and Nting, 2021).

Notwithstanding advancements in acknowledging the significance of women's property rights and economic involvement, considerable obstacles persist. Legal reforms, cultural transformations, and focused empowerment initiatives are crucial for improving women's access to resources and economic prospects. The enduring presence of cultural norms and economic obstacles underscores the intricacy of attaining gender equality in this area. Resolving these difficulties necessitates a comprehensive strategy that considers the cultural, legal, and economic aspects of gender inequality. Meaningful progress in improving women's economic empowerment and closing the gender gap in property ownership and entrepreneurship can only be achieved by addressing these fundamental concerns.

Entrepreneurial opportunities are crucial for women's economic progress, but gender-based obstacles frequently restrict their access to these chances. Female entrepreneurs encounter various obstacles, including restricted access to initial funding, marginalization from business networks, and societal norms that emphasize their domestic responsibilities over professional ambitions. These impediments substantially obstruct women's capacity to establish and expand firms, resulting in their underrepresentation in the entrepreneurial domain. Mitigating these obstacles through specific regulations, financial inclusion initiatives, and mentorship is essential for promoting economic equality and empowering women as entrepreneurs.

The economic influence of women's entrepreneurship is significant, especially in emerging nations, where it acts as a crucial catalyst for economic growth and wealth. Female entrepreneurs facilitate a more equitable allocation of wealth and enhance the optimization of human resources. In Nigeria, women-owned enterprises constitute a substantial segment of the economy, offering stability and autonomy to several women. Nonetheless, these organizations are frequently modest with restricted expansion potential, underscoring the want for supportive initiatives to assist women in expanding their businesses and enhancing their economic contributions (Gulvira et al., 2024). Notwithstanding the economic advantages of women's business, gender-specific obstacles persist in hindering advancement. Female entrepreneurs encounter financial limitations, societal and cultural obstacles, and challenges in market access. In Serbia, female entrepreneurs face prejudice and inequality, despite existing legislation and institutions designed to promote them (Stanisavljević, 2023). In India, societal norms that condition women for household responsibilities instead of entrepreneurial endeavors constrain their economic participation, hindering their capacity to create and expand firms (Sharma, 2023).

To overcome these obstacles, many measures can be implemented to assist women entrepreneurs. Targeted financial initiatives, mentorship programs, and campaigns designed to alter adverse stereotypes of women in business are essential for surmounting challenges. In Indonesia, female entrepreneurs have effectively utilized networking and social capital to overcome hurdles, highlighting the significance of mentorship and support networks in promoting entrepreneurial success (Hayati and Arini, 2023). Attaining gender parity in entrepreneurship can substantially enhance creativity and economic development. This necessitates confronting gender biases and enhancing women's access to markets, technology, and financial resources. In Visakhapatnam, India, governmental programs and promotions have been crucial in fostering women entrepreneurs, highlighting the significance of policy assistance in enabling women's participation in the entrepreneurial domain (Veckalne and Tambovceva, 2023). Moreover, the experiences of industrialized nations, which provide women with specific incentives and assistance for entrepreneurship, present significant insights for developing countries aiming to cultivate a more favorable climate for female entrepreneurs (Stanisavljević, 2023).

Promoting women's engagement in entrepreneurship as a legitimate professional path and equipping them with essential skills and financial resources can substantially enhance their economic involvement. In India, initiatives aimed at fostering women's entrepreneurship through education and financial inclusion are dismantling entrenched barriers, enabling a greater number of women to participate in economic activities and achieve their professional aspirations (Sharma, 2023). The obstacles encountered by women entrepreneurs are significant; however, the prospects for economic advancement and gender equality via entrepreneurship are considerable. Overcoming these obstacles necessitates a multifaceted strategy encompassing policy reforms, financial assistance, and cultural transformation. By cultivating an atmosphere that promotes women's advancement and development as entrepreneurs, societies can progress towards enhanced economic equality and empowerment for women worldwide.

Economic indicators, including labor force participation, the gender wage gap, access to financial resources, and property ownership, are crucial for comprehending the intricacies of gender inequality in the economic domain. These metrics demonstrate how societal norms, structural obstacles, and discriminatory behaviors restrict women's economic possibilities and sustain financial inequality. Resolving these difficulties necessitates comprehensive policies that foster gender equality in education, employment, entrepreneurship, and financial inclusion, guaranteeing that women possess equal opportunity to prosper in the economy.

4.2.3 Political and Legal Indicators of Gender Equality

Political and legal indicators are essential for evaluating the degree of women's representation and involvement in decision-making processes, as well as determining the efficacy of legislative frameworks in advancing gender equality.

Political representation serves as a vital indicator of gender equality, emphasizing the proportion of women in parliaments, municipal governments, and leadership positions within public organizations. The augmented participation of women in politics is not merely a question of equity; it also fosters more inclusive and balanced policies. Women in positions of political authority often prioritize issues that disproportionately impact them, including childcare, reproductive rights, healthcare accessibility, and violence against women. Research consistently demonstrates that female lawmakers are more inclined to support legislation that advances social welfare and gender equality, resulting in more equitable societal results. The appearance of women in political leadership acts as a potent symbol, contesting conventional gender roles and motivating future generations to participate in political leadership. Nonetheless, structural and cultural impediments persist in restricting women's comprehensive involvement in politics, necessitating focused measures to eradicate these barriers and augment women's political agency.

In numerous global countries, women's political representation is severely lacking due to a confluence of structural, cultural, religious, and economic issues. This underrepresentation is frequently caused by ingrained patriarchal attitudes, political nepotism, and financial obstacles that impede women's full participation in politics. Initiatives are underway to deliver political education, enhance awareness, and create support networks for women seeking political positions. Young women encounter supplementary obstacles that further restrict their political engagement, highlighting the necessity for political parties and institutions to proactively endorse and facilitate their participation. The influence of female political representation is beyond mere symbolic importance. Empirical research associates increased female involvement in politics with enhanced policy results, especially in health and social welfare sectors. In Europe, elevated female political representation is associated with reduced geographical disparities in newborn mortality and improved health outcomes for all genders (Reeves et al., 2022). In Bavaria, Germany, the involvement of female councilors expedited the growth of public childcare by 40%, illustrating the direct impact women can exert on policies that advantage families and children (Baskaran and Hessami, 2023). In India, female leaders have enhanced gender-specific outcomes via legislative reforms and improved public goods provision; however, this progress has occasionally resulted in unintended consequences, including a rise in reports of intimate partner violence as traditional gender dynamics evolve (Anukriti et al., 2022).

Beyond their concrete policy effects, women's political representation serves a symbolic and inspirational function in countries where women have traditionally been marginalized. Women in political leadership roles confront gender stereotypes and enhance the societal status of women, facilitating the implementation of legislation that safeguards women's rights and advances gender equality. This symbolic depiction is especially significant in areas where conventional gender roles are firmly established and women's involvement in public life has been restricted. Research indicates that greater female political representation can boost the delivery of public goods in developing nations and improve institutional quality by mitigating corruption. This highlights the extensive socioeconomic advantages of gender-balanced political leadership and the necessity of fostering women's involvement in decision-making at all levels.

Although the beneficial effects of women's political representation are clear, substantial obstacles persist in attaining genuine gender equality in political engagement. Structural and cultural impediments, including patriarchal attitudes and unequal access to political networks, persist in restricting women's access to political power. Furthermore, the impact of enhanced representation may differ based on the context, with certain areas encountering sluggish advancement due to ingrained societal standards. Nevertheless, the advancements achieved in numerous nations highlight the necessity of ongoing initiatives to assist and empower women in political positions, both to advance gender equality and to cultivate more inclusive and effective administration.

Legal rights are essential for achieving gender equality, since they establish the basis for frameworks that safeguard and advance women's rights in areas such as education, employment, and property ownership. Robust legal frameworks not only protect women from discrimination and assault but also enable them to exercise their rights and engage freely in both public and private spheres. Strong legal safeguards are essential for guaranteeing that women possess equal chances in society, allowing them to contest discriminatory acts and bridge gender disparities. Countries possessing robust legal frameworks, exemplified by those in the European Union, generally exhibit elevated levels of gender equality, as these statutes furnish essential legal support for women to assert their rights (Protosavitska, 2023).

Legal reforms are crucial for abolishing discriminatory practices, including inequitable inheritance rules and limitations on women's access to specific professions. These reforms facilitate the rebalancing of both economic participation and political representation, so that women may contribute equitably to their societies. The European Union has established legal frameworks to guarantee gender equality in employment and access to goods and services, illustrating the beneficial effects of these changes on narrowing the gender gap (Kaganovska and Syroid, 2023). Rectifying legal disparities in domains such as property ownership and electoral representation is crucial for fostering a more inclusive and equitable society.

Notwithstanding the presence of legislation safeguards, the execution of these frameworks encounters obstacles in numerous nations owing to socio-cultural impediments. Patriarchal influences and negative stigmas towards women frequently inform the viewpoints of law enforcement officials, resulting in discriminatory practices and inequitable treatment. These impediments obstruct the efficient implementation of legislative protections and restrict women's comprehensive access to their rights. In numerous areas, women encounter considerable obstacles in securing justice and legal support, frequently attributable to insufficient gender-sensitive training within law enforcement and inadequate lobbying from non-governmental groups. Workplace discrimination endures, with female employees often relegated to subordinate positions relative to their male colleagues, underscoring the necessity for more robust implementation of gender equality laws.

International and regional organizations are essential in advancing gender equality by establishing standards and offering frameworks for member governments to adhere to. The European Union has established strategic pledges and action plans to promote gender equality, encompassing provisions for maternity protection and parental leave. These international frameworks provide a paradigm for national governments to implement, enhancing accountability and advancement in gender equality initiatives (Kaganovska and Syroid, 2023). The United Nations' Sustainable Development Goal (SDG) five aims to attain gender equality by eradicating discrimination against women and girls, highlighting the necessity of international collaboration in tackling global gender inequality.

Although legal frameworks are essential for advancing gender equality, they are insufficient on their own. The efficacy of these frameworks is significantly influenced by the socio-cultural setting and the dedication of governments and societies to adopt and enforce them. Legal improvements must be complemented by comprehensive initiatives to alter societal attitudes and practices that sustain gender disparity. Moreover, global collaboration and the proactive engagement of civil society are essential in tackling the fundamental causes of gender inequality and ensuring that legislative safeguards result in meaningful enhancements in the lives of women and girls. In the absence of these collaborative endeavors, legal reforms may be only symbolic, lacking the capacity to effectuate the substantial social and economic transformations requisite for genuine gender equality.

Access to justice is an essential element of legal equality, enabling women to pursue remedies when their rights are infringed upon, especially in instances of gender-based violence, including domestic violence, sexual harassment, and human trafficking. Although legal frameworks are established to safeguard women, the efficient execution of these laws is essential to guarantee both the protection of women and the accountability of perpetrators. The availability of legal aid and accessible court processes is essential for enabling women, particularly from underprivileged areas, to seek justice. In the absence of these procedures, legal safeguards are rendered ineffective, as numerous women lack the finances, knowledge, or support necessary to traverse intricate legal systems. Consequently, access to justice is a fundamental component in safeguarding women's rights and preserving their security.

The legal system designed to safeguard women from gender-based violence frequently falls short, as socio-cultural elements, including patriarchal norms and adverse stigmas, affect law enforcement, resulting in discrimination and injustice for victims. Women frequently encounter substantial obstacles in legal proceedings, exacerbated by the insufficient gender sensitivity of law enforcement personnel. Effective law enforcement necessitates extensive reforms, encompassing gender-sensitive training for law enforcement personnel and the establishment of victim-centric legal procedures.

Ensuring justice accessibility transcends the mere existence of laws; it necessitates an egalitarian, efficient judicial system capable of producing equitable outcomes. Accessibility and affordability are essential to this process, as marginalized communities frequently encounter difficulties in obtaining legal assistance due to exorbitant charges or intricate bureaucratic procedures. Legal aid is crucial for assisting women from underprivileged backgrounds in navigating the court system; nonetheless, its availability is insufficient in numerous areas. Administrative obstacles and inadequate assistance from legal aid organizations intensify the difficulty women encounter in obtaining justice. Also socio-economic and cultural hurdles are substantial impediments that hinder women, especially those from marginalized groups, including ethnic minorities and individuals in poverty, from obtaining justice. Women in these communities frequently lack knowledge, information, and monetary resources, rendering them susceptible to exclusion from the judicial system. In numerous low- and middle-income nations, initiatives designed to enhance access to justice must confront these socio-economic and cultural obstacles to guarantee that judicial systems are inclusive and egalitarian. Addressing these overarching challenges is essential; otherwise, legal reforms will probably fail to effectively promote justice and gender equality.

Legal institutions and civil society organizations are essential in enhancing access to justice by increasing legal information, empowering victims, and offering vital support. Non-judicial dispute resolution procedures and legal assistance programs are vital for guaranteeing equitable access to justice for individuals of all socio-economic backgrounds. Efforts by civil society are crucial in fostering trust in the legal system, particularly among vulnerable populations who may view the system as biased or unattainable. Enhancing this trust is essential for advancing socio-economic integration and augmenting the overall efficacy of the justice system.

Although legal frameworks and enforcement procedures are crucial for safeguarding women's rights, wider socio-economic and cultural variables substantially influence women's access to justice. Overcoming these obstacles necessitates a comprehensive strategy encompassing legislation reform, enhancement of law enforcement capabilities, and civil society backing. Moreover, enhancing the accessibility and efficacy of legal aid is essential for empowering women to pursue justice and hold offenders accountable. In the absence of these comprehensive efforts, legal protections are inadequate, and numerous women will persist in encountering significant obstacles in obtaining justice, perpetuating gender inequities.

International legal obligations, such as the Convention on the Elimination of All Forms of Discrimination Against Women (CEDAW), are essential for promoting worldwide gender equality. These frameworks necessitate that ratifying countries harmonize their national legislation with international standards, thereby advancing women's rights and improving accountability. These international agreements establish a systematic framework for tracking progress, compelling governments to uphold their gender equality pledges, so guaranteeing they create legislation and implement essential reforms to safeguard and empower women. The ramifications of these legal obligations permeate all spheres, affecting the incorporation of gender equality within comprehensive human rights frameworks and fostering worldwide democratic advancement.

International legal commitments such as CEDAW significantly influence the mobilization of legal activism and accountability. Non-governmental organizations (NGOs) and human rights advocates utilize the endorsement of such treaties as a mechanism to ensure state accountability, frequently prior to formal ratification. This mobilization can result in significant advancements in women's rights, as seen by the favorable correlation between CEDAW and improved women's rights in several nations (Comstock, 2024). The advancement of international legal norms for women's rights has not only solidified these rights but has also promoted an inclusive framework that establishes equality for both men and women. The progression of these international agreements demonstrates an increasing dedication to gender equality, striving for fairer treatment in social, economic, cultural, and political spheres. International legal frameworks are essential for the systematic implementation of gender equality, mandating that nations include these concepts in their constitutions and legal systems. This procedure has resulted in more comprehensive regulations on women's rights and has aided in combating systematic discrimination. Nonetheless, obstacles remain in attaining the complete achievement of gender equality, notwithstanding these pledges.

Although international standards have significantly promoted gender equality, the advancement towards complete equality can be gradual, particularly in environments influenced by external variables like socio-political turmoil or pandemics. Notwithstanding the plethora of worldwide guidelines and frameworks, the practical execution of these standards frequently proceeds at a languid pace, as nations grapple with fulfilling their commitments. The sluggish advancement highlights the necessity for ongoing lobbying and more robust measures to guarantee that gender equality promises are converted into tangible changes. The wider ramifications of these international legal obligations pertain to their incorporation within essential human rights frameworks. Gender equality is fundamentally connected to global human rights, as evidenced by treaties such as the Universal Declaration of Human Rights, which acknowledges gender equality as a core element of human dignity and democratic advancement. This integration highlights the significance of gender equality as a fundamental aspect of broader human rights initiatives. International law has evolved from gender-neutral methods to more gender-sensitive frameworks, demonstrating a refined comprehension of the distinct issues encountered by women. This transition is seen in numerous international and European legislative frameworks that currently offer enhanced safeguards against gender-based discrimination.

Although frameworks such as CEDAW have significantly contributed to the promotion of global gender equality, the path to realizing these objectives continues to be intricate. Continued discrimination and the sluggish advancement of legal and societal reforms underscore the necessity for vigorous enforcement of international standards. Ongoing lobbying and the enhancement of gender-sensitive policies are crucial for maintaining progress in this domain. The incorporation of gender equality into comprehensive human rights frameworks underscores its essential function in fostering democratic and egalitarian communities. The promotion of gender equality must remain a primary objective of international legislative and political initiatives to guarantee ongoing progress and the complete realization of women's rights globally.

Political and legal indicators are mutually dependent, mirroring the comprehensive context of women's empowerment. Political representation empowers women to impact decision-making processes that define the legal structures regulating their lives, while strong legal rights and access to justice guarantee their full societal participation. Collectively, these indicators offer an extensive insight into the advancements achieved in gender equality and the persistent obstacles that exist.

4.2.4 Cultural and Normative Indicators of Gender Equality

Cultural and normative variables are essential for comprehending the cultural values and attitudes that influence gender roles and expectations. These indicators demonstrate the profound impact of entrenched gender norms on daily life, encompassing familial relationships, occupational responsibilities, and overarching social frameworks. Surveys of popular opinions of gender roles offer significant insights into societal expectations and assist in identifying enduring preconceptions that obstruct progress towards gender equality. These prejudices, grounded in conventional gender views, influence various aspects like profession selections, educational prospects, and leadership positions, so perpetuating societal inequality. By emphasizing these cultural and normative obstacles, these indicators direct initiatives to transform attitudes towards more equitable gender relations, facilitating the development of more inclusive social behaviors.

A fundamental element of cultural and normative indicators is media portrayal, which significantly influences and mirrors social opinions towards gender. The representation of women in media, including advertising, television, film, and news, profoundly influences societal conceptions of gender roles. In Indian advertising, women are frequently portrayed in conventional roles such as homemakers and caregivers, hence perpetuating antiquated gender norms and constraining perceptions of women's potential (Shivam Gaur and Ananta Sarkar, 2024). Research on media portrayals continually underscores the dominance of gender stereotypes, objectification, and sexualization, which sustain detrimental gender norms and promote sexism in society (Santoniccolo et al., 2023). Educational curriculum frequently mirrors these cultural and normative factors, often perpetuating conventional gender norms. English textbooks in China often portray women as accountable for domestic responsibilities, so perpetuating gender hierarchies that prioritize masculine roles over feminine ones (Huang and Liu, 2024). In Colombia, educational institutions frequently endorse binary gender representations that fail to acknowledge sexual variation, so maintaining conventional beliefs and constraining the evolution of more egalitarian viewpoints (Ramírez et al., 2023). These instructional portrayals reinforce gender stereotypes, complicating efforts to alter society ideas regarding gender equality. The media's impact transcends mere representations. It can either promote conventional gender norms or contest them, providing opportunity for cultural transformation. The media can challenge cultural attitudes by subverting traditional gender norms and bringing attention to issues like gender-based violence (Shivam Gaur and Ananta Sarkar, 2024). Nevertheless, when women are persistently portrayed in conventional positions or are inadequately represented in critical areas such as leadership or professional environments, it perpetuates detrimental norms that restrict women's involvement in many fields. Such depictions foster a limited comprehension of gender roles, hindering women's ability to transcend cultural expectations that restrict them to specific vocations or duties.

The economic and social consequences of these cultural norms are significant. The substantial economic value of unpaid domestic labor, primarily performed by women, emphasizes the inequality in family duties and reinforces the necessity for policies that tackle gender disparity in both domestic environments and the labor market (Li, 2022). Favorable attitudes towards women and progressive value orientations correlate with more egalitarian gender perceptions, underscoring the necessity of fostering values like self-direction and universalism to transform societal attitudes. Societal perceptions of gender roles are influenced by factors including class, education, and cultural standards. A study conducted in eight nations revealed that women received increased acceptability for seeking higher education and income, whereas men had diminished support for lower income and occupational status, particularly in conservative society (Weißflog et al., 2023). In Europe, educational and professional environments are linked to more egalitarian perspectives, whereas conventional family structures and elevated religious observance often reinforce patriarchal gender roles. Such views establish obstacles for women in professional and leadership roles, perpetuating conventional standards that restrict their progression chances.

Confronting the matter of media representation is essential in contesting these cultural conventions and fostering a more equitable depiction of gender roles. The media can facilitate social change by portraying diverse and equitable depictions of women and men, prompting society to reevaluate antiquated norms. Enhancing the visibility of women in non-traditional roles—such as leaders, innovators, and decision-makers—facilitates the dismantling of cultural obstacles that restrict women's prospects. When media presents a wider array of female experiences and competencies, it can motivate young girls to aspire to objectives in areas where female role models are frequently lacking, while simultaneously contesting cultural standards that depict males as fundamentally superior in authority and leadership positions.

Cultural and normative variables offer essential insights into societal attitudes and stereotypes that sustain gender inequity. Comprehending these cultural characteristics allows legislators, activists, educators, and media professionals to formulate methods that confront detrimental attitudes and advance gender equality. By confronting preconceptions and advocating for diverse and honest portrayals of all genders, cultures may dismantle the barriers that perpetuate gender imbalance and progress towards a more equitable and inclusive future.

4.2.5 Environmental and Sustainability Indicators of Gender Equality

Gender equality is essential for environmental sustainability, especially regarding climate change and resource management. The interplay between gender and the environment is intricate, grounded in societal norms, economic frameworks, and imbalanced power dynamics. This relationship is prominently evident in women's roles in managing natural resources and their increased susceptibility to environmental deterioration and climate change. Women, despite their essential contributions, are frequently marginalized in decision-making processes within environmental governance, hence perpetuating gender disparities and compromising the sustainability of resource management techniques.

In numerous regions globally, access to clean water and sanitation is a gender-specific concern. Women and girls frequently assume the principal obligation of collecting water, which adversely impacts their time and prospects for school, employment, and personal growth. In areas with limited water resources or inadequate clean water infrastructure, women may dedicate several hours daily to gathering water for their households. This obligation perpetuates conventional gender roles and constrains women's economic and social empowerment by denying them the time necessary for education, income-generating endeavors, or political engagement. The absence of suitable sanitation facilities exacerbates these issues, especially for adolescent girls, who may be compelled to miss school during menstruation due to inadequate provisions. These gendered obstacles underscore the necessity of tackling environmental issues via a gender-sensitive lens to foster both equity and sustainability.

In addition to these daily responsibilities, gender equality is closely connected to overarching concerns of environmental governance and resource management. Women, particularly in rural and indigenous communities, frequently act as the principal custodians of natural resources, overseeing water, woods, and agricultural fields. Their profound understanding of local ecosystems designates them as essential participants in environmental conservation and sustainability initiatives. In Zimbabwe, women have been instrumental in recognizing environmental threats and alleviating crises through indigenous knowledge systems, highlighting the significance of their efforts (Mukurazhizha and Matanga, 2023). Nevertheless, despite their critical importance, women are typically marginalized in leadership roles and excluded from decision-making bodies, which are predominantly male-dominated. This exclusion results in policies that neglect women's demands and the sustainable management of resources, hence reinforcing gender disparities and obstructing effective environmental governance.

The exclusion of women from environmental decision-making is influenced by various issues, including cultural, social, and institutional obstacles. In the Niger Delta, Nigeria, women experience heightened vulnerability owing to environmental degradation; however, they are frequently marginalized in decision-making processes due to cultural beliefs and a lack of faith in the system (Sam and Zibima, 2023). Gendered views and social conventions in Africa marginalize women from substantive involvement in sustainable forest management, confining their participation to public meetings rather than recognizing them as stakeholders or experts (Duguma et al., 2022). These instances exemplify the pervasive exclusion women encounter in environmental governance, which not only compromises gender equality but also diminishes the efficacy of environmental legislation.

Institutional and structural disparities intensify these issues. Women in conservation professions frequently encounter structural obstacles, such as gender bias, discrimination, and inadequate representation in leadership positions. Within organizations such as The Nature Conservancy, women encounter greater challenges than men in shaping conservation decisions, underscoring the necessity for structural reform in these institutions (James et al., 2021). The deficiency of resources and committed initiatives to overcome these obstacles also restricts women's participation in conservation and natural resource management. Notwithstanding these limitations, the involvement of women in environmental stewardship is essential. Research indicates that gender equality and environmental sustainability can mutually enhance one another. Pro-environmental activities, including sustainable food consumption, are more common in nations with elevated gender equality, indicating that the empowerment of women may result in improved environmental outcomes (De Boer and Aiking, 2023).

To augment women's participation in environmental governance, various initiatives may be implemented. Capacity-building programs and the formulation of gender-sensitive policies are crucial for the empowerment of women in this domain. Establishing forums for women to cultivate networks and partake in decision-making processes might enhance their involvement in environmental governance (Sam and Zibima, 2023). Furthermore, incorporating gender concerns into green policies and finance can enhance women's empowerment and foster sustainable environmental results (Supporting women's empowerment through green policies and finance, 2022). These initiatives must transcend superficial representation, guaranteeing that women's perspectives are actively incorporated into decision-making processes and tackling intersectional concerns such as race, ethnicity, and socioeconomic status (Duguma et al., 2022).

Integrating women into environmental governance poses considerable problems; however, there is increasing acknowledgment of their contributions to sustainable resource management. Empowering women and tackling gender inequities can enhance the efficacy and inclusivity of environmental policy, hence fostering long-term sustainability. Achieving this necessitates coordinated efforts to eliminate the cultural, institutional, and structural obstacles that impede women's participation. By fostering inclusivity and advancing equity in environmental governance, societies may leverage the distinct perspectives and expertise that women provide, so maintaining the interdependence of environmental sustainability and gender equality.

Climate change disproportionately impacts women, particularly in underdeveloped nations where they depend significantly on natural resources for their sustenance. As climate-related phenomena such as droughts, floods, and storms increase in frequency and intensity, women encounter heightened difficulties in obtaining vital resources like food, water, and shelter for their family. Environmental strains intensify pre-existing gender disparities, as women frequently have diminished access to essential resources such as land, credit, and education, which are vital for adapting to the effects of climate change. In some instances, the exacerbating effects of climate change amplify the home work responsibilities of women, thereby constraining their prospects for economic and social progress. This additional burden complicates women's ability to participate in activities that could enhance their economic status, education, or political involvement.

The susceptibility of women is particularly pronounced in regions such as Africa and South Asia. Women in these regions frequently participate in agricultural labor and resource management, although their capacity to adapt to climate change is impeded by restricted access to essential resources, like land and education, vital for enhancing resilience to climatic effects (Adeola et al., 2024a). In Bangladesh, the mental health challenges faced by women have escalated due to climate-induced stressors such as food insecurity and water scarcity, underscoring the extensive impact of climate change on women's well-being (Goudet et al., 2023). The economic strain is a notable result observed in the Indian Sundarbans, where climate change has heightened domestic labor obligations for women while concurrently diminishing their economic prospects (Ghosh and Dutta, 2024). These instances highlight the necessity for climate policy that addresses the specific vulnerabilities encountered by women.

Given these constraints, women's participation in environmental decision-making is essential for alleviating the gender-specific impacts of climate change. Research indicates that the inclusion of women in leadership roles within environmental governance frameworks provides distinct viewpoints that contribute to more sustainable and equitable results. Women frequently have a greater propensity to endorse policies that prioritize community resilience, sustainable resource management, and enduring environmental protection. Their inclusion guarantees that climatic and environmental policies are more thorough and attuned to the requirements of at-risk groups. Grassroots women's networks have successfully engaged in climate mitigation measures, showcasing the efficacy of women-led environmental projects (Dhal, 2024). Moreover, the inclusion of women in water management initiatives improves the sustainability and efficacy of water utilization, thereby impacting both ecosystems and local populations (Mukurazhizha and Matanga, 2023). These instances underscore the crucial role women may have in promoting sustainability when afforded leadership chances.

Nonetheless, substantial obstacles to women's complete involvement in environmental governance remain. Societal and cultural conventions frequently inhibit women's involvement in decision-making processes. In some areas, like East Africa, men predominantly influence decisions on climate adaptation, even if women play a crucial role in local resource management and environmental stewardship (Niemann et al., 2024). Patriarchal frameworks, established gender roles, and cultural prejudices frequently diminish women's perspectives and exclude them from critical dialogues that directly impact their lives and communities. Moreover, legislative constraints and restricted access to education further impede women's participation in environmental governance. In numerous developing nations, women encounter obstacles to land ownership, finance access, and education—elements essential for enhancing resilience to climate change and engaging in governance (Spitzer, 2023). Overcoming these obstacles is crucial for empowering women and guaranteeing their participation in climate-related decision-making.

The incorporation of women in environmental governance is both an issue of justice and pragmatism. The distinct viewpoints of women in leadership positions frequently lead to more thorough and inclusive climate policies. Women's participation primarily emphasizes enhancing community resilience and ensuring that the most vulnerable groups, like children and the elderly, are included in adaptation and mitigation initiatives. In Africa, policy proposals have highlighted the necessity of including gender equality in climate action plans and embracing technological breakthroughs that bolster women's contributions to climate adaptation (Adeola et al., 2024a). Such initiatives can markedly improve the efficacy of climate policies by ensuring they are more rooted in the realities of those most impacted by climate change.

The integration of women in environmental governance is essential; yet comprehensive societal reforms are required to effectively tackle the fundamental problems that sustain gender disparities. Initiatives to empower women within climate governance should be integrated into a broader movement aimed at contesting patriarchal practices and advancing gender equity across all sectors. This necessitates a comprehensive strategy that integrates legal reforms, education, and capacity-building efforts. International organizations, including UNESCO, are crucial in offering the assistance and frameworks required to augment women's involvement in mitigating climate change effects (Spitzer, 2023). Moreover, technical advancements and global assistance can enhance women's empowerment by equipping them with the necessary tools and resources for participation in environmental decision-making.

Tackling the gendered effects of climate change necessitates a comprehensive and diversified strategy. Empowering women, altering policies, and questioning societal norms are critical measures for fostering resilient communities and guaranteeing sustainable development for all. Encouraging women's engagement in environmental governance and tackling the systemic obstacles that restrict their participation can lead to more equitable and effective climate policies. This will result in more sustainable outcomes, benefiting not only women but also the wider community and environment.

Gender-responsive climate policies are widely seen as vital for attaining both gender equality and environmental sustainability. These policies seek to incorporate gender concerns into climate adaptation and mitigation efforts, ensuring that women's perspectives are acknowledged and their distinct needs are handled amid environmental challenges. Recognizing the various roles women fulfill in environmental stewardship, gender-responsive policies advance equality and improve the efficacy of climate action. This approach underscores the imperative of incorporating women into decision-making processes about climate change, therefore promoting more inclusive and sustainable solutions.

The convergence of gender and climate change is particularly pronounced in places such as Africa, where the effects of climate change disproportionately hit women, intensifying existing gender disparities and adding to the feminization of poverty. Patriarchal frameworks frequently constrain women's capacity to respond to climate change, rendering it essential for policies to tackle these inequities. Gender-responsive climate policies that promote inclusivity and equality can strengthen climate change mitigation efforts by addressing the concerns of women, who frequently serve as primary environmental stewards. Policies that emphasize gender equality and social inclusion can promote sustainable development and offer more holistic answers to climate concerns (Adeola et al., 2024b). The Gender Action Plan (GAP) serves as a fundamental foundation for these initiatives, delineating measures for gender-sensitive climate actions. This framework underscores the transition from mere commitments to actual strategies that promote women's involvement in climate adaptation and mitigation initiatives. When adopted and effectively implemented, GAP can substantially enhance sustainable development through participatory engagement, particularly by addressing the needs of marginalized women in at-risk areas (Adeola et al., 2024b).

Technological innovation constitutes a vital element of gender-responsive climate policies. In Africa, gender-sensitive policies that incorporate technical solutions might mitigate the effects of climate change on women and girls, ensuring their inclusion in resilience-building initiatives. Access to technology and training in climate-smart farming methods might enable women to adapt to evolving climatic conditions, so securing their livelihoods and contributing to broader climate mitigation initiatives (Adeola et al., 2024b). In China, it has been suggested that collaborative strategies combining climate action, rural revitalization, and gender equality are essential, highlighting the necessity of incorporating gender considerations into technological and environmental solutions (Jeffs, 2023). The interplay of gender in climate adaptation is intricate, especially in rural regions, where conventional gender norms frequently influence decision-making processes. In rural East African homes, men are generally seen as the principal decision-makers; nevertheless, women wield considerable influence in some areas concerning climate adaptation, including water and agricultural resource management. Gender-responsive climate initiatives must account for these factors to be effective, guaranteeing the engagement of both men and women in adaption strategies (Niemann et al., 2024).

Notwithstanding the increasing acknowledgment of the significance of gender-responsive climate policies, obstacles persist. Opposition to more intersectional feminist methodologies, which examine the interplay between gender and other types of injustice, has been noted within civil society and policy arenas. This reluctance frequently arises from systemic obstacles that segregate intertwined movements for gender and climate justice (Flavell, 2023). Furthermore, although a positive correlation exists between gender equality and environmental performance, the interplay between both domains is intricate and contingent upon circumstance.

Gender-responsive climate policies are essential for mitigating the unequal effects of climate change on women and for advancing gender equality and environmental sustainability. These policies acknowledge the distinct viewpoints and contributions of women in environmental stewardship, while simultaneously addressing the systemic obstacles that restrict their involvement. Nonetheless, obstacles such as opposition to intersectional methodologies and the intricacies of gender relations in climate adaptation underscore the necessity for ongoing study, activism, and inclusive policy-making. By tackling these obstacles, gender-responsive climate policies can enhance their efficacy, guaranteeing the attainment of climate justice in conjunction with gender justice.

Gender equality and environmental sustainability are fundamentally interconnected. The participation of women in environmental decision-making is crucial due to their roles in resource management and their heightened sensitivity to climate change. Attaining gender equality in this context necessitates confronting both the immediate obstacles confronting women—such as access to water and sanitation—and the overarching structural impediments that restrict their involvement in leadership. By advocating for gender-inclusive environmental governance and gender-responsive climate policy, societies can achieve more sustainable and equitable results for all.

4.2.6 Technological Indicators of Gender Equality

Technology represents a significant domain where gender gaps are frequently pronounced, and in the digital era, access to information and communication technology (ICT) has emerged as a crucial metric of gender equality. The **digital gender divide** denotes the disparity between men and women over access to and utilization of digital technology, including the internet, mobile devices, and digital literacy initiatives. This divide pertains not only to women's access to the internet but also to their interaction with technology, the digital competencies they develop, and their opportunity to utilize technology for education, employment, and entrepreneurship. This topic is complex, influenced by various factors such as socioeconomic status, cultural norms, and educational possibilities.

Women globally encounter substantial obstacles in obtaining access to digital technologies. Economic constraints, particularly in low-income areas, hinder numerous women from acquiring personal devices like cellphones, which are crucial for internet access and digital services. In India, women are less inclined to possess smartphones and frequently depend on communal home gadgets, hence constraining their access to digital tools and the duration of their usage (Islam and Manchanda, 2023). The disparity is particularly evident in rural regions, where infrastructural inadequacies and cultural norms further restrict women's access to mobile phones and the internet, as seen in Pakistan (Amber and Chichaibelu, 2023). The rural-urban divide underscores the interplay of gender, location, and socioeconomic variables in sustaining the digital gender disparity.

In addition to access, digital literacy and the cultivation of digital skills are essential elements of the divide. Gender differences in these domains commence early in life, as girls frequently experience diminished exposure to technology-related disciplines relative to males. This disparity continues in higher education and the workforce, where women are still underrepresented in disciplines such as software development, data science, and artificial intelligence. In Germany, gender disparities in ICT literacy are affected by socioeconomic position, with girls from underprivileged families encountering heightened obstacles to technology participation (Passaretta and Gil-Hernández, 2023). Notwithstanding these limitations, technology possesses considerable potential to empower women by facilitating access to education and chances for skill enhancement. E-learning platforms are essential in addressing educational disparities and enhancing women's employability in the digital economy (Hasni et al., 2023).

The digital gender divide adversely affects women's involvement in the labor market and their economic potential. Access to digital technologies is intricately associated with enhanced female workforce participation. In Pakistan, mobile phone ownership correlates with increased female labor force participation, highlighting the economic advantages of bridging the digital divide (Amber and Chichaibelu, 2023). Moreover, digital platforms present women with prospects for entrepreneurship and participation in the gig economy, offering flexible employment arrangements that might align with familial obligations (Hasni et al., 2023). Technology can disrupt conventional gender norms and foster economic empowerment by facilitating women's engagement in alternative work opportunities.

Nevertheless, societal and cultural conventions frequently provide substantial obstacles to women's participation in technology. In numerous societies, cultural norms and preconceptions inhibit women from employing digital tools, perpetuating the digital gender barrier. In certain areas, women encounter societal pressure to prioritize domestic duties over digital participation, hence restricting their capacity to leverage technology (Ortiz Cobo and Jeri Levano, 2023). Confronting these cultural obstacles is crucial for empowering women to properly utilize digital technology and engage in the digital economy. Furthermore, the design of digital technology significantly influences the reinforcement or challenge of gender inequality. Participatory design methodologies that engage women in the production and development of technologies can effectively address gender-specific requirements and promote inclusivity in digital tools. Incorporating women into the design process enables the creation of technologies that more effectively address varied user demographics and mitigate the reinforcement of gender stereotypes (Rießenberger and Fischer, 2023).

The digital gender barrier poses considerable obstacles but also provides potential for dramatic change. Mitigating the social, educational, and cultural obstacles that perpetuate this disparity can facilitate the reduction of the gap and advance gender equality. The design of inclusive technology, along with policies that enhance access to digital tools and literacy initiatives, is essential for preventing the marginalization of women in the digital revolution. Moreover, bridging the digital gender difference will enhance not just gender equality but also overall social and economic advancement. It is crucial to acknowledge that the digital gender divide is a component of a more extensive system of inequities, and actions to mitigate it must be aligned with wider efforts to advance gender equity and social inclusion.

The digital gender barrier goes beyond mere access to technology; it also includes the **quality of digital engagement** and the cultivation of skills essential for women to succeed in the contemporary technology-driven landscape. This disparity profoundly affects women's economic prospects, political engagement, and personal safety, as numerous contemporary services, including banking, healthcare, and social networking, necessitate computer literacy. Nevertheless, women frequently lack representation in digital literacy programs, which are essential for acquiring the skills necessary to effectively engage with and utilize modern technologies. In the absence of these abilities, women face a significant disadvantage, hindering their capacity to engage comprehensively in the digital economy.

The digital gender difference is interconnected with other inequalities, including wealth, education, and geography. Women in rural regions or from underprivileged socio-economic origins encounter far bigger obstacles in obtaining technology than their urban or affluent peers. This layered inequality perpetuates existing social and economic disadvantages, making it harder for women to escape cycles of poverty or marginalization. In India, rural women are less inclined to possess smartphones, frequently depending on communal home gadgets, so constraining their digital participation and prospects for economic progress (Islam and Manchanda, 2023). Women in lower-income households in Canada encounter substantial obstacles to accessing digital services, especially in rural regions, underscoring the necessity for infrastructure enhancement and policies that foster digital equity (Singh and Chobotaru, 2022).

The digital gender difference significantly affects women's economic prospects. In numerous regions globally, access to digital tools and competencies is intricately connected to enhanced work opportunities and labor force participation. In Pakistan, mobile phone ownership has been demonstrated to enhance female labor force participation, highlighting the economic advantages of granting women access to technology and digital skill development possibilities (Amber and Chichaibelu, 2023). The digital divide in Europe is intricately linked to gender and income disparities, with educational attainment significantly impacting digital participation and economic results (Picatoste et al., 2023).

Socioeconomic and geographic disparities also intensify the digital gender barrier. Women from low-income households or rural regions encounter further obstacles in obtaining technology, hence constraining their participation in the digital realm. These obstacles not only limit economic prospects but also exacerbate educational inequality. Studies indicate that students from low-income families frequently experience reduced internet access, adversely impacting their educational performance (Afzal et al., 2023). Creating technology resource centers and promoting public-private collaborations can mitigate these discrepancies and provide access to digital technologies.

Addressing the digital gender disparity necessitates an emphasis on digital literacy and skill development. Research indicates that girls frequently excel over males in digital competencies; nonetheless, enduring gender disparities in attitudes towards technology may still affect results, especially regarding profession selections and sustained involvement with technology (Consoli et al., 2023). Rectifying these attitudinal disparities and guaranteeing equitable access to digital literacy initiatives is essential for advancing gender equality in the digital domain. Technology significantly contributes to the political and social empowerment of women. Digital tools can enhance women's representation and impact in decision-making by facilitating their participation in political debate and challenging traditional conventions. Nonetheless, digital inequality persists as an obstacle in specific circumstances, particularly among migratory women. Peruvian migrant women in Italy and Spain encounter digital disparities that hinder their social inclusion and intensify their vulnerability (Afzal et al., 2023).

Tackling the digital gender barrier necessitates a thorough and diversified strategy that accounts for the interaction of gender with other inequalities. By prioritizing the enhancement of digital literacy, expanding technological access, and tackling socioeconomic and geographic inequality, we may strive for a more equitable digital environment for women. It is essential to remain alert to the possibility that digital systems may exacerbate existing disparities, ensuring that technology functions as a means of empowerment and inclusion rather than perpetuating exclusion and disparity. Mitigating the digital gender difference necessitates focused interventions, such as enhancing affordable internet access, advancing digital literacy initiatives for women and girls, and fostering female engagement in STEM (science, technology, engineering, and mathematics) disciplines. These initiatives can bridge the divide and guarantee that women are not marginalized in the digital revolution. Empowering women with the necessary tools and abilities to traverse the digital realm enables nations to foster more inclusive economies and advance gender equality in both public and commercial sectors.

Gender equality constitutes a fundamental human right and a pillar of social justice, providing extensive advantages for sustainable development and societal advancement. The examination of social, economic, political, legal, cultural, and environmental factors highlights the necessity of deconstructing systemic obstacles and confronting ingrained gender inequality. Attaining gender equality in these areas empowers individuals, enhances communities, stimulates economic progress, and fortifies social cohesiveness. By adopting comprehensive and inclusive policies, communities can establish an equitable future in which all genders flourish and contribute equally to development and well-being.

4.3 DRIVERS OF AND BARRIERS TO GENDER EQUALITY

The issue of gender equality has gained prominence as nations strive to establish inclusive and equitable environments that afford equal opportunity to all individuals, irrespective of gender. Attaining gender equality is not merely a social objective but also an economic and political imperative, as it directly impacts national development, economic advancement, and societal welfare. The interaction between facilitators and obstacles to gender equality is intricate, and comprehending this dynamic is essential for formulating policies that advance equity in many aspects of life.

The principal catalysts of gender equality include economic development, education, and legal structures. Economic growth can promote the integration of women into the workforce, offering avenues for financial autonomy and professional advancement. Education, especially at the secondary and tertiary levels, empowers women with the competencies and information necessary for engagement in the workforce, governance, and civic life. Government programs that promote education and provide labor market protections are particularly effective in advancing gender equality. Moreover, legislative frameworks and international obligations, including the United Nations' Sustainable Development Goals (SDGs), establish a foundation for countries to enhance gender equality via policies such as equitable remuneration, anti-discrimination legislation and safeguard against gender-based violence.

Nonetheless, substantial obstacles persist that hinder gender equality. Cultural norms and societal expectations frequently sustain conventional gender roles, constraining women's involvement in education, employment, and decision-making. Economic inequality, exemplified by the enduring wage gap and disparate access to resources, perpetuates the disadvantage of women, especially in developing areas. Moreover, institutional prejudices inside businesses and governments can obstruct the implementation of legislation designed to foster equality. The ramifications of conflict and instability in specific places further intensify these concerns, disproportionately impacting women through violence, displacement, and poverty.

The present research examines the economic determinants influencing gender equality, recognizing that although other factors contribute to this issue, the analysis is confined to economic drivers and obstacles. A restriction of the research pertains to the specification of the model. This research omits parameters encompassed by the Gender Inequality Index (GII)—such as secondary and tertiary education attainment and women's labor force participation—while utilizing the GII as a proxy for gender inequality to preserve a reasonably straightforward framework. The analysis encompasses broader economic issues that affect these dimensions, including government expenditure on education and labor market laws.

Notwithstanding these constraints, the study seeks to elucidate the fundamental factors that can promote gender equality, especially via education, economic policies, and labor market reforms, as well as the enduring obstacles that hinder advancement, including societal norms and economic inequality. This research enhances the comprehension of achieving gender equality among economic and socio-cultural obstacles by assessing these elements.

To empirically examine the determinants and obstacles to gender equality, this study is grounded in theoretical assumptions, establishing a framework for the approach while concentrating on the principal elements affecting gender inequality:

1. Higher economic development is linked with lower gender inequality since the availability of material and social resources promotes gender-specific preferences, and gender equality is associated with greater gender differences in preferences (Falk and Hermle, 2018).
2. Economic growth promotes gender inequality since growth is positively related to gender wage inequality. This should be especially evident in export-oriented economies, where women's lower wages stimulate growth through increased exports (Seguino, 2000). Therefore, it can be expected that the size of exports moderates the impact of economic growth on gender inequality.
3. Earnings for women lag behind those of men in nearly every industrialized country, with motherhood significantly affecting women's employment and earnings (Gornick, 2004). A higher gender pay gap is expected to have a positive correlation with gender inequality. Moreover, the ability of labor to bargain collectively may substantially impact the reduction of gender inequality. Stronger labor unions, for instance, can help address income disparities, including the gender earnings gap (Rosenfeld and Kalleberg, 1991).
4. Cultural norms, such as patrilocality and concerns for women's purity and not the rule of law, exacerbate gender inequality in many countries and increase male-skewed sex ratios (Jayachandran, 2014). Informal institutions and societal perceptions, particularly regarding women in business executive roles, significantly hinder gender equality (Kim, 2021). Therefore, it can be expected that a more equal gender distribution in the population, stronger rule of law, and a higher percentage of firms with female participation in ownership will have a negative effect on gender inequality.
5. Cross-national variation in tax-and-transfer policies explains a substantial share of the variation in gender inequality outcomes, with stronger social protection policies leading to better economic status for women (Gornick, 2004). Therefore, higher expenditures on social protection are expected to reduce gender inequality. Policies that improve demand-side factors—such as employment opportunities—and reduce the gender gap in economic activity are crucial for achieving gender equality (Altarawneh, 2020).

Consistent with the theoretical framework and to empirically evaluate the facilitators and obstacles to gender equality, Table 4.1 presents a summary of the principal variables utilized in the study, along with their descriptive statistics. This data overview allows for a thorough empirical analysis of the elements affecting gender inequality, enhancing the understanding of both the catalysts and obstacles to attaining gender equality.

The general equation for the empirical analysis of factors affecting gender inequality is specified for panel data and includes an interaction term to analyze the moderating effect of export intensity on the relationship between economic growth and gender inequality.

$$\begin{aligned} lnGII_{i,t} &= \alpha_i + \theta_t + \beta_1 lnY_{i,t} + \beta_2 \Delta lnY_{i,t} + \delta_{23} lnY_{i,t} \cdot EXPint_{i,t} + \\ &\beta_3 EXPint_{i,t} + \beta_4 MFr * 100_{i,t} + \beta_5 LMR_{i,t} + \beta_6 RoL_{i,t} + \beta_7 GS_{i,t} + \\ &\beta_8 \mathrm{FwFPiO}_{i,t} + \beta_9 GPG_{i,t} + \varepsilon_{i,t} \end{aligned} \tag{4.1}$$

Data for 27 EU countries from 1995 to 2022 was collected, though certain variables, such as the Gender Pay Gap (GPG) and Firms with Female Participation in Ownership (FwFPiO), contain significant gaps. To mitigate the risk of statistically insignificant results due to a small sample

TABLE 4.1
Descriptive Statistics and Variable List for Empirical Analysis

No.	Variable	Unit of measure	Years	Source
1.	Gender Inequality Index (GII)	GII measures gender inequality across three areas: health, assessed by maternal mortality and adolescent fertility; empowerment, measured by parliamentary seats and education levels; and labor, reflecting women's workforce participation	1990–2022	United Nations Development Programme
2.	GDP	Gross domestic product at market prices, million euro	1995–2022	Eurostat
3.	Population	Total, number	1995–2022	Eurostat
4.	Males	Total, number	1995–2022	Eurostat
5.	Export	Exports of goods and services, million euro	1995–2022	Eurostat
6.	Gender pay gap	Age class from 35 to 44 years, percentage	2007–2022	Eurostat
7.	Social protection	Total general government expenditure, percentage of gross domestic product (GDP)	1995–2022	Eurostat
8.	Education	Total general government expenditure, percentage of gross domestic product (GDP)	1995–2022	Eurostat
9.	Sickness and disability	Total general government expenditure, percentage of gross domestic product (GDP)	1995–2022	Eurostat
10.	Family and children	Total general government expenditure, percentage of gross domestic product (GDP)	1995–2022	Eurostat
11.	Unemployment	Total general government expenditure, percentage of gross domestic product (GDP)	1995–2022	Eurostat
12.	Social exclusion n.e.c.	Total general government expenditure, percentage of gross domestic product (GDP)	1995–2022	Eurostat
13.	Labor market regulations	Index	1995–2021	Fraser Institute
14.	Firms with female participation in ownership (% of firms)	Firms with female participation in ownership are the percentage of firms with a woman among the principal owners.	2004–2021	World Development Indicators, World Bank
15.	Rule of Law: Estimate	Rule of Law captures perceptions of the extent to which agents have confidence in and abide by the rules of society, in particular the quality of contract enforcement, property rights, the police, and the courts, as well as the likelihood of crime and violence. Estimate gives the country's score on the aggregate indicator, in units of a standard normal distribution, i.e. ranging from approximately −2.5 to 2.5.	1996–2021	World Development Indicators, World Bank Development Research Group

size, several estimations were conducted, both with and without these variables. Since government spending for various objectives (GSsp, GSedu, GSs&d, GSf&ch, GSunem, GSsocex) is positively intercorrelated, with Pearson correlation coefficients ranging from 0.3 to 0.7, these variables were included one by one in different estimations to avoid multicollinearity. The first two estimations aggregate all government spending under investigation (GS).

A two-way fixed effects model was used to control for country-specific factors (α_i) that do not change over time and common time-specific changes (θ_t) across all EU27 countries. Diagnostic tests revealed heteroscedasticity, minor serial correlation, and slight cross-sectional correlation in the error term. To correct for potentially inaccurate standard errors, adjustments recommended by Driscoll and Kraay were applied. Estimation results are provided in Table 4.2.

TABLE 4.2
Two-Way Fixed Effects Estimates of Eq. 1. Dependent Variable—lnGII

Variable		(1)	(2)	(3)	(4)	(5)	(6)	(7)	(8)
Per capita GDP (lnY)	β_1	−0.2694***	−0.2639***	−0.2500***	−0.4531***	−0.4128***	−0.3419***	−0.3197***	−0.3536***
		(0.03382)	(0.03449)	(0.04835)	(0.04794)	(0.03060)	(0.05496)	(0.05526)	(0.04267)
Growth (ΔlnY)	β_2	−2.340**	−1.831**	−1.563*	−0.9891	−1.010*	−0.6053	−0.6446	−0.7184
		(1.035)	(0.7812)	(0.9004)	(0.6778)	(0.5336)	(0.6933)	(0.7793)	(0.7340)
ΔlnY * EXPint	δ_{23}	0.007674							
		(0.008874)							
Export intensity (EXPint)	β_3	0.003666***	0.003751***	0.004356***	0.006748***	0.004864***	0.006654***	0.005842***	0.006343***
		(0.0003851)	(0.0003086)	(0.0003626)	(0.0004494)	(0.0005487)	(0.0003887)	(0.0004079)	(0.0003580)
Male-to-female ratio (MFr) *100%	β_4	0.02130***	0.02133***	0.01934***	0.03136***	0.02644***	0.02357***	0.02210***	0.02091***
		(0.002708)	(0.002740)	(0.002937)	(0.003462)	(0.004017)	(0.002850)	(0.003218)	(0.002523)
Labor market regulation (LMR)	β_5	−0.01119	−0.01114	−0.02820***	−0.04720***	0.04477***	−0.01970*	−0.02182**	−0.02147*
		(0.01128)	(0.01100)	(0.009311)	(0.01374)	(0.01027)	(0.01152)	(0.01042)	(0.01258)
Rule of Law (RoL)	β_6	−0.1493***	−0.1498***	−0.3028***	−0.1320***	−0.04551	−0.2850***	−0.3044***	−0.2987***
		(0.04943)	(0.04992)	(0.04123)	(0.03823)	(0.04689)	(0.03452)	(0.04369)	(0.03770)
Government spending on social protection, education, unemployment, etc. (GS)	β_7	−0.03656***	−0.03689***						
		(0.001694)	(0.001698)						
Government spending on Social protection (GSsp)				−0.03954***					
				(0.004645)					
Government spending on education (GSeduc)					−0.1596***				
					(0.007217)				
Government spending on sickness and disability (GSs&d)						−0.2105***			
						(0.009684)			
Government spending on family and children (GSf&ch)							−0.08709***		
							(0.01506)		
Government spending on unemployment (GSunem)								−0.07672**	
								(0.02826)	
Government spending on social exclusion n.e.c. (GSsocex)									−0.1199***
									(0.03346)
Sample size		556	556	556	556	556	556	556	556
Adjusted R^2		0.7541	0.7542	0.7206	0.7325	0.7971	0.7014	0.6939	0.6980

All estimates include country- and time-specific dummies. Driscoll-Kraay standard errors are presented in parentheses * significant at the 10% level, ** significant at the 5% level, *** significant at the 1% level.

Estimated coefficients on per capita GDP are negative and statistically significant. They ranges from −0.25 to −0.45 depending on the slight change in the specification or using government spending for different objectives. Nevertheless, results clearly show that a higher development level of a country is associated with lower gender inequality. Specifically, other factors being fixed, in countries that have 1% higher per capita GDP, it is estimated that the gender inequality index is lower by 0.25–0.45%.

In estimations that include total government spending (regardless of whether there are interactions or not), growth shows a significant effect in reducing gender inequality. However, the statistical significance of this effect disappears, even though the direction remains the same, when government spending on specific objectives is included. Therefore, robust evidence on the effect of economic growth on gender inequality is lacking.

The interaction term between growth and export intensity was found to be statistically insignificant. Thus, there is no evidence that exports mediate the effect of growth on gender inequality. However, there is robust evidence across estimations that export intensity is positively associated with gender inequality. Export intensity higher by 1% is estimated to be associated with 0.36–0.66% higher gender inequality.

Higher male-to-female ratios are positively linked to gender inequality. This ratio higher by one percentage point is associated with 1.9–3.1% higher gender inequality. Despite the fact that size of the effect varies, its statistical significance remains robust.

Rule of law is estimated to be negatively related to gender inequality. This finding is robust across all estimations except one where government spending on sickness and disability is included. Rule of law higher by one point is estimated to be linked to gender inequality lower by 13–30%. It is worth noting that rule of law estimate theoretically varies from minus 2.5 to plus 2.5. Thus change of this variable by one point should be considered extremely huge.

All government spendings under the investigation seem to have a negative and statistically significant effect on gender inequality. The size of the estimated effect greatly varies, but the tendency remains the same—the higher the government spending on social protection, education, unemployment, etc., the lower the gender inequity.

The empirical findings presented offer significant economic insights that correspond with the facilitators and obstacles to gender equality. The pronounced inverse correlation between per capita GDP and gender disparity highlights that economic development is essential for advancing gender equality. As nations become more affluent, they frequently provide greater resources to social programs, education, and healthcare, which typically advantage women and other underprivileged populations, therefore diminishing inequality. This corresponds with the economic notion that augmented resources in affluent nations facilitate greater options for women's participation in the job market and access to superior services.

Nonetheless, the results indicate obstacles. The relationship between economic development and export intensity did not exhibit a notable moderating influence on gender inequality. This indicates that, although growth can diminish inequality, export-driven growth may not inherently result in improved gender results. Conversely, export-driven sectors may be predominantly male-oriented or foster gender-segregated labor markets, so perpetuating or intensifying inequality. This insight underscores the necessity for specific policies that guarantee women reap the benefits of global trade, including education and training in high-growth areas.

The correlation between the male-to-female ratio and gender inequality underscores a systemic obstacle: nations with a higher male-to-female ratio frequently encounter more gender inequality. This is likely attributable to labor market dynamics in which women's work is undervalued or where societal norms and expectations restrict women's engagement in higher-paying occupations. The enduring nature of this inequality indicates that economic strategies may be inadequate unless cultural or institutional biases are also addressed.

The strong inverse correlation between the rule of law and gender disparity highlights that effective governance is crucial for achieving equality. Legal safeguards, implementation of anti-discrimination statutes, and an impartial judiciary are essential in advancing gender equality by guaranteeing women equal access to economic opportunities and rights.

The persistent evidence that government expenditure on social protection, education, and other welfare services mitigates gender inequality indicates that public investment in social infrastructure is a crucial factor in diminishing obstacles to gender equality. This underscores the significance of governmental actions, especially in domains that directly influence women's capacity to participate in the workforce, including childcare, healthcare, and educational possibilities. By reallocating resources to social objectives, nations can cultivate more inclusive economic conditions, diminish the structural obstacles confronting women, and promote enduring gender equality.

The findings underscore that economic growth alone is insufficient to advance gender equality. Targeted government expenditure, robust institutional frameworks, and initiatives to mitigate structural barriers—such as gender disparities in the labor market and legal inequities—are crucial for attaining lasting advancement in gender equality.

In summary, the quest for gender equality is both a moral and social need, as well as a vital catalyst for economic progress and political stability. The approach emphasizes the complex interaction of economic development, legal structures, and cultural norms in influencing gender results. Although economic growth helps mitigate gender disparity by broadening opportunities and resources, it is not a comprehensive solution. The findings emphasize that focused policies, especially in education, labor market changes, and social protection, are crucial for addressing deep-seated inequality. The importance of strong legal systems and governance structures is paramount, since they establish the basis for implementing gender parity and safeguarding women's rights. Achieving gender equality necessitates a comprehensive strategy that integrates economic, structural, and cultural reforms, guaranteeing equal chances for both women and men to engage and prosper in all facets of life.

4.4 POLICIES AND MEASURES TO ENSURE GENDER EQUALITY

Attaining gender equality is crucial for cultivating a harmonious, successful, and sustainable global society. It is fundamental to sustainable development, as it fosters equitable engagement in economic activities, enhances societal resilience, and elevates outcomes in health, education, and overall well-being. Furthermore, when both sexes are afforded equal opportunity, communities witness augmented economic progress, more efficient governance, and heightened social cohesion. To advance gender equality, it is essential to enact specific policies and actions that tackle the fundamental economic, social, and institutional obstacles.

Fostering inclusive economic growth by increasing female labor participation, especially in high-growth industries such as technology and finance, is crucial for attaining gender equity and economic prosperity. This entails enacting laws that promote women's entrepreneurship, facilitate access to financing, and provide vocational training in industries traditionally dominated by males.

The advancement and empowerment of women substantially enhance inclusive economic growth. Research indicates that women's education and income positively influence economic growth, underscoring the significance of equitable development strategies (Mafida Amalia et al., 2023). Investing in women across diverse areas, such as technology and finance, propels economic growth and creativity. This investment represents both a social justice imperative and a strategic initiative for economic progress. Female entrepreneurship represents an unexploited opportunity for economic advancement. Establishing advantageous conditions for women entrepreneurs, encompassing access to financial resources and supportive policies, is essential for realizing this potential (Gulvira et al., 2024). National Financial Inclusion Strategies are progressively acknowledging the necessity of advancing women's financial inclusion. Nevertheless, additional measures are required to guarantee women equitable access to finance and financial services, which are essential for entrepreneurship (Bin-Humam et al., 2023). Vocational training and skills development are essential for advancing gender diversity in high-growth sectors. Promoting women's participation in science, technology, engineering, and mathematics (STEM) through specialized education and training initiatives can stimulate innovation and economic development. Mitigating gender gaps in these areas necessitates comprehensive policies that facilitate women's participation. Although these strategies are vital, it is equally important to acknowledge the wider societal and cultural variables that affect women's workforce involvement. Mitigating gender-based obstacles, like disproportionate caregiving duties and prevailing societal norms, is essential for establishing a genuinely inclusive economic landscape. This comprehensive approach guarantees that women may entirely participate in and gain from economic progress.

Enhancing legal and institutional frameworks to advance gender equality in the workplace is crucial for mitigating the institutional obstacles faced by women. This necessitates the augmentation of legal safeguards, the steadfast enforcement of gender equality legislation, and the broadening of anti-discrimination policies in both the commercial and governmental sectors. Legal frameworks frequently inadequately address more intricate kinds of gender discrimination, including indirect and environmental discrimination. Enhancing existing regulations by instituting more explicit rules and broadening their applicability is essential for addressing these concerns (Liu, 2024). International human rights standards establish a basis for developing national legal systems; yet, their efficacy is frequently obstructed by insufficient awareness and cultural obstacles (Dhahri, 2023). Despite the existence of extensive legal frameworks, enforcement deficiencies and the absence of strategic initiatives can enable the continuation of gender inequities. Institutional systems are essential in combating workplace discrimination. Implementing efficient complaint and redress mechanisms is essential for the prompt and equitable resolution of discrimination claims (Liu, 2024). Furthermore, cultivating a culture of corporate social responsibility and advancing gender equality inside firms can aid in dismantling enduring prejudices and biases (Liu, 2024). Workplace discrimination against women adversely affects equity and significantly harms their physical and emotional well-being, resulting in heightened stress, diminished job satisfaction, and decreased motivation. Enhancing legal safeguards is crucial for alleviating these negative consequences (Shaheen et al., 2023). Moreover, gender imbalance in the workplace impacts organizational creativity, productivity, and financial performance. A comprehensive approach encompassing governments, organizations, and society is essential for achieving gender equality (Du, 2024). Enhancing legal and administrative frameworks is essential, yet it is equally vital to confront the cultural and societal norms that sustain gender discrimination. Increasing awareness and strengthening institutional capacity will guarantee the proper execution of gender equality legislation and policies (Dhahri, 2023). Legal safeguards that advocate for equal remuneration for equivalent work and shield against workplace discrimination are essential, and the execution of these statutes will aid in diminishing the systemic obstacles women encounter across various sectors.

Enhance social protection policies. Government expenditure on social welfare initiatives, including childcare, family assistance, healthcare, and education, is crucial for augmenting women's involvement in the labor market, especially among low-income households. These rules not only offer direct advantages but also facilitate women's ability to harmonize job and familial obligations more efficiently. Gender-responsive social protection systems are essential for tackling the unique issues encountered by women, especially in low- and middle-income nations. Analysis of 52 national social protection policies indicates that although numerous nations recognize gendered vulnerabilities, a considerable disparity frequently occurs between policy language and actual execution (Cookson et al., 2024). This disconnection highlights the necessity for more robust pledges to convert policy into practice. The COVID-19 epidemic highlighted the imperative for gender-sensitive social protection measures. Women encountered an augmented weight of caring duties and confronted intensified work obstacles, underscoring the necessity for more inclusive policies that cater to gender-specific requirements (Cookson et al., 2024). The pandemic underscored the necessity of flexible and adaptive social protection policies to ensure the protecting of women's involvement in the labor market, particularly during emergencies.

Social policies aimed at disadvantaged populations, especially women, are essential for advancing social justice and equity. These policies promote equitable opportunities and facilitate inclusive growth, which is vital for establishing a balanced and just society. In unregulated areas, such as the informal labor market, understanding and accessibility of social welfare programs are essential for female workers. Nonetheless, discrepancies in social security coverage persist, requiring more efficiency and outreach to guarantee equitable access to these programs. Health-related social protection services, including social insurance and assistance programs, are essential for enhancing healthcare coverage for at-risk populations. These services are essential for attaining universal healthcare, which subsequently promotes women's health and economic involvement by diminishing employment barriers and improving overall well-being (Yokobori et al., 2023). The insights gained during the COVID-19 pandemic underscore the necessity of robust and adaptable social protection policies in managing crises. Countries that enacted comprehensive social protection measures were more effective in alleviating the adverse effects of the pandemic, underscoring the necessity for systems that can adjust to the requirements of women and other marginalized groups during crises (Abdoul-Azize and El Gamil, 2021). Although social protection measures are crucial for improving women's engagement in the labor market, considerable obstacles persist in guaranteeing their efficient implementation and accessibility for all women. Connecting policy formulation and implementation is essential for ensuring that social protection systems are really inclusive and sensitive to gender considerations. To tackle these difficulties, it is essential to enhance the accessibility, efficiency, and responsiveness of social welfare programs, guaranteeing that they are available to all women, irrespective of their socio-economic situation.

Encourage gender-responsive budgeting. Gender-responsive budgeting is a strategy framework aimed at ensuring equitable distribution of public expenditures, benefiting both men and women, while addressing gender imbalances in essential areas such as infrastructure, healthcare, and education. Numerous countries globally have implemented gender-responsive budgeting, demonstrating its capacity to advance gender parity. A principal advantage of gender-responsive budgeting is its capacity to convert national budgets from impartial resource distribution systems into instruments for promoting social fairness. Focusing on gender equality enables governments to meticulously evaluate their expenditures and ensure that resources are allocated to mitigate gender inequities in sectors such as health, education, and economic involvement (Gupta et al., 2024).

Nonetheless, the execution of gender-responsive budgeting faces numerous hurdles. Despite advancements in its implementation, numerous countries encounter substantial obstacles in attaining the intended gender equality results. These challenges frequently arise from cultural and institutional issues that impede the comprehensive implementation of gender-responsive policies. Despite acknowledging the need for gender equity, a considerable disparity sometimes exists between policy formulation and actual execution, highlighting the intricacies of incorporating gender considerations into public budgeting procedures (Koirala, 2023). Despite the potential of gender-responsive budgeting to mitigate gender inequities, numerous institutional and cultural obstacles frequently impede its efficacy. Notwithstanding these limitations, the experiences of nations that have used this strategy underscore its significance as a mechanism for advancing gender equity. Gleaning insights from these experiences enables governments to enhance their strategies regarding public expenditures, guaranteeing inclusivity and equity for all genders.

Governments ought to persist in implementing gender-responsive budgeting to guarantee that public expenditures meet the needs of both men and women. This methodology is especially crucial in sectors like infrastructure, healthcare, and education, where gender inequality endures. Through meticulous analysis and alignment of expenditures aimed at diminishing these disparities, gender-responsive budgeting can serve as an effective instrument for advancing enduring gender equity and social inclusion.

Close the gender pay gap. Enforcing stringent measures, including pay transparency legislation and obligatory reporting, is essential in mitigating the gender wage disparity. The implementation of these measures, alongside incentives for compliant organizations and penalties for non-compliance, can generate substantial progress towards attaining gender pay equality. Legislation on pay transparency and obligatory reporting has demonstrated the capacity to improve accountability and compel organizations to rectify gender pay inequities. The Gender Pay Gap Information Regulations in the UK mandate that big firms annually disclose gender pay statistics, fostering public scrutiny that promotes policy changes to address the gap (Benedi Lahuerta et al., 2024). Nonetheless, research from Austria suggests that wage transparency by itself may be insufficient. A study on Austria's legislation requiring internal reporting of gender pay discrepancies revealed no substantial effect on diminishing wage inequality, indicating that transparency initiatives need to be complemented by further actions to effectuate significant change (Gulyas et al., 2021).

Incentives for organizations that diligently strive to eliminate the gender wage gap might serve as a significant motivator. Examples encompass tax incentives or public acknowledgment for corporations that achieve significant advancements in female pay equity (Du, 2024). Conversely, sanctions for non-compliance, such as monetary fines or limitations on government contracts, may deter corporations from disregarding or dismissing gender pay inequities (Salman, 2023). The enforcement of these sanctions introduces an additional level of accountability that motivates corporations to adhere to female pay equity measures. Closing the gender wage gap involves more than just transparency and adherence to regulations. An integrated strategy is crucial, encompassing the rectification of institutional gender biases, occupational segregation, and social gender preconceptions. Promoting work-life balance and cultivating inclusive leadership practices are essential elements of comprehensive strategies to attain gender equality in the workplace (Du, 2024). Legal frameworks must be comprehensive, addressing both immediate salary difficulties and the broader cultural and structural prejudices that perpetuate inequality (Salman, 2023).

Although pay transparency rules and compulsory reporting are significant advancements, they do not constitute comprehensive solutions. The efficacy of these strategies is largely contingent upon the quality of the data gathered and the context in which these initiatives are implemented. A comprehensive strategy, integrating legislative, organizational, and cultural reforms, is crucial for attaining enduring gender pay equity.

Support for female participation in leadership. Advancing women into leadership positions is crucial for mitigating gender imbalance in both the public and commercial sectors. Organizations can substantially enhance female presence in decision-making roles by instituting quotas and mentorship programs. This strategy not only confronts and deconstructs prevailing gender biases but also cultivates a more inclusive atmosphere that acknowledges and develops women's leadership potential. Empowering women in leadership roles is essential for promoting gender equality. Research demonstrates that offering women leadership chances can effectively confront gender prejudices and preconceptions, resulting in more fair workplace dynamics (Lambert et al., 2023). Research indicates that female leadership in the business sector enhances economic performance, as companies with a greater percentage of female executives exhibit approximately two-percentage-points higher economic value added than those with fewer or no female executives, underscoring the essential role of gender diversity in corporate governance (Roessle et al., 2024).

Notwithstanding these advantages, women in leadership encounter substantial obstacles, including bias, discrimination, and societal conventions that favor familial responsibilities over professional pursuits. Nonetheless, there is an increasing tendency towards establishing supportive workplaces via mentorship programs and leadership development opportunities (The influence of the professional environment on women leadership, 2023). Such programs are essential for delivering the requisite social support and personal development opportunities that empower women to fully actualize their leadership potential within corporate contexts (Nugroho, 2023). Beyond business environments, public leadership is essential in reshaping societal and institutional standards concerning gender inequity.

The advancement of women into leadership positions has accelerated, while obstacles persist. Ongoing preconceptions and societal expectations impede women's progress; nonetheless, the growing adoption of supportive initiatives indicates a favorable movement towards gender equality. By overcoming these obstacles and promoting inclusive settings, sustainable advancement in gender equality within leadership can be realized.

Investment in education and training. Investing in education, especially for girls, is essential for promoting gender equality and stimulating economic development. Governments must prioritize access to high-quality education for females, particularly in disciplines such as science, technology, engineering, and mathematics (STEM), where female representation is markedly deficient. These expenditures not only mitigate gender inequities but also promote societal advancement and innovation. Investing in girls' education yields enduring societal advantages, encompassing economic expansion, enhanced healthcare results, and increased gender equality. Educated women significantly contribute to diverse areas, promoting innovation and enhancing inclusive decision-making processes.

Gender imbalance persists in STEM education due to cultural preconceptions and a dearth of prominent female role models. Notwithstanding advancements in educational accessibility in certain areas, female participation in STEM-related disciplines, including technical and vocational education and training (TVET), remains insufficient. Gender-responsive techniques and female role models are crucial for motivating more girls to engage in STEM education (Ngugi and Muthima, 2017). Scholarships and targeted initiatives that offer financial support and incentives for females are essential to closing the gender gap in STEM. Moreover, augmenting the representation of female lecturers and role models in STEM can motivate young women to pursue analogous career trajectories. Policies ensuring a specific ratio of female students in technical disciplines can help foster gender equity (Ngugi and Muthima, 2017). Investment in girls' education and STEM is essential; however, broader socio-economic variables also affect educational attainment. Guaranteeing student safety, enhancing educational quality, and matching training programs with socio-economic demands are essential measures to optimize the effectiveness of these investments. Governments should prioritize establishing an environment that facilitates these activities and promotes enduring change.

Addressing cultural and institutional impediments to gender equality necessitates a holistic strategy that confronts prejudices, fosters inclusive norms, and advocates for active male involvement in caring duties. Educational initiatives and media campaigns are crucial in transforming social perceptions towards more gender-equitable standards. Confronting gender stereotypes is crucial for transforming cultural views on gender roles. Media portrayals profoundly influence the comprehension and reinforcement of gender roles. Women are frequently depicted in conventional or subordinate roles, reinforcing antiquated stereotypes. Nevertheless, media possesses the capacity to instigate societal transformation by offering more progressive and equitable portrayals of gender roles, therefore contesting stereotypes and promoting enhanced gender equality (Shivam Gaur and Ananta Sarkar, 2024). In the professional environment, these prejudices frequently result in salary inequities and the underrepresentation of women in executive positions. Confronting gender biases and implementing diversity and inclusion measures can mitigate inequality (Du, 2024).

Advancing inclusive norms necessitates addressing cultural views and confronting structural disparities within businesses and societies. Thorough actions, including the enactment of affirmative policies and the promotion of inclusive leadership, are essential for eradicating systemic gender discrimination and harassment. Furthermore, socio-legal frameworks emphasize the necessity of dismantling patriarchal gender roles and advancing gender-sensitive education to surmount these structural impediments. Promoting male involvement in caring is a vital component of fostering gender equality. Although numerous campaigns emphasize education and awareness, these endeavors frequently yield restricted success. To achieve significant change, it is essential to incorporate men in caring roles and confront the emotional and societal obstacles that hinder their full participation. This transition can contest conventional gender stereotypes and foster a more equitable distribution of labor in both professional and domestic spheres.

Although these tactics are essential, efforts to date have frequently failed to effect substantial macro-level change. This underscores the necessity for more holistic strategies that concurrently tackle micro, meso, and macro societal levels. Integrating legal, cultural, and structural changes can facilitate enduring advancements in gender equality, guaranteeing the efficacy and permanence of initiatives aimed at dismantling cultural and structural obstacles.

Promoting export-related gender equality initiatives provides a strategic approach to mitigating gender inequality in export-driven businesses. Research indicates a favorable correlation between export intensity and gender inequality, underscoring the necessity for focused efforts in these areas. These activities correspond with the overarching endeavor to incorporate gender concerns into trade policies, ensuring equitable benefits from economic advancement for both men and women.

Historically, trade practices have generated imbalances that frequently disadvantage women. The World Trade Organization (WTO) is crucial in developing gender-responsive trade norms and assessing progress via processes such as the Trade Policy Review Mechanism. This method guarantees that trade agreements effectively promote gender equality (Der Boghossian, 2023). Gender-responsive trade policies are crucial for eliminating obstacles encountered by women in commerce. These policies can offer both financial and non-financial incentives while enhancing access to essential trade-related infrastructure. Addressing these hurdles might generate additional opportunities for women entrepreneurs and female workers in export sectors, hence promoting a more inclusive trading environment (Der Boghossian, 2023). Women-owned SMEs in underdeveloped nations encounter distinct problems that necessitate customized governmental solutions. Gender-disaggregated data is essential for comprehending the differences between women-owned exporting and non-exporting SMEs, facilitating more effective policy interventions (Samy et al., 2023).

Advancing gender equality is essential for fostering entrepreneurship and creativity, both of which are vital for economic development. Women frequently encounter barriers to market access, technology, and financing, intensified by enduring gender bias and inequality (Veckalne and Tambovceva, 2023). Enacting regulations that address these biases, enhance resource accessibility, and cultivate supportive networks and mentorship opportunities for female entrepreneurs might mitigate these barriers, hence promoting women's advancement in the entrepreneurial ecosystem.

Integrating gender equality programs in export-linked industries is a viable technique; nonetheless, it is essential to acknowledge that overarching cultural norms and biases may impede development. An integrated strategy encompassing policy interventions and cultural transformations is essential to foster an environment that promotes gender equality in all sectors. Addressing these systemic concerns enhances the feasibility of achieving more inclusive and fair economic growth.

Achieving gender equality necessitates an integrated, comprehensive strategy that confronts both institutional and cultural obstacles, rather than relying on single solutions. By integrating strategic policy measures with endeavors to transform societal norms, we can facilitate the development of a more inclusive and equitable global society. Governments, organizations, and communities must cooperate to eliminate institutional barriers, advance educational and leadership possibilities for women, and guarantee that economic development equitably benefits all genders.

Gender equality is both a moral obligation and a driver of sustainable development and innovation. Nations pursuing advancement should invest in women's education, encourage their involvement in high-growth sectors, advocate for equitable remuneration, and enhance female leadership to achieve extensive societal and economic advantages. These approaches foster more resilient economies, equitable communities, and robust governance frameworks. By addressing entrenched biases and aligning policies with a gender-inclusive framework, we can guarantee that all persons, irrespective of gender, possess equal chances to contribute to and benefit from global development. This comprehensive approach ensures that gender equality is realized, promoting a harmonious, wealthy, and sustainable world for future generations.

4.5 CONCLUSIONS

Attaining gender equality is not merely a moral obligation but also an essential catalyst for economic advancement, social well-being, and political stability. The complex interplay of economic growth, legal systems, cultural norms, and structural obstacles significantly influences gender outcomes. Although economic development can facilitate chances for enhanced gender equality, it is not an all-encompassing answer by itself. This intricacy underscores the imperative of implementing specific policies, institutional reforms, and governmental expenditures to promote gender equality.

Economic development has a distinct inverse correlation with gender disparity, as affluent nations possess greater capacity to commit resources towards social initiatives that support women, encompassing healthcare, education, and employment engagement. The empirical evidence indicates that an increase in per capita GDP correlates with a reduction in gender inequality. This is primarily due to wealthier nations allocating larger resources to social infrastructure, hence fostering women's empowerment and enhancing their participation in economic activity.

Nonetheless, economic growth alone does not inherently result in gender equality. The study indicates that export-oriented economies, frequently dependent on labor-intensive sectors, may indeed sustain gender disparities. Export-oriented sectors are typically male-dominated or promote gender-segregated labor markets, so constraining the advantages women can get from economic growth. Consequently, growth must be supported by targeted policies that guarantee women's access to education, training, and employment in high-growth industries such as technology and renewable energy.

A primary barrier to attaining gender equality is the male-to-female ratio. Nations with elevated male-to-female ratios typically have more gender disparity, attributable to societal norms that devalue women's contributions in the workforce. These traditions frequently inhibit women's access to higher-paying employment, constraining their financial autonomy and perpetuating gender inequality. The endurance of these gender prejudices underscores the necessity for policies that tackle not only economic but also cultural and institutional obstacles.

Alongside economic issues, proficient governance is essential for promoting gender equality. The pronounced negative link between the rule of law and gender disparity highlights the significance of legal frameworks in advancing equality. Countries with strong legal frameworks that implement anti-discrimination laws and protect women's rights in the workplace are more likely to diminish gender inequality. Governance systems that foster accountability, openness, and impartiality are crucial for the effective implementation of law provisions.

Furthermore, governmental spending on social safety, education, and welfare services is essential in alleviating gender inequality. Investments in childcare, healthcare, and educational opportunities facilitate women's fuller participation in the workforce, ultimately diminishing gender inequities. Government expenditure on education directly affects women's access to higher-paying employment, while social protection programs facilitate the reconciliation of professional and familial obligations for women. The persistent evidence that social spending mitigates gender inequality underscores the necessity for governments to prioritize these expenditures in their budgets.

Notwithstanding these advancements, institutional and cultural obstacles persist in hindering progress. Cultural standards, especially in patriarchal societies, frequently prescribe conventional gender roles that restrict women's involvement in school, employment, and leadership responsibilities. These conventions are well established and pose significant obstacles to the execution of gender equality programs. Moreover, even in nations with progressive legislation, institutional prejudices inside corporations and governmental bodies may obstruct the implementation of laws, permitting gender inequality to endure.

The research underscores that a comprehensive strategy is vital for attaining enduring advancements in gender equality. Economic growth must be accompanied by specific social and labor market changes, as well as robust legal frameworks that promote gender equality. Cultural and institutional reforms are equally vital for eliminating the structural impediments that sustain gender disparity. Governments must invest in social safety and education while aggressively transforming societal beliefs that limit women's potential.

REFERENCES

Abd Majid, N., Jaaffar, A.H., 2023. The Effect of Women's Leadership on Carbon Disclosure by the Top 100 Global Energy Leaders. *Sustainability* 15, 8491. https://doi.org/10.3390/su15118491

Abdoul-Azize, H.T., El Gamil, R., 2021. Social Protection as a Key Tool in Crisis Management: Learnt Lessons from the COVID-19 Pandemic. *Glob. Soc. Welf.* 8, 107–116. https://doi.org/10.1007/s40609-020-00190-4

Adeola, O., Evans, O., Ngare, I., 2024a. Explaining Gendered Vulnerability to Climate Change: The Contextual Conditions. In: *Gender Equality, Climate Action, and Technological Innovation for Sustainable Development in Africa*, Sustainable Development Goals Series. Springer International Publishing, Cham, pp. 59–79. https://doi.org/10.1007/978-3-031-40124-4_3

Adeola, O., Evans, O., Ngare, I., 2024b. Gender Equality, Climate Action, and Technological Innovation for Sustainable Development in Africa: A Background. In: *Gender Equality, Climate Action, and Technological Innovation for Sustainable Development in Africa*, Sustainable Development Goals Series. Springer International Publishing, Cham, pp. 3–34. https://doi.org/10.1007/978-3-031-40124-4_1

Adeola, O., Evans, O., Ngare, I., 2024c. Gender and Climate Issues in Africa: Regional Insights, Challenges, and Recommendations. In: *Gender Equality, Climate Action, and Technological Innovation for Sustainable Development in Africa*, Sustainable Development Goals Series. Springer International Publishing, Cham, pp. 35–58. https://doi.org/10.1007/978-3-031-40124-4_2

Afzal, A., Khan, S., Daud, S., Ahmad, Z., Butt, A., 2023. Addressing the Digital Divide: Access and Use of Technology in Education. *J. Soc. Sci. Rev.* 3, 883–895. https://doi.org/10.54183/jssr.v3i2.326

Alamneh, T.S., Teshale, A.B., Yeshaw, Y., Alem, A.Z., Ayalew, H.G., Liyew, A.M., Tessema, Z.T., Tesema, G.A., Worku, M.G., 2022. Socioeconomic Inequality in Barriers for Accessing Health Care Among Married Reproductive Aged Women in Sub-Saharan African Countries: A Decomposition Analysis. *BMC Womens Health* 22, 130. https://doi.org/10.1186/s12905-022-01716-y

Altarawneh, Y., 2020. Explaining the Gender-Gap in Economic Activity: A Cross-Country Study. *Res. World Econ.* 11, 16. https://doi.org/10.5430/rwe.v11n5p16.

Althaber, A., Leuze, K., Künzel, R., 2023. Financial Solidarity or Autonomy? How Gendered Wealth and Income Inequality Influence Couples' Money Management. *Soc. Incl.* 11. https://doi.org/10.17645/si.v11i1.6005

Amber, H., Chichaibelu, B.B., 2023. Narrowing the Gender Digital Divide in Pakistan: Mobile Phone Ownership and Female Labor Force Participation. *Rev. Dev. Econ.* 27, 1354–1382. https://doi.org/10.1111/rode.12994

Amir-ud-Din, R., Naz, L., Ali, H., 2023. *Relationship Between Asset Ownership and Women's Empowerment: Evidence from DHS Data from 18 Developing Countries*. Cambridge University Press, Cambridge.

Antón, J.-I., Grande, R., Muñoz De Bustillo, R., Pinto, F., 2023. Gender Gaps in Working Conditions. *Soc. Indic. Res.* 166, 53–83. https://doi.org/10.1007/s11205-022-03035-z

Anukriti, S., Erten, B., Mukherjee, P., 2022. *Women's Political Representation and Intimate Partner Violence*. IZA Discussion Paper No. 15395. https://docs.iza.org/dp15395.pdf

Asongu, S., Nting, R., 2021. The Comparative Economics of Financial Access in Gender Economic Inclusion. *Afr. J. Econ. Manag. Stud.* 12, 193–207. https://doi.org/10.1108/AJEMS-06-2020-0268

Assaduzzaman, M., Filatova, T., Lovett, J.C., Coenen, F.H.J.M., 2023. Gender-Ethnicity Intersectionality in Climate Change Adaptation in the Coastal Areas of Bangladesh. *Sustainability* 15, 3744. https://doi.org/10.3390/su15043744

Baba, A., Iddrisu, T.I., Doke, E.A., 2024. The Cry of Landless Smallholder Women Farmers and Agro-Processors in Northern Ghana-Evidence from West Mamprusi Municipality. *Int. J. Multidiscip. Res.* 6, 25229. https://doi.org/10.36948/ijfmr.2024.v06i04.25229

Bagade, T., Chojenta, C., Harris, M., Oldmeadow, C., Loxton, D., 2022. The Human Right to Safely Give Birth: Data from 193 Countries Show That Gender Equality Does Affect Maternal Mortality. *BMC Pregnancy Childb.* 22, 874. https://doi.org/10.1186/s12884-022-05225-6

Baskaran, T., Hessami, Z., 2023. Women in Political Bodies as Policymakers. *Rev. Econ. Stat.*, 1–46.

Benedi Lahuerta, S., Rejchrt, P., Patrick, A., 2024. The UK Pay Transparency Regulations: Apparent Transparency Without Accountability? *Leg. Stud.* 44, 21–44. https://doi.org/10.1017/lst.2023.12

Bin-Humam, Y., Braunmiller, J.C., Elsaman, M., 2023. *Emerging Trends in National Financial Inclusion Strategies That Support Women's Entrepreneurship*. Washington, DC: World Bank. https://doi.org/10.1596/39659

Börkan, B., Bakış, O., 2016. Determinants of Academic Achievement of Middle Schoolers in Turkey. *Educ. Sci. Theory Pract.* https://doi.org/10.12738/estp.2016.6.0227

Carneiro, C.M.M., Pinho, P.D.S., Teixeira, J.R.B., Araújo, T.M.D., 2023. Trabalho doméstico não remunerado: persistência da divisão sexual e transtornos mentais. *Rev. Saúde Pública* 57, 31. https://doi.org/10.11606/s1518-8787.2023057004502

Castán Broto, V., 2024. Gender Equality and Social Inclusion in Community Energy: An Intersectional Perspective. In: Castán Broto, V. (Ed.), *Community Energy and Sustainable Energy Transitions*. Springer Nature Switzerland, Cham, pp. 71–93. https://doi.org/10.1007/978-3-031-57938-7_4

Chandrasekaran, M., Krishnapriya, P.P., Jeuland, M., Pattanayak, S.K., 2023. Gender Empowerment and Energy Access: Evidence from Seven Countries. *Environ. Res. Lett.* 18, 045003. https://doi.org/10.1088/1748-9326/acc2d3

Chowdhuri, P.D., Kundu, K., Meyur, S., 2023. Change in Women's Work Burden and Gender Norms of Unpaid Care Work after the Second Wave of the COVID-19 Pandemic in India. *Work* 76, 897–905. https://doi.org/10.3233/WOR-220047

Comstock, A.L., 2024. Signing CEDAW and Women's Rights: Human Rights Treaty Signature and Legal Mobilization. *Law Soc. Inq.* 49, 1222–1255. https://doi.org/10.1017/lsi.2023.24

Consoli, D., Castellacci, F., Santoalha, A., 2023. E-Skills and Income Inequality within European Regions. *Ind. Innov.* 30, 919–946. https://doi.org/10.1080/13662716.2023.2230222

Cookson, T.P., Sandoval, R., Staab, S., Tabbush, C., Bitterly, J., Mathew, M., 2024. Do Governments Account for Gender When Designing Their Social Protection Systems? Findings from an Analysis of National Social Protection Strategies. *Soc. Policy Adm.* 58, 78–92. https://doi.org/10.1111/spol.12944

Dahal, P., Joshi, S.K., Swahnberg, K., 2022. A Qualitative Study on Gender Inequality and Gender-Based Violence In Nepal. *BMC Public Health* 22, 2005. https://doi.org/10.1186/s12889-022-14389-x

De Boer, J., Aiking, H., 2023. Pro-environmental Food Practices in EU Countries Strongly Suggest Mutually Reinforcing Improvements in Gender Equality and Environmental Sustainability. *Appetite* 180, 106350. https://doi.org/10.1016/j.appet.2022.106350

Der Boghossian, A., 2023. Gender-Responsive WTO: Making Trade Rules and Policies Work for Women*. In: Bahri, A., López, D., Yves Remy, J. (Eds.), *Trade Policy and Gender Equality*. Cambridge University Press, pp. 21–43. https://doi.org/10.1017/9781009363716.004

Deshpande, A., Kabeer, N., UNU-WIDER, 2021. *Norms That Matter: Exploring the Distribution of Women's Work Between Income Generation, Expenditure-Saving, and Unpaid Domestic Responsibilities in India* (WIDER Working Paper No. 2021). WIDER Working Paper. UNU-WIDER. https://doi.org/10.35188/UNU-WIDER/2021/070-2

Dhahri, I., 2023. The Role of International Human Rights Norms in Combating Discrimination and Promoting Equality. *Easta J. Law Hum. Rights* 1, 137–146. https://doi.org/10.58812/eslhr.v1i03.90

Dhal, S., 2024. Rural Women's Response to Climate Change: An Exploratory Study of Women's Grassroot Network in Odisha. *Cosmop. Civ. Soc. Interdiscip. J.* 15, 131–141. https://doi.org/10.5130/ccs.v15.i3.8755

Diendéré, A.A., Ouédraogo, D., 2023. Women Farmers and Climate Change: Empirical Evidence from Burkina Faso. *Agric. Resour. Econ. Rev.* 52, 451–475. https://doi.org/10.1017/age.2023.12

Du, J., 2024. Advancing Gender Equality in the Workplace: Challenges, Strategies, and the Way Forward. *J. Theory Pract. Soc. Sci.* 4, 46–50. https://doi.org/10.53469/jtpss.2024.04(04).11

Duguma, L.A., Nzyoka, J., Obwocha, E., Minang, P., Wainaina, P., Muthee, K., 2022. The Forgotten Half? Women in the Forest Management and Development Discourse in Africa: A Review. *Front. For. Glob. Change* 5, 948618. https://doi.org/10.3389/ffgc.2022.948618

Duvendack, M., Sonne, L., Garikipati, S., 2023. Gender Inclusivity of India's Digital Financial Revolution for Attainment of SDGs: Macro Achievements and the Micro Experiences of Targeted Initiatives. *Eur. J. Dev. Res.* 35, 1369–1391. https://doi.org/10.1057/s41287-023-00585-x

Evans, D.K., Akmal, M., Jakiela, P., 2021. Gender Gaps in Education: The Long View. *IZA J. Dev. Migr.* 12, 20210001. https://doi.org/10.2478/izajodm-2021-0001

Falk, A., Hermle, J., 2018. Relationship of Gender Differences in Preferences to Economic Development and Gender Equality. *Science* 362. https://doi.org/10.1126/science.aas9899.

Flavell, J., 2023. Lessons from the Women and Gender Constituency: Interrogating Civil Society Strategies for Organising in the UNFCCC. *Int. J. Polit. Cult. Soc.* 36, 385–403. https://doi.org/10.1007/s10767-023-09448-z

Fosch-Villaronga, E., Poulsen, A., Søraa, R.A., Custers, B.H.M., 2021. A Little Bird Told Me Your Gender: Gender Inferences in Social Media. *Inf. Process. Manag.* 58, 102541. https://doi.org/10.1016/j.ipm.2021.102541

Galos, D.R., Kulic, N., 2023. Variations of Gender Gaps in the Labour Market Outcomes of Graduates Across Fields of Study: A (Combined) Test of Two Theories. *Sociology* 57, 882–903. https://doi.org/10.1177/00380385221122400

Gaur, S., Sarkar, A., 2024. Deconstructing Gender Stereotypes in Indian Advertising: Insights from Media and "Bulbbul". *Int. J. Multidiscip. Res.* 6, 23751. https://doi.org/10.36948/ijfmr.2024.v06i03.23751

Ghosh, A., Dutta, K., 2024. Health Threats of Climate Change: From Intersectional Analysis to Justice-Based Radicalism. *Ecol. Soc.* 29, art15. https://doi.org/10.5751/ES-14045-290215

Głodowska, A.E., 2018. Gender Differences in Educational Attainment: The Evidence from the European Union Countries. *Horyzonty Wych.* 221–230. https://doi.org/10.17399/HW.2018.174319

Gornick, J., 2004. Women's Economic Outcomes, Gender Inequality and Public Policy: Findings from the Luxembourg Income Study. *Socio-Econ. Rev.* 2, 213–238. https://doi.org/10.1093/SOCECO/2.2.213

Goudet, J.-M., Arif, F.B., Owais, H., Ahmed, H.U., Ridde, V., 2023. Intersecting Vulnerabilities: Unraveling the Complex Relationship between Women, Climate Change, and Mental Health in Bangladeshi Communities. *MedRxiv*. https://doi.org/10.1101/2023.06.06.23290721

Gulvira, A., Ainash, M., Sagynysh, M., Meiramgul, A., Aliya, K., 2024. The Impact of Female Entrepreneurship on Economic Growth in Developing and Developed Economies. *Economics* 12, 145–162. https://doi.org/10.2478/eoik-2024-0016

Gulyas, A., Seitz, S., Sinha, S., 2021. *Does Pay Transparency Affect the Gender Wage Gap*? Evidence from Austria.

Gupta, I., Ranjan, A., Barman, K., 2024. Are Gender Budgets Necessary for Reducing Inequality in Health Outcomes? An Exploratory Analysis. In: Dev, S.M., Ganesh-Kumar, A., Pandey, V.L. (Eds.), *Achieving Zero Hunger in India, India Studies in Business and Economics*. Springer Nature Singapore, Singapore, pp. 243–259. https://doi.org/10.1007/978-981-99-4413-2_9

Hasni M.N.N., Fernando, P.V., Silva, D.I.D., Ahamed, Vidhanaarachchi, S., 2023. *Empowering Women through Technology: Addressing Gender Disparities*. https://doi.org/10.22541/au.168296462.20128634/v1

Hayati, F.A., Arini, R.E., 2023. Exploring the Challenges and Opportunities Faced by Women Entrepreneurs in Sukabumi District: A Qualitative Study. *West Sci. Interdiscip. Stud.* 1, 77–86. https://doi.org/10.58812/wsis.v1i05.76

Hoi, N.K., Xuan, W.J., 2023. Breaking Barriers: An Examination of Gender Inequality and Discrimination in Malaysia's Workplace. *Int. J. Acad. Res. Bus. Soc. Sci.* 13, 1113–1127. https://doi.org/10.6007/IJARBSS/v13-i5/16791

Huang, P., Liu, X., 2024. Challenging Gender Stereotypes: Representations of Gender Through Social Interactions in English Learning Textbooks. *Humanit. Soc. Sci. Commun.* 11, 819. https://doi.org/10.1057/s41599-024-03293-x

The influence of the professional environment on women leadership: A Quantitative Investigation, 2023. *PsychologyandEducation* 55. https://doi.org/10.48047/pne.2018.55.1.73

Ingole, C.K., 2023. Economic Impact of Productive Use of Renewable Energy: A Case of a Women-Collective from Rural Maharashtra (India). *Eur. J. Sustain. Dev.* 12, 467. https://doi.org/10.14207/ejsd.2023.v12n4p467

Islam, A., Manchanda, P., 2023. Gender Inequality in Digital India: A Survey on Digital Literacy, Access, and Use. *Digital Futures at Work Research Centre (Digit)*. https://doi.org/10.20919/MCUU2363

Islam, Q.S., Akter, N., 2024. Gender-based Violence (GBV) on Bangladeshi Women and Girls during COVID-19 in Toronto: Forms, Causes, and Mental Health Impacts. *J. Adv. Med. Med. Res.* 36, 307–319. https://doi.org/10.9734/jammr/2024/v36i75505

James, R., Gibbs, B., Whitford, L., Leisher, C., Konia, R., Butt, N., 2021. Conservation and Natural Resource Management: Where Are All the Women? *Oryx* 55, 860–867. https://doi.org/10.1017/S0030605320001349

Jayachandran, S., 2014. The Roots of Gender Inequality in Developing Countries. *Political Behavior: Race*. https://doi.org/10.1146/ANNUREV-ECONOMICS-080614-115404

Jeffs, N., 2023. Climate Action in China. *Royal Institute of International Affairs*. https://doi.org/10.55317/9781784135591

Kaganovska, T.E., Syroid, T.L., 2023. The European Union Organizational and Legal Mechanism for Ensuring Gender Equality. *Uzhhorod Natl. Univ. Her. Ser. Law* 2, 232–239. https://doi.org/10.24144/2307-3322.2022.76.2.37

Keränen, T., Enwald, H., 2023. Everyday Energy Information Literacy and Attitudes Towards Energy-Related Decisions: Gender Differences among Finns. *Resources* 12, 70. https://doi.org/10.3390/resources12060070

Kim, S., 2021. Determining Critical Factors of Gender Inequality: Evidence from 34 OECD and non-OECD Countries. *World Dev. Perspect.*, 21, 100284. https://doi.org/10.1016/J.WDP.2020.100284.

Koirala, B., Harvey, S., Bollinger, R.C., Davidson, P.M., 2024. A Call for Action: Need to Expand the Scope of Women's Health Assessment. *J. Adv. Nurs.* 80, 407–409. https://doi.org/10.1111/jan.15788

Koirala, P., 2023. Practices of Gender-Responsive Budgeting in Nepal. *Bouddh. Abhiyan* 8, 163–173. https://doi.org/10.3126/bdkan.v8i01.57805

Kowalewska, H., 2023. Gendered Employment Patterns: Women's Labour Market Outcomes Across 24 Countries. *J. Eur. Soc. Policy* 33, 151–168. https://doi.org/10.1177/09589287221148336

Kucharski, L., Rutkowska, A., 2021. The Situation of Women on the Polish Labour Market Compared to the Situation of Men. *Wiad. Stat. Pol. Stat.* 66, 46–67. https://doi.org/10.5604/01.3001.0015.2306

Kumari, N., 2024. Politics of Education in the USA: Deconstructing Gender Dynamics. *World J. Adv. Res. Rev.* 22, 116–123. https://doi.org/10.30574/wjarr.2024.22.1.1040

Lambert, E., Deyganto, K.O., Yure, M., Abebe, A., Sherbo, H.H., 2023. *Empowering Women in Leadership and Management Positions to Maintain Gender Equality: A Case Study on Sidama Region*. https://doi.org/10.32388/AH0O94

Leite, B.O., De Medeiros, D.S., Magno, L., Bastos, F.I., Coutinho, C., De Brito, A.M., Cavalcante, M.S., Dourado, I., 2021. Association between Gender-based Discrimination and Medical Visits and HIV Testing in a Large Sample of Transgender Women in Northeast Brazil. *Int. J. Equity Health* 20, 199. https://doi.org/10.1186/s12939-021-01541-z

Leoncini, R., Macaluso, M., Polselli, A., 2024. Gender Segregation: Analysis across Sectoral-Dominance in the UK Labour Market. *Empir. Econ.* https://doi.org/10.1007/s00181-024-02611-1

Leshchenko, O., Chung, H., 2023. *Homeworking and Division of Domestic Work: The Role of Gender Role Attitudes in Germany*. https://doi.org/10.31235/osf.io/85b23

Li, Y., 2022. Analyzing Gender Roles in Housework Norms from a Quantitative Perspective. *Stud. Soc. Sci. Humanit.* 1, 89–92. https://doi.org/10.56397/SSSH.2022.10.09

Liu, S., 2022. Gender Disparities in Rural Education Attainments and Agricultural Landownership from the Perspective of Sustainable Development Goals (SDGs): Evidence from 16 Sub-Sahara African Countries. *RAUSP Manag. J.* 57, 401–412. https://doi.org/10.1108/RAUSP-01-2022-0031

Liu, S., 2024. Research on the Legal Dilemma of Workplace Gender Discrimination and Its Feminist Countermeasures. *J. Humanit. Arts Soc. Sci.* 8, 841–845. https://doi.org/10.26855/jhass.2024.04.004\

Mafida Amalia, A., Rahmawati, F., Nuraini Dwi Putri, I., 2023. The Influence of Women's Development and Empowerment on Inclusive Economic Growth. *J. Appl. Bus. Tax. Econ. Res.* 2, 546–559. https://doi.org/10.54408/jabter.v2i5.198

Magda, I., Cukrowska-Torzewska, E., Palczyńska, M., 2024. What If She Earns More? Gender Norms, Income Inequality, and the Division of Housework. *J. Fam. Econ. Issues* 45, 1–20. https://doi.org/10.1007/s10834-023-09893-0

Maria Giammarioli, A., Longo, E., Bucciardini, R., 2024. Gender-Based Violence Is a Never to be Forgotten Social Determinant of Health: A Narrative Literature Review. In: Kabir, R., Davod Parsa, A., Lakhno, I. (Eds.), *Women's Health Problems—A Global Perspective*. IntechOpen. https://doi.org/10.5772/intechopen.110651

Markan, M., Dhingra, R., Segan, M., Dabla, V., Sagar, M., Neogi, S., Dey, S., Chakravarty, N., 2022. Gender-based Violence Programming in Times of COVID-19: Challenges, Strategies and Recommendations. *Front. Glob. Womens Health* 3, 952688. https://doi.org/10.3389/fgwh.2022.952688

Masenya, M., 2024. The Implications of Gender Inequality and Land Allocation South African Rural Areas: A Theoretical Perspective. *Int. J. Bus. Ecosyst. Strategy 2687–2293* 6, 190–195. https://doi.org/10.36096/ijbes.v6i3.485

Moreira, L.R., Blumenberg, C., Caicedo Velasquez, B.E., Ewerling, F., Balandrán, A., Vidaletti, L.P., Varela, A.R., Hellwig, F., Ponce De Leon, R.G., Barros, A.J.D., Silveira, M.F., Wehrmeister, F.C., 2023. The Role of Gender Inequality and Health Expenditure on the Coverage of Demand for Family Planning Satisfied by Modern Contraceptives: A Multilevel Analysis of Cross-Sectional Studies in 14 LAC Countries. *Lancet Reg. Health Am.* 19, 100435. https://doi.org/10.1016/j.lana.2023.100435

Morgan, A.U., Chaiyachati, K.H., Weissman, G.E., Liao, J.M., 2018. Eliminating Gender-Based Bias in Academic Medicine: More Than Naming the "Elephant in the Room." *J. Gen. Intern. Med.* 33, 966–968. https://doi.org/10.1007/s11606-018-4411-0

Mukurazhizha, R.M., Matanga, S.Y., 2023. The Role of Women in Managing the Environmental Crisis: A Case Study of Cyclone Idai in Chipinge, Zimbabwe. *HTS Teol. Stud. Theol. Stud.* 79. https://doi.org/10.4102/hts.v79i3.8042

Ndep, A.O., Obidile, V.C., Akwafuo, S., Ekwebene, O.C., Azubike, C.P., Umechinedu, R.-D., Obidile-Ikwegbu, L.E., 2022. Knowledge and Incidence of Gender-Based Violence among Young Adults in the South-South Region of Nigeria: A Comparative Study. *Res. Square*. https://doi.org/10.21203/rs.3.rs-1821610/v2

Ndimele, G.C., Xin, S., Vefemsi, V.E., 2023. Addressing Time Poverty among Women in Nigeria Through Building of Infrastructures. *J. Int. Coop. Dev*. 6, 1. https://doi.org/10.36941/jicd-2023-0018

Nevo, C.M., Egenti, S., 2019. A Disaggregated Analysis of Wealth Status and Educational Attainment in Nigeria Using the Multinomial Logit Approach. *Economies* 7, 39. https://doi.org/10.3390/economies7020039

Ngugi, M., Muthima, P., 2017. Female Participation in Technical, Vocational Education and Training Institutions (TVET) Subsector: The Kenyan Experience. *Public Policy Adm. Res.* 7(4), 9–23. https://www.iiste.org/Journals/index.php/PPAR/article/view/36735/37749

Niemann, J., El-Mahdi, M., Samuelsen, H., Tersbøl, B.P., 2024. Gender Relations and Decision-Making on Climate Change Adaptation in Rural East African Households: A Qualitative Systematic Review. *PLoS Clim.* 3, e0000279. https://doi.org/10.1371/journal.pclm.0000279

Nugroho, W.S., 2023. The Effect of Social Support on Women-Inclusive Leadership Mediated with Personal Growth Initiative. *RSF Conf. Ser. Bus. Manag. Soc. Sci.* 3, 494–501. https://doi.org/10.31098/bmss.v3i3.714

Ortiz Cobo, M., Jeri Levano, R.I., 2023. Gender Digital Divide in Migratory Contexts: The Case of Peruvian Migrant Women in Italy and Spain. *J. Educ. Cult. Soc.* 14, 575–596. https://doi.org/10.15503/jecs2023.1.575.596

Pasha, H.K., 2024. Gender Differences in Education: Are Girls Neglected in Pakistani Society? *J. Knowl. Econ.* 15, 3466–3511. https://doi.org/10.1007/s13132-023-01222-y

Passaretta, G., Gil-Hernández, C., 2023. The Early Roots of the Digital Divide: Socioeconomic Inequality in Childrens ICT Literacy from Primary to Secondary Schooling. In: Skopek, J. (Ed.), *Research Handbook on Digital Sociology*. Edward Elgar Publishing, pp. 307–327. https://doi.org/10.4337/9781789906769.00026

Picatoste, X., Mesquita, A., González-Laxe, F., 2023. Gender Wage Gap, Quality of Earnings and Gender Digital Divide in the European Context. *Empirica* 50, 301–321. https://doi.org/10.1007/s10663-022-09555-8

Pinem, M.L., 2023. Applying an Intersectionality Approach to Multiple Dimensions of Social Life. *J. Ilmu Sos. Dan Hum.* 12, 228–235. https://doi.org/10.23887/jish.v12i2.59906

Popoola, N.I., Ojetunde, I., Sule, A.I., Okoh, S.A., 2023. Factors Influencing Gender Inequality in Land and Property Ownership Among Peri-Urban Women in Minna, Nigeria. *Ghana J. Dev. Stud.* 20, 58–81. https://doi.org/10.4314/gjds.v20i1.4

Protosavitska, L., 2023. Legal Aspects of Gender Equality and Their Legislative Consolidation. *Law Hum. Environ.* 14. https://doi.org/10.31548/law/1.2023.88

Quadlin, N., 2022. Striving for Equality in an Unequal World. *Contemp. Sociol. J. Rev.* 51, 89–92. https://doi.org/10.1177/00943061221076190a

Rafferty, Y., 2013. International Dimensions of Discrimination and Violence against Girls: A Human Rights Perspective. *J. Int. Women's Stud.* 14(1), 1–5.

Ramírez, L., Palacios-Espinosa, X., Ferroni, P., Beltrán, C.A., 2023. Social Representations of Gender: A Contextual Construction in Schools. *Rev. Colomb. Psicol.* 32, 65–84. https://doi.org/10.15446/rcp.v32n2.102351

Reeves, A., Brown, C., Hanefeld, J., 2022. Female Political Representation and the Gender Health Gap: A Cross-National Analysis of 49 European Countries. *Eur. J. Public Health* 32, 684–689. https://doi.org/10.1093/eurpub/ckac122

Reich-Stiebert, N., Froehlich, L., Voltmer, J.-B., 2023. Gendered Mental Labor: A Systematic Literature Review on the Cognitive Dimension of Unpaid Work Within the Household and Childcare. *Sex Roles* 88, 475–494. https://doi.org/10.1007/s11199-023-01362-0

Rießenberger, K.A., Fischer, F., 2023. Age and Gender in Gerontechnology Development: Emphasizing the Need for an Intersectional Approach. *Z. Für Gerontol. Geriatr.* 56, 189–194. https://doi.org/10.1007/s00391-023-02183-2

Roessle, F., Fleischmann, C., Roessle, K., 2024. Gender Diversity and Financial Performance in Executive Positions in German Companies. *Probl. Perspect. Manag.* 22, 571–581. https://doi.org/10.21511/ppm.22(2).2024.44

Rohmatilah, D.A., 2023. The Role of Gender Equality on Poverty Alleviation: Case of Indonesia. *J. Perenc. Pembang. Indones. J. Dev. Plan.* 7, 272–287. https://doi.org/10.36574/jpp.v7i2.450

Rosenfeld, R., Kalleberg, A., 1991. Gender Inequality in the Labor Market. *Acta Sociol.*, 34, 207–225. https://doi.org/10.1177/000169939103400304.

Sahu, V.K., Baral, S.K., Singh, R., 2024. Financial Empowerment of Tribal Women: An Inquiry into Sustainable Economic Justice Initiatives and Pathways towards Inclusive Development. *Asian J. Econ. Bus. Account.* 24, 182–194. https://doi.org/10.9734/ajeba/2024/v24i41272

Salman, M., 2023. *Addressing Gender Pay Disparity: A Closer Look at Wage Inequality for Women.* https://doi.org/10.31219/osf.io/n2ke5

Sam, K., Zibima, T., 2023. *Inclusive Environmental Decision-Making in Sub-Saharan Africa: Insights from the Ogoni Remediation Project.* Niger Delta, Nigeria. https://doi.org/10.21203/rs.3.rs-2827128/v1

Samy, Y., Adedeji, A., Iraoya, A., Dutta, M.K., Fakmawii, J.L., Hao, W., 2023. Trade and Women's Economic Empowerment: Survey Results for SMEs Across Six Developing Countries. In: *Trade and Women's Economic Empowerment.* Springer Nature Switzerland, Cham, pp. 21–57. https://doi.org/10.1007/978-3-031-39039-5_2

Santoniccolo, F., Trombetta, T., Paradiso, M.N., Rollè, L., 2023. Gender and Media Representations: A Review of the Literature on Gender Stereotypes, Objectification and Sexualization. *Int. J. Environ. Res. Public. Health* 20, 5770. https://doi.org/10.3390/ijerph20105770

Seguino, S., 2000. Gender Inequality and Economic Growth: A Cross-Country Analysis. *World Dev.*, 28, 1211–1230. https://doi.org/10.1016/S0305-750X(00)00018-8

Sen, S.K., Karmakar, P., Adhikari, S., 2023. Empowerment and Women's Empowerment—A Theoretical Basis. *Int. J. Multidiscip. Res.* 5(3). https://doi.org/10.36948/ijfmr.2023.v05i03.4111

Shaheen, S., Naz, M., Ali, B., Gillani, S.M.A., Abbas, T., Ali, W., 2023. Discrimination against Working Women and Its Impact on Their Work Performance in District Sahiwal. *J. Educ. Soc. Stud.* 4, 59–69. https://doi.org/10.52223/jess.20234105

Sharma, M., 2022. Gendered Dimensions of Educational Premium Disadvantage in Earnings in USA, 2019. *GeoJournal* 87, 847–868. https://doi.org/10.1007/s10708-022-10623-6

Sharma, M., 2023. A Study on Women Entrepreneurship in India. *J. Surv. Fish. Sci.* 10, 3721–3723. https://doi.org/10.53555//sfs.v10i1.1715

Singh, V., Chobotaru, J., 2022. Digital Divide: Barriers to Accessing Online Government Services in Canada. *Adm. Sci.* 12, 112. https://doi.org/10.3390/admsci12030112

Singh, V., Patel, S., Singh, R., 2023. Analyzing Gender Disparities in Land Ownership and Wage Rates in Indian Agriculture: An Empirical Study. *Asian J. Agric. Ext. Econ. Sociol.* 41, 652–660. https://doi.org/10.9734/ajaees/2023/v41i92087

Spitzer, P.A., 2023. The Case for Gender Equity. In: *Empowering Female Climate Change Activists in the Global South: The Path Toward Environmental Social Justice.* Emerald Publishing Limited, pp. 19–41. https://doi.org/10.1108/978-1-80382-919-720231003

Stanisavljević, B., 2023. Perspectives and Problems of the Development of Women's Entrepreneurship in Serbia. *Knowl. Int. J.* 60, 119–124. https://doi.org/10.35120/kij6001119s

Suleiman, K.A., Akinrefon, A.A., Olakiitan, A.I., 2024. Variation in Educational Attainment Levels: Statistical Evidence from the Nigeria Demographic Health Surveys (2008, 2013 and 2018). *Int. J. Dev. Math. IJDM* 1, 191–199. https://doi.org/10.62054/ijdm/0102.15

Supporting Women's Empowerment Through Green Policies and Finance (OECD Environment Policy Papers No. 33), 2022. OECD Environment Policy Papers. https://doi.org/10.1787/16771957-en

Tehseem, T., Sibtain, M., Obaid, Z., 2018. Exploring Gender Stereotypes in Media Adverts: A Multimodal Analysis. *J. Soc. Sci. Humanit.* 57, 155–175. https://doi.org/10.46568/jssh.v57i2.42

Thwala, P., 2023. Women Representation in the Electoral Office: Progress and Challenges. *Curr. Res. J. Soc. Sci. Humanit.* 6, 112–121. https://doi.org/10.12944/CRJSSH.6.1.09

Veckalne, R., Tambovceva, T., 2023. The Importance of Gender Equality in Promoting Entrepreneurship and Innovation. *Mark. Manag. Innov.* 14, 158–168. https://doi.org/10.21272/mmi.2023.1-14

Weißflog, M.I., Grigoryan, L., Hofmann, W., 2023. The Interplay of Gender and Class in Social Judgment Across Different Cultural Levels of (Gender) Inequality. https://doi.org/10.31219/osf.io/sqdu5

Wielgoszewska, B., Bryson, A., Costa Dias, M., Foliano, F., Joshi, H., Wilkinson, D., 2023. Exploring the Reasons for Labour Market Gender Inequality a Year into the COVID-19 Pandemic: Evidence from the UK Cohort Studies. *Longitud. Life Course Stud.* 14, 180–202. https://doi.org/10.1332/175795921X16751166213852

Williams, A., Wood, S.N., Stuart, H.C., Wamue-Ngare, G., Thiongo, M., Gichangi, P., Devoto, B., Decker, M.R., 2022. Gendered Time Use During COVID-19 Among Adolescents and Young Adults in Nairobi, Kenya. *eClinicalMedicine* 49, 101479. https://doi.org/10.1016/j.eclinm.2022.101479

Yokobori, Y., Kiyohara, H., Mulati, N., Lwin, K.S., Bao, T.Q.Q., Aung, M.N., Yuasa, M., Fujita, M., 2023. Roles of Social Protection to Promote Health Service Coverage among Vulnerable People Toward Achieving Universal Health Coverage: A Literature Review of International Organizations. *Int. J. Environ. Res. Public. Health* 20, 5754. https://doi.org/10.3390/ijerph20095754

Zhang, A.Y., Gao, G., 2024. Intersectionality in Educational Psychology and Curriculum Studies: A Critical Review. *Extensive Rev*. 4, 1–14. https://doi.org/10.21467/exr.4.1.5387

Zhu, Z., 2024. The Impact of Gender Differences on Years of Education. *J. Educ*. 29.

5 Energy Efficiency

Mindaugas Butkus and Raimonda Valasevičienė

5.1 THE ENERGY EFFICIENCY CONCEPT

Energy efficiency is a complex concept that is represented in diverse fields including engineering, economics, energy, computer sciences, environmental sciences, mathematics, and physics (Dunlop, 2019). Energy efficiency is an evolving research topic for scholars, businesses, and policymakers for its undeniable role in meeting increasing energy demand, reducing CO_2 emissions, and tackling climate change.

Optimal energy use is globally recognized as the most crucial, cost-effective, immediate, underutilized, neglected, and misunderstood method to ensure future energy supply (Lovins, 2004). Energy efficiency is a broad concept that lacks a specific quantitative indicator. Energy efficiency has evolved since the first energy crisis in 1973. It was defined as the concept of achieving better economic output by consuming less energy (Yang and Yu, 2015). Various indicators are necessary to measure changes in energy efficiency. Energy efficiency is often quantified by the ratio of energy output to energy intake, represented as a percentage ranging from 0% to 100%. Energy efficiency refers to the practice of reducing energy use while maintaining the same level of service or production. Energy efficiency is often described as the proportion of usable outputs of a process to the energy input into that process. The challenge is in accurately determining the beneficial result and the energy required, leading to many significant methodological factors that are often ignored in the literature (Patterson, 1996).

Energy efficiency is, in principle, a simple idea: an output, for example, vehicle-km travelled, divided by the needed input energy. Efficiency improvements are regarded as an important means of mitigating not only climate change but also other environmental problems (Moriarty and Honnery, 2023).

Useful energy encompasses several technical kinds of energy needed by consumers to provide energy services, such as heat, mechanical energy, light, electrical energy, magnetic field energy, and electromagnetic radiation. Useful energies must generally be generated at the time and place of demand from final energy by employing energy converters such as motors, boilers, and light sources. Energy inputs refer to the fuels and electricity used in the production of products and services.

Energy efficiency may be assessed at several levels, ranging from the overall energy intensity of the economy and specific sectors to individual activities. An energy system may refer to a single energy conversion unit like a boiler, a building, an industrial process, a company, a sector, an entire economy, or a global energy network (Yang and Yu, 2015). Efficient energy use involves maximizing benefits by using energy wisely and possibly adjusting consumption to times when energy is plentiful and inexpensive, such as during off-peak hours or when renewable sources are generating more power. Efficiency has distinct interpretations for the two professions that prioritize it the most. Efficiency for engineers is defined as the physical ratio of output to input. For economists, efficiency is defined as the ratio of monetary output to input. However, many economists utilize the ratio of monetary output to physical input for practical reasons (Lovins, 2004).

Energy efficiency can also be defined as the ratio of effort, services, commodities, or received energy to energy prices. For example, thermal comfort in a building represents performance, transporting people or information is a service, and a smartphone is a product that needs energy for manufacture. GDP is often compared to energy use on a broad scale to calculate energy intensity,

DOI: 10.1201/9781003517481-5

typically measured in kilowatt-hours per euro. Energy efficiency is a top priority in the national energy strategy and is seen as a fundamental aspect of efforts to address climate change and promote sustainable development. Energy statistics are often categorized based on the economic sector where energy is used, rather than the specific physical end-use techniques or services being used (Dunlop, 2019).

End uses were first comprehensively examined in 1976. However, they are currently seldom included in official data and may be challenging to quantify precisely. End-use analysis is beneficial as it helps align energy suppliers in terms of quality, size, and quantity with end-use requirements, resulting in significant energy and cost savings. Providing excessive-quantity energy, rather than just the necessary amount for a job, is inefficient and costly (Lovins, 2004).

Energy efficiency inevitably has an ethical and technical dimension; conventional efficiency measures ignore important elements of energy costs, particularly in mining materials needed for renewable energy plants (Moriarty and Honnery, 2023).

5.1.1 Energy Conservation

Energy conservation is crucial when assessing energy efficiency. Energy saving is the act of decreasing energy use without considering the resulting output. Energy conservation involves reducing energy usage while maintaining comfort and productivity levels (Yang and Yu, 2015). Being conscious of shutting off lights when natural light is enough into buildings helps preserve energy. Another way to enhance energy efficiency is by replacing incandescent lamps with light-emitting diode (LED) lamps, which need less energy to generate the same number of lumens.

Energy savings are outcomes of investments and management practices that enhance energy efficiency. The sources of greenhouse gas emissions include several sectors, including homes, transportation, buildings, industry, agriculture, and energy supply and conversion systems including coal mining, power production, and oil and gas extraction and refining. The savings can be categorized as follows: a) savings resulting from high energy prices impacting demand for energy services and consumption; b) savings from switching fuel and energy technologies, such as replacing electric heaters with heat pumps to reduce electricity consumption; c) savings from changes in economic structure, like transitioning from heavy and manufacturing industries to the service sector; d) savings from enhancing energy efficiency, such as upgrading single-glazed windows to double- or triple-glazed windows in buildings; e) savings from utilizing information technology and smart grid management, such as global positioning systems and smart grid technologies, to save energy in transportation and power generation systems (Trianni et al., 2014).

Energy efficiency is rapidly increasing in size, speed, and affordability due to advancements in technology, delivery methods, and integrated design. Some individuals do not see the continual growth of energy services as a valuable objective, instead opting to lead a more simple lifestyle, beautifully thrifty rather than chaotic, fulfilling intangible wants with intangible resources. Such decisions may result in greater energy savings than technological advancements alone, even though they are often seen as outside the scope of energy efficiency (Lovins, 2004).

Energy efficiency is often considered an easily achievable goal by many observers. McKinsey (2009) identified 38 distinct greenhouse gas (GHG) reduction strategies, assessing each based on the amount of CO_2 equivalent emissions saved and the societal cost per ton averted (excluding taxes and subsidies). Out of the 38 energy savings options, 20 had a negative cost, indicating that the savings from reduced energy use would exceed the extra capital expenses of implementing more efficient equipment. The cost reductions were a result of transitioning to LED lighting, using hybrid automobiles, and producing power from landfills. Geothermal energy and some land management strategies have negligible replacement costs. Approaches, including nuclear, wind, solar, and technologies like CCS, achieved cost-effective reductions in greenhouse gas emissions (Huaman and Jun, 2014). We should implement steps to the point where marginal cost is equivalent to marginal gain from an economic perspective. It is contended that employing cost-cutting initiatives would decrease carbon emissions and lead to cost savings. Imagine a home aiming to optimize its winter comfort by adjusting the thermostat setting of its current heating system. The cost of the comfort unit is determined by its energy efficiency. Enhanced energy efficiency significantly decreases the expenses associated with certain comforts. Households choosing more insulation primarily want to save energy costs, but they may also seek to enhance their comfort.

Appliance manufacturers are motivated to promote significant energy savings, but customers are unable to immediately see these benefits and may be skeptical of the manufacturer's assertions. Electric providers promote energy-saving measures via advertising. Electric utility forecasts of energy savings surpass actual savings. Utility reports show greater predicted savings compared to projections from independent third parties. The findings align with the motivation utilities providers have to exaggerate reported savings when regulators reward utilities providers based on such savings.

5.1.2 Energy Intensity

Energy intensity is the antithesis of energy efficiency. Energy intensity is a commonly used metric for assessing energy efficiency, often calculated as the amount of primary energy used per unit of economic output (Yang and Yu, 2015). Energy intensity is often computed on a macroeconomic level, using energy consumption to gauge the whole economy and gross domestic product (GDP) to assess production (Andrieu et al., 2022). Energy intensity is the reciprocal of energy efficiency; however, it may vary due to factors other than changes in energy efficiency. Energy intensity is influenced by both energy efficiency within individual enterprises and the composition of industries. A transition in the industrial sector towards sectors with lower energy use will decrease energy intensity. Shifting the country's focus to services would also decrease energy intensity (Voigt et al., 2014). The cost of energy is another element that influences energy intensity. As energy intensity rises, it is expected to subsequently drop, but this change will be transient. If energy costs stay elevated for an extended duration, they will stimulate technical advancements that will continue to decrease energy intensity in the long run.

5.1.3 Energy Efficiency in Consumption

The main goal of energy efficiency is to reduce energy loss from the primary source to the secondary source or from the secondary source to the end usage. Implementing energy practices is essential for decreasing losses between infrastructure requirements and energy consumption. Energy is vital for enhancing productivity and maintaining a high standard of living, making the connection between energy and economic development significant (Moriarty and Honnery, 2023).

Energy efficiency is the practice of using minimal energy for heating, cooling, lighting, and equipment necessary to maintain suitable conditions in a structure. The user aims to maximize value under budget restrictions from an economic standpoint. Energy efficiency decreases the amount of energy needed from the power provider. The decrease indirectly impacts the utility's required capacity. By purchasing a programmable thermostat to regulate the temperature during winter nights, homeowners may reduce the amount of energy needed from utilities while they are asleep. The thermostat's energy-saving function does not always decrease the utility's capacity requirements, since they are determined by peak energy consumption, which usually happens around noon and sundown in a winter peak system. The distinction is in the units used to measure energy consumption and capacity demand: kilowatt-hours (kWh) for energy consumption and kilowatts (kW) for capacity demand. Energy may be defined as the capacity to do labor. Energy is the result of multiplying power, measured in kilowatts (kW), by time, measured in hours (Babuta et al., 2021). A 60 W bulb burning for one hour consumes 60 Wh (Lovins, 2017). Energy efficiency refers to the practice of reducing the amount of energy required to accomplish a certain job. Electricity efficiency should not decrease the kilowatt capacity needed to provide the electricity. A 60 W bulb running for 30 minutes out of 60 will use 30 Wh, but the electrical system will continue to draw 60 W (Trianni et al., 2014).

The efficiency of the overall energy system is another concern. Peak electricity demand necessitates backup power plants and extra transmission capacity, reducing the efficiency of the energy system.

There are some key terminological differences in energy efficiency. Five types of energy efficiency can be measured at different stages of energy conversion chains: a) conversion efficiency from primary to secondary energy; b) distribution efficiency from conversion point to end use; c) end-use efficiency in converting secondary energy to desired services; d) extractive efficiency in converting fuel or renewable sources to primary energy; e) hedonic efficiency in converting energy services to human welfare (Heun et al., 2018).

Considering all five efficiencies allows for comparing the main energy used to achieve the desired outcomes, such as enjoyment or economic well-being. Concentrating only on intermediate methods and goals overlooks the ultimate human objectives that an energy system should fulfil. Most cultures prioritize three aspects of energy efficiency: extraction, conversion, and distribution. Customers have the greatest potential to benefit from end-use and hedonic efficiency since they are not fully used (Ramos et al., 2015).

5.1.4 Efficiency of Energy in Production

Energy demand is driven by production demand. Companies use cost efficiency to lower production costs and boost revenue. Production is the first stage in cost evaluation. To consider energy efficiency, energy must be included as a raw element in the production function. Economic efficiency is determined by the Marginal Energy Product and relying on the average product may lead to wasteful outcomes. The corporation aims to optimize output within a certain budget to reduce manufacturing expenses. Companies aim to optimize resource use by minimizing the overall cost of generating a certain level of output. The correlation among energy, labor, and capital is crucial for evaluating energy efficiency in manufacturing.

It was discovered that when energy and capital are interdependent, increased energy costs lead to a decrease in the use of both energy and capital, resulting in a deceleration of the economy. An energy-capital substitution occurs when increased energy costs result in a higher use of capital rather than energy, hence mitigating the adverse impacts of elevated energy prices. The cross-price elasticity is a more straightforward measure of input substitutability (Frondel, 2004). A rise in energy prices causing a reduction in energy consumption would prompt the business to raise capital equipment to sustain production. This indicates a positive cross-price elasticity and suggests that the two inputs are substitutes. When a company decreases both capital and energy use in response to rising energy prices, it indicates that energy and capital are complementary (Todorov et al., 2019).

5.1.5 Benefits of Energy Efficiency

Supporters of energy efficiency often mention various advantages that could result from enhancing it, including energy conservation, environmental enhancement, energy security, reduced energy expenses, improved economic competitiveness, and job creation. Whole-system design may achieve significant energy reductions at a lower cost compared to smaller-scale efforts. To fully harness the significant potential of energy efficiency, it is essential to use precise language, prioritize effectively, pay attention to technical specifics and market shortcomings, and be open to acknowledging empirical evidence even if it contradicts economic theory (Lovins, 2004). Efficient energy practices may replace expensive and undesirable energy sources, improve security and wealth, accelerate global progress, and safeguard the earth's environment while still being profitable (Asif and Muneer, 2007). Energy efficiency is often the most significant, cost-effective, preferred, quickly adopted, least apparent, least comprehended, and most overlooked method for delivering energy services.

The United States Environmental Protection Agency (EPA) stresses that improving energy efficiency is a quick, cost-effective method to save money, lower greenhouse gas emissions, generate employment, and address increasing energy needs. Energy efficiency offers various advantages, including environmental benefits such as reducing greenhouse gas emissions and pollutants, as well as decreasing water consumption. It also provides economic benefits by lowering utility bills, generating employment opportunities, and stabilizing electricity costs. Additionally, energy efficiency contributes to utility system benefits by reducing overall electricity demand, thereby minimizing the necessity for new electricity infrastructure investments. Moreover, it serves as a risk management strategy by diversifying utility resource portfolios and mitigating uncertainties related to fluctuating fuel prices (Fuss et al., 2012).

It is emphasized that prioritizing energy efficiency is crucial for achieving the greatest reduction in carbon emissions at the most cost-effective rate. The IEA states that improving energy efficiency, primarily via cost-effective measures, has the most significant potential for decreasing CO_2 emissions by 2050. The greatest emphasis in the immediate term should be given to it (International Energy Agency, 2009).

5.1.6 Energy Efficiency Losses

Many subtleties of defining and measuring energy efficiency merit but seldom get rigorous treatment, such as the following losses, services, or metrics: a) distribution losses downstream of end-use devices (an efficient furnace feeding leaky ducts yields costlier delivered comfort); b) undesired or useless services, such as leaving equipment on all the time (as many factories do) even when it serves no useful purpose; c) misused services, such as space-conditioning rooms that are open to the outdoors; d) conflicting services, such as heating and cooling the same space simultaneously (wasteful even if both services are provided efficiently); e) parasitic loads, as when the inefficiencies of a central cooling system reappear as additional feedback cooling loads that make the whole system less efficient than the sum of its parts; f) misplaced efficiency, such as applying energy using equipment, however efficiently, to a task that does not need it, say, cooling with a mechanical chiller when groundwater or ambient conditions can more cheaply do the same thing; g) incorrect metrics, such as measuring lighting by raw quantity (lux or footcandles) unadjusted for its visual effectiveness, which may actually decrease if greater illuminance is improperly delivered (Lovins, 2004).

The energy-to-GDP ratio is a crude and oversimplified metric that may be deceptive since it encompasses variations in technological efficiency, human behavior, and the GDP composition (Lovins, 2004). Energy use should be cost-effective. Minimal energy use may not always align with the least expensive overall cost. Opportunity costs must be taken into account to pursue possibilities that provide the highest net gain. To address the negative externalities associated with energy usage, such as sulfur, NOx, and carbon dioxide emissions, it is necessary to align the marginal social benefit with the marginal social cost. This ensures that energy consumption remains below the market equilibrium level to promote social welfare. Effectiveness is the capacity to accomplish a desired outcome, closely related to efficiency, which measures how well something generates the intended result. The term "efficiency" might be positioned as the ratio of the achieved outcome to the theoretically optimal outcome, representing the near to ideal limit. Efficiency is enhanced when a service is delivered using fewer resources or when a higher level of service is attained with the same energy usage.

5.1.7 Energy Efficiency Gap and Rebound Effect

Government involvement in energy decision-making is justified by market failure resulting in socially inefficient consequences. The energy efficiency gap is the disparity between the actual energy efficiency level and the most cost-effective optimum energy efficiency level (Backlund et al., 2012). When a gap like this occurs, customers do not accurately assess the cost of energy saved, preventing the market from reaching an economically efficient outcome. Although incandescent lighting may be more cost-effective, people may still choose it (Reames et al., 2018). Consumers may discern disparities in quality and convenience. Consumers may not be inclined to invest more now to save future energy expenses. In 1865, British economist William Stanley Jevons observed that advancements in energy-efficient technology lead to increased use of current appliances at a higher intensity (intense margin) and the adoption of new appliances (high margin) (Lovins, 2017). The Jevons paradox states that energy-efficient appliances may not always lead to a decrease in energy use. The paradox of energy efficiency arises when the increased use of energy-saving equipment results in saving less energy compared to if the devices were not used more intensely. This conduct does not constitute a market failure. When energy prices are low, it is cost-effective to use more energy. Various factors may influence people's choices:

Significant discount rate. The energy efficiency gap originated from research that studied the acquisition and use of room air conditioners and the acquisition and use of water and space heaters. It was discovered that consumers tend to use a greater implicit discount rate than the market discount rate when making choices related to energy efficiency. Investigations identified an implicit discount rate of around 20%, which exceeds the market discount rate (Lovins, 2017).

Constrained reason. Bounded rationality is a cognitive theory that explains why customers do not choose energy-efficient options. Consumers utilize heuristics, which are simplified decision principles when they encounter challenging options instead of analyzing benefits and costs. Consumers often aim for a payback time of three years or less when considering an energy investment. Behavioral economics provides different approaches, such as limited rationality, to analyze consumer decision-making. Behavioral explanations pertain to psychological motives that result in deviation from rational utility maximizing. Framing may impact consumer decision-making. Decreasing energy expenses from $200 to $190 or from $20 to $10 results in a $10 savings (Lovins, 2017). However, the customer can see the second decrease as more significant if she considers it a 50% rather than a 5% save. Bounded rationality acknowledges the considerable expenses associated with acquiring more precise information, and thus it is not considered illogical conduct or a market failure. If customers find energy efficiency selections challenging, they could rely on different criteria other than the market interest rate to make such judgments (Gillingham and Palmer, 2013).

Principal-agent conflicts. Energy inefficiencies, such as the energy efficiency gap, may result from conflicting interests between principals and agents. The principal-agent issue occurs when the principal and agent have conflicting objectives, leading to potential failure in achieving the principal's goals.

Jevons identified the rebound effect as a situation where improved technological efficiency leads to an increase in energy consumption. This intense rebound phenomenon is now identified as the "backfire" effect. Jevons suggested that the use of coal would rise rather than fall with a more effective steam engine, leading to the Jevons Paradox. Examine energy efficiency initiatives designed to decrease carbon emissions. The impact of energy efficiency on reducing carbon emissions relies on the presence and extent of the feedback effect. The opposite impact diminishes energy savings by increasing energy efficiency, which is unfavorable for energy efficiency proponents (Gillingham and Palmer, 2013). From a cost-effectiveness perspective, this may be advantageous. The rebound effect enhances customer satisfaction and exemplifies a well-operating market rather than a market failure. The direct rebound refers to the impact of improvements in energy efficiency on pricing, including impacts on both substitutability and revenue. This outcome may be achieved by using the price elasticity of energy demand. An indirect rebound refers to the impact of increased income spent on other commodities after the initial purchase. Gillingham and Palmer (2013) differentiate between the net feedback impact caused by a decrease in the perceived energy price owing to improved energy efficiency and the overall feedback effect, which encompasses additional changes related to a shift in energy efficiency. If energy efficiency improves without affecting the product's qualities or manufacturing costs, there will be a net influence on pricing. If customers see changes in product characteristics as unfavorable, the overall rebound will be lower than the pure rebound.

A minor feedback effect will result in a minimal energy-saving adjustment. When the feedback effect is significant or, in extreme cases, present, the energy savings resulting from improved energy efficiency will be little or perhaps negative (Colmenares et al., 2020).

Cost-effective energy efficiency upgrades lower the cost of energy services like heating and lighting, leading to higher demand for these services, which partially counteracts the energy savings achieved per unit of service (Arimura et al., 2012). The immediate rebound effect is well recognized and is now the focus of a substantial and expanding body of empirical research, particularly on efficiency enhancements made by consumers. Energy efficiency improvements may lead to indirect and macroeconomic reactions, including rebound effects that affect energy consumption throughout the economy (Brockway et al., 2021).

For instance, the reduction in gasoline use achieved by using fuel-efficient vehicles might be redirected towards purchasing other products and services that also need energy for production and operation, known as indirect rebound (Walnum et al., 2014). The widespread use of energy-efficient automobiles might lower gasoline demand, leading to decreased gasoline prices (Burke and Nishitateno, 2013). This could result in increasing gasoline consumption and other energy sources, known as macroeconomic rebound, with subsequent effects on other markets. Both direct and indirect rebound effects are analyzed within a partial equilibrium framework (Thomas and Azevedo, 2013). This means that the approaches used to estimate them, such as input-output models, maintain input and commodity prices constant throughout the economy and focus only on changes in the effective price of the energy service. The macroeconomic rebound effects are general equilibrium because the approaches used to evaluate them, such as computable general equilibrium models, enable input and commodity prices to fluctuate throughout the whole economy (Wei, 2010). The combined effects occur simultaneously, resulting in the economy-wide rebound effect, typically expressed as a percentage of the anticipated economy-wide energy savings. This is calculated based on a scenario where no adjustments take place (Brockway et al., 2021).

Estimating economy-wide recovery effects is difficult, but there is increasing evidence indicating that they might be substantial. It was discovered that economic development in emerging nations is just as reliant on energy as historical growth in developed nations, even with significant advancements in the energy efficiency of certain technologies. The balance in energy intensity indicates that the energy conservation achieved via advancements in certain technologies has been counteracted by other developments, including a transition towards more energy-demanding consumer habits. It is demonstrated that the sustained decrease in regional and global energy intensity is mostly attributed to nations becoming wealthier, rather than achieving certain levels of income with reduced energy use.

Energy efficiency often involves achieving more performance with less energy, rather than reducing performance or doing nothing at all (Lovins, 2004). An example is the mechanical energy output of an electric motor relative to the electrical input. Efficiency refers to achieving more services with the same energy input or obtaining the same services with less energy use. Technological advancements in energy efficiency, behavioral modifications, and pricing or regulatory incentives are necessary to attain efficient results. However, only a combination of three things—technological improvements in energy efficiency (such as thermal insulation), behavioral changes (such as resetting thermostats), and price or policy measures to encourage compensation for these energy changes—may help to achieve efficiency outcomes.

***

Energy efficiency is a fundamental idea that is significant in a variety of fields, including engineering, economics, and environmental sciences, among others. Its primary objective is to maximize the usage of energy to boost economic production while simultaneously lowering energy consumption. This rapidly evolving area of research is very necessary to address global issues such as the growing need for energy, the effects of climate change, and the reduction of carbon dioxide emissions.

During the 1970s, Lovins was a well-known advocate for the notion of energy efficiency. He emphasized the significance of this concept in terms of ensuring sustainable energy supply and economic productivity via the reduction of energy use. In contrast to energy conservation, which just involves reducing the amount of energy that is used, energy efficiency is centered on the goal of maintaining the same level of service while using less energy. Both the optimization of energy inputs—such as fuels and electricity—and the use of efficient energy converters, such as motors and boilers, are required for the process to reach its full potential.

Although there are many different ways to assess energy efficiency, most of them revolve around the ratio of the amount of usable energy generated to the amount of energy that is used. Because of the complex process of estimating both the precise amount of energy required and the amount of energy that may be used, quantifying this ratio may prove to be challenging. As a consequence of this, there are several levels of efficiency assessment that include not only individual devices but also whole economic sectors and even global energy networks. Energy intensity is defined as the amount of energy utilized per unit of economic production and is thus negatively connected with energy efficiency. These assessments help in grasping and improving energy intensity, thereby contributing to the improvement of energy efficiency.

In addition, energy efficiency in consumption and production highlights the relevance of minimizing energy loss in energy conversion processes and making the most of the use of resources in manufacturing to improve economic efficiency. The development of new technologies and the adoption of new behaviors both have a significant influence on the enhancement of energy efficiency. On the other hand, this is occasionally complicated by phenomena such as the rebound effect, which is a situation in which improvements in energy efficiency may result in greater energy consumption due to behavioral or economic responses.

As a result of the many benefits it provides, including the reduction of emissions of greenhouse gases, the enhancement of energy security, and the promotion of economic competitiveness, energy efficiency is a major priority for governments and politicians across the world. On the other hand, it is not yet being used to its full capacity, and its comprehension is not all that strong. Because of this, there is a need for more focused research, better methods of evaluation, and the widespread adoption of environmentally friendly products and behaviors.

Efficiency in energy use is a strategic approach to thoughtfully managing our energy resources. This strategy ensures that we can meet our present demands without severely influencing the ability of future generations to meet their obligations. To achieve sustainable development, all aspects of society need to make efforts to comprehend, implement, and improve energy-efficient solutions.

5.2 ENERGY EFFICIENCY INDICATORS

Energy efficiency is a widely utilized concept in several disciplines, including engineering, architectural design, manufacturing, management, organization, economics, policy-making, and development (Chen et al., 2020). Bosseboeuf et al. (1997) analyze efficiency using two fundamental concepts: cost-effectiveness and technical efficiency.

Cost-effectiveness refers to achieving increased output and improved living standards while using the same or less energy content. This may result in savings in terms of both CO_2 emissions and costs. Economic relations are used when measuring energy efficiency at a macro level, which cannot be captured by technical or physical measures of economic activity. This is done at the level of the whole economy or sector. Energy intensity refers to the economic links that exist between final energy consumption and performance indicators, such as GDP or value added. It is quantified by calculating the ratio of energy consumption (measured in units like joules) to the monetary worth of the performance indicator. It is necessary to assess energy intensity at a constant price to prevent the influence of inflation.

Technical and economic efficiency refers to the decrease in energy consumption resulting from advancements in technology, changes in behavior, improved governance, and other factors. This decrease may be measured using both economic and technical indicators. Technical and economic connections are evaluated without being separated into different categories (such as subsectors or final consumption). This evaluation links energy use to a performance measure that is expressed in physical terms (such as a ton of cement or the number of passenger-kilometers) or to a specific unit of consumption (such as a car or home). The term used to refer to the average energy consumption per unit is "unit consumption" (Bosseboeuf et al., 1997).

These indicators may be used to measure economic activity at the product, sector, or national levels, as well as primary energy consumption and secondary final energy consumption (Filippini and Hunt, 2015). In his research on energy efficiency, Patterson (1996) highlights that many indicators may be used to track developments in this domain. The entities are categorized into four primary classifications:

- Thermodynamic indicators are assessments of energy efficiency that rely only on data collected from thermodynamics, where energy consumption and production are quantified in thermodynamic units. Thermodynamics is often described as the scientific study of energy and its many processes.
- Physical-thermodynamic indicators are a combination of measurements that assess energy use in thermodynamic units and yield in physical units. These physical units attempt to quantify the delivery of process services, such as the amount of product in tons.
- Economic and thermodynamic indicators are hybrid measures that quantify the supply of process services. These indicators are quantified in terms of market pricing for production and energy consumption in thermodynamic units.
- Economic metrics only assess the efficiency of energy transformations based on their market worth. Put simply, the measurement of production and expenses is limited to monetary terms alone.

Energy efficiency indicators may be stated either as ratios—where energy consumption is divided by a performance indicator—or as quantities, which capture changes in consumption caused by certain explanatory variables (Velasco-Fernández et al., 2020). There are two sorts of indicators that have been recognized based on their purpose: descriptive indicators and explanatory indicators (Bosseboeuf et al., 1997). Descriptive indicators provide a detailed account of the condition and progress of energy efficiency. They pertain to economic or technical-economic relationships that aim to represent some characteristics of efficiency and cannot be understood in isolation. Nevertheless, it is possible to integrate many descriptive indicators to elucidate energy efficiency trends (Bosseboeuf et al., 1997). This category encompasses several measures of energy intensity, including the monetary worth of energy in terms of GDP and value added, as well as unit consumption and particular consumption connected to physical value such as homes, vehicles, steel, cement, and employment (Tanaka, 2008). Descriptive indicators are often shown as historical time series, typically sourced from official statistical entities (Eichhammer and Mannsbart, 1997).

Explanatory indicators are indicators that elucidate the elements influencing the state and progress of energy efficiency, as well as the significance of energy efficiency in the advancement of energy consumption. Explanatory indicators are specifically developed to elucidate the reasons behind the variations seen in the descriptive indicators, such as the advancement or decline of energy efficiency in a certain country. One of the primary objectives in this context is to ascertain the respective contributions of technical development, structural changes, behavioral changes, and other factors. Furthermore, these indicators will serve as valuable tools for elucidating the disparities across nations, such as variations in climatic conditions and housing dimensions (Haas, 1997). Explanatory indicators may be obtained either by directly deriving them from other economic indicators or by calculating them as new indicators. An example of the latter is when there is a structural influence on the industry, resulting in various technical and economic repercussions (Bosseboeuf et al., 1997). These indicators are not easily accessible via statistical data, particularly every year. They may include specialized surveys, research, and interviews conducted with industry and equipment providers. Typically, they are chosen by proficient professionals with extensive knowledge of technologies associated with a certain manufacturing process (García-Quevedo and Jové-Llopis, 2021).

Explanatory indicators have the potential to be transformed into descriptive indicators and included in the research by using more advanced statistical methods. However, in reality, it would be prohibitively costly. Hence, explanatory indicators are included with descriptive indicators to enhance comprehension of variations in international comparisons or comparisons across time (Eichhammer and Mannsbart, 1997). Typically, each indicator is computed for three categories of energy: fossil fuels, heat, and electricity. In addition, the majority of indicators are presented in both absolute and relative terms, either as indices or as a percentage of consumption for the reference year, to assess the technical and economic consequences (Bosseboeuf et al., 1997). The inefficient use of energy or the existence of a technical disparity is often classified into two groups: market failures—which include issues related to information, incentives, and behavior—and behavioral failures, which involve decision-making based on heuristics, lack of attention, and limited reasoning (Filippini and Hunt, 2015). It is stated that both a parametric and non-parametric limit analysis may be used to evaluate the total production efficiency, the efficiency in using a single input, and the amount of technological development. Parametric approaches are preferable for assessing energy efficiency levels due to the presence of subtle variations in the provision of energy services (Filippini and Hunt, 2015).

According to Filippini and Hunt (2015), there are three approaches available for evaluating a resource-related technical measure of efficiency, such as energy efficiency.

Alternatively, there is a method to determine the precise technical efficiency of the input by evaluating the input demand function. This function calculates the minimal input needed to achieve a certain level of output, considering the technology and the quantity of other production components.

Finally, Zhou et al. (2012) suggested assessing the Shephard energy input distance function and demonstrated that this particular form of input distance function is synonymous with the input requirement function.

It is important to highlight that all three approaches may alone evaluate the degree of technical efficiency in input use. Nevertheless, considering the economic perspective, it is crucial to possess data about the extent of total resource efficiency. This metric encompasses both technical and distribution inefficiencies (Filippini and Hunt, 2015).

Energy efficiency may be defined more accurately by considering the economic outlook and cost-effective energy efficiency indicators, rather than depending just on basic energy intensity indicators (Lovins, 2017). The definition of energy efficiency should be grounded on an unconventional notion, namely, a quantification of cost-related technology. This measure is defined as the ratio of the minimum amount of energy required to the actual energy consumption, considering factors such as production technology, level of output, and other costs. Furthermore, Filippini and Hunt (2011) introduced a modified version of this metric, which calculates the minimal energy required to achieve a certain level of output, including both technological and economic factors. Actual energy demand in this particular situation deviates from the projected marginal energy demand because of the existence of distribution and technological inefficiencies (Boogen et al., 2021). Three ways that are more appropriate for assessing the parametric methods of non-traditional energy efficiency include the energy demand function and the shepherd's energy distance function (Filippini and Hunt, 2015).

5.2.1 Comparative Analysis of Energy Efficiency Indices across Different Countries

There are several justifications for investing effort in evaluating energy efficiency by conducting a national ex-post assessment and comparing it to other nations. The primary cause is the implementation of diverse energy policies, particularly those related to austerity measures and energy pricing, among European nations with varying demographic, economic, and socio-economic conditions. Typically, the variations in energy policy across various nations are more significant than the changes that occur over time inside a certain country (Baldwin et al., 2019). Politically, the discussion over the greenhouse effect would necessarily include comparing the magnitude and rate of CO_2 emissions increases across different countries. However, the issues develop as a result of the constraints on the accessibility of data (Bosseboeuf et al., 1997).

To evaluate energy efficiency across borders, it is essential to examine the accomplishments of each country in the past and establish national energy goals for the sector based on previous performance and ongoing initiatives in other nations. Additionally, it is important to consider the European Commission's recommendations and directives regarding energy efficiency and the environment.

Energy intensity variations are influenced by several elements, including the economic organization and structure, as well as the actual degree of energy. The demand for energy is determined by the desire for various outputs, which might include goods, services, or energy services such as heating, cooling, and lighting. Households and businesses use energy, labor, and capital to manufacture goods. From an economic perspective, it is crucial to achieve efficiency in production by selecting a mix of expenses that minimizes the overall cost of production (Filippini and Hunt, 2015).

The energy efficiency indicators used to assess a country's energy efficiency are broad indicators that can be applied to the entire economy or specific sectors such as industry, agriculture, etc. They can also be used to measure energy efficiency in subsectors—such as the household sector—or specific end-uses, like space heating (Bosseboeuf et al., 1997). An ongoing challenge in energy efficiency policy has been the process of evaluation: determining the outcomes accomplished, rating the objectives, and comparing the relative positions of different nations.

To assess the energy efficiency of nations, the first step is to establish the necessary prerequisites. Therefore, the indications being analyzed must be able to be compared to one another. When considering disparities in the size of a nation, it is more advantageous to use relationships and indices rather than quantity. Furthermore, it is crucial to prioritize the establishment of a consistent definition of indicators across different nations. Also, many matters that may not have much significance at a national level gain considerable importance when compared internationally. Examples encompass the issue of climatic disparities among nations, which holds particular significance when assessing energy efficiency for heating spaces, the geographical expanse of a country (particularly for transportation), transit for transportation, and the influence of variations in trade composition (Berardi, 2017; Busch and Ortner, 2019). To calculate comparable indicators, it is essential to have technical and economic data. This includes information on energy consumption by final consumers, macroeconomic variables such as GDP, value added by sector, and energy prices. Additionally, socio-economic data reflecting various activities and levels of service, such as production, traffic, housing stock, vehicles, and electrical appliances, are necessary. Technical data, such as specific consumption, is also required (Bosseboeuf et al., 1997).

The systematic selection may influence the interpretation of the findings. When comparing energy intensities, which refers to the ratio of energy consumption to GDP, it is important to consider whether the energy consumption refers to primary or final energy and whether any structural changes in the economy should be taken into account or not. This comparison yields different types of information and presents various challenges (Ang and Goh, 2018). Comparing energy intensity directly, primary energy intensity is a more suitable measure for overall energy requirements compared to GDP. This is because it considers the final energy consumption, self-consumption, and losses in the energy transformation system. However, the comparison may be biased due to the primary electricity metering rules adopted, which result in significant differences in the quantity and nature of electricity among different parties (Bosseboeuf et al., 1997).

Final energy intensities are a more suitable metric for comparing the actual energy consumption model in relation to GDP. They specifically measure the energy used in the final stages of production and consumption. However, this comparison can be misleading if the proportion of electricity in final consumption varies significantly between countries. This is because electricity consumption tends to be more efficient in terms of end-use, while thermal power plants have lower transformation efficiency (Tanaka, 2008). Undoubtedly, the advancements made in energy efficiency are somewhat counteracted by the growing losses incurred during the process of generating nuclear power (Bosseboeuf et al., 1997).

The overall energy intensity, including both primary and final energy, encompasses all aspects that contribute to the alteration in the amount of energy needed to generate one unit of GDP, including technological, managerial, and economic considerations. Changes in the economic structure have an impact on the total energy intensity, although this effect is not directly linked to energy efficiency. For instance, the process of tertiarization in the economy or the shift of output to less resource-intensive sectors (such as electronics and computers) would result in an equivalent reduction in total intensity across all areas. To effectively track energy efficiency trends in energy pricing and energy management strategies, it is essential to disregard the impact of structural changes. The process involves determining the energy intensity while keeping the economy's structure constant, including the distribution of GDP among agriculture, industry, and services, as well as the distribution of added value among the main industries. The evaluation of structural changes specifically focuses on final energy consumption and has the potential to alter the impression of relative energy efficiency trends across nations, but not their relative levels (Bosseboeuf et al., 1997).

5.2.2 Indicators of Energy Efficiency in the Industrial Sector

The primary goal of energy indicators in the industrial sector is to comprehend the principal technical and economic influences on the overall final energy consumption in both the industry as a whole and specific industrial subsectors or branches (Brodny and Tutak, 2022). An examination of energy indicators can assist in establishing connections between alterations in energy consumption and energy efficiency policies or environmental policies, fluctuations in energy prices, variations in external trade of energy-intensive intermediate or final products, and structural impacts resulting from business cycles. This analysis aims to enhance the adjustment of energy efficiency measures (Eichhammer and Mannsbart, 1997).

The influence of a combination of product/process and efficiency may be partially elucidated by technical advancements, such as alterations in recycling rates, the composition of raw materials or fuels, or modifications in the process (Fu et al., 2018). The primary cause of changes in energy consumption is the improvement in efficiency, considering the variations in structure and the influence on international commerce. The increase in efficiency may be attributed to certain key technical advancements, along with other significant aspects, specifically modification in process technology, such as the implementation of a membrane process for mercury or a diaphragm process for chlorine generation. The phenomenon of being able to modify numerous industrial processes is referred to as the Process Replacement Effect. Occasionally, the outcomes of several procedures, such as the chlorine process, may provide equivalent levels of quality. Additional modifications in the procedure, such as alterations in cement or steel manufacturing procedures, might impact the caliber of the result, hence restricting the extent of substitution. The descriptive indicator that is most affected by process substitutability is the mean unit of consumption of the process blend. The explanatory indication refers to the alteration in certain stages of the manufacturing processes of a particular product, such as chlorine, cement, or raw steel. When it comes to replacement, if the products have the same quality and are used equally in both the old and new processes, the replacement should be considered an enhancement in energy efficiency. In other situations, it can be seen as a shift between industries (Eichhammer and Mannsbart, 1997).

5.2.3 Residential Sector Energy Efficiency Indicators

The home sector is intriguing because of its substantial customer base, which has comparable equipment, making it relatively homogenous when compared to the industrial and service sectors.

Because of the consistent and uncomplicated nature of demand patterns, this sector is the primary target of the majority of energy policy. Comprehending the elements that contribute to fluctuations in home energy use is crucial for developing a strategy for future energy consumption. This enables the differentiation between growth elements that may reach a maximum limit (such as equipment ownership and interior temperature) and those that can potentially keep growing with rising earnings or declining energy costs (Haas, 1997).

Energy efficiency refers to the ability to either offer more services using the same amount of energy or provide the same services using less energy. These two definitions, which are considered authoritative and traditional, consist of two primary elements for assessing energy efficiency. The first aspect is comparing the result and input. The other is a comparison between the output (service) resulting from a certain input (more or less) or a comparison between the input required to get a specific output (service). To discuss techniques for assessing energy efficiency, it is logical to categorize the issue into two primary divisions: the Energy Efficiency Indicator (ECEI) and the Energy Cost-Efficiency Assessment (EEE). The assessment of energy efficiency is comprised of two components: an energy efficiency indicator (ECEI) and an energy cost-efficiency estimate (EEE) (Chen et al., 2020).

An analysis of energy efficiency distribution by end-use is crucial since advancements in enhancing efficiency in one specific sector might be obscured by structural modifications and rises in other sectors. Without segmenting by distinct end-uses, we may fail to discern the underlying variables that explain the shifts in energy consumption patterns within households. Another concern is economic variables: specifically, prices and salaries have a direct impact on energy use. Nevertheless, the degree and consistency of the impact remain uncertain. Furthermore, the impact may differ depending on the ultimate purpose. Furthermore, the demand for services is contingent upon a multitude of elements that are contingent upon cultural norms and human behavior. The variability of these parameters is also highly dependent on the ultimate purpose of their utilization (Haas, 1997).

When determining long-term demand, several important factors need to be considered. First, the general lifestyle and demographic factors play a significant role, such as the occupancy patterns of the house and the size of the family. Second, the impact of energy and electricity services saturation should be taken into account (Haas, 1997). Third, the fuel mix and its relationship to efficiency and the structure of energy services are crucial. Additionally, it is important to differentiate between long-term and short-term changes in structure, service, and intensity. Last, it is difficult to accurately assess the effectiveness of equipment or the house itself without conducting expensive studies (Haas, 1997).

The primary factors of significance are the structure, behavior, and technological efficiency. The structure indicates a persistent need for services, such as housing, gadget use, and the quantity of water heating systems. Behavior encompasses the immediate requirements for services, such as adjusting the interior temperature, selecting the ventilation rate, and determining the operating duration of gadgets and lighting. Technical efficiency refers to the amount of energy required to provide a certain quantity of services (Filippini and Hunt, 2015). Furthermore, other factors, such as climate and population, have an impact on the demand for energy (Haas, 1997).

Certain factors may be succinctly stated using specific terminology. The structure and behavior of a society are contingent upon the autonomous choices made by its consumers and their respective lifestyles, therefore rendering them a manifestation of one's way of life. Income and attitudes have the most significant impact on lifestyle. The consumption of a unit of equipment is influenced by technical efficiency and customer behavior, which may be described as intensity (Duflou et al., 2012). Examples of intensity include the measurement of kilowatt-hours (kWh) per unit of the device and kilowatt-hours per square meter of living space. The relationship between efficiency and behavior is reciprocal, as increased efficiency results in reduced service pricing, which in turn might impact customer behavior when selecting service levels. The rebound effect, also known as the indirect rebound effect, is a significant factor in several end uses such as space heating, water heating, and lighting. The primary determinants of enhanced efficiency are prices and standards. Income, energy costs, and attitudes are the primary indirect factors that impact energy usage. Energy policy measures, such as taxes, norms, preferences, and information, may directly or indirectly affect energy use (Haas, 1997).

The energy consumption in residential areas is often categorized into five distinct end uses: space heating, water heating, cooking, lighting, and electrical appliances (Swan and Ugursal, 2009). Appliances may be categorized as refrigerators (including combination units), freezers, clothes washing machines, clothes dryers, dishwashers, and even televisions (Cabeza et al., 2018). Explanatory variables may be regarded as the factors that determine energy efficiency and CO_2 indicators. They may be used to standardize energy usage and obtain intensity. The use of energy for a defined purpose may be employed to standardize carbon dioxide emissions, resulting in the measurement of carbon dioxide intensity. The primary determinant for space heating is the size of the living area. Larger houses need more energy for heating compared to smaller ones, as stated by Haas (1997).

To modify growth rates, it is necessary to implement multiple actions simultaneously. These actions include enhancing thermal efficiency, not only for buildings but also for heating systems, in response to stricter standards. Additionally, improving energy management systems is crucial, with a specific focus on enhancing simplicity and convenience for consumers. Lighting is the household end use with the most potential for enhancing efficiency. The widespread use of CFLs may have a substantial effect since they utilize just 20–25% of the power consumed by regular incandescent bulbs to offer the same level of service output (Haas, 1997). The stagnation and decline in end-use energy consumption are determined by the balance between the gain in efficiency and the rise in demand for services caused by the widespread use of traditional appliances and the introduction of innovative gadgets.

5.2.4 Challenges in Evaluating the Efficiency of the Economy

Given the multidisciplinary nature of energy efficiency, definitions and measures may be, at best, imprecise and, at worst, incomprehensible. Undoubtedly, energy efficiency is a crucial concern for the long-term viability of civilization. The area of science and engineering encompasses many difficulties and employs diverse methodologies. However, the idea of energy efficiency might vary significantly based on the specific goals and objectives being addressed (Chen et al., 2020).

These two authoritative and traditional definitions of an effective measurement of energy efficiency consist of two basic components. There are two types of comparisons: one involves comparing the output and input, while the other involves comparing the output (service) for a given input (to some extent) or comparing the inputs for a given output (service). By now, the proposal should have unequivocally centered on the concept of "comparison," with the issue being the subject of comparison, specifically, "who to compare with." To present the techniques used to measure energy efficiency, it is logical to categorize the topic into two primary divisions: energy-consumption-efficacy-indicator (ECEI) and energy-economic-efficiency-estimate (EEEE) (Chen et al., 2020).

Initially, it is important to note that energy in isolation is insufficient for generating output. This indicates that the use of energy as a single input must be extended to include a situation with numerous inputs. Furthermore, the determination of a technological boundary building is required.

The suggested EEEE aims to build an estimating approach. The drawback is that although the indicator displays average energy efficiency, it lacks clarity and independent criteria for discerning between high and bad performance. Therefore, there is a lack of definitive criteria for making comparisons. Thus, this indicator often depends on its historical data, alternative objectives, or established benchmarks for evaluating performance. The availability and dependability of data have many effects that impede the effective study of energy efficiency indicators in the industrial business. Several of these impacts are linked to statistical issues that restrict the decomposition and, therefore, the precision of such analyses. For instance, several industry statistics fail to include tiny firms (Eichhammer and Mannsbart, 1997). One issue with statistical accuracy is that companies are decreasing their reported energy usage by outsourcing it to other plants through contracting companies. This problem of shifting energy consumption to different sectors occurs when certain parts of energy consumption, originally attributed to one industry, are often moved and become a transportation issue (Eichhammer and Mannsbart, 1997).

***

The concept of energy efficiency is extensively useful in a variety of domains, including engineering, economics, and the formulation of public policy. Essentially, it comprises determining the cost-effectiveness as well as the technical efficiency of the use of energy. To achieve cost-effectiveness goals, it is necessary to raise living standards and productivity without increasing energy consumption. This might result in a reduction in CO_2 emissions and expenditures. On the other hand, technical efficiency is a specialized goal that aims to reduce energy consumption via the implementation of technological advancements, behavioral changes, and improved political administration.

The evaluation of energy efficiency involves the use of a wide range of indicators at several levels, such as product, sector, or national, while taking into consideration the utilization of both primary and secondary energy sources. The thermodynamic, physical-thermodynamic, economic-thermodynamic, and purely economic indicators are the four primary kinds that are used to classify the indicators. Quantifying energy use in terms of thermodynamic or economic value—or a combination of the two—is the focus of each category, which offers a unique perspective on energy efficiency.

Different sorts of indicators may be broken down into two primary categories: descriptive and explanatory. Indicators that are descriptive provide specific information on the existing state of energy efficiency as well as its future growth. Measurements such as energy intensity in proportion to GDP or consumption per unit are examples of the types of indicators that are often given in the form of a series of data that spans time. Explanatory indicators, on the other hand, are a kind of indicator that analyzes the factors that affect these patterns. These variables include things like structural adjustments or technological advancements. These indicators are often tailored to shed light on disparities across nations or between various periods.

At the general level, energy intensity is an important statistic to consider. The formula for calculating it is to divide the amount of energy used by an economic performance metric such as GDP, which is then adjusted for inflation to reduce the impact of inflation. This ratio makes it easier to understand the relationship between energy consumption and economic production. It also provides significant insights into the efficiency with which energy is used in a variety of industries or the economy as a whole.

To measure energy efficiency, it is essential to conduct comparisons between countries, especially to take into consideration the various energy policies and socio-economic considerations. When considering energy efficiency, these comparisons often highlight the influence of external factors, such as the structure of the economy and the policies that govern it. In addition, to have a complete understanding of these indicators, it is necessary to grasp a mix of technical and economic information. This includes specialized information on energy consumption, general economic data, and socio-economic measures.

The lack of clarity in language and the varied range of metrics that are employed as a result of the engagement of numerous disciplines are both factors that contribute to the challenges that occur when evaluating energy efficiency. To conduct an accurate review, it is necessary to have trustworthy data; however, this may be often hampered by statistical errors or a lack of data, particularly in the case of small businesses or sectors that outsource their energy use.

For monitoring and improving energy usage efficiency, influencing policy decisions, and promoting sustainable development, energy efficiency indicators are essential tools. Having a complete understanding of the technical and economic elements of the complexity and variety of energy efficiency in diverse sectors and national contexts is very necessary to correctly evaluate and enhance energy efficiency in these settings.

5.3 DRIVERS OF AND BARRIERS TO ENERGY EFFICIENCY

Review of previous publications analyzing energy efficiency aspects across different countries or regions, such as Petrović et al. (2018), Guevara et al. (2021), Duro et al. (2010), Sun et al. (2022), Liddle (2010), Díaz et al. (2019), Gales et al. (2007), Yu et al. (2022), Andrieu et al. (2022), Ajayi and Reiner (2020), Samargandi (2019), Gyamfi et al. (2023), Deichmann et al. (2019), Recalde and Ramos-Martin (2012), Jin (2022), Zhou et al. (2012), Cao et al. (2020), and Antonietti and Fontini (2019) conducted studies that primarily focused on identifying the key factors and obstacles affecting energy efficiency. The following topics are arranged in order of their level of discussion, starting with the most discussed: industrial structure, economic development, energy costs, technical innovation, energy policy and regulations, trade openness, urbanization, renewable energy usage, energy efficiency measures, and foreign direct investment.

The impact of **industrial structure** on energy intensity is based on the makeup and effectiveness of various economic sectors. The industrial structure is a measure of the proportion of energy-intensive sectors, such as manufacturing and mining, compared to less energy-intensive sectors like services and agriculture, in an economy. An increased share of industrial activity often leads to a greater total demand for energy, resulting in a rise in energy intensity. The scientific discourse highlights the shift from basic and secondary sectors to tertiary sectors as essential for decreasing energy intensity. This transition is linked to technical advancements, enhancements in efficiency, and a move towards knowledge-intensive occupations that need less energy per unit of economic production. Structural decomposition analysis is often used in theoretical models and empirical research to separate the influence of industrial structure from other variables that affect energy intensity.

The relationship between industrial structure and energy intensity is extensively studied in the subject of environmental economics. The structure and productivity of various economic sectors are pivotal in defining the total energy requirements and efficiency of an economy. Ang and Zhang (2000) argue that the industrial structure is a reflection of the proportion of energy-intensive sectors, such as manufacturing and mining, in contrast to less energy-intensive sectors like services and agriculture. They contend that a greater share of industrial activity generally leads to an increase in the total energy demand, thereby raising energy intensity.

The shift from primary and secondary sectors to tertiary sectors is emphasized as a crucial route for decreasing energy intensity. This viewpoint is corroborated by much research, which highlights the significance of technical advancements, enhancements in efficiency, and a transition towards knowledge-intensive activities that need less energy per unit of economic production. This change is often linked to the emergence of a service-oriented economy, which typically exhibits a reduced energy intensity in comparison to industrial economies.

In addition, both theoretical models and empirical research often use structural decomposition analysis to investigate the influence of industrial structure on energy intensity. This approach enables researchers to separate the impacts of industrial structure from other variables that influence energy intensity. Voigt et al. (2014) illustrate this method by showing how changes in the composition of industries have influenced the differences in energy efficiency across various nations.

The impact of **economic development** on energy intensity is based on the theoretical framework of the environmental Kuznets curve (EKC) hypothesis. This concept suggests that energy intensity does not follow a linear trajectory as economies progress. During the first phases of economic growth, there is a rise in energy intensity as a result of a strong dependence on energy-intensive industries. As economies progress and develop technologically, there is a transition towards service-oriented industries and high-value, less energy-consuming industrial sectors, resulting in a reduction in energy intensity. The shift is made easier by technological advancements, improved energy efficiency, and a stronger focus on environmental sustainability in industrialized countries.

The EKC hypothesis posits a curvilinear relationship between economic growth and environmental concerns, such as energy intensity. The concept suggests that energy intensity often rises in the first phases of economic growth when nations significantly depend on energy-intensive industries. As economies expand and develop, there is a clear trend towards increased focus on service-oriented industries and high-value, less energy-intensive manufacturing. This transition leads to a decrease in energy intensity.

Extensive examination of the environmental Kuznets curve (EKC) theory and its impact on energy intensity emphasizes the significant contributions of technical innovation and enhanced energy efficiency in facilitating this shift. As economies progress, they often embrace more sophisticated technology that enhances productivity while lowering energy use, resulting in a decrease in energy intensity.

Furthermore, the significance of environmental sustainability in industrialized economies cannot be underestimated. Examining the relationship among policy interventions, sustainability, and the adoption of energy-efficient technologies and practices, one can argue that these factors have a significant role in reducing energy intensity in mature economies. These changes are a result of a larger movement in society's attitudes towards taking care of the environment and are a sign of more advanced economic growth, as described by the EKC hypothesis.

The relationship between **energy costs** and energy intensity is based on the economic theory that elevated prices stimulate efficiency and decrease consumption. Theoretical models propose that when energy prices rise, both producers and consumers strive to reduce energy expenses, resulting in investments in energy-efficient technology and a transition towards less energy-demanding activities. The price elasticity of energy demand quantifies the degree to which energy consumption reacts to variations in price. The increase in energy costs incentivizes the use of more effective technologies and methods, resulting in a decrease in the energy required for economic production.

The literature broadly acknowledges the economic concept that higher energy costs result in enhanced efficiency and decreased use. Gillingham et al. (2009) examine the impact of increased energy costs on the behavior of consumers and producers. They argue that higher energy prices provide a strong motivation for individuals and businesses to invest in energy-saving devices and adopt activities that need less energy. The statement emphasizes the importance of price elasticity of energy demand, which quantifies how energy consumption reacts to variations in energy costs. It explores how elevated energy costs stimulate the adoption of more efficient technology and practices, as in, when energy costs increase, the economic advantages of investing in energy efficiency become more evident, resulting in a reduction in the energy intensity of economic production. This phenomenon is especially noticeable in industries where energy bills make up a substantial proportion of overall costs.

In addition, Linares and Labandeira (2010) examine the consequences of changes in energy prices on the energy intensity of various sectors in the economy. Their investigation verifies that elevated energy costs have a crucial role in promoting energy efficiency and transitioning economic activity towards less energy-intensive models. It is emphasized that the price elasticity of energy demand differs across sectors and changes over time, which affects how quickly and to what degree adjustments are made in energy intensity.

The impact of **energy innovations** on energy intensity is based on the concept of technical advancement enabling more effective energy use. Advancements in energy technology, which include renewable energy sources, energy-efficient appliances, and industrial processes, directly aid in decreasing the energy intensity of economic production. The link is highlighted by the diffusion of innovation theory, which elucidates the process by which new technologies get widely adopted in society, leading to improved energy efficiency and decreased total energy intensity. Innovations enhance both the energy efficiency of industrial processes and the promotion of cleaner energy sources. This leads to a change in the energy balance towards choices that have lower carbon emissions, resulting in a reduction in energy intensity.

The idea that advancements in technology, especially in the field of energy, result in a more effective use of energy is a fundamental principle in the studies on sustainable development and energy policy. Jaffe et al. (2002) provide a thorough analysis of how advancements in energy technology, such as improvements in renewable energy sources and energy-efficient appliances, have a major role in decreasing the amount of energy used in economic activities. The significance of the diffusion of innovation theory in elucidating the widespread acceptance of novel technologies and the resulting enhancements in energy efficiency is emphasized. The diffusion of innovation theory explains how new technologies are gradually adopted across society, shedding light on how energy innovations can lead to reduced energy intensity. By promoting more efficient energy use and facilitating the shift toward cleaner energy sources, this process supports the transition to lower energy consumption and improved sustainability over time.

In addition, Popp et al. (2010) examine how energy innovations might improve the efficiency of manufacturing processes and promote the use of energy sources with lower carbon emissions. Their research emphasizes the role of technology improvements in enhancing energy efficiency and enabling a shift in the energy mix. This shift leads to a reduction in energy intensity and supports endeavors to address climate change.

The impact of **energy policy** and regulations on energy intensity is based on the theoretical principles of correcting market failures and influencing behavior via nudges. Energy policies, such as carbon taxes, subsidies for renewable energy, and efficiency requirements, aim to address market failures by including the external costs associated with energy usage and emissions. These policies provide financial benefits for investing in energy efficiency and adopting cleaner energy technology, resulting in a decrease in energy consumption per unit of output. Behavioral economics concepts indicate that policies have the potential to influence consumers and enterprises to adopt more energy-efficient behaviors. This may enhance the effectiveness of technology improvements and market-based instruments in decreasing energy intensity.

Goulder and Parry (2008) conducted a comprehensive examination of how carbon taxes and other market-based instruments effectively address market failures by internalizing the external costs linked to energy usage and emissions. Advocates assert that these regulations incentivize investments in energy efficiency and the use of greener technology, resulting in a decrease in energy intensity.

Renewable energy subsidies and efficiency regulations play a critical role in promoting a shift toward less energy-intensive economic activities. By encouraging technological advancements and facilitating the broader adoption of energy-efficient solutions across various industries, these measures have a significant impact on reducing energy intensity.

Behavioral economics also offers valuable insights into energy policy by highlighting how subtle incentives can guide consumers and businesses toward more energy-efficient practices. Even small policy interventions can meaningfully influence energy consumption patterns, complementing direct regulations and technological advancements in reducing energy intensity.

The impact of **trade (as well as financial) openness** on energy intensity is based on the theoretical principles of comparative advantage and technology transfer. Trade liberalization enables nations to focus on areas where they possess a competitive advantage, perhaps resulting in a transition towards sectors that use less energy. Financial openness promotes the movement of money towards energy-efficient technologies and practices, which in turn encourages the adoption of innovations that decrease energy use per unit of GDP. Furthermore, the process of global integration may stimulate competitiveness and enhance efficiency, resulting in a decrease in energy intensity. This is achieved by sharing and implementing the most effective techniques and technology across different countries. Trade openness is closely linked to economic development, with comparative advantage playing a key role in shaping energy intensity. Free trade allows nations to specialize in sectors where they hold a competitive edge, potentially leading to a shift toward less energy-intensive industries as economies grow and evolve.

Trade openness can positively influence environmental outcomes, particularly in reducing energy intensity. By promoting efficiency and specialization through comparative advantage, trade liberalization has the potential to enhance energy efficiency, especially when coupled with effective environmental regulations. Similarly, financial openness facilitates the flow of capital toward environmentally friendly and more productive technologies. This access to global capital markets supports the adoption of advanced technologies that reduce energy consumption relative to GDP, highlighting its importance for improving energy efficiency.

Global integration influences competitiveness and efficiency, contributing to a reduction in energy intensity. Exposure to international markets and competition encourages organizations to adopt better practices and more advanced technologies. This drive for competitiveness can lead to significant reductions in energy consumption as companies strive to stay efficient.

Urbanization has a theoretical basis for its impact on energy intensity, which is explained by the increasing demand for energy and the development of infrastructure in urban areas. Urbanization results in increased energy consumption as a result of the clustering of economic activity, transportation demands, and household energy usage. Nevertheless, metropolitan regions have the potential for enhanced energy efficiency as a result of economies of scale and more efficient infrastructure. The total influence of urbanization on energy intensity is determined by the delicate equilibrium between the rising demand and the efficiency improvements.

Urbanization is linked to higher energy consumption due to the concentration of economic activities, increased transportation needs, and household energy demands. However, metropolitan areas also hold the potential to improve energy efficiency by leveraging economies of scale and adopting advanced infrastructure and technologies.

The relationship between urbanization and energy intensity reveals a dynamic interplay between rising energy demand in metropolitan areas and the potential for efficiency gains through concentrated energy use. Initially, urbanization drives an increase in energy consumption, but over time, implementing energy-efficient technologies and infrastructure in urban regions may offset this effect, potentially leading to a decrease in energy intensity.

A comprehensive review of urbanization and the global environment shows that urban areas play a dual role in both escalating energy demand and enabling enhancements in energy efficiency. The need to implement efficient urban planning and policy interventions to fully use the potential for energy efficiency in urban settings, which will have a favorable impact on total energy intensity, can be stressed.

The impact of using **renewable energy** on energy intensity is based on the substitution of traditional, energy-intensive sources with cleaner and more efficient alternatives. The integration of renewable energy sources into the energy mix decreases the total energy needed per unit of GDP by using energy sources that have lower operational energy demands. This shift is backed by the theory of innovation diffusion, which proposes that as renewable technologies become more prevalent, their impact on reducing energy intensity grows due to advancements in efficiency, cost reductions, and improved energy security. As a result, reliance on fossil fuels and the associated energy inefficiencies are reduced.

In their study, Dagoumas and Koltsaklis (2019) provide a fundamental comprehension of how including renewable energy sources into the energy mix may effectively diminish the total energy used per unit of GDP. This process entails replacing traditional, more energy-consuming sources with cleaner, more efficient alternatives, hence leading to a decrease in energy intensity. The research emphasizes the significance of using renewable energy sources due to their reduced operational energy needs and the environmental advantages they provide.

Renn and Marshall (2016) examine the intricacies of innovation diffusion theory within the realm of renewable energy technology. They contend that as these technologies grow more prevalent, their impact on lowering energy intensity becomes more noteworthy. The ongoing gains in efficiency, cost reductions, and greater security of the energy supply provided by these sources jointly decrease dependence on fossil fuels and alleviate related energy inefficiencies.

Examining the capacity of renewable energy technologies to significantly reduce energy intensity, the focus is on the importance of advancements in renewable energy, which not only enhance efficiency but also have a vital role in tackling climate change and resolving problems related to energy security. The authors demonstrate how the spread of renewable energy technology, facilitated by policy measures and investments, speeds up the shift towards a less energy-consuming and more environmentally friendly energy system.

The impact of **energy efficiency measures** on energy intensity is theoretically derived from the notion of reducing the energy consumption needed to generate a unit of economic output. These methods, which include advancements in technology, modifications in behavior, and adherence to regulations, directly increase the effectiveness of energy use in industrial, commercial, and residential areas. The theoretical basis is rooted in the notion of the energy efficiency gap, which suggests that market inefficiencies and knowledge asymmetries hinder the complete achievement of potential energy savings. Through the use of energy efficiency measures, economies may close this disparity, resulting in a reduction in energy intensity as more output is produced using less energy.

The energy efficiency gap is defined as the disparity between the economically ideal degree of energy efficiency and the level that is achieved. Market inefficiencies, such as imbalances in knowledge and external costs, hinder the complete achievement of potential energy conservation. By tackling these market inefficiencies through advancements in technology, modifications in behavior, and implementation of regulatory standards, it is feasible to increase the effectiveness of energy use in different industries, therefore lowering energy intensity.

Gillingham et al. (2009) examine the direct impact of energy efficiency measures, including technology advancements and governmental interventions, on reducing energy intensity. The authors provide a thorough examination of the obstacles to energy efficiency and demonstrate how overcoming these obstacles may result in substantial reductions in energy consumption per unit of economic production. Their study substantiates the notion that the implementation of energy efficiency measures allows economies to overcome the energy efficiency gap. They also investigate the consequences of the energy efficiency gap on policy and economic results. The importance of behavioral interventions and regulatory norms is highlighted in conjunction with technical advancements to enhance energy efficiency. Through the efficient implementation of a combination of strategies, proponents say that it is feasible to achieve significant reductions in energy intensity, resulting in increased productivity while using less energy.

The impact of **foreign direct investment (FDI)** on energy intensity is based on the theoretical concepts of the "pollution haven" hypothesis and the technology transfer argument. Foreign direct investment (FDI) may result in a rise in energy intensity when it is focused on sectors that use a lot of energy, particularly in countries that do not have strict environmental rules. This supports the pollution haven concept. On the other hand, foreign direct investment (FDI) may decrease the amount of energy used per unit of output by sharing sophisticated, energy-efficient technologies and management practices with the nations where the investment takes place. This supports the idea of transferring technology. The impact of foreign direct investment (FDI) on energy intensity is determined by the interplay of contrasting factors, which are impacted by the regulatory environment, the sectoral distribution of FDI, and the host country's ability to adopt new technology.

Cole (2004) presents an empirical study that supports the pollution haven hypothesis. The study suggests that foreign direct investment (FDI) tends to go to countries with less strict environmental rules. This may lead to an increase in energy intensity in such countries since the investments are focused on companies that use a lot of energy. This phenomenon is especially noticeable in emerging nations where environmental restrictions are less strict, therefore enticing foreign investments in industries like manufacturing and mining, which often use a significant amount of energy.

In contrast, foreign direct investment (FDI) may work as a channel for spreading sophisticated, energy-efficient technologies and management practices to the nations where it takes place. Foreign direct investment (FDI) is capable of decreasing energy intensity, particularly when investments are focused on sectors that use advanced technology or when the host nation has the ability to adopt and utilize these new technologies efficiently. The impact of foreign direct investment (FDI) on reducing energy intensity is greatly affected by the host country's regulatory framework, the specific industries that FDI is involved in, and the current infrastructure and human resources. These factors determine the country's capacity to adopt and make use of new technologies.

Furthermore, research publications in the fields of economics, environmental science, sociology, and political science also examine little-recognized issues that impact energy intensity:

Climate and geographic conditions: the underlying principle here is to the direct influence of climate on the energy requirements for heating and cooling. Colder temperatures need more energy consumption for heating, while hotter locations see increased energy use for cooling, hence directly impacting a region's energy intensity. It is found that places experiencing more severe temperatures have higher energy intensity since there is a greater need for temperature adjustment.

Population density: a high concentration of people may result in reduced energy consumption per capita due to the implementation of more efficient public transportation systems and the reduced distances required for the movement of products and services. Theoretical models often include urban planning notions to elucidate these processes. Empirical data demonstrates that increased population density is associated with decreased energy intensity. This relationship is primarily attributed to the effectiveness of public transportation systems and the reduced distances required for the delivery of products and services. Their study emphasizes the significance of urban design in fostering energy-efficient living conditions.

Sectoral energy strategies are based on a focused strategy to enhance energy efficiency in certain sectors. Theoretical framework often entails studying the specific influence of customized policies on sectors such as industry, transport, or residential areas, evaluating their efficacy in decreasing the energy intensity within such sectors.

Cultural factors, such as cultural tastes and practices, have a significant impact on energy use patterns. The theoretical foundation in this context might entail socio-economic models that integrate cultural factors to explain variations in energy consumption patterns across different cultures. Dunlop (2019) examines the impact of cultural preferences and habits on energy consumption patterns, offering a socio-economic framework that integrates cultural factors. Investigation demonstrates that cultural influences have a substantial influence on energy consumption patterns among diverse civilizations.

International environmental commitments: the influence of these commitments is conceptually elucidated by international relations and environmental economics theories, indicating that global commitments might motivate national policies and measures aimed at decreasing energy intensity. The theoretical foundations of international environmental accords form the perspectives of international relations and environmental economics. He contends that these agreements might incentivize states to embrace policies and practices aimed at reducing energy intensity, emphasizing the interdependence between global commitments and domestic measures.

Resource endowment refers to the presence of natural resources, which may impact the selection of energy sources. Countries that possess abundant fossil fuels may have greater energy intensity. The link may be understood by examining the theoretical underpinning provided by the resource curse hypothesis and economic models of comparative advantage. Countries that possess ample natural resources, particularly fossil fuels, often have greater levels of energy intensity as a result of their dependence on these resources. The use of economic models that analyze comparative advantage may assist to clarify the continued existence of this pattern and its consequences for energy policy.

Political stability and governance: A stable government may allow the effective implementation of an enduring energy policy. The theoretical foundation includes political economics models that establish a connection between the quality of governance and the efficacy of policies in attaining energy efficiency objectives. The use of political economics models illustrates that stable governance institutions are essential for effectively implementing long-term energy efficiency initiatives. They establish a clear connection between the quality of governance and policy results.

***

There is a substantial relationship between the structure of industries and the quantity of energy that is used for each unit of production. When compared to industries that use less energy, such as agriculture and services, industries that require more energy, such as manufacturing and mining, contribute more to the overall demand and intensity of energy. An industry that requires more energy includes mining and manufacturing. To reduce the amount of energy that is used, it is essential to shift toward enterprises that are primarily concerned with the provision of services and depend significantly on information. This shift may be made feasible via the development of new technologies and the enhancement of operational efficiency. It is common practice to use structural decomposition analysis to disentangle the effect of industrial structure from that of other factors that have an impact on it.

The environmental Kuznets curve (EKC) theory, which suggests a non-linear pattern of energy intensity as economies grow, is used to investigate the relationship between economic development and energy intensity. It is common for there to be an increase in energy intensity during the early phases of economic growth. This is because of the degree to which the economy is dependent on firms that use a significant quantity of energy. Economies tend to shift toward sectors that are service-oriented and high-value as they expand, which ultimately results in a reduction in the amount of energy that is used. This change is made possible by advancements in technology as well as an increased awareness or concern for the environment.

When energy prices are higher, there is often a greater incentive for increased efficiency and lower consumption. This is because both customers and producers are driven to invest in energy-saving technologies. The price elasticity of energy demand is a measurement that determines the extent to which consumption shifts in reaction to changes in price. This demonstrates the significance of energy pricing in terms of promoting energy conservation.

Through the enhancement of the effectiveness of industrial processes and the promotion of the use of more environmentally friendly energy sources, technological innovation plays a significant part in the reduction of energy intensity. An explanation for the broad adoption of innovative technologies, which ultimately leads to a drop in energy intensity, is provided by the idea of the diffusion of innovation.

Policies and regulations on energy are formulated to address market failures by taking into account the external costs that are connected with energy use and pollution. Carbon taxes, subsidies for renewable energy sources, and efficiency standards are all examples of policies that play a significant part in encouraging investments in energy-efficient technologies and practices. These policies also emphasize the influence that policies have on the patterns of energy use.

There is a possibility that urbanization may increase energy consumption; nevertheless, it also gives opportunities for increased efficiency via economies of scale and improvements in infrastructure. At the same time, the insertion of renewable energy sources into the energy mix helps to minimize dependency on traditional energy-consuming sources. This is made possible by technology advancements and incentives provided by the government.

There is a fluctuating relationship between the effect of foreign direct investment (FDI) and energy intensity. Even though it may result in increased energy consumption in industries that require a great deal of energy, particularly in regions with lax environmental regulations (a phenomenon that is referred to as the pollution haven theory), it may also make it easier for host nations to adopt contemporary technology that is efficient in terms of energy consumption, which would lead to a reduction in energy intensity.

5.4 POLICIES AND MEASURES TO INCREASE ENERGY EFFICIENCY

Enhancing energy efficiency and achieving cost savings have always been significant factors in the worldwide energy equilibrium. Recently, energy policy has gained more significance due to growing worries about global climate change and energy security.

The oil crisis of the 1970s marked the inception of economic efficiency policy, as governments implemented several initiatives to enhance energy efficiency across all sectors of their economies, according to Geller et al. (2006). Since 1973, developed nations have seen a substantial increase in energy efficiency. The implementation of this strategy resulted in a reduction in energy intensity (Geller et al., 2006).

The next significant milestone in terms of energy efficiency occurred in 1986. Anderson (2005) designates 1986 as the Year of Energy Efficiency, during which the UK government disseminated detailed information on energy consumption, technologies, standards, and more. An energy audit was conducted to enable consumers to conduct a comprehensive analysis of their energy usage and receive guidance on reducing consumption and, consequently, costs. Governments have the authority to establish mandatory minimum energy efficiency requirements for industry, buildings, and transportation.

Undoubtedly, the energy industry requires political leadership due to its strategic nature. To enhance cost-effectiveness, policies must effectively mitigate energy efficiency failures, while ensuring that the benefits derived from these policies outweigh the costs of their implementation (Gillingham et al., 2009).

The essence of the energy efficiency policy is the deliberate effort to bring about a significant change in the market. Market transformation projects are deliberate interventions that bring about enduring changes in the composition or operation of all markets for energy-efficient goods, services, and behaviors. An efficient market transformation program should incorporate the alignment of policy measures and interventions tailored to market conditions. These measures should aim to enhance energy efficiency and be of interest to all market participants, including the government, private sector, consumers, equipment manufacturers, service providers, and financial institutions. The program should also promote energy efficiency improvements by facilitating the interaction between supply and demand through competitive market forces.

Consequently, it is essential to do a market evaluation before formulating an energy efficiency strategy. This statement emphasizes the significance of considering various measurements that may have distinct impacts and are therefore appropriate for different stages of market development (Birner and Martinot, 2005). Before any action, it is important to carry out thorough market research to identify market characteristics that need reinforcement via incentives or reduction through punishments. Policy actions should be meticulously crafted and synchronized to eliminate identified obstacles in the market. The energy efficiency policy cycle often starts with strategic planning and goal setting, followed by the formulation of particular measures for various target groups, such as market players. Policy measures are still being put into action, and at the end of each cycle, a policy effect evaluation is conducted. After the policy review phase, the outcomes are integrated into the planning, design, and implementation processes. This cycle then repeats. The subsequent sections will analyze each phase of this dynamic cycle.

5.4.1 Policies Aimed at Improving the Efficiency of Energy Use

A single policy alone is unlikely to have a significant influence on all customers with varying spending habits. According to Anderson (2005), it is more effective to use a range of policy areas than to concentrate on a single policy instrument. Often, many policies seem to enhance each other's efficacy. Consumers often lack knowledge of energy prices, technology, and energy management systems. Therefore, information plays a crucial role in supporting various policies (Anderson, 2005).

The overarching extent of the policy has more significance than the inquiry into which particular policy domains are most efficacious. The primary objective of policy-making is not to choose among various policies but to enhance the overall magnitude at which almost all forms of energy efficiency programs are implemented. Certain policies, such as the dissemination of information, may be implemented based on prevailing energy costs and the required payback periods at present. The effectiveness of information and publicity campaigns relies on the extent to which governments can effectively appeal to the public's interests in energy efficiency, which in turn depends on whether the prices are sufficiently high. However, it is important to note that not all policies have a cumulative effect on each other's effectiveness. For instance, regulations that establish minimum efficiency standards for electrical appliances may not have as significant an impact if there are already existing measures such as labeling and consumer information in place. In the absence of such information, these regulations would already contribute to improving efficiency levels. The precise form of policy, including the allocation of funds, the grant amount, the criteria employed, the calculation method, and the administration process, significantly impacts the outcome. Additional policy measures, such as shifting the financial burden of energy efficiency investments from consumers to energy efficiency companies (which may also be energy suppliers), can alter the necessary time for recovering the initial investment. However, these measures can be implemented while considering the prevailing energy prices. The implementation of this strategy is influenced by unique conditions, such as fluctuations in energy market pricing.

5.4.2 Political Regulations and Standards

Political regulation refers to the process of creating and enforcing rules and guidelines by the government to control various aspects of society. These regulations and standards are put in place to ensure compliance and maintain order.

Governments influence the organization of energy efficiency promotion, both via their institutions and by promoting change in the corporate and volunteer sectors. According to Anderson (2005), the energy supply business has the potential to contribute to the adoption of energy efficiency by incorporating it into its structure, instead of just concentrating on supplying energy and leaving efficiency and preservation efforts to other groups. Policy formulation may be achieved by implementing price and tax policies. Governments exert substantial control over energy efficiency by leveraging their authority to tax, manage, and regulate energy-supplying corporations. Additionally, governments may employ antitrust, anti-inflationary, regional, and trade policies to further shape energy prices. As an example, oil often incurs more taxation compared to other forms of energy. The primary objective of this taxation is often to enhance income efficiency rather than energy consumption. However, taxes that result in higher pricing of petroleum products generally encourage their efficient use, irrespective of their original intention (Anderson, 2005).

The cost of energy consumption is determined by several factors, including the price of electricity, insulating materials, energy-efficient heating systems, energy studies and audits, and any interest payments incurred if funds are borrowed for these expenditures. The pricing may be impacted by government programs that attempt to subsidize and encourage energy efficiency.

The primary categories of incentives are grants (provided to both industries and residents who use energy), tax incentives, and loans. They are often integrated with information and publicity initiatives to guarantee that energy customers are knowledgeable about the diverse incentives that are accessible. Grants and tax incentives are more prevalent than loans. Grants are often more economical than tax incentives and loans, although they may be complex and expensive to manage.

Combining financial incentives with knowledge and publicity has been shown to enhance the effectiveness of energy efficiency legislation (Anderson, 2005).

Regulations and standards policies are generally beneficial, particularly when it pertains to construction regulations. Regulating the technologies utilized by the sector has become complex due to the vast array of technologies that need to be considered. Enforcing minimal standards enables the resolution of issues stemming from insufficient consumer knowledge and skills. From this perspective, they might be seen as a substitute for the dissemination of information, but it is possible to integrate both strategies. Standards have the additional benefit of imposing a minimal financial burden on governments (Anderson, 2005).

Empirical evidence demonstrates that a meticulously crafted policy has the potential to yield substantial energy conservation. However, it is crucial to consider the aspects of commercialization, market development, and technology advancements while creating such a policy (Koronen et al., 2020). Implementing minimum efficiency criteria may be a very successful approach to encourage a significant and widespread improvement in energy efficiency, particularly when they are regularly revised (Geller et al., 2006).

Policymakers must guarantee that efficiency criteria are both technically and economically viable. Monetary incentives have the potential to enhance the adoption of energy-saving techniques. Removing subsidies for fossil fuels may facilitate the advancement of energy efficiency. Energy or CO_2 taxes may be raised as well, even if the energy demand is typically unresponsive to price changes in the near term. Labelling, distribution, and training may enhance awareness of energy-saving measures and enhance expertise in energy management (Geller et al., 2006).

Energy efficiency policies and programs are most effective when they are included in market transformation plans that specifically target the impediments present in a given country. Energy efficiency rules should be enforced over 10 years or more to guarantee the systematic growth of energy efficiency markets. Simultaneously, it is necessary to regularly evaluate regulations such as performance goals and objectives, product labeling, and financial incentives. The effectiveness of devices has consistently been enhanced by dynamic rules (Geller et al., 2006).

5.4.3 Strategies to Enhance Energy Efficiency

Energy efficiency should be seen not as a goal in itself but as a method of attaining a more effective distribution of resources (Patterson, 1996). Improved energy efficiency may also contribute to other key public policy goals, namely economic growth and climate change mitigation.

Reducing the energy demand is considered essential by many advocates and policymakers to attain energy efficiency (Anderson, 2005). According to economic theory and the general literature, rising costs motivate the move to alternative items. The energy efficiency market is a heterogeneous collection of diverse enterprises with varying interests that operate in the field of energy efficiency. The energy efficiency market includes suppliers of energy-saving equipment and services, as well as organizations engaged in the financing and execution of energy efficiency projects, such as banks, investment funds, design engineers, and builders. The demand in the energy efficiency market includes individuals and organizations that advocate for energy efficiency enhancements, such as end-users (building owners, renters, and managers), public sector entities, local authorities, and industry representatives.

Policy measures should be customized to suit the particular market conditions. The energy efficiency market is comprised of several stakeholders with diverse backgrounds, making it very susceptible to the effect of behavioral, social, economic, and psychological aspects that shape the decision-making of market players. To address obstacles to both supply and demand, it is necessary to use a variety of policy tools concurrently. Policymakers must recognize that the significance of policy measures fluctuates with time. The appropriateness of various instruments depends on the market's level of development and the timing of their implementation. Consequently, the policy has to be modified to accommodate fluctuating market circumstances.

It is crucial to comprehend that some policy tools may not be appropriate for every market or all players in the market. To effectively implement policy measures, it is necessary to employ targeted approaches to identify certain groups. For example, voluntary agreements are not appropriate for domestic customers, and the labeling of devices will have little impact on big industrial users. However, it is important to note that not all policy initiatives are appropriate for every sector of energy consumption, such as homes, public services, commercial services, industry, and transport. This consideration takes into account the unique characteristics of each sector. Occasionally, it can be advantageous to implement a package of measures, which involves combining two or more measures. For example, this could include pairing a building code with subsidies to demonstrate adherence to higher standards or launching a campaign to encourage cleaner transportation along with subsidies for purchasing hybrid cars. The purpose of implementing such a package is to enhance the effectiveness and efficiency of the policy.

5.4.4 Causes of Policy Failure

The existing national energy efficiency policies are consistently deficient or failing to attain the intended outcomes. We can identify several factors contributing to this policy failure, but fundamentally, the issue can be distilled into three main components:

- Policymakers fail to adequately address all parties involved in energy efficiency. Specifically, not all actors in the market are targeted with suitable policy measures to alleviate market inefficiencies and promote sustainability. We need a comprehensive and customized policy that is specifically designed to address the evolving market circumstances.
- Policy-making should consider the unique characteristics of the implementation environment and the time limitations of implementation. This means that the emphasis should be on developing enough acceptable implementation capabilities that are suited to attaining the goals.
- Politics is a dynamic field, necessitating the implementation of robust policy monitoring and assessment protocols. These processes are essential for identifying effective policies and pinpointing areas that need improvement, facilitating more effective policy change. The development of a unified energy efficiency strategy requires a novel methodology, characterized primarily by its dynamism.

For an energy efficiency strategy to achieve success, its creation must be approached as a process of learning, drawing from both theoretical knowledge and practical evidence. The learning process can be accurately described as a closed-loop process, comprising the following stages: policy formation, policy implementation, and policy evaluation. Policy formation involves setting goals, objectives, methods, and legal frameworks for different target groups. Policy implementation includes establishing the institutional framework, involving stakeholders, managing human resources, and building capacity. Policy evaluation involves monitoring the results achieved through energy statistics and energy efficiency indicators, as well as assessing the impact of policy measures both qualitatively and quantitatively.

5.4.5 The Process of Formulating and Executing Energy Efficiency Policies

The conventional method of formulating energy policies was finding a middle ground between energy demand and supply and gradually changing policy goals, combinations, and targets in response to external changes and variables.

The process of initiating energy management starts with the determination to establish an energy management policy statement. Subsequently, at the highest level of leadership, a strategic plan for energy management is implemented. The strategy outlines specific objectives that can be quantifiably accomplished. The plan including the indicated goals will be disclosed to the public. This measure guarantees the unwavering backing of senior executives and all staff members in the execution of the energy management initiative. Subsequently, an organizational infrastructure is established to facilitate the execution of the strategy. An energy management team is designated to take on the responsibility of coordinating energy management efforts at either the municipal or business level. Furthermore, it is necessary to designate someone, often with technical or maintenance expertise, to oversee the administration of local energy in every department or facility held by the city. Ultimately, every member of the energy management team must get comprehensive education and training to effectively carry out their assigned responsibilities. This guarantees the ability and capability to carry out energy efficiency initiatives. Furthermore, the energy management team is accountable for conducting educational and promotional initiatives to modify the behavior and mindset of employees regarding energy usage in the workplace. They also play a role in initiating green public procurement activities to drive market transformation by leveraging the substantial purchasing influence of the public sector.

The policy should be enforced in all settings where energy is regularly used, including both workplaces and homes. It is an indisputable reality that all energy provided is used by individuals, either directly or indirectly via the many institutional and commercial entities that individuals establish, in both their professional and personal lives. Hence, to effectively execute energy efficiency measures and successfully implement legislation, it is essential to mobilize and collaborate with all stakeholders. The implementation environment is governed by international institutions and initiatives, which provide the framework for policy formation and execution. At the national level, four primary stakeholder groups have distinct roles in implementing energy efficiency policies. These groups, known as vertical social structures, have activities that directly affect the energy efficiency market. Public sector entities primarily exist to guarantee the execution of national policies throughout many sectors, including households, services, industry, and transport. However, the public sector, similar to the commercial sector, is where policy is effectively put into action. However, civil society groups and the media have a crucial role in disseminating knowledge and advocating for widespread energy efficiency. This will ultimately result in a transformation of consumer attitudes and a transition towards more effective energy use.

Most policymakers aim to integrate the demands and standards of global politics into the domestic strategic and legal structure, without fully considering national conditions, such as the level of development of the country's energy efficiency market. Furthermore, there is a notable deficiency in the focus on executing the policy, coupled with a prevailing assumption that implementation is straightforward and would occur naturally, thereby diminishing the need for significant exertion. A fresh strategy is necessary in the development of a shared energy efficiency policy.

In conclusion, the effectiveness of the energy efficiency strategy relies on it being approached as a learning process that incorporates both theoretical knowledge and empirical data.

5.4.6 Assessment of the Efficacy of Energy Efficiency Initiatives

Evaluating the overall efficacy of several individual energy efficiency programs is a very challenging undertaking. As policies and energy patterns evolve, it is crucial to identify cause-and-effect links to make accurate judgments regarding performance. According to Anderson (2005), a method to do this is by computing the hypothetical outcomes that would have occurred if the policy had not been implemented, comparing them to the actual outcomes, and attributing any disparities to the influence of politics. Consequently, it can be inferred that the disparity cannot be ascribed to any additional variable that was not taken into account during the computations. Issues may also arise when implementing policies in specific situations. It may be unclear if certain aspects of these situations are essential for the policy to work or whether they are irrelevant to the program's efficacy (Tanaka, 2008). It is important to be cautious when reading research that specifically addresses the concept of "cost-effectiveness." Cost-effectiveness, in this context, refers only to the financial aspect. It focuses on the monetary savings that consumers may achieve via improved energy efficiency. However, it does not take into account non-financial environmental impacts.

The assessment methods are simultaneously an essential component of the policy-making phase and aligned with the concurrent and coherent activities of policy implementation. The first stage in policy formation involves the formulation of a dependable theory on how a policy instrument (or a set of policies) might effectively enhance energy efficiency. A mix of policy measures is constructed based on sound assumptions from established theories. The term "well argued" indicates a firm conviction that this action would effectively enhance energy efficiency in the market at a reasonable cost. An evaluation will be conducted to determine the anticipated effects of reduced energy use and cost efficiency resulting from the implementation of the measures. These effects will be compared to the situation in which the measures are not implemented.

Cost-effectiveness refers to the ratio of the extra expenses borne by the end-users, society, or the government sector as a whole, to the overall impact of the policy being studied. Government expenditure includes the execution, management, enforcement, supervision, and assessment of regulations, as well as the provision of subsidies and tax advantages. Cost-effectiveness is a measure used to assess the efficiency of public funding in achieving socially advantageous objectives. The expenses incurred by end users will be calculated based on the energy price, incremental investments, and incremental costs associated with running and maintaining the energy efficiency measure.

Typically, many indicators are used to evaluate the efficacy:

- Energy intensity refers to the relationship between energy consumption (measured in units of energy) and an activity indicator defined in monetary units (such as gross domestic product or value added) (Bosseboeuf et al., 1997).
- Unit consumption, also known as particular consumption, is the measurement of energy consumption based on an activity indicator expressed in physical terms (such as steel tons or number of vehicle-kilometers) or a specific unit of consumption (such as vehicles or houses).
- The Energy Efficiency Index (ODEX) offers a comprehensive evaluation of the sector's energy efficiency developments. The indicators are computed by taking a weighted average of certain subsectoral factors, such as end-use and mode of transport. Reduction denotes a rise in energy efficiency. An index that measures changes in energy efficiency is more crucial for understanding the true nature of these changes than just looking at the intensity of energy use.
- Diffusion rates refer to the measurement of how widely certain indicators are adopted. These indicators can be categorized into three types: the market penetration of renewable energy sources, the market penetration of efficient technologies, and the dissemination of energy efficiency practices. Examples of these indicators include the number of solar water heaters, the percentage of wooden boilers used for heating, the number of efficient lamps sold, the percentage of label A when selling new electrical appliances, the percentage of public and non-motorized passenger transport, the percentage of rail freight and combined rail and road freight, and the percentage of efficient processes in the industry.

5.4.7 Strategies to Minimize Energy Efficiency Deficits

A significant factor contributing to the failure of energy efficiency policy is the tendency of policymakers to favor generic solutions and blindly adopt policies from other countries, without considering the unique characteristics of their energy efficiency market. While some typical obstacles in the market hinder energy efficiency, they alone are insufficient to stimulate a market transformation and accomplish the ultimate objective of establishing a self-sustaining energy efficiency market.

The availability of all types of energy at reasonable rates is the driving force behind the economic and social progress of society. Enhancing energy efficiency is undeniably beneficial for the general public, particularly in the context of combating climate change. To fully use the potential for cost-effective efficiency increases, it is imperative to implement legislative policies that eliminate current market obstacles. Policies designed to enhance energy efficiency should facilitate the transition of the market towards higher efficiency, with the ultimate objective of achieving a cleaner environment, an improved quality of life, a more competitive industry, and enhanced energy supply security, evaluating energy efficiency policies by transitioning from a techno-economic approach to a broader societal view on the energy efficiency market. Furthermore, these products must be designed with careful consideration of the actual demands of the market, particularly tailored to meet those needs. They should also possess the ability to be adaptable and responsive to evolving market demands, to effectively accomplish objectives most efficiently. Energy efficiency policy-making necessitates the adoption of novel and inventive strategies, characterized primarily by their dynamism. Dynamic policy-making refers to a process that is characterized by continual learning and feedback. It involves the integration and balancing of policy development, implementation, and assessment.

Energy efficiency is linked to the reduction of energy losses. Losses occur throughout the conversion, transmission, and distribution of energy, as well as during its ultimate use. While technology plays a significant role in minimizing losses in the first three areas of activity, addressing the last problem requires a combination of technological and non-technical approaches. Improved organization of activities, enhanced energy management, and modifications in consumer behavior may effectively prevent the wasteful use of energy. The latter, in particular, is a significant challenge when it comes to making lifestyle adjustments. The most challenging aspect is coordinating activities, optimizing energy use, and modifying customer behavior, which is becoming more integrated into lifestyle transformations. Energy efficiency should be seen as an ongoing process that encompasses not just isolated measures to prevent excessive energy use and reduce energy waste but also includes the monitoring and regulation of energy consumption to attain a sustained minimal level of energy usage.

To maintain consistent progress in enhancing energy efficiency, it is important to see energy use as a regular task that requires management. Energy management may be defined as the foundation for assuring the ongoing prevention of excessive energy use and the minimization of energy losses by the application of knowledge, suitable metering, and ICT.

It is important to consider not just the technical and economic aspects of energy use but also the reality that energy efficiency is an ongoing social process. Furthermore, the price of energy must be determined in a manner that appropriately represents its true costs, including factors such as environmental, health, and geopolitical effects. Additionally, it is crucial to educate people on the repercussions of energy usage.

In policy-making, it is crucial to consider these fundamental principles to enhance energy efficiency. The design of policies should prioritize the avoidance and promotion of reducing energy consumption. Additionally, it is essential to ensure that there are sufficient resources and infrastructure in place for effective monitoring and management of policy implementation.

***

Particularly in light of concerns over climate change and energy security, energy efficiency has emerged as an essential component of every energy policy implemented around the globe. The oil crisis of the 1970s marked the beginning of the deployment of energy efficiency measures, which marked a significant movement toward increasing energy efficiency in a variety of businesses all over the globe. Major advancements have been made, which have led to major reductions in energy intensity and the promotion of behaviors that are ecologically responsible in terms of energy use.

This history of energy efficiency policies has achieved several important milestones, with considerable interventions aiming to modify the market dynamics of energy-efficient products and services. These interventions have been effective in achieving their goals. To encourage a more energy-efficient economy, governments have implemented a variety of regulatory measures and incentives, such as mandatory energy requirements for businesses, buildings, and transportation.

These interventions are specially designed to synchronize with the current market conditions, to maximize the interaction between supply and demand via the use of competitive market forces.

The first step in developing an effective strategy for energy efficiency is to conduct a thorough market analysis to tailor energy efficiency techniques to the different stages of market development. All stakeholders, including governmental bodies, corporate sectors, and consumers, are encouraged to engage in energy-saving behaviors via the implementation of this approach, which ensures that programs are targeted to successfully address specific market constraints and promote energy-saving behaviors. It is common practice for the policy cycle to involve strategic planning, the formation of specific metrics, and continual assessment to integrate the outcomes of the previous planning stages into succeeding stages.

In order to address the many diverse aspects of energy efficiency, several different policy initiatives and regulatory measures have been implemented. It is vital to disseminate information to support these policies and close the knowledge gap among customers about energy pricing and technology that is efficient. In addition, the incorporation of financial incentives such as grants, tax incentives, and loans is sometimes accompanied by promotional efforts to maximize the adoption of policies and their effectiveness.

In spite of the fact that there are strategic measures in place, energy efficiency projects often face challenges and failures. These include inadequate market targeting, limited implementation capabilities, and a lack of appropriate monitoring and assessment methods. It is of the utmost importance to gain knowledge from these problems, which calls for a flexible approach to policymaking that can adapt to changing market conditions and absorb continuing feedback.

In general, the process of developing policies that promote energy efficiency is an ongoing process of gathering information that combines theoretical understanding with data from the actual world to create an environment that encourages behaviors that are sustainable and energy conservation. To accomplish the overarching goals of energy efficiency, economic growth, and climate mitigation, it is very necessary to continually improve policies, as well as to properly execute and analyze them.

5.5 CASE STUDY

A case study to explore drivers and barriers to energy efficiency in the EU countries during 2000–2022 is based on regression analysis aiming to find which factors are indicated in the literature (see Subchapter 4.3) and in which direction they affect energy efficiency. The general equation linking energy efficiency with its factors for panel data could be specified as follows:

$$\ln EE_{i,t} = \alpha + \mathbf{X}\beta + \mu_i + \theta_t + \varepsilon_{i,t}, \tag{1}$$

where: *EEi,t* is energy efficiency in country *i* over the period *t*. **X** is a matrix of time- and across country-varying factors that affect energy efficiency. μ_i is the unobserved time-invariant countries heterogeneity. θ_t is time-dummies and $\varepsilon_{i,t}$ is the idiosyncratic error term. To proxy energy efficiency, here we will use energy Intensity (*EI*). To account for industrial structure, we will use the size of different sectors like manufacturing (*MAN*), services (*SER*), etc. in the economy. To proxy the level of economic development, we will use per capita GDP (*Y*) and for economic development speed—economic growth, i.e., per capita GDP growth (*ΔlnY*). To proxy dynamics of energy prices, we will use change in the energy price index (*ΔE_CPI*). To account for the effect that innovations might have on energy efficiency, we will use two alternative input and output approaches. For the input approach, we will use investment in research and development (R&D). For the output approach, we will use change in capital (ΔlnCP) and labor (ΔlnLP) productivity as the possible outcome of investment in R&D that could affect energy use directly. For trade and financial openness, we will use a sum of imports and exports to GDP (*TRADE*) and stock inward FDI to GDP (*FDI*). For urbanization, we will use the percentage of the urban population to total population (*URB*), for the renewable energy use—energy use from renewable energy sources (*REN*). To proxy energy policy and energy efficiency measures we will use several alternative variables—per capita emissions of greenhouse gases in CO_2 equivalent (GHG), energy imports dependency (EID), energy taxes (TAX), and per capita energy use (FEC). For the descriptive statistics of the variables, please see Table 5.1.

The energy intensity of GDP, measured in kilos of oil equivalent per thousand euros, has an average value of 200 and a large standard deviation of 116, showing considerable variation within nations or over time. The energy efficiency of economic operations varies significantly across EU nations, as seen by the range spanning from a minimum of 32.9 to a high of 775. The industrial structure statistics, shown as a proportion of the total value added, emphasizes the variety in the

TABLE 5.1
Summary Statistics of the Variables

Full name		Abbrev.	Unit of measurement	Mean	Median	S.D.	Min	Max
Energy intensity of GDP in chain-linked volumes (2015)		EI	Kilograms of oil equivalent per thousand euro	200	166	116	32.9	775
Industry structure (share in total value added)	Agriculture	AGR	%	3.20	2.60	2.51	0.210	22.4
	Industry	IND	%	21.1	21.4	5.94	5.68	41.4
	Manufacturing	MAN	%	17.2	17.0	5.48	4.21	39.9
	Construction	CON	%	6.10	5.91	1.71	1.08	12.7
	Services	SER	%	69.6	69.5	7.27	41.9	88.3
Gross domestic product per capita at market prices. Chain linked volumes (2015)		Y	euro	25644	20055	18754	3183.5	102130
Economic growth		ΔlnY	*100%	2.34	2.46	3.90	−15.6	21.0
CPI energy change compared to the previous year		ΔE_CPI	%	5.52	4.20	9.91	−14.9	70.6
R&D all sectors. Percentage of gross domestic product		R&D	%	1.45	1.24	0.880	0.200	3.87
Real labor productivity per person. Index, 2015=100		ΔlnLP	*100%	1.74	1.48	3.10	−12.1	18.5
Gross value added per unit of net fixed assets, Index, 2015=100		ΔlnCP	*100%	0.482	0.732	3.82	−21.1	19.6
Exports and imports of goods and services, % of GDP		TRADE	%	116.	100.	63.1	37.1	393
FDI stock inward % of GDP		FDI	%	122.	39.6	316	2.19	2062.1
Urban population (% of total population)		URB	%	71.8	69.6	12.5	50.6	98.2
Renewable energy sources		REN	%	18.6	16.7	11.9	0.102	66.0
Per capita emissions of greenhouse gases		GHG	Kilograms per capita in CO_2 equivalent	8372.9	7404.6	3304.2	3620.3	20340
Energy imports dependency		EID	%	57.1	58.2	26.3	−50.6	109
Energy taxes. Percentage of GDP		TAX	%	1.94	1.88	0.542	0.460	4.81
Final energy consumption		FEC	Kilograms per capita of oil equivalent	2460.8	2158.0	1352.4	937.56	9713.2

economic composition of the EU. The services sector is the most prominent, accounting for an average of 69.6% and displaying a comparatively low standard deviation of 7.27. This suggests a continuous focus on services across the European Union. On the other hand, the agricultural sector has a much lower average value of 3.20% but with a rather large level of variability (standard deviation = 2.51). This indicates the diverse significance of agriculture in the economies of various nations. The GDP per capita in EU nations varies significantly, ranging from 3,183.5 to 102,130 euros, with an average of 25,644 euros. The vast variety of economic indicators highlights the substantial economic inequalities present within the European Union. The average economic growth rate (ΔlnY) is 2.34%, with a significant variation ranging from –15.6% to 21.0%. This demonstrates the influence of economic cycles and the diverse situations of different nations. The mean of R&D investment as a proportion of GDP is 1.45%, suggesting that there is room for governmental attention to boost innovation and development. ΔlnLP represents variations in labor productivity per individual. The data exhibits a mean of 1.74% and a range spanning from –12.1% to 18.5%. These figures suggest that labor productivity has seen oscillations, which might perhaps be attributed to technology advancements, labor market reforms, or economic cycles. The standard deviation of 3.10% signifies significant diversity, implying distinct economic situations and varying phases of development throughout the EU. The gross value added per unit of net fixed assets (ΔlnCP) has a moderate average growth of 0.482% and a very broad range spanning from –21.1% to 19.6%. The substantial variety (standard deviation = 3.82) highlights the influence of economic downturns, capital investments, and depreciation policies, which may have a considerable effect on asset productivity in various locations. The trade figures, which include both exports and imports as a proportion of GDP, have an average of 116% with a wide range from 37.1 to 393%. The high maximum value indicates that these economies are highly dependent on international trade, but the wide range reveals that there are different levels of trade dependency across EU nations. These variations may be ascribed to factors such as geographical location, economic magnitude, and industry composition. The FDI as a percentage of GDP has a very high average of 122% with an exceptional level of variability of 316, indicating a highest value of 2,062.1%. The data emphasizes the important impact of foreign investments on the economy of EU members, while there is considerable fluctuation and inequality, which may be influenced by policy conditions, market size, and economic stability. The average proportion of urban population in relation to the total population is 71.8%, with a range of values from 50.6% to 98.2%. This indicates that the majority of the population in the EU resides in urban areas, and there is a significant variation in urbanization levels across different nations, as seen by the relatively high standard deviation of 12.5%. Urbanization is often associated with increased economic growth and improved access to services, which may have an impact on several other economic and social indices. The average energy taxation, expressed as a percentage of GDP, is 1.94%, ranging from 0.46% to 4.81%. The standard deviation of 0.542% indicates a moderate level of heterogeneity in the way various EU nations use fiscal policy to control energy consumption and environmental effects. Increased energy tariffs may signal more robust environmental policies, which can impact energy consumption patterns and promote the use of renewable energy sources.

The environmental indicators, such as the amount of greenhouse gases (GHG) emitted per person and the proportion of renewable energy sources (REN), provide valuable information on the EU's impact on the environment and its efforts towards sustainability. The average GHG emissions per capita is 8,372.9 kg, with a broad range of values (3,620.3 to 20,340), indicating differing degrees of environmental strain across nations. Renewable energy sources, on average accounting for 18.6% of total energy production, indicate the continued shift towards cleaner energy. However, there is significant variation in this percentage, with a standard deviation of 11.9. The EU's energy imports dependence (EID) of 57.1%, which varies from –50.6% to 109%, demonstrates its significant reliance on energy supplies from outside the region. Additionally, several nations inside the EU are net energy exporters. The data on final energy consumption (FEC) underscore the significant energy use across the European Union (EU), underscoring the need of using energy-efficient practices and implementing measures to decrease reliance on imported energy.

To estimate Eq. (1), we will use fixed-effect estimators to be able to control unobserved cross-country heterogeneity (μ_i). Some variables make sense just in the dynamic form due to the nature of the data collected to proxy them. For example, indexes used to proxy energy price level or labor (capital) productivity, etc. Also, some variables like economic growth are dynamic. Thus, we specify an alternative dynamic equation:

$$\Delta \ln EE_{i,t} = \alpha + \Delta \mathbf{X}\beta + \theta_t + \Delta\varepsilon_{i,t}, \tag{2}$$

where all variables enter the equation in first-differences (Δ). μ_i does not appear in the equation anymore. Since it represents time-invariant factors, this term is "differenced away" because $\Delta\mu_i = 0$. We estimate Eq. (2) using POLS as the unobserved cross-country heterogeneity will be removed by first-difference transformation.

TABLE 5.2
Fixed Effects Estimates of Eq. (1)

Regressors	**(1)**	**(2)**	**(3)**	**(4)**	**(5)**	**(6)**	**(7)**
Const.	6.583***	14.73***	15.31***	14.69***	14.61***	11.66***	12.36***
	(1.551)	(0.3527)	(0.4679)	(0.3553)	(0.3485)	(0.3470)	(0.4609)
IND	0.0335***	0.0295***	0.0398***	0.0293***	0.0319***	0.0077**	0.0105**
	(0.0042)	(0.0042)	(0.0052)	(0.0042)	(0.0042)	(0.0037)	(0.0051)
SER	−0.0307***	−0.0258***	−0.0347***	−0.0252***	−0.0304***	−0.0089**	−0.0111**
	(0.0040)	(0.0041)	(0.0051)	(0.0041)	(0.0042)	(0.0035)	(0.0051)
lnY	0.8265***	−0.8579***	−0.8630***	−0.8537***	−0.8226***	−1.0750***	−1.0710***
	(0.3132)	(0.0207)	(0.0246)	(0.0211)	(0.0222)	(0.0218)	(0.0304)
(lnY)^2	−0.0832***						
	(0.0155)						
R&D	0.0201	0.0217	0.0212	0.0215	0.0209	0.0145	0.0140
	(0.0165)	(0.0167)	(0.0173)	(0.0168)	(0.0165)	(0.0143)	(0.0156)
TRADE	0.0029***	0.0025***	0.0025***	0.0026***	0.0025***	0.0014***	0.0014***
	(0.0001)	(0.0002)	(0.0002)	(0.0002)	(0.0002)	(0.0001)	(0.0002)
FDI	0.0000	0.0000	0.0001*	0.0001*	0.0001**	0.0002**	0.0002**
	(0.0000)	(0.0000)	(0.0000)	(0.0000)	(0.0000)	(0.0000)	(0.0000)
URB	0.0081***	0.0074***	0.0070***	0.0072***	0.0080***	0.0093***	0.0094***
	(0.0008)	(0.0008)	(0.0009)	(0.0009)	(0.0008)	(0.0007)	(0.0008)
REN	−0.0021**	−0.0028***	−0.0027***	−0.0030***	−0.0029***	−0.0031***	−0.0036***
	(0.0008)	(0.0008)	(0.0009)	(0.0009)	(0.0008)	(0.0007)	(0.0008)
lnGHG			0.0415				0.0657**
			(0.0269)				(0.0281)
EID				−0.0004			—0.0005
				(0.0004)			(0.0004)
TAX					−0.0634***		−0.0823***
					(0.0157)		(0.0165)
lnFEC						0.4935***	0.5072***
						(0.0313)	(0.0394)
Sample size	512	512	433	512	512	512	433
Adj. R^2	0.8932	0.8871	0.8825	0.8871	0.8905	0.9252	0.9178

Notes: all estimates include time dummies. To account for serial autocorrelation, heteroscedasticity, and cross-sectional correlation of the error term we compute Beck-Katz standard errors (PCSE are presented in parentheses). *, **, and *** indicate significance at the 10, 5, and 1% levels respectively.

Results suggest that industry structure has a significant effect on energy efficiency, which is, decreasing the size of the industry sector and increasing the size of the service sector in the economy is associated with smaller energy intensity. According to the estimations, which do not include per capita energy use, which is (1)—(5), other factors being fixed, a one percentage point increase in the share of the industry and services sector in the economy is associated with on average 3% higher and lower energy intensity, respectively.

Manufacturing operations sometimes need substantial energy consumption for equipment, heating, cooling, and chemical procedures. On the other hand, the service sector, which encompasses industries such as banking, education, healthcare, and retail, depends less on large equipment and more on human resources and digital technology, which typically need lower energy use. The service sector often has a higher degree of adaptability to embrace novel technologies and methodologies that might effectively diminish energy use. For instance, the use of digital technology in the service industry may result in increased efficiency that greatly reduces energy usage, such as by promoting telecommuting, online shopping, and digital media. Industrial sectors have the option to incorporate new technologies in order to decrease energy consumption. However, due to the large amount of energy needed for physical manufacturing processes, their total energy intensity may still be greater in comparison to the services sector. The transition from a mostly industrial to a primarily service-based economy is often a component of a more comprehensive economic growth and maturation process. As economies develop, they often go towards higher levels of the value chain, prioritizing knowledge-intensive and high-value services. This transition is characterized by enhancements in energy efficiency across the economy, along with a shift away from industries that use higher amounts of energy.

In our analysis, we aimed to examine the EKC hypothesis, which posits a curvilinear link between a nation's degree of development and its energy intensity, characterized by an inverted U-shape. The EKC theory posits that, during the first phases of economic progress, nations tend to mainly depend on agriculture, which is relatively less energy-intensive as compared to industrial pursuits. As these nations embark on the path of industrialization, they adopt more energy-intensive technology and methods. The transition from agricultural to basic manufacturing is characterized by a higher energy consumption per unit of economic output, resulting in an increase in energy intensity. As development advances, economies undergo a period of rapid industrialization. During this era of economic expansion, industries that use a significant amount of energy, such as steel, cement, and chemicals, tend to have a strong presence. At this level, the energy intensity keeps rising since these sectors need substantial energy inputs for the production of commodities. As nations attain greater levels of development, there is often a transition towards more effective technology and procedures. Emphasizing innovation and implementing greater efficiency standards leads to a decrease in the amount of energy used for each unit of output. Furthermore, there is often a legislative inclination towards cleaner and more sustainable methods that also promote energy efficiency. During the most advanced phases of economic progress, there is a discernible transition from the production of goods to the provision of services. Service sectors, like banking, education, and information technology, often need less energy compared to industrial sectors. This transition leads to a reduction in the total energy intensity as the economy expands while using less energy per unit of GDP. The progression of technology and growing environmental consciousness have resulted in a wider acceptance and use of renewable energy sources. Renewable energy sources, such as solar and wind, are often more efficient and less resource-intensive compared to conventional energy sources. As nations increasingly prioritize sustainability, the level of energy intensity may further decrease. The inverted U-shape represents the shift from an agrarian economy to an industrial economy (with a rise in energy intensity), followed by a transition to a service-oriented, technology-driven economy (with a decline in energy intensity).

The positive coefficient on Y and negative on Y-squared, both being statistically significant, provide empirical evidence on EKC. However, the estimated turning point lies around 143–144 euros of per capita GDP, which is 22 times smaller than the minimal observed value in the sample. Thus, we can conclude that all EU countries have already reached the development level to the point where further development is associated with decreasing energy intensity and, at the same time, increasing energy efficiency. All other estimations, except (I), are based on a linear relationship between *Y* and *EI* and show that with every increase of per capita GDP by 1%, other factors being fixed, the energy intensity is estimated to decrease by around 0.8–0.9%, which is a clear energy efficiency increasing effect of economic development (this size of the effect is found in the estimations without per capita final energy consumption).

We do not find empirical evidence to support the relationship between R&D spending and energy intensity. It might be related to the approach used here to proxy technological advancements. Since R&D, as the proxy for the input approach, is a prerequisite to technological advances, it is not necessarily channeling to higher energy efficiency.

We find strong evidence that trade is positively related to energy intensity. In estimations without per capita final energy consumption (I)–(V), other factors being fixed, higher openness to trade by 1 percentage point is associated with 0.25% higher energy intensity.

The influence of increasing trade openness on energy intensity and energy efficiency may vary based on factors such as the economy's structure, the kind of commodities being traded, and the efficiency of the sectors involved. If a nation has a competitive advantage in energy-intensive sectors such as steel, aluminum, or chemical manufacture, engaging in trade may result in a concentration of expertise and focus in these areas. As the nation increases its export of commodities that need a large amount of energy, the energy intensity of the economy, which is the ratio of energy consumption to economic production, may rise. Increased trade openness may result in expanded markets and heightened demand, which in turn encourages enterprises to expand their production capacity. If there is an expansion in energy-intensive industries, it might lead to a rise in the total energy intensity of the economy. This phenomenon is most noticeable when the improvements in energy efficiency fail to match the growth in output. Trade may result in the acquisition of technology and practices that can either enhance or diminish energy efficiency. If a nation chooses to import outdated and less effective technologies due to their lower cost or if imported items are manufactured using less efficient technologies, it might lead to a decrease in the overall energy efficiency of the economy. On the other hand, the adoption of advanced, energy-efficient technology might enhance overall efficiency. Nations possess varying environmental and energy legislation. Trade openness may result in "regulatory arbitrage," a phenomenon where industry relocates to nations with less strict restrictions, possibly causing a rise in global energy intensity. This is especially important when energy-intensive manufacturing moves from nations with stringent environmental regulations to those with lenient regulations. Greater commerce often results in the transportation of a larger quantity of commodities across extended distances. This kind of transportation is very energy intensive, especially when it depends on methods such as shipping, trucking, or air freight, which heavily rely on the use of fossil fuels. As the amount of commerce increases, the energy used in transportation may have a substantial impact on the total energy intensity of the nations involved in the trade. Although trade rivalry might incentivize enterprises to adopt more efficient technologies to maintain their competitiveness, it can also result in cost-cutting tactics that may include the use of cheaper but less energy-efficient manufacturing methods.

We have mixed results of FDI impact on energy intensity ranging from insignificant to 0.01–0.02% higher energy intensity, which almost negligible, due to one percentage point higher FDI-to-GDP level. The estimated effect of financial openness is less stable and 10–20 times smaller compared to trade openness. These results can be explained by several reasons.

If foreign direct investment (FDI) primarily targets sectors that have lower energy consumption, such as services or technology, as opposed to heavy industries like manufacturing or mining, the impact on a country's energy intensity might be minimal. Investments in industries such as finance, software, or retail have a negligible impact on energy consumption patterns when compared to investments in industrial sectors. Frequently, global businesses engaged in foreign direct investment (FDI) introduce cutting-edge technology and streamlined methods. These technologies have the ability to decrease the amount of energy used in the processes they are implemented in. By extensively implementing these practices, the energy efficiency of the host economy may be enhanced, which may result in a reduction rather than a rise in energy intensity. Foreign investors often adhere to stringent environmental legislation in their home countries and use these elevated standards in their operations abroad. This might result in the adoption of cleaner and more energy-efficient technology and practices in the nation where the event is taking place. The influence of foreign direct investment (FDI) on energy intensity might also be contingent upon the current stage of economic growth in the nation where the investment is taking place. In advanced economies with already high levels of energy efficiency, more foreign direct investment (FDI) may have no impact on further enhancing energy intensity. On the other hand, in developing nations, while foreign direct investment (FDI) might initially result in increased energy use owing to industrial expansion, it can eventually promote the adoption of more energy-efficient industrial methods. Foreign direct investment (FDI) has the potential to replace outdated and ineffective local technology with contemporary and highly energy-efficient alternatives. This substitution often results in a decrease in the energy intensity of manufacturing operations.

Results of estimations without per capita final energy consumption (I)–(V) suggest that level of urbanization is positively and significantly related to energy intensity. Other factors being fixed, urban-to-total population ratio higher by one percentage point is associated with 0.7–0.8% higher energy intensity.

Urbanization necessitates substantial infrastructure development, including the construction of roads, buildings, public transit systems, and utilities. The process of building and upkeeping this infrastructure requires substantial energy, particularly during the first phases of urban growth. The construction industry exhibits a significant need for energy. As metropolitan areas grow, there is a higher need for residential and commercial structures, leading to an increasing need for construction materials such as steel, cement, and glass. However, the manufacture of these materials requires a significant amount of energy. Urban lives often entail more energy usage in comparison to rural living. This encompasses increased use of electrical appliances, heating and cooling systems, and other consumer devices. Urban regions also exhibit a positive correlation between improved standards of life and increasing energy consumption per person. Urbanization generally results in an increased dependence on motor vehicles, such as automobiles, buses, and trains, leading to a rise in total energy consumption. While public transportation may really be more energy efficient per person, the overall energy consumption may nevertheless rise due to the increased demand for travel in densely populated regions. Urban locations serve as centers for commercial and industrial operations, which often need more energy compared to the mostly agricultural employment seen in rural areas. Urban centers have a notable increase in energy consumption due to the concentration of industry, services, and retail activities. In metropolitan locations, there is a need to carry energy across extended distances from power plants to users. The inefficiencies in energy distribution networks, particularly when the infrastructure is not properly maintained or modernized, may result in increased energy losses in this transmission. Urban areas produce substantial quantities of trash, necessitating energy-intensive procedures such as waste collection, transportation, and treatment. Trash-to-energy facilities not only provide electricity but also demonstrate the substantial energy input needed to handle urban trash. Urbanization often results in alterations to land use patterns, whereby green spaces are transformed into urban environments. The decrease in green areas might diminish the natural cooling benefits of flora, resulting in an escalation in energy consumption for cooling structures.

Results of estimations without per capita final energy consumption (I)–(V) suggest that increase in the renewable energy use (renewable-to-total energy) by one percentage point, other factors being fixed, reduces energy intensity by 0.2–0.3%.

Renewable energy technologies often exhibit higher levels of efficiency compared to conventional fossil fuel systems. Modern solar panels and wind turbines exhibit exceptional efficiency in turning renewable energy into electrical power. The enhanced efficiency results in a reduction in energy wastage, hence decreasing the energy intensity of power production. Renewable energy solutions often enable localized power generation, which means producing energy closer to where it is used. For example, solar panels installed on roofs minimize energy losses in the process of transmitting and distributing electricity, a typical occurrence in centralized power systems that depend on transmitting electricity across long distances from huge power plants. Conventional energy generation via the use of fossil fuels is characterized by a significant use of energy, which includes the activities of extracting, transporting, and processing fuels like coal, oil, and natural gas. In contrast, renewable energy sources often need less rigorous extraction and processing. Once solar panels or wind turbines are deployed, the energy they provide does not need continuous extraction or refining of raw materials. Renewable energy sources provide a sustained and inexhaustible supply of power that can be tapped again without running out. The sustainability of this element implies that the energy produced may be maintained or grown over time without requiring a proportional increase in energy input. As a result, the total energy intensity is reduced compared to limited and depleting sources such as fossil fuels. Utilizing renewable energy sources decreases the release of greenhouse gases and pollutants that are linked to the combustion of fossil fuels. This environmental advantage, while not directly reducing energy intensity, promotes a transition towards more sustainable and possibly less energy-demanding economic activity. For instance, improved air quality may decrease healthcare demands and minimize productivity declines associated with pollution. Investing in renewable energy often coincides with improvements in energy efficiency. As part of a wider shift towards sustainable energy, buildings and businesses might use more efficient practices and technology, resulting in a decrease in energy consumption per unit of production. Countries that make significant investments in renewable energy sources frequently see a transition towards service industries that need less energy use. This transition is partially attributed to the evolving skill sets and infrastructural requirements of an economy that prioritizes sustainability and technology, as opposed to heavy industry and manufacturing. Government policies sometimes endorse the deployment of renewable energy and concurrently encourage energy efficiency via measures such as building rules, appliance standards, and subsidies for energy-saving devices. These strategies effectively decrease the amount of energy used per unit of economic output.

We used several variables to proxy energy policy and energy efficiency measures like greenhouse gas emissions, energy import dependency, energy taxes and final energy consumption intensity. Estimation results are rather mixed, but it is clear that with higher taxation and thus increasing energy prices, we can achieve lower energy intensity. Also, promoting policies that reduce per capita energy consumption is a powerful tool to reduce energy intensity and thus increase energy efficiency. We find no evidence that energy import dependency would affect energy intensity. The variable of greenhouse gas emissions becomes significant if only final energy consumption is introduced in the equation. Moreover, the inclusion of final energy consumption changes the effect of most other factors considered as influencing energy intensity. The effect of trade decreases while the effect of others increases.

POLS estimates of Eq. (2) also suggest that increasing share of service sector and decreasing share of industry service reduces energy intensity and thus positively affects energy efficiency. The same is true considering economic growth—increasing per capita GDP is related to lower energy intensity.

Estimates reveal that increasing energy price levels are negatively related to energy intensity, which is, other factors being fixed, increase in energy price level by 1% is associated with decreasing energy intensity by 0.1–0.13%.

The negative link arises from the fact that higher costs provide a strong motivation for individuals and businesses to adopt more efficient energy consumption practices and invest in the development of energy-saving technology. Rising energy prices prompt individuals and companies to alter their behavior in order to save expenses. This might include reducing energy use or transitioning to more energy-efficient methods and equipment. For instance, a rise in power costs might incentivize homes to switch off unnecessary lights or prompt enterprises to invest in energy-efficient

TABLE 5.3
POLS Estimates of Eq. (2)

Regressors	(1)	(2)	(3)	(4)	(5)	(6)
Const.	−0.0409*	−0.0612***	−0.0414**	−0.0625***	−0.0348*	−0.0161
	(0.0219)	(0.0226)	(0.0210)	(0.0228)	(0.0198)	(0.0186)
ΔIND	0.0036***	0.0035***	0.0040***	0.0032***	0.0035***	0.0035***
	(0.0007)	(0.0009)	(0.0008)	(0.0009)	(0.0008)	(0.0006)
ΔSER	0.0064***	0.0070***	0.0078***	0.0080***	0.0074***	0.0072***
	(0.0027)	(0.0029)	(0.0032)	(0.0039)	(0.0024)	(0.0029)
ΔlnY	−0.6074***	−0.6314***	−0.6764***	−0.6422***	−0.9654***	−0.8816***
	(0.1423)	(0.1485)	(0.1420)	(0.1509)	(0.1326)	(0.1284)
ΔE_CPI	−0.0010**	−0.0012**	−0.0012***	−0.0012**	−0.0013***	−0.0013***
	(0.0005)	(0.0005)	(0.0004)	(0.0005)	(0.0004)	(0.0004)
ΔlnLP	−0.0548***	−0.0551***	−0.0523***	−0.0498***	−0.0525***	−0.0579***
	(0.0151)	(0.0157)	(0.0149)	(0.0158)	(0.0137)	(0.0130)
ΔlnCP	−0.0247***	−0.0238***	−0.0216***	−0.0259***	−0.0260***	−0.0287***
	(0.0094)	(0.098)	(0.0094)	(0.0098)	(0.0085)	(0.0082)
ΔTRADE	0.0004**	0.0004**	0.0003**	0.0004**	0.0004**	0.0004**
	(0.0002)	(0.0002)	(0.0002)	(0.0002)	(0.0002)	(0.0002)
ΔFDI	0.0000**	0.0000**	0.0001***	0.0000**	0.0000	0.0000
	(0.0000)	(0.0000)	(0.0000)	(0.0000)	(0.0000)	(0.0000)
ΔURB	0.0060***	0.0057***	0.0061***	0.0060***	0.0062***	0.0062***
	(0.0023)	(0.0017)	(0.0026)	(0.0018)	(0.0025)	(0.0024)
ΔREN	−0.0021***	−0.0022***	−0.0019***	−0.0019**	−0.0023***	−0.0023***
	(0.0006)	(0.0007)	(0.0005)	(0.0007)	(0.0005)	(0.0003)
ΔlnGHG		0.4244***				0.5356***
		(0.0408)				(0.0654)
ΔEID			−0.0024			−0.0033
			(0.0034)			(0.0036)
ΔTAX				−0.0016***		−0.0017***
				(0.0005)		(0.0003)
ΔlnFEC					0.7198***	0.6820***
					(0.0665)	(0.0625)
Sample size	384	321	384	384	384	337
Adj. R2	0.2916	0.4892	0.3499	0.2899	0.4659	0.2933

Notes: all estimates include time dummies. To account for serial autocorrelation, heteroscedasticity, and cross-sectional correlation of the error term we compute Beck-Katz standard errors (PCSE are presented in parentheses). *, **, and *** indicate significance at the 10, 5, and 1% levels respectively.

equipment. As the cost of energy increases, the potential benefits of investing in energy-saving technology and practices become increasingly appealing. This may result in increased acceptance of energy-efficient appliances, construction materials, and industrial procedures, all of which decrease energy intensity (the quantity of energy used per unit of production). Increased energy costs may incentivize innovation by making energy-efficient technology more financially feasible. Research and development in fields such as renewable energy, energy storage, and energy efficiency are more prone to securing financing during periods of elevated energy prices. Rising energy costs may expedite the transition from energy-intensive sectors to less energy-intensive ones. As an example, economies may transition from being reliant on manufacturing to focusing more on service-oriented industries, which generally use less energy per unit of economic production.

There is a favorable correlation between the improvement of labor and capital productivity and the enhancement of energy efficiency. The impact of labor productivity on lowering energy intensity is about twice as significant as the impact of capital productivity. Assuming all other conditions remain constant, a 1% increase in productivity is linked to a 0.05–0.06% decrease in energy intensity for labor productivity and a 0.02–0.03% decrease for capital productivity.

When labor and capital increase their productivity, they often do so by using methods that also improve energy efficiency. Technological innovation is a key factor in boosting productivity, for both labor and capital. The same technologies that enhance the productivity of both labor and capital often result in energy conservation as well. For instance, the implementation of a novel manufacturing technology that enhances production efficiency while reducing power consumption per unit of output results in simultaneous improvements in labor and capital productivity, as well as a decrease in energy consumption. Enhancing productivity often entails improving manufacturing processes to achieve more output with the same resources. This optimization often involves enhancing energy efficiency. For example, optimizing logistics or enhancing supply chain management may decrease both the energy used in transportation and the duration that labor and capital are engaged in less efficient tasks. When companies raise their output as a result of improved productivity, they often attain economies of scale. Increasing production volume may enhance energy efficiency by reducing the energy expenditures per unit of output. This is especially apparent in sectors such as manufacturing, where extensive operations have the ability to maximize energy use more effectively than smaller-scale activities. Increased productivity may result in more earnings, enabling companies to allocate more resources towards enhancing their equipment. This newly acquired equipment often exhibits higher efficiency, using less energy while achieving the same or superior output. Investing in contemporary, energy-efficient machinery not only increases capital productivity but also improves energy efficiency. Enhancing productivity often entails enhancing the quality of the product and minimizing waste. Efficient processes not only reduce waste but also save energy by preventing the manufacture of unneeded or faulty items that would demand extra energy for remaking or repairing.

The estimation results of Eq. (2) provide very comparable findings about the direction and statistical significance of other variables related to energy intensity such as trade, foreign direct investment (FDI), and so on.

In conclusion, our regression study demonstrates that the economic structure has a substantial impact on energy efficiency. Specifically, transitioning from industrial sectors to service sectors leads to a significant reduction in energy intensity. More precisely, if the industrial sector share increases by 1%, the energy intensity also increases by 3%. On the other hand, if the service sector share increases by the same amount, the energy intensity reduces by 3%. This may be attributed to the greater energy requirements of manufacturing in comparison to services, which mostly depend on digital technology and human resources. Furthermore, our findings provide evidence in favor of the environmental Kuznets curve (EKC) theory, which posits that energy intensity tends to rise in the early stages of economic growth but declines as economies progress and shift their emphasis towards services and high-value industries.

Our data suggest that a rise of 1% in trade openness results in a 0.25% increase in energy intensity, perhaps because energy-intensive businesses play a significant role in trade dynamics. On the other hand, when renewable energy is used more, there is a decrease in energy intensity, highlighting the effectiveness of these energy sources. Urbanization is shown to have a positive correlation with energy intensity, with a 0.7–0.8% rise in energy intensity for every 1% increase in urban population. This may be attributed to the significant energy requirements of metropolitan infrastructure and lifestyles.

Our research also examines the effects of financial dynamics, namely foreign direct investment (FDI) and financial openness, on energy intensity. We show that these factors have few or unpredictable consequences on energy intensity. In terms of policy consequences, increased energy costs, which serve as a motivation for efficiency, are linked to a decrease in energy intensity. Furthermore, there is a direct relationship between improvements in labor and capital productivity and a decrease in energy intensity, highlighting the importance of technical progress in improving energy efficiency. These observations should direct policymakers and stakeholders in formulating policies that use economic structure, trade, urbanization, and technology to promote energy efficiency.

5.6 CONCLUSIONS

The purpose of this case study is to investigate the factors that have had an impact on energy efficiency in the European Union between the years 2000 and 2022. To analyze the impacts of several variables on energy intensity (EI) and efficiency, it employs regression analysis using panel data. The industrial mix, the level and pace of economic development, changes in energy costs, investments in innovation, and the degree of openness in trade and finance are the primary elements that are taken into consideration. In addition, the analysis encompasses the investigation of urbanization, the use of renewable energy sources, the implementation of energy policy measures, and the pursuit of improvements in energy efficiency.

One of the most significant findings is that the transition from industrial economies to service-based economies results in a significant improvement in energy efficiency. As a result of this study, the environmental Kuznets curve (EKC) is shown to be correct. This curve proposes that as economies mature, energy intensity initially increases and then decreases. Evidence suggests that there is a correlation between an increase in GDP per capita and a matching drop in energy intensity. This suggests that there is a direct correlation between economic development and an improvement in energy efficiency.

There is a positive correlation between the openness of commerce and the intensity of energy consumption, which suggests that companies that use a significant amount of energy are more dominant in the trade sector. On the other hand, urbanization is linked to increased energy consumption due to the need for city infrastructure and methods of life. This is because urbanization is related to greater energy consumption. There is a correlation between the use of renewable energy and a decrease in energy intensity, which highlights the effectiveness of sustainable energy sources.

Different types of investments in technology, such as those made via research and development (R&D) and foreign direct investment (FDI), have different effects on the amount of energy that is used. Through this, the complex link that exists between economic activity and energy use is brought into focus. The ever-increasing costs of energy serve as a driving force behind the adoption of more efficient processes for energy consumption, which is under legislation that promotes energy efficiency.

To summarize, the findings of the study highlight how important it is to take into consideration a wide range of economic and policy concerns to effectively enhance energy efficiency across the European Union.

REFERENCES

Ajayi, V., & Reiner, D. (2020). European industrial energy intensity: Innovation, environmental regulation, and price effects. *The Energy Journal*, *41*(4), 105–128. https://doi.org/10.5547/01956574.41.4.vaja

Anderson, V. (2005). *Energy efficiency policies*. Routledge.

Andrieu, B., Vidal, O., Le Boulzec, H., Delannoy, L., & Verzier, F. (2022). Energy intensity of final consumption: The richer, the poorer the efficiency. *Environmental Science & Technology*, *56*(19), 13909–13919. https://doi.org/10.1021/acs.est.2c02015

Ang, B. W., & Goh, T. (2018). Bridging the gap between energy-to-GDP ratio and composite energy intensity index. *Energy Policy*, *119*, 105–112. https://doi.org/10.1016/j.enpol.2018.04.038

Ang, B. W., & Zhang, F. Q. (2000). A survey of index decomposition analysis in energy and environmental studies. *Energy*, *25*(12), 1149–1176. https://doi.org/10.1016/S0360-5442(00)00039-6

Antonietti, R., & Fontini, F. (2019). Does energy price affect energy efficiency? Cross-country panel evidence. *Energy Policy*, *129*, 896–906. https://doi.org/10.1016/j.enpol.2019.03.022

Arimura, T. H., Li, S., Newell, R. G., & Palmer, K. (2012). Cost-effectiveness of electricity energy efficiency programs. *The Energy Journal*, *33*(2), 63–100. https://doi.org/10.5547/01956574.33.2.4

Asif, M., & Muneer, T. (2007). Energy supply, its demand and security issues for developed and emerging economies. *Renewable and Sustainable Energy Reviews*, *11*(7), 1388–1413. https://doi.org/10.1016/j.rser.2005.12.004

Babuta, A., Gupta, B., Kumar, A., & Ganguli, S. (2021). Power and energy measurement devices: A review, comparison, discussion, and the future of research. *Measurement*, *172*, 108961. https://doi.org/10.1016/j.measurement.2020.108961

Backlund, S., Thollander, P., Palm, J., & Ottosson, M. (2012). Extending the energy efficiency gap. *Energy Policy*, *51*, 392–396. https://doi.org/10.1016/j.enpol.2012.08.042

Baldwin, E., Carley, S., & Nicholson-Crotty, S. (2019). Why do countries emulate each other's policies? A global study of renewable energy policy diffusion. *World Development*, *120*, 29–45. https://doi.org/10.1016/j.worlddev.2019.04.022

Berardi, U. (2017). A cross-country comparison of the building energy consumptions and their trends. *Resources, Conservation and Recycling*, *123*, 230–241. https://doi.org/10.1016/j.resconrec.2016.04.010

Birner, S., & Martinot, E. (2005). Promoting energy-efficient products: GEF experience and lessons for market transformation in developing countries. *Energy Policy*, *33*(14), 1765–1779. https://doi.org/10.1016/j.enpol.2004.01.012

Boogen, N., Cattaneo, C., Filippini, M., & Obrist, A. (2021). Energy efficiency and the role of energy-related financial literacy: Evidence from the European residential sector. *Energy Efficiency*, *14*, 1–17. https://doi.org/10.1007/s12053-020-09874-5

Bosseboeuf, D., Chateau, B., & Lapillonne, B. (1997). Cross-country comparison on energy efficiency indicators: The on-going European effort towards a common methodology. *Energy Policy*, *25*(7–9), 673–682. https://doi.org/10.1016/S0301-4215(97)00056-5

Brockway, P. E., Sorrell, S., Semieniuk, G., Heun, M. K., & Court, V. (2021). Energy efficiency and economy-wide rebound effects: A review of the evidence and its implications. *Renewable and Sustainable Energy Reviews*, *141*, 110781. https://doi.org/10.1016/j.rser.2021.110781

Brodny, J., & Tutak, M. (2022). Analysis of the efficiency and structure of energy consumption in the industrial sector in the European Union countries between 1995 and 2019. *Science of the Total Environment*, *808*, 152052. https://doi.org/10.1016/j.scitotenv.2021.152052

Burke, P. J., & Nishitateno, S. (2013). Gasoline prices, gasoline consumption, and new-vehicle fuel economy: Evidence for a large sample of countries. *Energy Economics*, *36*, 363–370. https://doi.org/10.1016/j.eneco.2012.09.021

Busch, S., & Ortner, A. (2019). From national to cross-border support of renewable electricity in the European Union. In *The European Dimension of Germany's Energy Transition: Opportunities and Conflicts* (pp. 207–225). Springer. https://doi.org/10.1007/978-3-030-28636-2_11

Cabeza, L. F., Ürge-Vorsatz, D., Ürge, D., Palacios, A., & Barreneche, C. (2018). Household appliances penetration and ownership trends in residential buildings. *Renewable and Sustainable Energy Reviews*, *98*, 1–8. https://doi.org/10.1016/j.rser.2018.09.017

Cao, W., Chen, S., & Huang, Z. (2020). Does foreign direct investment impact energy intensity? Evidence from developing countries. *Mathematical Problems in Engineering*, *2020*, 1–11. https://doi.org/10.1155/2020/9645391

Chen, K. H., Cheng, J. C., Lee, J. M., Li, L. Y., & Peng, S. Y. (2020). Energy efficiency: Indicator, estimation, and a new idea. *Sustainability*, *12*(12), 4944. https://doi.org/10.3390/su12124944

Cole, M. A. (2004). Trade, the pollution haven hypothesis and the environmental Kuznets curve: Examining the linkages. *Ecological Economics*, *48*(1), 71–81. https://doi.org/10.1016/j.ecolecon.2003.09.007

Colmenares, G., Löschel, A., & Madlener, R. (2020). The rebound effect representation in climate and energy models. *Environmental Research Letters*, *15*(12), 123010. https://doi.org/10.1088/1748-9326/abc9c7

Dagoumas, A. S., & Koltsaklis, N. E. (2019). Review of models for integrating renewable energy in the generation expansion planning. *Applied Energy*, *242*, 1573–1587. https://doi.org/10.1016/j.apenergy.2019.03.194

Deichmann, U., Reuter, A., Vollmer, S., & Zhang, F. (2019). The relationship between energy intensity and economic growth: New evidence from a multi-country multi-sectoral dataset. *World Development*, *124*, 104664. https://doi.org/10.1016/j.worlddev.2019.104664

Díaz, A., Marrero, G. A., Puch, L. A., & Rodríguez, J. (2019). Economic growth, energy intensity and the energy mix. *Energy Economics*, *81*, 1056–1077. https://doi.org/10.1016/j.eneco.2019.05.020

Duflou, J. R., Sutherland, J. W., Dornfeld, D., Herrmann, C., Jeswiet, J., Kara, S., . . . & Kellens, K. (2012). Towards energy and resource efficient manufacturing: A processes and systems approach. *CIRP Annals*, *61*(2), 587–609. https://doi.org/10.1016/j.cirp.2012.05.002

Dunlop, T. (2019). Mind the gap: A social sciences review of energy efficiency. *Energy Research & Social Science*, *52*, 145–152. https://doi.org/10.1016/j.erss.2019.05.026

Duro, J. A., Alcantara, V., & Padilla, E. (2010). International inequality in energy intensity levels and the role of production composition and energy efficiency: An analysis of OECD countries. *Ecological Economics*, *69*(12), 2468–2474. https://doi.org/10.1016/j.ecolecon.2010.07.008

Eichhammer, W., & Mannsbart, W. (1997). Industrial energy efficiency: Indicators for a European cross-country comparison of energy efficiency in the manufacturing industry. *Energy Policy*, *25*(7–9), 759–772. https://doi.org/10.1016/S0301-4215(97)00049-8

Filippini, M., & Hunt, L. C. (2011). Energy demand and energy efficiency in the OECD countries: A stochastic demand frontier approach. *The Energy Journal*, *32*(2), 59–80. https://doi.org/10.5547/ISSN0195-6574-EJ-Vol32-No2-4

Filippini, M., & Hunt, L. C. (2015). Measurement of energy efficiency based on economic foundations. *Energy Economics*, *52*, S5–S16. https://doi.org/10.1016/j.eneco.2015.09.004

Frondel, M. (2004). Empirical assessment of energy-price policies: The case for cross-price elasticities. *Energy Policy*, *32*(8), 989–1000. https://doi.org/10.1016/j.enpol.2003.11.010

Fu, Y., Kok, R. A., Dankbaar, B., Ligthart, P. E., & van Riel, A. C. (2018). Factors affecting sustainable process technology adoption: A systematic literature review. *Journal of Cleaner Production*, *205*, 226–251. https://doi.org/10.1016/j.jclepro.2018.08.258

Fuss, S., Szolgayová, J., Khabarov, N., & Obersteiner, M. (2012). Renewables and climate change mitigation: Irreversible energy investment under uncertainty and portfolio effects. *Energy Policy*, *40*, 59–68. https://doi.org/10.1016/j.enpol.2010.09.061

Gales, B., Kander, A., Malanima, P., & Rubio, M. (2007). North versus South: Energy transition and energy intensity in Europe over 200 years. *European Review of Economic History*, *11*(2), 219–253. https://doi.org/10.1017/S1361491607001967

García-Quevedo, J., & Jové-Llopis, E. (2021). Environmental policies and energy efficiency investments: An industry-level analysis. *Energy Policy*, *156*, 112461. https://doi.org/10.1016/j.enpol.2021.112461

Geller, H., Harrington, P., Rosenfeld, A. H., Tanishima, S., & Unander, F. (2006). Policies for increasing energy efficiency: Thirty years of experience in OECD countries. *Energy Policy*, *34*(5), 556–573. https://doi.org/10.1016/j.enpol.2005.11.010

Gillingham, K., Newell, R. G., & Palmer, K. (2009). Energy efficiency economics and policy. *Annual Review of Resource Economics*, *1*(1), 597–620. https://doi.org/10.1146/annurev.resource.102308.124234

Gillingham, K., & Palmer, K. (2013). Bridging the energy efficiency gap: Policy insights from economic theory and empirical evidence. *Review of Environmental Economics and Policy*, *8*(1), 18–38. https://doi.org/10.1093/reep/ret021

Goulder, L. H., & Parry, I. W. H. (2008). Instrument choice in environmental policy. *Review of Environmental Economics and Policy*, *2*(2), 152–174. https://doi.org/10.1093/reep/ren005

Guevara, Z., Henriques, S., & Sousa, T. (2021). Driving factors of differences in primary energy intensities of 14 European countries. *Energy Policy*, *149*, 112090. https://doi.org/10.1016/j.enpol.2020.112090

Gyamfi, B. A., Kwakwa, P. A., & Adebayo, T. S. (2023). Energy intensity among European Union countries: The role of renewable energy, income and trade. *International Journal of Energy Sector Management*, *17*(4), 801–819. https://doi.org/10.1108/ijesm-05-2022-0018

Haas, R. (1997). Energy efficiency indicators in the residential sector: What do we know and what has to be ensured? *Energy Policy*, *25*(7–9), 789–802. https://doi.org/10.1016/S0301-4215(97)00057-7

Heun, M. K., Owen, A., & Brockway, P. E. (2018). A physical supply-use table framework for energy analysis on the energy conversion chain. *Applied Energy*, *226*, 1134–1162. https://doi.org/10.1016/j.apenergy.2018.05.109

Huaman, R. N. E., & Jun, T. X. (2014). Energy-related CO2 emissions and the progress on CCS projects: A review. *Renewable and Sustainable Energy Reviews*, *31*, 368–385. https://doi.org/10.1016/j.rser.2013.12.002

International Energy Agency. (2009). Energy technology perspectives 2010. In *Scenarios and Strategies to 2050*. https://doi.org/10.1787/energy_tech-2010-en

Jaffe, A. B., Newell, R. G., & Stavins, R. N. (2002). Environmental policy and technological change. *Environmental and Resource Economics*, *22*, 41–70. https://doi.org/10.1023/A:1015519401088

Jin, T. (2022). Impact of heat and electricity consumption on energy intensity: A panel data analysis. *Energy*, *239*, 121903. https://doi.org/10.1016/j.energy.2021.121903

Koronen, C., Åhman, M., & Nilsson, L. J. (2020). Data centres in future European energy systems—energy efficiency, integration and policy. *Energy Efficiency*, *13*(1), 129–144. https://doi.org/10.1007/s12053-019-09814-8

Liddle, B. (2010). Revisiting world energy intensity convergence for regional differences. *Applied Energy*, *87*(10), 3218–3225. https://doi.org/10.1016/j.apenergy.2010.03.024

Linares, P., & Labandeira, X. (2010). Energy efficiency: Economics and policy. *Journal of Economic Surveys*, *24*(3), 573–592. https://doi.org/10.1111/j.1467-6419.2009.00609.x

Lovins, A. B. (2004). Energy Efficiency, taxonomic overview. In *Elsevier eBooks* (pp. 383–401). https://doi.org/10.1016/B0-12-176480-X/00167-4

Lovins, A. B. (2017). Energy efficiency. *Energy Economics*, *1*, 234–258. https://doi.org/10.1016/j.eneco.2017.05.014

McKinsey, G. I. (2009). *Pathways to a Low-Carbon Economy-Version2 of the Global Greenhouse Gas Abatement Cost Curve*. McKinsey Company.

Moriarty, P., & Honnery, D. (2023). Rethinking notions of energy efficiency in a global context. *Energies*, *16*(12), 4706. https://doi.org/10.3390/en16124706

Patterson, M. G. (1996). What is energy efficiency? Concepts, indicators and methodological issues. *Fuel and Energy Abstracts*, *37*(4), 314. https://doi.org/10.1016/0140-6701(96)82844-x

Petrović, P., Filipović, S., & Radovanović, M. (2018). Underlying causal factors of the European Union energy intensity: Econometric evidence. *Renewable and Sustainable Energy Reviews*, *89*, 216–227. https://doi.org/10.1016/j.rser.2018.02.036

Popp, D., Newell, R. G., & Jaffe, A. B. (2010). Energy, the environment, and technological change. In B. H. Hall & N. Rosenberg (Eds.), *Handbook of the Economics of Innovation* (Vol. 2, pp. 873–937). https://doi.org/10.1016/S0169-7218(10)02005-8

Ramos, A., Gago, A., Labandeira, X., & Linares, P. (2015). The role of information for energy efficiency in the residential sector. *Energy Economics*, *52*, S17–S29. https://doi.org/10.1016/j.eneco.2015.08.022

Reames, T. G., Reiner, M. A., & Stacey, M. B. (2018). An incandescent truth: Disparities in energy-efficient lighting availability and prices in an urban US county. *Applied Energy*, *218*, 95–103. https://doi.org/10.1016/j.apenergy.2018.02.135

Recalde, M. Y., & Ramos-Martin, J. (2012). Going beyond energy intensity to understand the energy metabolism of nations: The case of Argentina. *Energy*, *37*(1), 122–132. https://doi.org/10.1016/j.energy.2011.10.025

Renn, O., & Marshall, J. P. (2016). Coal, nuclear and renewable energy policies in Germany: From the 1950s to the "Energiewende". *Energy Policy*, *99*, 224–232. https://doi.org/10.1016/j.enpol.2016.05.004

Samargandi, N. (2019). Energy intensity and its determinants in OPEC countries. *Energy*, *186*, 115803. https://doi.org/10.1016/j.energy.2019.07.158

Sun, X., Jia, M., Xu, Z., Liu, Z., Liu, X., & Liu, Q. (2022). An investigation of the determinants of energy intensity in emerging market countries. *Energy Strategy Reviews*, *39*, 100790. https://doi.org/10.1016/j.esr.2021.100790

Swan, L. G., & Ugursal, V. I. (2009). Modeling of end-use energy consumption in the residential sector: A review of modeling techniques. *Renewable and Sustainable Energy Reviews*, *13*(8), 1819–1835. https://doi.org/10.1016/j.rser.2008.09.033

Tanaka, K. (2008). Assessment of energy efficiency performance measures in industry and their application for policy. *Energy Policy*, *36*(8), 2887–2902. https://doi.org/10.1016/j.enpol.2008.03.032

Thomas, B. A., & Azevedo, I. L. (2013). Estimating direct and indirect rebound effects for US households with input–output analysis Part 1: Theoretical framework. *Ecological Economics*, *86*, 199–210. https://doi.org/10.1016/j.ecolecon.2012.12.003

Todorov, G., Volkova, E., Vlasov, A., & Nikitina, N. (2019). Modeling energy-efficient consumption at industrial enterprises. *International Journal of Energy Economics and Policy*, *10*(4). https://doi.org/10.32479/IJEEP.7376

Trianni, A., Cagno, E., & Donatis, A. (2014). A framework to characterize energy efficiency measures. *Applied Energy*, *118*, 207–220. https://doi.org/10.1016/j.apenergy.2013.12.042

Velasco-Fernández, R., Dunlop, T., & Giampietro, M. (2020). Fallacies of energy efficiency indicators: Recognizing the complexity of the metabolic pattern of the economy. *Energy Policy*, *137*, 111089. https://doi.org/10.1016/j.enpol.2019.111089

Voigt, S., De Cian, E., Schymura, M., & Verdolini, E. (2014). Energy intensity developments in 40 major economies: Structural change or technology improvement? *Energy Economics*, *41*, 47–62. https://doi.org/10.1016/j.eneco.2013.10.015

Walnum, H. J., Aall, C., & Løkke, S. (2014). Can rebound effects explain why sustainable mobility has not been achieved? *Sustainability*, *6*(12), 9510–9537. https://doi.org/10.3390/su6129510

Wei, T. (2010). A general equilibrium view of global rebound effects. *Energy Economics*, *32*(3), 661–672. https://doi.org/10.1016/j.eneco.2010.01.002

Yang, M., & Yu, X. (2015). Energy efficiency becomes first fuel. In *Energy Efficiency. Green Energy and Technology* (pp. 11–18). Springer. https://doi.org/10.1007/978-1-4471-6666-5_2

Yu, S., Liu, J., Hu, X., & Tian, P. (2022). Does development of renewable energy reduce energy intensity? Evidence from 82 countries. *Technological Forecasting and Social Change*, *174*, 121254. https://doi.org/10.1016/j.techfore.2021.121254

Zhou, P., Ang, B., & Zhou, D. (2012). Measuring economy-wide energy efficiency performance: A parametric frontier approach. *Applied Energy*, *90*, 196–200. https://doi.org/10.1016/J.APENERGY.2011.02.025.

6 Renewable Energy

Ahmad Bathaei

6.1 RENEWABLE ENERGY SOURCES

6.1.1 Sources

Renewable energy is described by Lund as energy generated from natural resources that reestablish themselves organically in a matter of years. The technical properties of renewable energies can be examined in addition to their sustainability. Managers are thus assisted in selecting a specific renewable energy source by factors such as the ability to meet market demand, technical attributes like integration with other sources, and economic feasibility derived from energy efficiency and operating costs (Verbruggen et al., 2010).

The pursuit of renewable energy has a long history, dating back centuries, as people have looked for sustainable substitutes for conventional, limited energy sources. The roots of this quest can be found in ancient civilizations that used the power of wind, water, and sunshine for a variety of reasons, even though our understanding and application of renewable energy have greatly changed over time (Kalair et al., 2021).

The idea of renewable energy originated in the earliest human societies, when windmills were used for mechanical chores like grinding grain. In order to harness the power of flowing water for industrial and agricultural purposes, water wheels were also essential. The modern period of renewable energy was defined by the later development of more advanced technologies, which were made possible by these early innovations (Md Momtazur Rahman et al., 2022).

Due to concerns about the environmental effects of conventional fossil fuels and their limited supply, the 20th century saw a shift in the research and development of renewable energy sources. The energy crisis of the 1970s accelerated efforts to find sustainable and alternative solutions and stimulated research and development of renewable energy technology. The first important wind and solar energy projects started at this time and paved the way for the world to transition to cleaner and more sustainable energy sources (Sen and Ganguly, 2017).

Fossil fuel has been the main energy resource since the early days of industrialization, and nowadays the renewable resource has a diversified setup that takes advantage of each natural process replenished over time. The solar energy harvested via sunlight can get into the photovoltaic cells or the solar thermal systems. Air stream energy that has a kinetic motion in the wind is captured using wind turbines. Hydropower captures electricity from moving water, whereas geothermal energy operates on Earth's generated heat. Biomass has the residing organic substances for production of energy. It sucks the energy of the tides and the waves, which are spread by the strength of waves (Tovar-Facio et al., 2022).

The utilization of renewable resources dates back to ancient civilizations and it has been continuing through the 20th century by both technologies and communities. Hence it is possible to use almost anything in terms of renewable energy. For today renewable energy includes all varieties of power sources that we call different names like hydro, wind, tidal, geothermal, and biomass depending on natural processes in which the energy is embedded and that are replenished recurrently over time (Montoya et al., 2021).

Being one of the famous renewable energy sources, sunlight is transformed to solar energy through the process of the sun's shining. Cells of photovoltaic utilize light to power electricity generation and thermal systems using sun for heating objectives as other forms are two means of

DOI: 10.1201/9781003517481-6

use of this energy. The next component is wind power, which uses wind turbines to harness moving air's kinetic energy, and hydropower becomes prevalent due to dams and other infrastructure built to mobilize water flow (Li and Zhu, 2023).

Other than hydropower that is known for the stability and reliability of electricity, there is geothermal energy that converts the heat inside the earth to electricity. Use of organic agents like wood, plant waste, and waste products from animals results in new, green, and regenerated sources of energy production, which are independent and sustainable. An energy source that is renewable with great potential, tidal and wave power, is presented by the rolling and swelling motions of water (Suwaileh et al., 2023).

Thus, it is seen that renewable energy has certain features that makes its implementation on a wide range again a dynamic one. The expensive options range of renewable energy sources is aroused in a conjunction of ongoing technological progress in the field and the growth of our understanding of these resources. While today there are ample choices for the technological solution to move beyond conventional and restricted energy sources, the moves we make toward a low-carbon economy and social system are intended to be low-carbon and sustainable (Sovacool et al., 2022).

Utilizing renewable energy represents a huge step towards the realization of a sustainable society as this process is directly derived from organic sources. Installations with solar term and photovoltaic components are among the solar energy resources that are harnessed for production due to sunlight. With the stream, the hydropower operates using the potential energy of accelerating water, and by means of the wind turbine the wind power captures kinetic energy. Than biomass energy that comes from biological material and geothermal energy takes energy from inside Earth. Oceans waves and tides are primarily driven by the critical energy. Collectively, these abundant energy sources add to the vast and ever-changing pool of renewable energy, which is essential to our transition to a low-carbon future (Sovacool et al., 2023).

The global effort to shift towards a low-carbon and more sustainable energy system must include renewable energies. Renewable energy sources, which exploit natural processes that are continuously replenished, provide a cleaner and more environmentally friendly source of energy than finite fossil fuels (Hoang and Nguyen, 2021). The variety of ways in which we may access these sustainable resources is reflected in the wide range of renewable energy sources. Here, we describe and group several types of renewable energy sources.

- Solar energy: energy obtained by photovoltaic cells or solar thermal systems from the sun's rays.
- Wind energy: produced by wind turbines, which transform the kinetic energy of moving air into electrical power.
- Hydropower: produced by utilizing the force of flowing water, frequently by building dams.
- Geothermal energy: derived via heat pumps or geothermal power plants from Earth's interior heat.
- Biomass: produces heat or power by using organic materials like wood, crop waste, and animal dung.
- Tidal energy: this type of energy is obtained from the moon's gravitational pull on ocean tides.
- Wave energy: uses the force of the ocean's waves to generate electricity.

6.1.1.1 Solar Energy

Based on the sun's limitless power, solar energy is a crucial and quickly growing part of the renewable energy scene. Photovoltaic cells and solar thermal systems are the two main ways to use this clean and sustainable energy source. Unlike traditional power plants that convert fossil fuels to electricity, solar panels made of semiconductor materials convert sunlight into electricity directly using photovoltaic cells, while for the solar thermal systems there are applications such as heating, water heating, and power production, which make use of focused solar radiation. Due to the many ways it can be used, solar energy has a broad application, for instance, solar farms to small home installations for general use (Bandaru et al., 2021). We are witnessing an increase in solar power worldwide as a consequence of ongoing technological improvements and reduced costs, which make it one of the elements crucial to building a future with a low-carbon and sustainable energy source. Today, the sun still serves as a source of inspiration for scientists who tap its power for the prosperity and existence of a planet that is kind to itself because the sun provided the world an unlimited supply of clean energy (Caineng et al., 2022).

Suffice it to say that solar energy has an array of advantages, and the prospect of it becoming one of the major players in the energy sources worldwide is an actual potential. Its eco-friendliness stands out as a key perk. With solar power, there is very low emission of pollutants into the environment. So, it helps in climate change mitigation, hence increasing the air quality of the atmosphere. One noteworthy aspect is the presence of an unlimited supply of sunlight, which renewable energy is close to, which lowers the risk of using conventional energy sources and ensures energy security and independence at the same time (Kabeyi and Olanrewaju, 2022).

If the financial incentives that exist for solar power are explored, a strong case can be made for it. As manufacturing increases and technology advances, panels are making solar power more cheaply and become just as competitive as, if not more so than, traditional energy sources. Job creation in the solar business is a huge consequence of that, including the cost reduction mechanism, government support, and subsidy. The solar energy sector creates jobs from manufacturing of the equipment, installation, maintenance, and detecting as well as economic development (Gebreslassie, 2021).

Also, the solar energy causes the population to have their own grid and generate their power without having to rely on power authorities. Decentralization is thus a measure that serves to remedy problems associated with dependency on centralized systems and more importantly empowers people to get actively involved in administering their energy consumption patterns. Local area grids enjoyed more stability and residents got more energy independence through their own installations of solar at their houses or offices (Cantoni et al., 2022).

Though there are distinct disadvantages of solar energy, in spite of its benefits, it is worth a try. The single major blunder is intermittency; collecting solar power depends on the sunshine and this means that once the sun shines, then it stops and disappears as well during the night and poor weather conditions. Obstacles like weather unreliability are being taken care of and stable and reliable generation of solar power is brought about by the improvement of energy storage systems and smart grid technologies (Kalair et al., 2021).

The next thing to mention is that there are environmental issues concerning the fabrication and recycling of solar panels. When talking about some of the materials and chemicals used in the production, it is important to note that disposing of currently outdated solar panels that accumulate is not an option, but instead careful recycling is required. Going forward with research and innovations, some of the challenges arising such as recycling and manufacturing approaches being more sustainable can be tackled (Jurasz et al., 2020).

Briefly, solar energy is closely connected with the switch to a low-carbon world thanks to solar power being clean and cheaper. It gives a lot of economic opportunities, allows distributed energy generation, and is extremely compatible with the environment. These flaws still exist, for example unpredictability of power and other environmental concerns, but the persistent research and development of solar technologies will surely improve both their effectiveness and accessibility and even will bring overall positive changes to the world's energy systems (Huang et al., 2021).

6.1.1.2 Wind Energy

The world of renewable energy evolves rapidly and wind energy resulting from aerodynamic motion is a powerful and speedily developing source that is used to generate electric power. It is environmental sustainability that is probably the most precious advantage of wind energy. Wind power manufacturing has next to zero greenhouse gas emissions as compared to the conventional types of fossil fuels that cause a tremendous amount of air pollution and, hence, such power generation is in a fight against climate change. Wind power stands out as a renewable and green energy to dump the present fossil fuel-based power systems because of large scale availability and global resource distribution (A. Rahman et al., 2022).

The cost-effectiveness of wind energy also contributes to its allure since it brings economic advantages. With the increase in the level of technology, wind turbines thus have become much more efficient and cheaper than they used to be. Because of a competitive pricing that has become active, wind power is now amounting to be environmentally sound and in its majority cheaper than fossil fuels. In addition, the wind power industry is an important job producer creating and shaping manufacturing, installation, servicing, and R&D jobs, leading to an increase in number of workers. Wind farms situated both onshore and offshore have a very important positive effect resulting in job creation and a stronger economy not only in energy generation but also in communities and locations (Glasson et al., 2022).

Besides that, the wind energy process entails decentralization. Although you have the option to build wind farms in different geographical areas like coastal regions and rural areas, this, however, provides for creating a more spread-out energy generation strategy. Through the involvement of more localized power units and weakening the impact of power shortages on energy resilience, this decentralization is going to increase the resilience of energy infrastructure. Moreover, the wind turbines are proving to be of different types, suitable to any scale, of a micro installation or a multi-megawatt wind farm. As a result, the wind energy contributes to satisfying energy demands from a small household all the way to industrial scale use (Kalair et al., 2021).

Wind energy, however, does create a certain level of discontentment. Intermittency—the wind being variable in its strengths—and consistency are only two of many issues that should be considered, however. The fluctuating nature of this energy production causes the grid's reliability to decline and an electricity storage technique must be introduced or complementary technology must be developed in order to fulfil supply and demand. Along with deployment, wind energy projects planning and implementation face the problems of local populace anxiety, aesthetic and noise risks, and eventually influence on the wildlife as well (Huang et al., 2021).

To summarize, many community wind projects are becoming the suitable choice for supplying energy due to a decentralized generation capability, economic boost, and the fact that they are environmentally friendly. Navigating the intermittency and environmental constraints, constant technological advancements and increasing financial investment in the wind energy sector raise the expectation that those problems will be resolved to give the wind energy source more space to play the role of a pivotal factor in the worldwide transition to energy sources that are clean and sustainable.

6.1.1.3 Hydropower

The kinetic energy of the flowing water is utilized for hydropower's generation, which is still a key part of the luminary global energy. Hydropower is a competent source and overlooked, but what outshines all its pros is reliability. The hydropower energy is a stable and precisely controlled source of energy that has the ability to consistently meet demand on a reliable basis compared to some energy sources such as solar and wind, which are not equally available at times. As it is very reliable and dependable, hydropower is an integral element of a diverse and resilient system, which ensures constant operation with different energy generators (Phillips et al., 2020).

The hydropower industry is not without economic advantages of a high order. By their nature, these projects are generally cheaper than their conventional counterparts once the infrastructure installation and other related expenses are taken into consideration. Economic viability of hydropower facilities and the return on investment will be further stabilized by the prolonged operation time, which often exceeds 50 years of the installation. In addition to that, hydropower plant development and construction create local jobs in the areas where the plants are installed; thus, the local economies are stimulated (Enevoldsen and Jacobson, 2021).

The environmental non-toxicity in hydropower production is one of the major factors that makes it in so much more demand compared to other renewable energy sources. The use of hydropower equals a very small rate of emission of greenhouse gas and the carbon footprint of electricity generation is tremendously reduced. Beyond that, the construction of a lake can also manifest the multifaceted nature of the hydroelectric projects in the sense of a synergy between flood management, agriculture, and recreation (Davies et al., 2021).

Although some environmental benefits can be seen in the uses of hydropower, still, there are problems accompanying this mechanism. The degradation or loss of aquatic habitats and ecosystem of rivers is one of the most critical concerns for aquatic ecosystems. The water current circulation, sediment discharge, and quality of water may be affected by the operation and the development of reservoirs. On the other hand, substantial hydropower stations have a major impact on the population's displacement and the environment transformation. These issues that raise questions of social and environmental aspects have to be treated with profound strategic thinking and disaster risk management measures (Phillips et al., 2020).

In summary, this energy solution is characterized by its guaranteed reliability, economic feasibility, and ecological viability, which helped it to take a prized place in the intentions and plans of the countries that decided to alter their energy structure. Although social and environmental issues have been tackled through technological breakthroughs and sustainable practices initiatives, new ways of finding renewable and clean energy will still remain important, hence hydropower will continue to be fundamental.

6.1.1.4 Geothermal Energy

Leveraging geothermal energy is a unique source of renewable energy that exploits the heat that is stored in the earth. The key benefit of this form of energy is its capacity to be unchanged and reliable in this way. The demand for power is a continuous operation, without fluctuations in daylight hours or weather, a power base that provides a constant supply. Geothermal energy, in addition to being universal and flexible in design, also acts as a dependable and balancing force within the framework of the power system that supports grid stability and balances out intermittent renewable sources (Mlilo et al., 2021).

Getting geothermal power could be significantly more attractive due to its financial gains or benefits. The installation of plants that tap energy from the earth, heat geothermal plants can remain in good order for about 25 years and always come in handy to supply sustainable and affordable heat for a long period. Additionally, the geothermal projects increase the local economies and employment growth by creating jobs around various stages, for example in exploration, drilling, plant construction and already-existing plant operation (Settembre-Blundo et al., 2021).

Furthermore, there are no emissions involved in power generation, being instead a very eco-friendly form of energy. Contrasting fossil fuel-based power plants with geothermal facilities, the latter utilize less land area and incur a less drastic ecological impact; thus they prevent much disturbance to natural habitats. All those features are finally achieved by direct application of geothermal resources for heating and cooling of houses and commercial structures, which becomes a direct and effective way to take advantage of the heated earth (Soltani et al., 2019).

Geothermal energy does have disadvantages despite many of its benefits. The occurrence of geothermal energy in a very limited geographical region limits its development and application. High generating volume of geothermal power should be prioritized with the regions where there is a strong geothermal potential such as in the USA and Iceland. Besides, the exploration and drilling operation escalates the correlation with environment, and devotion to the proper management of resources is unavoidable to prevent resource depletion or changes in subsurface pressure (Davies et al., 2021).

In general, the features of geothermal energy that make it favorable are saving, being economically feasible, and being environmentally friendly. Even errands regarding capital and environmental management, current geothermal energy, which is powered from research and technology, has been developed for the purpose of increasing accessibility and a lower degree of environmental damage, meaning that this would still be used in the future as the world searches for environmentally friendly and clean sources of energy.

6.1.1.5 Biomass

Biomass energy, which comes from carbon-containing substances including wood, crop residue, and animal waste, occupies the niche energy space filled up by the renewable energy sources. The main advantage of it lies in its various customization and in its widespread abundance. Accommodation to biomass energy resources is easy and within reach when you can access resources from nearby sources thus minimizing dependency on centralized sources of energy. Moreover, biomass may be employed for diverse reasons like producing electricity, supplying heat, and producing biofuels. This allows for its use as a substitution for the power generation across the grid usage spectrum (Sikiru et al., 2024).

Economically, the biomass energy gains increasing demand for attention. Biomass projects, one way or another, create local economy opportunities by supporting the development of centers for agriculture, forestry, and bioenergy. The use of leftover residues, for example agricultural or forest waste, to produce an added income source for farmers and forest industries is the optimal option. Moreover, inherently green energy systems such as biomass are able to be inserted into existing power grid structures either by repurposing facilities or by installing equipment developed for conventional fossil fuels. This way, you do not have to fund the project on a large scale from the beginning (Settembre-Blundo et al., 2021).

From an environmental point of view, biomass energy is understood to be neutral if its processing is controlled as well as possible. On the one hand, the process of combusting biomass results in the disposal of the residual carbon in the form of carbon dioxide. But, on the other hand, plants that are burned as fuel to produce energy are known to absorb carbon dioxide, which is on par with the carbon dioxide emitted. Through its closed carbon cycle, biomass can be a renewable energy source coupled with using carbon capture and storage, which ultimately leads to fewer emissions (Chu et al., 2024).

Additionally, there are some problems associated with biomass. One of the cons is the potential of deforestation and the switch to different forms of land use when biomass production is conducted on a large scale. Unsustainable farming practices, like excessive land clearance, can lead to losing habitats and lowering biodiversity as well as make more carbon emissions in our environment. Other than releasing air pollutants, although at lower levels compared to the amount produced by methane emissions, biomass also emits air pollutants as it burns. Regular monitoring, policy, and protocol guidelines that move biomass energy forward will therefore be essential in ensuring the sustainability and ecological benefits that are brought by biomass energy (Siwal et al., 2021).

The main achievements of biomass energy are: it allows for flexibility in usage, offers a cost efficient way of meeting energy needs, and is considered a carbon-free energy option. The environmental impact should be thoroughly assessed, whereas we must adopt sustainable sourcing methods to be sure that the benefits generated by the biomass energy usage could be fully implemented for a scalable transition to a clean and sustainable future (Black et al., 2022).

6.1.1.6 Tidal Energy

The energy generated from the push of the moon on the oceans' water (tidal energy) has great prospects in the energy sector, namely renewable sources. The outstanding quality of it is in the possibility to be predicted. Celestial factors cause interplay, which leads to a stable tide of habits for which predicting tides in advance is possible. The green energy precisely contributes to consistently and reliably generating the energy and hence becomes an important asset to play a role in maintaining an energy that is stable and resilient. Additionally, tides show a large and somehow regular bi-daily pattern, which creates a stable and affordable power source (Ayaz et al., 2022).

The economic benefits of tidal streams increase the available options for using tidal energy. One of the benefits that may be highlighted regarding this perpetual and reliable source of energy is that it may achieve price stability for a long period of time. Through this, the rate at which the universe of commodities and services is affected by the changes in the prices of energy shall be reduced if not eliminated so that consumers and enterprises have fewer headaches of irregular prices of energy (Settembre-Blundo et al., 2021). There lies a great scope of employment throughout the various stages of tidal energy projects, right from design and construction to after sales and operations, hence also swelling the local economy and fueling the pursuit of green energy industry development. Furthermore, the technological progress in tidal energy technologies will possibly result in the increased production volumes that may, ultimately, bring down the price and a further progress towards competitiveness compared to the current energy-generating sources (Khare and Bhuiyan, 2022).

Producing power through tidal flow is generally viewed as a means of having no harmful effects on the environment and being renewable. Because it gives off very few greenhouse gases, tidal power is ideal for instances where you would like to mitigate climate change and reduce fossil fuels as one of the major sources of energy. Second, tidal energy installations tend to have very little environmental impact as opposed to what conventional systems of energy generation would. Besides that, to diminish the amount of habitat disruption and marine ecosystems protection is this case (A. Rahman et al., 2022).

Nevertheless, tidal energy owns its hurdles. One of the major problems is the location of the existing resources to expand tidal energy sites that can be specific for a particular region. The problem of usable coastal sites for tidal generation is high, which in turn reduces the expansion of tidal energy technologies on a large scale. In addition, in some cases the potentially negative environmental consequences caused by the creation and operation of tidal power plants in their vicinity including disturbances of marine habitat and navigation are noted. The fact that a healthy balance between the tide energy production and marine life protection is one of the key factors of how much tide energy is going to develop is purely obvious (Jurasz et al., 2020).

In summary, there are three major pluses of using tidal energy, and they are the regularity of its release, the economic rationale of this source, and the environmental safety of its use. Still, conflicting with the restrictions on specific sites and effecting the environment, break-throughs and research are being made annually to track all the problems. This is a starting point for the expansion of tidal power that as a whole can potentially resolve the global problem of fossil fuel dependency on green and renewable energy sources.

6.1.1.7 Wave Energy

Wave energy, coming from the movements of waves in the oceans, has proven to have great potential as a renewable and sustainable source of power. The main strength of wave energy is its reliability; this is one of the biggest advantages of the power plant. Differently from a few reasonable sources of the renewable energy sector including sun or wind, waves are characterized with an accurate and predictable pattern, making a reliable and uninterrupted supply of energy. The advantage of wave power lies in its predictability, which is a requirement for stabilizing the power systems and, thus, presents wave energy as an essential item in the diverse and robust power mix (Ouramdane et al., 2021).

The economical side gains it popularity and acceptance. Technology has constantly been evolving together with the industry maturing, and the realistic probability of wave energy being feasible looks far better than ever. Wave energy projects possess a very long lifespan and are becoming more competitive than other sustainable sources, so it is imperative to consider these sources for long-term energy planning. On the other hand, part of the process that entails the choice of wave energy technologies can also stimulate economic growth by providing employment opportunities among those employed in the manufacture, installation, and maintenance, as well as research and development sectors (Khare and Bhuiyan, 2022).

In a word, wave energy can be regarded as an alternative that is eco-friendly and environmentally friendly too. Because wave energy projects primarily involve not introducing large amounts of greenhouse gases into the environment during operations, this also aids in reducing levels of climate change and the reliance on fossil fuels. Besides that, it is usually the case that these projects have a smaller scale in terms of the pollution they cause, compared to power generation, which often has a negative impact on the environment, particularly disrupting the habitats of marine wildlife and destroying ecosystems (A. Rahman et al., 2022).

Despite the numerous benefits of wave energy, it presents some bottlenecks in development. Although the turbulent ocean region is one of the drawbacks, it also guarantees the stability of the wave conditions. Long-lasting and environmentally resistant equipment that would only need minimal maintenance and effectively exploit the power of the water to produce energy remains complex in terms of the technical aspects. Additionally, rigorous sea conditions, which will be encountered in the course of putting and keeping the wave energy amenities in place, might create logistical and operational trouble (Soltani et al., 2019).

The alternative production of energy from a wave mainly ensures reliability, feasibility, and ecological sustainability. Reaching an end in technology development and the maritime environment obstacles is not a simple task (Clemente et al., 2021). Hence, research and development in the field, creative thinking, and strategic planning are the main elements that have been undertaken to address these challenges. This is likened to a struggle that is aimed at ensuring that wave energy becomes a dominant alternative in the rise of clean and sustainable energy.

6.1.2 Benefits and Features

6.1.2.1 Solar Energy

Solar energy, because of its numerous merits and one-of-a-kind characteristics, has a significant value when considering the path towards low-carbon renewables. An important advantage of this is its ample and extensive availability, harnessing an almost infinite resource, which is the source of all life. Solar power is a renewable energy source for the future that generates no pollution and is minimal in carbon footprint. The installation of solar panels can be scaled up or down; from a small home system to a gigantic solar farm providing a varying degree of its power to cater to the different needs of a household. The renewable character of solar power allows communities to generate electricity that is not transmitted through the power grid which saves energy that is locally produced, thus leading to energy independence, which reduces reliance disadvantages. Also importantly, the solar energy technology, including the development of incredibly efficient photovoltaic cells materials and new solar panel designs, is advancing every day to make it more and more useful for a wider consumer base (Soltani et al., 2019).

6.1.2.2 Wind Energy

One of wind energy's many advantages is its capacity to harness the kinetic energy of the wind for free and without restriction, as well as to generate clean, renewable electricity. Wind turbines have incredible scalability and adaptability characteristics in them, which gives them the chance to easily fit into different climates. Their use may embody individual turbines for homes and power units in enormous wind farms. Wind power can therefore be reliably taken into account for supply planning and also implies a decrease of fossil fuel dependency. The production of jobs is only one aspect of this economy, and it includes multiple sectors such as manufacturing, installations, and ongoing maintenance. Moreover, wind energy facilities often have a fast payback time during which the operation costs are low. Therefore, these enterprises are economically viable actors of changeable energy profiles (Chapman et al., 2021).

6.1.2.3 Hydropower

Riding the waves of water, hydropower is one of the recognized components in the process of switching to a renewable energy system that may be utilized because of the unique features of the technology. On the positive side, this plays the role of teacher and trainer to the other natural sources, which are intermittent in their cut-off power supply. Hydroelectric projects tend to be tied to long investment intervals, which further increases their economic and stability prospects. The advantages to the environment include a reduction in greenhouse gas emissions during operation as well as the ability to use pools for a variety of purposes, such as recreation and swimming, as well as flood control. However, in order to correct the growth of this type of power plant in a sustainable manner, we must carefully consider the environmental effects, such as the disruption of habitats and the transformation of river ecosystems (Kuriqi et al., 2021).

6.1.2.4 Geothermal Energy

Geothermal energy utilizes Earth's internal heat to generate a steady and dependable power supply. The main benefit of this is its ability to provide a consistent and uninterrupted supply of electricity, regardless of the weather conditions. Geothermal power facilities have a limited environmental impact and release negligible amounts of pollutants while in operation, contributing to a low-carbon energy combination. The utilization of geothermal resources for heating and cooling purposes directly boosts its adaptability. Although geothermal energy may not be as readily available as other sources, ongoing technological improvements and research techniques are continuously broadening its potential contributions to the move towards sustainable energy (Guzović et al., 2022).

Biomass, which is energy obtained from organic sources, provides distinct advantages in facilitating a shift to low-carbon energy. The main benefit of this is its adaptability and capacity for several uses, such as power production, heating, and biofuel production. Biomass resources are plentiful and may be obtained from nearby sources, hence decreasing reliance on centralized energy sources. The closed carbon cycle of biomass, in which the carbon emissions from combustion are balanced by the carbon uptake during growth, renders it a theoretically carbon-neutral energy source. Nevertheless, to fully optimize the environmental advantages of biomass in a sustainable energy transition, it is crucial to implement sustainable sourcing procedures and take into account the implications of land-use change (Chapman et al., 2021).

6.1.2.5 Tidal and Wave Energy

Tidal and wave energy are novel methods for harnessing the ocean's power. The regularity of tide patterns and the steady behavior of ocean waves enhance the dependability of these sources. Tidal and wave energy projects have a reduced environmental impact compared to certain land-based renewable sources, hence conserving marine habitats. Their capacity for consistent and foreseeable power generation improves the stability of the grid, hence contributing to a robust energy system. Despite the presence of obstacles like site uniqueness and potential environmental consequences, continuous research and development efforts are focused on harnessing the complete capabilities of tidal and wave energy in a sustainable transition to renewable energy (Clemente et al., 2021).

6.1.2.6 Worldwide Patterns and Acceptance

According to the most recent data, there is a notable global shift towards adopting sustainable and low-carbon options in the field of renewable energy. Various influential areas and countries are taking on crucial roles in spearheading this transformation, demonstrating a wide range of ideas and plans to attain a more sustainable energy environment (Hassan et al., 2024).

China: the role of China cannot be underestimated being the main global powerhouse in the use of renewable energy sources. China, by building its reputation as having the highest solar panel and wind turbine manufacturing and utilizing levels, has been able to reach this by setting out and defending very aggressive targets and making enormous investments. A country is at present implementing its renewable energy program intensively and it is evident that solar, wind and hydraulic power take the biggest role. As China is the country that set the goal of switching to the renewable energy, it is the world's indisputable leader in global renewable energy deployment (Hou et al., 2023).

Europe: renewable energy has already been the greenest foothold of renewable power in the countries of Europe for more than one quarter of a century now. Germany is known for its "energy Wende" program. This program is implemented to enable Germany to prioritize the importance of the environment and sustainability in its energy systems. Denmark, Sweden, and Spain that have realized impressive advancements in the area of renewables, which is mainly associated with the following rapid expansion of wind power, solar energies, and smart gird technology (Jurasz et al., 2020).

The United States: under the regime of US, renewable energy used for the last couple of years has been green intensifying. The states of California, Texas, and others work towards the supply of renewable energy through an application of tough rules regarding the implementation of renewable energy. The boom in wind and solar programs along with advancements in energy storage have brought a trend towards increased diversification in the energy mix (Wakiyama and Zusman, 2021).

India: Indian participation in renewable energy adoption is gaining more and more attention. This growing need for electricity is behind this move as well as commitment to lowering pollution levels. The country allows for substantial economic cooperation with the clean energy projects aimed at transition towards solar, wind, and hydropower use. India's renewable energy aspirations not only plan the country's energy system but also involve provision of energy services to its people, hence addressing the issue of energy accessibility (Elavarasan et al., 2020).

Latin America: many countries of this region have already implemented considerable advancements with regard to renewable energy uptake. There is a possibility that Brazil has further possibilities for getting electricity from hydropower, and the potential of wind and sun is starting to be seen with the passage of time. Chile is likewise in the lead, the countries' main focus being predominantly the development of solar power. This undoubtedly represents a sign that the region is committed to a cleaner and greener future (Murshed et al., 2021).

The Middle East and North African region: formerly great oil and gas importers, the move of so many countries in the Middle East and North Africa towards renewable energy has changed that situation. At first look, the United Arab Emirates has installed what is described as the largest solar panel globally. These countries are fully aware of the necessity of increasing the energy diversity for the sake of long term sustainability and security (Esmaeili Shayan et al., 2022).

Worldwide patterns: in general, there is a worldwide movement towards greater use of renewable energy, which is influenced by factors like decreasing costs, government incentives, and a rising awareness of environmental issues. The expansion of solar and wind energy has been particularly rapid. The incorporation of cutting-edge technology, such as energy storage and smart grids, is augmenting the dependability and adaptability of renewable energy systems on a global scale (Khare and Bhuiyan, 2022).

Notwithstanding these favorable patterns, there are still obstacles that remain, such as problems with irregularity, the requirement for strong infrastructure, and ambiguities in policies. Nevertheless, the clear fact that countries are equally committed to gaining a greener and carbon-free future is a reflection of the far-reaching influence and impact that renewables are having on the global energy system.

6.2 RENEWABLE ENERGY INDICATORS

Renewable energy indicators are the numerical metric employed for tying up the elements of the implementation, effectiveness, and impact of renewable energy. These indicators have helped to paint a comprehensive picture of how renewable energy initiatives are gradually taking root and growing in strength within countries and regardless of their level. A variety of indicators that measure renewable output range from capacity, to generation, to efficiency, to investment, to environmental implications (Jurasz et al., 2020).

In addition, the reliability of academic research, and peer-reviewed journals in elucidating the indicators of renewable energy, is too important to be disregarded. Scientists and researchers conduct experiments, studies, and assessments on renewable energy implementation, and among the many elements they try to understand are technology improvement, efficiency of the policy, and socio-economic effects. The discoveries of world-renowned scientists solely accumulate the present information base on sustainable energy so as to supply the policymakers and industry experts with sustainable solutions (Guzović et al., 2022).

Besides, international treaties and activities contribute to the attribution of renewable energy indices. Being extremely detailed, the paragraphs of the Paris Agreement and the Sustainable Development Goals (SDGs) define the targets and infrastructure for the development of renewable energy sources and the endorsement of sustainable living. A set of indicators, including those addressing the accessibility of renewable energy, investment in renewable energy, as well as the environmental impacts of the transition, takes place in the context of policy-making carried out both nationwide and internationally (Mulder et al., 2023).

In particular, renewable energy indicators are imperative to observe, pin-point the problems, and inform solutions needed in the transition process to sustainable energy. It is the case that the indicators are a source of influence over the policies, investments, and efforts that are goal orientated toward developing the renewable energy sector faster by including clean, comprehensive, and co-operative data of renewable energy deployment and performance (Pandey and Asif, 2022).

Kumar et al. discussed multitude criteria decision-making (MCDM) strategies to get a better answer of renewable energy systems in complex conditions that are specified in all possible variables and criteria. Afterwards, other writers have moved on to discover how often social aspects have on a regular basis remained vital in the process of energy planning. Energy projects, considered by developers, are often plagued with new technologies and a cost-effective energy for all, yet they are often unsuccessful because of the three components of the social factors. In consequence, while designing an energy system, the softer aspects must be treated along with the core ones (Kumar et al., 2017).

Taking Strantzali and Aravossis' research for reviewing relevant literature, which was based on the last 31 years and covered 183 research articles, the aim went beyond the patterns in energy (RES) assessment and even to analyzing the structures or tools applied in energy planning project evaluation. The approach instead was which were left (94%), eligibility in the labor being the most important consideration. After that was social amenity with 28%, economic advantages with 15%, visual effects at 14%, neighborhood rectification at 13%, effects on health at 10%, and enhancing job opportunities at 8% (Strantzali and Aravossis, 2016).

Liu identified 11 fundamental sustainability metrics for renewable energy systems, categorizing them as follows: environmental indicators (such as CO_2 emissions and renewable fractions), economic indications (including costs, return on investment, and payback), and social indicators (such as job generation and benefits for inhabitants) (Liu, 2014).

Beccali et al. established a set of criteria for conducting multi-criteria decision-making in energy planning. These criteria were divided into three categories: technological (such as reliability and cost of saved primary energy), energy and environmental (including land requirement and greenhouse pollutant emissions), and social and economic (such as labor impact, market maturity, and compatibility with political situation) (Beccali et al., 2003).

Rahman et al. outlined various factors that decision-makers should consider when selecting the most suitable option for sustainable rural electrification in developing nations. These factors were categorized into five groups: technical (such as capacity factor, annual resource availability, compatibility with existing infrastructure), economic (including capital cost, system lifespan, reliance on fossil fuels), social (considering public and political acceptance, potential for local employment, public awareness and willingness, conflicts with other applications), environmental (taking into account CO_2 emissions and local impact), and policy/regulation (considering land requirements, emphasis on local resources, and tax incentives) (Md Mizanur Rahman et al., 2013).

Upadhya and Sharma compiled a comprehensive list of factors and criteria to consider when designing a hybrid energy system for a specific location. These factors include socio-political considerations such as alignment with national energy policy goals, political and social acceptance (including factors like land use, visual impact, electromagnetic interference, acoustic noise, shadow flicker, ecosystem disturbance, etc.), the risk associated with the energy system's fuel prices, and the impact on labor (Upadhyay and Sharma, 2014).

Štreimikienė et al. took an additional step by assembling a panel of 25 experts to identify qualitative and quantitative criteria for evaluating various electricity generation technologies (such as nuclear, gas, biomass, geothermal, hydro, wind, etc.) with regard to their economic, technological, environmental, social, and political factors. The social dimensions encompassed the impact on social welfare (employment, economic stability), the impact on the sustainable advancement of societies (education, research, culture), and public reception/attitude (Štreimikienė et al., 2016).

Niu and Wei presented a simulation of an optimization algorithm in a hybrid thermal/wind power system. The algorithm aimed to achieve a social-environmental-economic dispatch. The objective function of the algorithm was to calculate the risk cost associated with power outages caused by the instability of wind energy generation. These outages can lead to socio-economic losses (Niu and Wei, 2013). Rojas-Zerpa and Yusta offered a general tip for making a decision for rural and distance areas that don't have modern infrastructure. The Fourth Pillar does not only implement technical, economic, and environmental factors but social factors as well. They surveyed 16 different professionals around the world and on this basis defined the set of criteria, which included social acceptance of power, job creation, and human development index (HDI) among others (Rojas-Zerpa and Yusta, 2015).

Firma Dufo-Lopez et al. summon off-grid HRESs using a new method in combination with a multi-objective genetic algorithm to achieve the best system design decisions. The goal is to achieve this by taking into account both, minimizing the total discounted cost (NPC) and maximizing the HDI and job creation at the same time (Dufo-López et al., 2016). This implementation is developed through of Rojas-Zerpa and Yusta observations (Rojas-Zerpa and Yusta, 2015). A study by Al-falahi et al. reported mainly contained summaries that then explained and illustrated the criteria using economic, environmental, and social elements. Moreover, it integrated energy indicators like the "human development index," the creation factor of jobs, the cost of social carbon, and the characterization factor of socio-demographic patterns in the determination of the energy load pattern of a household in a special place (Al-Falahi et al., 2017).

Lyden et al. conducted a thorough examination of 51 tools that simulated general energy systems. From this list, they chose and assessed 15 tools that specifically focused on modeling community-scale hybrid renewable energy systems (HRESs). These selected tools solely took into account storage and demand-side management. The study compiled a set of design optimization factors and outputs, some of which were correlated with social indicators such as Human Development Index (HDI) and employment generation (Lyden et al., 2018). Pfenninger et al. examined how energy models tackle the current system's difficulties, including handling issues of time and space, managing uncertainty and transparency, dealing with increasing complexity, and incorporating human behavior, societal hazards, and possibilities (Pfenninger et al., 2014).

Hilpert et al. stated that modelers bear a significant societal obligation due to the widespread utilization of their models in supporting policies and choices. Adhering to scientific standards in energy system modeling is crucial for the construction of accurate models. Ultimately, models must adhere to scientific criteria, but the significance of public acceptance is becoming paramount (Hilpert et al., 2017).

According to Wiese et al., model-based analyses are now essential in the transformation process. These analyses are used to address various questions related to grid control and planning, dispatch and unit commitment, expansion planning, energy market design, and environmental and social analysis of highly integrated energy systems (Wiese et al., 2018).

When compared to other factors, energy has the biggest impact on changing environmental conditions. When it comes to the negative effects of energy on the environment, Myers and Kent think that burning fossil fuels is mostly responsible for greenhouse gas emissions, which primarily consist of carbon dioxide (CO_2). As such, there is a close relationship between environmental sustainability and energy sector policies. Because renewable energy is thought to be more environmentally benign than fossil fuels, countries that now rely on fossil fuels must switch due to the negative effects of energy on the environment. It's also important to note that using renewable energy sources reduces CO_2 emissions and aids in environmental protection (Soltani et al., 2019).

6.3 DRIVERS OF AND BARRIERS TO RENEWABLE ENERGY USAGE

6.3.1 Drivers

Renewable resources such as solar radiation, wind, tidal waves, and heat from biomass are widely available across the globe. Certain forms of energy, such as solar power, have incredible potential and are practically boundless. It is worth noting that the solar radiation reaching the earth's surface in one year is more than sufficient to meet the world's yearly energy needs by a factor of over 10,000. In addition, utilizing only a fraction of the solar energy that reaches the world's paved surfaces could easily fulfill the current global energy demands. Technologies that convert these resources into electricity and energy are readily available and continually advancing (Azarpour et al., 2022).

There are certain factors that give positive influence to the use of renewable energy while on the other hand there are hindrance factors. Due to this, it is important to appreciate these drivers and barriers as an important first step toward propelling policies that will promote and support the shift to a sustainable energy future. Understanding the economic, technological, and policy-related factors can help us better grasp the realities of transitioning to a sustainability-focused model that incorporates renewable energy use (Wiese et al., 2018).

Generally, this is driven by several factors and prohibited by several forces. Enthusiasm for renewable energy as well as governmental, financial, and technological support drive the usage of renewables as follows:

Environmental concerns: accelerated acceptance of renewables to a significant degree; concerns concerning climate change and its implications have become well-recognized. Clean alternatives are more accepted as a result of worries about deforestation, air and water pollution, as well as other environmental consequences of fossil fuel activity (Hassan et al., 2024).

The current global focus on climate change control necessitates a push for renewable energy source use, as underscored by IRENA's REmap 2030. The report therefore stresses that there is a need to increase the usage of renewable energy sources so as to prevent further global warming and reduce the emission of greenhouse gases even more. The current IEA report predicts that the global share of renewables in total final energy will grow only slightly to 26% by 2030. Transition to renewable energy has many advantages apart from climate change public health, and social and economic opportunities are only a few of them. Green technologies as a means not only decreases emission of gases affecting the quality of the air, therefore improving the health of the population, but also minimize instances of environmental dangers like oil leakages and habitat loss linked with fossil fuel extraction and transport (Tham et al., 2020). In addition, renewable electricity can contribute to climate change mitigation; solar and wind power, for example, cool the earth and hydropower provided solutions to the problem of increase in temperature and water scarcity (A. Rahman et al., 2022).

Yet, there are challenges felt in the transition to renewable energy, which are categorized into the administrative challenge, the political challenge, the grid challenge, and the socioeconomic challenge. Policies, regulation, and approval from the authorities also act as major constraints in deploying renewable energy as it involves administrative procedures like environmental management plans and approval. Also, traditional and political barriers such as regulatory policies and policies that differ from one country to another slow down progress in the adoption of renewable energy (Lu et al., 2020). Technical issues include those related to grid integration and the capacities for allowing, for instance, wind power to balance fluctuating supply and demand, while other barriers include financing of renewables and the acceptance by the public of such technologies. To do this requires overcoming some barriers that are a hindrance to the shift towards sustainable energy systems as well as the achievement of climate objectives (Streimikiene et al., 2021).

Energy security: reliance on frequently imported fossil fuels exposes energy security and the impossibility of securing energy supplies. Renewable energy sources such as solar and wind do NOT increase dependence on imported energy but rather address energy security since they are more readily available domestically (Hassan et al., 2024).

Energy security remains a central precondition for ensuring economic stability especially for developing nations with large stakes on imported fossil energy sources. However, geopolitical issues explain the fluctuating nature of traditional forms of energy, meaning that from a portfolio perspective, there is a need to tap into non-traditional options. The establishment of local renewable energy resources is a sure way to enhance energy security since it reduces reliance on imported fuel. Focusing on manufactured goods over imported ones, especially within areas where domestic substitutes are available, reduces financial pressures driven by the use of foreign exchange for the importation of energy. Due to the economic development, there is an ongoing problem of energy scarcity, thus the need to explore avenues of diversifying the types of energy available in the industrial societies as well as in the domestic domains (Gatto, 2022).

Cost competitiveness: noteworthy cost reductions enabled by technological improvements and economies of scale have resulted in renewable energy technologies progressively becoming competitive with conventional energy sources (Hassan et al., 2024).

The cost structure of renewable energy is another crucial aspect that has witnessed a tremendous enhancement due to aspects such as technology changes and scale economies. Thus, the components of renewable energy technologies, for instance, PV solar systems and wind power systems, among others, have seen a drastic reduction in the manufacturing costs thereof over the years. For example, the efficiencies of solar panels have increased, meaning that for a similar area more power can be produced and the cost per kilowatt hour (kWh) has been reduced. As technology of wind turbines increased, so did the improvements in design, composite material used, and fabrication processes, thus resulting in bigger and more efficient turbines harnessing more wind energy at reduced cost (Mlilo et al., 2021).

Furthermore, it is important to appreciate the crucial advisory of economies of scale in bringing down the costs of renewable energy. Another factor is the scale effect for renewables production. Volumes have risen, and also since cost of equipment and the necessary materials have decreased with the scale of production, more operational advantages can be introduced in procurement and supply chains. Much of this has been seen in the utility scale solar PV industry, where manufacturing at high volume has meant volume pricing (Albrecht and Steinrücke, 2020).

This has meant the renewable energy technologies gradually achieve competitive prices and costs with other conventional energy sources especially due to the advancements in technology and scale economies. Today, in some instances, the cost of generating electricity from renewables has touched a new low to the point of being cheaper than the cost of generating electricity from conventional sources in new power plant investments. This trend is anticipated to persist, and with improvements in technologies, coupled with the increased adoption of renewable energy projects, costs are also expected to come down further, making renewable energy cost-effective and sustainable as a long-term substitute to fossil energy (Azarpour et al., 2022).

Policy support: policy support has greatly helped facilitate feedback and supportive mechanisms for using renewable sources of energy. It includes objectives for renewable energy, renewable portfolio requirements, feed-in tariffs, and tax credit (Rosales-Asensio et al., 2024).

Ensuring that government provides its full support to the projects is essential for investment in the renewable energy sector. Lack of clear long-term policies/strategies and regulatory policies affecting foreign investments provide a high degree of policy risk, which is costly to investors/developers. So to solve this problem government must establish proper policies for the development of the renewable energy that can offer good support to these renewable energy projects. Environmental support and acceptance are also critical measures: communities are urged to initiate their own renewable goals. Some of the regulatory authorities require changes to obtain a more favorable result for the development of renewable energy sources (Relva et al., 2021). To overcome these barriers, active integrated policy, public participation, legislation adaptations, finances, and infrastructure towards expansion and promoting health systems, successful policies, and technological advancements are needed. It is clear that through careful actions as well as the adoption of appropriate measures, national and other relevant authorities, parties and actors ought to foster the shift to the utilization of the renewable energy sources (Gungah et al., 2019).

Technology innovation: research and improvement in innovative technologies have increased the cost-effective efficiency and reliability of renewable energy systems, in many cases even surpassing installed demand forecasts (Mlilo et al., 2021).

The use of technology is therefore relevant in the shift towards utilizing renewable sources of energy especially with the growth in the economies and the related demand for energy. This is clearly seen in the transition from conventional sources of energy, such as fossil, to cleaner electrical sources especially for nations in the developing world that undergo the social and economic development process. Technologies in renewable energy sources happen to be an opportunity of the decentralized renewable source of power that may help in the progress of the substandard areas and in addition as a possibility to offer jobs in the new developed nations and the developing ones (Javid et al., 2021). Moreover, the governments formulate liberal tax incentives, favorable policies for renewable energy technology companies, and relaxed regulatory framework, which in turn promote the development of renewable energy technology businesses (Mirzania et al., 2019).

Although the implementation of renewable energy technologies is independent of geographical factors, they are commercially important based on geographical factors. For instance, wind-based systems involve harnessing wind velocity and aspects of geography; location of the wind turbines is fundamental in generation of electricity. Likewise, for determining solar energy opportunities, it is relevant to have precise irradiation data, while it is crucial to properly define geothermal reservoirs for geothermal energy facilities. Some of the issues have not been solved fully due to the difficulties that people face in acquiring wholly comprehensive data through advancing technology. Mitigating these challenges with enhanced quantitative approach in data gathering and analysis is crucial towards achieving the best outcomes with renewable energy technologies (Ahmad et al., 2022).

Job creation: renewable energy may create many more jobs than incumbent sectors of the global economy. Jobs have been created in research, manufacture, and installation, as well as in maintenance and service (Mlilo et al., 2021).

6.3.2 Barriers

Nevertheless, there are certain barriers and associated concerns that hinder the progress of clean energy development and its integration into contemporary economic systems. To remove these barriers and provide conditions that will facilitate the broad use of renewable energy, governments, business associations, financial institutions, and the general public must work together.

Intermittency and system integration: the uncertain nature of renewable energy sources like wind and solar drives this issue. It affects the stability and reliability of the system. The most common issue is the generation of adequate energy storage systems and ensuring that the grid's infrastructure can manage these fluctuations (Wiese et al., 2018).

The use of wind and solar power for generation of electricity is one of a very special kind since these sources of energy are intermittent in nature and hence present many difficulties when adopted into the power grid. This intermittency is hence defined as the feature that, different from fuel supplies or stored energy, these energy sources are variable and unpredictable in nature, making them a challenge to operate on a large scale and integrate with the current power systems (Albrecht and Steinrücke, 2020).

Reliability is always a big question mark where renewable energy is concerned and intermittency can be defined as the measure of how much renewable energy is actually available at any given time. Wind and solar power are unpredictable as they rely on the ability to capture forces of nature such as wind and heat from the sun. This variability often causes times of either low or no electricity output and can be very frustrating in attempting to guarantee constant energy availability. Some of the most significant challenges noted earlier include variability and unpredictability, such as the variation in wind power output as a result of turbulence affecting the wind energy generation, which has been cited as a major hindrance to ramp up wind energy integration due to the resultant system reliability issues and increased operational costs (Sovacool et al., 2023). Along the same vein, the predictable fluctuation of solar energy has implications for the grid in terms of reliability and control.

Several solutions and approaches have been developed to support intermittency to meet the challenges faced, such as progress in grid operators' services, in the area of energy storage products, and in the creation of more effective and adaptive electricity networks. The recent literature has gone further in analyzing the characteristics of intermittent renewables and in adopting different kinds of models in an attempt to improve the management of intermittent renewables and enhance their penetration in power systems. These approaches are vital for explaining system behavior under various levels of renewable energy mix and an important platform on which the results can be embedded for overall decarburization efforts (McPherson et al., 2022).

Initial capital costs: although renewable energy systems often have lower operating costs over the long term than traditional energy sources, some populations, entities, or the government may be unable to afford the up-front investment required to install them (Mirzania et al., 2019).

A major challenge for the deployment of renewable energy systems stems from the high up-front capital costs and often their operating costs are lower in the long-run than conventional energy suppliers. Although the renewable energy systems involve relatively minimal and manageable running costs compared to their non-renewable counterparts in the long run, the installation expenses often act as barriers to people, organizations, or even governments. Financial divides hinder the implementation of renewable energy technologies, especially in areas of low investor capacity or where conventional energy is subsidized. In order to tackle this problem, incentives, grants, or financing programs have to be provided in as many countries as possible so that the population can afford renewable energy systems and their integration into homes, businesses, and other buildings, which, in return, would create the framework for the shift to clean energy (Caineng et al., 2022).

Fossil fuel subsidies: financial incentives and subsidies for the development and use of fossil fuels skew market dynamics, making it more difficult for renewable energy solutions to compete fairly (Diallo et al., 2024).

Subsidizing of fossil fuels has indirect negative impacts on renewable energy solutions as it distorts the market structure and offers unfair competitive advantage by offering favorable prices for fossil fuel products. Such financial incentives and subsidies help in presenting a lower-cost aspect of fossil fuel exploration and use, hence making them more economically competitive to renewable power sources. Regarding this, there are challenges that renewable energy technologies encounter in growing market share and cost of technology. Furthermore, supporting fossil fuels from subsidies prolongs the continued use of carbon-intensive and environmentally unfriendly energy sources, which deepen climate shifts and pollution. Challenging fossil fuel subsidies remains key in understanding the absence of a level playing ground as a viable means to incorporate the increased use of renewable resources into the energy mix of climate change mitigation initiatives (Li and Zhu, 2023).

Infrastructure limitations: in certain areas, especially in isolated or rural areas, the development and integration of renewable energy projects are hampered by a lack of suitable transmission and distribution infrastructure (Elavarasan et al., 2020).

Remote or rural infrastructures are also a weakness within the development and integration of renewable energy projects due to infrastructure constraints. Most of these regions lack adequate transmission and distribution infrastructure, hence providing a challenge to the deployment of renewable energy technologies. Controlling and transporting renewable energy generated in remote areas also poses another problem due to insufficient transfer infrastructure, meaning it may not be utilized efficiently even when produced. This infrastructure deficit explains why there's perhaps no stronger priority for the enhanced deployment of renewable energy than to sink more resources into upgrading transmission and distribution networks (Castillo-Calzadilla et al., 2022).

Public perception and nimbyism: the opposition of local populations to renewable energy projects, known as nimbyism, can hinder the process of locating and obtaining permits for these projects, leading to delays and increased expenses. These concerns may pertain to issues such as noise, visual disturbances, or perceived devaluation of property (Kabeyi and Olanrewaju, 2022).

These projects entail risks associated with their development and implementation, such as the possibility of local people in specific places halting their progress, resulting in delays and subsequent cost escalation. Emotional responses are triggered by difficulties related to noise pollution, visual disruptions, or the potential decrease in real estate values. These challenges should be resolved by engaging in community consultations, clear and direct communication, and implementing efficient decision-making processes that take into account the demands of the local population. Furthermore, it is crucial to implement concrete measures to address community concerns and tackle the primary causes of people's hostility. This includes offering compensation for any negative consequences and providing tangible incentives to encourage support for the establishment of renewable energy facilities. These actions are essential in order to streamline the process of siting and permitting (Rosales-Asensio et al., 2024).

Policy and regulatory uncertainty: long-term investment in renewable energy projects is discouraged by inconsistent or shifting government policies, regulatory frameworks, and support mechanisms (Tham et al., 2020).

Reliability is another issue that has to be tackled as the policy and regulatory environment for investment in renewables remains less than certain. They include policy and regulatory uncertainties, which are evidenced by volatile policies and regulations and which encourage investors and developers. This uncertainty and lack of clear direction in the law makes investments in the renewable energy sector relatively risky and slows the industry's expansion. This problem, therefore, requires the governments to ensure that the policies that support investment are clear, harmonized, and supportive to induce certainty and confidence among investors. Through the adoption of long-term policies that offer stability and certainty to investors, governments can unlock new developments in the renewable energy field, as well as foster innovations in the renewable energy field to unlock a better future for sustainable energy (Phillips et al., 2020).

Other significant barriers include market failures, such as insufficient funding for research and development programs of renewable energy technologies due to the lack of exclusive property rights and environmental impacts without a price tag. Monopolies in the energy sector can limit competition and restrict new players from entering the market. High initial investment costs for renewable energy systems may make them unaffordable for many potential customers, especially in developing countries. Additionally, all power projects carry financial risks due to uncertainties in future electricity prices, and renewable energy projects often face higher costs due to the need for higher returns on risky investments in the private capital market compared to established technologies (Taghizadeh-Hesary and Yoshino, 2020).

6.3.3 Opportunities

There are four major sectors where renewable energy is well suited, and in this discussion we will explore these as areas of opportunity. These factors include socioeconomic development, energy access, energy security, and climate change mitigation.

6.3.3.1 Socioeconomic Development

Per capita income has a positive correlation with per capita energy use as well as the Human Development Index. It's clear that economic growth has been a major driver of rising energy demand and consumption in recent decades. As the economy expands, there is a growing need for more advanced and adaptable energy sources. There has been a clear trend towards transitioning from traditional fossil fuels to cleaner electricity as a means of promoting economic growth (Kabeyi and Olanrewaju, 2022). Advocating for clean energy is crucial for human development as it has the potential to improve income generation, enhance health and education quality, and reduce poverty levels. Given the decentralized nature of renewable energy technologies, they can play a crucial role in fostering rural development. The creation of new job opportunities is a significant long-term benefit of renewable energy in both developing and developed nations. Government bodies and policymakers have shown their support for the domestic market in renewable energy technology business by providing tax benefits, flexible norms, and investment-friendly policies (M. M. Rahman et al., 2022).

6.3.3.2 Access to Energy

Humanitarian development is constantly implementing different strategies to boost electricity as well as better methods of cooking, and the outcomes are positive. Renewable sources such as solar energy, hydropower, and bio power are some of the economic and eco-friendly power sources in the rural setting. For example, photovalic power offers lighting and clean water provision while biomass combustion offers the provision of an energy demand locally. Replacing cow dung or any traditional fuel with advanced biomass stoves or biogas systems will help in utilizing energy sources in a much eco-friendlier way (Bandaru et al., 2021).

Accessing energy is a very key aspect in carrying out humanitarian concern and several approaches are being deployed with the enhancements. Sources of energy such as the sun, water, biomass, etc. fare especially well in these rural areas since they have cost effective and environmentally friendly impacts. For instance, solar villages (SVs) provide light, which is also used as a source of power for water purification processes, two essential functions required in off-grid regions (Elavarasan et al., 2020). Similarly, the biomass combustion technologies also have the advantage of distributed energy generation systems against a number of local energy needs, which means we eliminate the conventional energy sources like cow dung etc. Conversion to the superior forms of biomass stoves or biogas systems adds more efficiency on use of energy, hence boosting the sustainability tag alongside enhanced standards of living. Therefore, it can be stated that the humanitarian approaches should focus on the promotion of renewable energy sources for Waska to support and transform the communities and provide a better future in terms of environmental impact reduction (Agyeman and Lin, 2022).

6.3.3.3 Energy Security

The issue of energy security is one of the major areas concerning economic stability, especially in the developing countries where any unpredictable standstill of energy availability could be detrimental to their economy (Relva et al., 2021). Today, we mostly depend on the fossil fuels that have high sensitivity to geopolitics and prices that are unpredictable. Stability of energy is always critical, hence backing the green energy types would strengthen the energy security agenda due to diversification and the reduction of reliance on conventional fuel. Thus, engaging diverse energy sources and the increase in renewable resources has advantages over the dependency on individual energy types or specific countries, due to mitigation of conflicts of interest and high energy uncertainties that lead to less economic stability and more sustainability (Elavarasan et al., 2020).

Apart from that, it is an effective way to wean the communities off fossil fuel while at the same time ensuring that they receive energy from various sources, not just from a sole source, thus improving energy security. Some sources of energy like solar, wind, and hydro are usually available locally and are not as affected by the global market pricing as compared to the traditional sources of energy. It is the ability to produce energy domestically and independently of fluctuations in the global market and political instability that can affect fossil fuel imports (Kabeyi and Olanrewaju, 2022). Moreover, utilizing renewable energy infrastructure for satisfaction increases employment opportunities and boosts electrification, making it more economically secure and dependable.

Also, the renewable energy systems need comparatively fewer investments on operation and maintenance as a rule, hence offering smoother economic gains and steadiness through the life cycle of such systems than that required for adopting the fossil fuel energy systems. In a nutshell, there is no doubt that prioritizing local renewable energy solutions has never been more critical for energy security but also as a pathway towards securing developmental and economic growth given looming global challenges (Hassan et al., 2024). By prioritizing the use of foreign exchange for importing goods that cannot be produced locally, we can reduce our reliance on energy imports, which could be highly advantageous for countries that rely on oil imports.

6.3.3.4 Climate Change Mitigation

Solar cooling of buildings can help mitigate the direct effects of increasing temperatures. Hydropower plants with dams play a crucial role in managing droughts and floods. Given the increasing frequency of global warming and climate changes, the construction of these dams is justified. Solar PV and wind power generation are ideal for dry and arid areas with ample solar insulation as they do not require water. These are crucial during dry seasons for running water pumps, thereby increasing agricultural productivity. Planting trees along coasts and riverbanks helps to prevent erosion and can serve as a source for biomass energy (A. Rahman et al., 2022).

Climate change adaptation measures include a diverse array of measures that have been developed with the specific goal of lowering emissions of greenhouse gases as well as preventing and responding to the consequences of global warming. Shading of buildings through application of solar technology presents an acceptable means of mitigating the impacts of heatwaves and high temperatures by creating impressionable indoor thermal conditions without risking energy utilization and emission of greenhouse gases into the atmosphere. Hydropower plants with dams also rely on altered climate impacts on water supply and demand; these structures act as multi-purpose facilities for water management responding to both drought and flood risks. For this reason, the construction as well as the maintenance of such structures is justified as the incidents of and the intensity of severe weather conditions increase in the future (Štreimikienė et al., 2016).

Renewable energy technologies provide new clean sources for electricity generation instead of depending on fossil fuels and increasing water scarcity. Furthermore, during dry or low-rainfall periods, the systems can help to pump water to help farmers carry out their activities and improve production in areas that receive little rainfall or experience dry spells. Further, afforestation measures, which include planting trees along the coastlines and riverbanks, not only address erosion problems but also harness carbon stocks and enhancement of carbon sinks besides acting as potential stock for biomass energy in the fight against climate change. Mainstreaming of climate change means that, through adopting various and multi-disciplinary strategies, societies can cope with impacts of climate change on various fronts and graduate to meet these impacts with sustainability and resilience across the sectors (Kabeyi and Olanrewaju, 2022).

In conclusion, developing the implementation of renewable energy sources is a complex but necessary issue. Tackling regulatory barriers, offering financial assistance, encouraging innovation, and raising public awareness can advance the path towards a more sustainable and environmentally friendly energy future. Governments and stakeholders have the opportunity to accelerate the transition to renewable energy sources through consistent policy measures, monetary incentives, creation of infrastructure, and active community participation.

6.4 POLICIES AND MEASURES TO PROMOTE RENEWABLES

The renewable energy sources promotion through policies and measures is the key element of all international initiatives that aim at climate change mitigation, energy security enhancement, and sustainable development promotion. In an attempt to advance the use of renewable energy technologies, governments, organizations, and communities have started many initiatives like wind power projects all over the world to address the urgent need for transition from fossil fuels to renewable energy. Such policies and measures range from various regulatory frameworks, financial incentives, research and development initiatives, and public awareness campaigns that are designed to speed up the expansion of generating capacities from renewables (Štreimikienė et al., 2016).

Frequently employed strategies to promote renewable energy sources utilize mandates and targets. These targets set objective parameters for the mix of renewable sources in the overall energy consumption for a specific period. Renewable energy quotas and requirements, such as Renewable Portfolio Standards and mandates, call for utilities and energy suppliers to supply a certain proportion of the electricity by renewable sources. Investors and stakeholders are effectively communicated with via these mandates and targets that serve to provide investment certainty for renewable energy projects (Sikiru et al., 2024).

Besides the vital role in developing a renewable energy infrastructure, it also tackles the financial setbacks that usually hinder deployment. Different processes using the likes of subsidies, tax credits, grants, and feed-in tariffs are often used to promote investments in renewable energy initiatives, hence making the alternatives that use fossil fuels cost ineffective. In other words, feed-in tariffs ensure firm prices for the energy generated and provided to the power system by renewable technologies, which, in turn, stimulate their adoption and offer revenue certainty to project investors (Bandaru et al., 2021).

Setting market mechanisms and regulatory frameworks is paramount in order to create a climate that will enable the deployment of renewable energy. As an example, net metering policies enable the electricity consumer who has onsite renewable energy systems to settle their electricity demand with the additional energy that they produce. This has the dual effect of encouraging both distributed generation and self-use. Furthermore, the internalization of the external costs of greenhouse gas emissions are 9g. Among them, carbon taxes or emissions trading systems increase the competitiveness of renewable energy compared to fossil fuel (Glasson et al., 2022).

Research, development, and innovation investments are the pillars in the economic and technological advancement of renewable energy technologies. Technology demonstration projects, public funding for research institutions, and industry-academia-government collaborative research partnerships are important in overcoming technological barriers and realizing the full promise of renewable energy sources. Additionally, the capacity building initiatives and the knowledge sharing platforms contribute to the development of expertise and skills in the renewable energy sector, facilitating a foundation for innovation and entrepreneurship (A. Rahman et al., 2022). Education and awareness programs are a good approach to developing societal understanding and acceptance of renewable energy sources. By sharing the benefits of renewable energy, fact-checking the misunderstandings, and encouraging behavioral modifications, the efforts serve to create a favorable environment for the implementation of renewable energy. Through the inclusion of stakeholders including entrepreneurs, civil society organizations (CSOs), and local communities in the decision-making process, the social and environmental sustainability of renewable energy projects can be guaranteed (Elavarasan et al., 2020).

Policies and initiatives that encourage the use of renewable energy sources are pivotal to the acceleration of worldwide transition to the low-carbon and sustainable clean energy future. Through the use of a sound blend of regulatory, financial, technological, and social principles, governmental authorities and interested parties can successfully tap into the immense potential in renewable sources of energy and harness their manifold benefits in terms of social equity, energy security, climate mitigation, and economic progress (Kabeyi and Olanrewaju, 2022). Despite the remaining obstacles and the increasing deployment of renewable energy to effectively meet the challenges of the 21st century, continued political commitment, increased international cooperation, and innovations are basically required. The following policies are some that could be implemented to support development of renewable energy sources.

6.4.1 Feed-in Tariffs (FiTs)

FiTs (Feed-in tariffs) are an efficient policy mechanism primarily aimed at speeding up the installation rate of renewable energy sources. They attain this objective by providing a sustainable and predictable financial incentive to the renewable energy producers (Liu, 2014).

1. Revenue stability stream: FiT grants renewable energy producers a guaranteed payment for power supplied to the grid for an agreed period of time, such as several years. Such is their key strength. This regular source of funding leads to financial stability and security, which is absolutely essential for renewable energy projects that have high upfront costs and long payback periods. FiTs address the financial risks related to power generation from renewable energy technologies by providing a predictable revenue stream, which strengthens investor confidence and project development (Sovacool et al., 2023).
2. Risk mitigation measures FiTs play an important role in removing the various financial risks that are integrally connected with renewable energy projects. These risks are technology risk, market risk, and policy uncertainty. FiTs help investors and developers by providing them with a stable rate of payment for renewable electricity generation and also hedging them against electricity price fluctuations, market demand fluctuations, and legislative changes. A steady institutional environment leads to sustained investments in infrastructure related to renewable energy and smooth implementation of new technologies thus promoting innovation and gradual cost reduction (Verbruggen et al., 2010).
3. Equalizing the competitive environment: FiT programs are needed to achieve equity in the usage of renewable energy technologies, which have higher initial costs than traditional fuel sources like coal, oil, and natural gas. In the absence of solid financial incentives, the substantial preliminary expenditure on infrastructure, equipment, and installation, which tends to be compulsory for renewable energy projects, may impede investors. FiTs are guaranteed income that clearly states the true value of renewable energy production, which includes not only the environmental but also the social benefits like cleaner air and lower carbon emissions. Such measures will close the cost gap (Štreimikienė et al., 2016).
4. Facilitation of market development: FiTs provide a basis for expansion for renewable energy sectors through the development of a stable and suitable environment for investment and implementation. FiTs promote market confidence and bring in many investment types: from researchers to corporations, utilities, and financials. The latter ensure that renewable electricity is properly priced against the competition. Investing more in renewal projects stimulates the low-carbon economy's transition through providing job opportunities, economic growth, and local development (Montoya et al., 2021).
5. Adaptability and flexibility: FiTs are therefore very adjustable to the specific conditions of any given market, technology, and region. Governments have the duty to change FiT rates, duration, and eligibility criteria in line with policy objectives, cost competitiveness, and technological maturity. The capacity of FiTs to ratify and modernize according to the market dynamics, technology developments, and policy objectives makes them an effective tool in the penetration of renewable energy (Kabeyi and Olanrewaju, 2022).

At a glance, feed-in tariffs emerge as an effective policy instrument to encourage the use of renewable energy sources by providing producers a predictable and long-term financial incentive. FiTs do just this: motivate investment, level the playing field for renewable energy technologies, reduce financial risks, and stimulate market development by providing a guaranteed payment for the generation of power using renewable sources. When it comes to the development of a more resilient and sustainable energy system, feed-in tariffs are going to continue in their core role of speeding up installations of renewable energy infrastructures and helping governments accomplish their climate and energy targets (Guzović et al., 2022).

6.4.2 Tax Incentives and Credits

Unquestionably, tax incentives and credits are the main instruments of policy that promote renewable energy investments because they give monetary benefits to interested entities and investors in such projects.

1. Initially, decrease initial expenditures: a main benefit of tax credits and incentives is their ability to lessen the initial expenses associated with renewable energy systems. A common disincentive for energy investors and developers is the need for substantial upfront investments in infrastructure, equipment, and installation, which can be an obstacle for potential investors and developers. The tax credits, for example investment tax credits (ITCs) or production tax credits (PTCs), offers a direct financial benefit by providing you with a fraction of the initial costs associated with renewable energy projects. This makes them more financially practical and attractive (Soltani et al., 2019).
2. Promoting investment: the use of tax incentives and credits helps in the deployment of clean energy technologies and the installation of renewable energy infrastructure by reducing the financial barriers of entry. These incentives not only facilitate capital flow but also encourage institutions, businesses, and individuals to invest in renewable energy initiatives, which in turn stimulates the sector's growth and innovation. Alongside job creation, economic growth, and local prosperity, environmental and climate objectives would be more effective in achieving their goals if more resources were assigned to renewable energy projects (Jurasz et al., 2020).
3. Market expansion: by elevating the competitiveness of renewable energy technologies in relation to conventional energy sources, tax credits and other incentives stimulate the growth of the market for such technologies. These incentives contribute to cost reduction of renewable energy systems, which in turn makes these renewable energy initiatives more financially feasible and thus pushes for the increased deployment of renewable energy in different sectors and applications. Tax incentives may be designed to facilitate particular technology or categories of projects such as geothermal energy, wind power, solar photovoltaic (PV), biomass, or wind power. This system encourages more variety of renewable energy sources (Lu et al., 2020).
4. Promotion of innovation: tax allowances and incentives are the leading factors of innovation and technological progress in this industry of renewable energy. These incentives aim at encouraging the search for new technology developments, process creation, and innovation through the giving of financial rewards to investment in renewable energy campaigns. It provides constant development and financial savings in the sphere of renewable energy, thus leading to reduced costs and implementation of renewable energy (Štreimikienė et al., 2016).
5. Policy flexibility: through tax incentives and credits, policy-makers have the opportunity to craft the renewable energy policies that align with their specific goals and preferences. The ability of the government to alter the conditions of tax incentives such as expanse, duration, and qualification to promote particular renewal investments and tackle new avenues and challenges is manifest. Different kinds of tax incentives ensure that they can adapt to the new technologies and to fluctuating market conditions and policy goals. In this way, renewable energy sources are continuously enhanced (Mulder et al., 2023).

In essence, tax credits and incentives are indispensable instruments for fostering investment in renewable energy through the following mechanisms: alleviating the need for large initial investments, sparking investment activity, expanding the market, bringing fresh ideas to the fore, and empowering the government to design renewable energy policies of its own. In the progress of the transition towards a sustainable and resilient energy future, nations worldwide will undoubtedly utilize tax incentives and credits as instruments for the timely installation of renewable energy facilities and the achievement of climate and energy objectives (Jurasz et al., 2020).

6.4.3 Renewable Energy Grants and Subsidies

Governments employ grants and subsidies as vital financial instruments for moving forward and implementing renewable energy technologies.

1. Provision of development and deployment support: subsidies and grants serve as the pillars of the financial assistance framework for renewable energy projects in their cradle, construction, operation, and maintenance stages. This assistance contributes to minimizing the risk of financial uncertainties and associated risks in the process of investment in renewable energy, especially those related to new technology and difficult market conditions. Subsidies and grants are instrumental in the process of installation of renewable energy infrastructure and the development of clean energy capacity by way of providing access to financial resources (Pandey and Asif, 2022).
2. Provision for capital expenses: Subsidies, grants, and assistance are normally provided to renewable energy mainly to directly assist the companies in constructing the renewable projects so that they complete the total investment of their overly high-capital intensive project. This outflow consists of the amount spent for: employees' remuneration, resources, equipment, land acquisitions, and infrastructure. These are divided into the ones for the installing and running of renewable energy systems. The initial financing expenditure required for projects is alleviated via subsidies and grants, which are a source of financial assistance toward a project's capital expense for both the investors and developers (Clemente et al., 2021).
3. Mitigating operational expenditures: capital campaigns, as well as bearings and savings for operations purposes related to renewable energy projects, may cover the cost of the regular administration, maintenance, repairs, and inspection. Governments as the major players in the long-term sustainability and reliability of renewable energy projects provide their compliance to operational expenses and increase performance and longevity. This kind of support is particularly significant during the foregoing reason about community-based or small initiatives that normally have limited funding (Chu et al., 2024).
4. Advancement of research and development: inevitably the projects that bring innovations to the development of green energy receive beneficiations like grants and allowances for the ease of operation in the interest of simplifying renewable energy technologies, effectiveness, cutting costs, and solving technical challenges. Universities, research institutes, private firms, and organizations primarily in areas of novel technology would be participating in the investment of these designated funds, which has the possibility of not only bringing on game-changing results but also of vast technological leaps in the renewable energy sector. It is setting up the collaboration and triggering productiveness that grants and subsidies achieve on their part, thus motioning the development and improvement of clean energy solutions (Settembre-Blundo et al., 2021).
5. Addressing the gap in affordability: the first and possibly the most important reason why grants and subsidies are being used in renewable energy is that they narrow the disparity that exists between the expenses that processors can incur and the cost of the machines that produce sustainable electricity. Subsidies from governments to the developers attached to the financial plans for renewable energy reduce the investors' expenses on clean power generation, thus competing with conventional energy sources that become gradually not cost effective. Thus, this would result in a lowering of electricity fees, hence making energy more affordable. On the other hand, it helps the business to transition toward green or renewable energy alternatives (Kabeyi and Olanrewaju, 2022).

In conclusion, grants and subsidies are essential for promoting renewable energy development and are therefore instrumental in the development, implementation, and operation of such projects. This is done through the funding of initial investments, running operational expenses, research and development promotion, and filling the gap of affordability. Financial incentive is an important factor that contributes to the extent to which governments can achieve their climate and energy objectives, stimulate investment in renewable energy infrastructure, and accelerate the transition towards a decarbonized energy future.

6.4.3.1 Renewable Portfolio Standards (RPS) and Renewable Energy Obligations (REOs)

Policies like Renewable Portfolio Standards (RPS) and Renewable Energy Obligations (REOs) that compel suppliers and utilities to obtain a certain percentage of their electricity from renewable sources are known as policy mechanisms.

1. The generation of market demand: Renewable Power Suppliers (RPS) and Energy Retailers (REO) are legally mandated to get a pre-designated portion of their electricity from renewable energy sources like wind, solar, biomass, and hydroelectricity. Formulating legislation that obligates the incorporation of renewable energy sources into the energy portfolio induces a stable and reliable market demand for renewable electricity. Consequently, the provision of a secure market allows developers of renewable energy projects to attract investment (Streimikiene and Kyriakopoulos, 2023).
2. Energy-providing investment: the incorporation of RPS and REOs acts as a clear message to the market that it is especially focused on the development of renewable energy sources. Such action makes investing in renewable energy projects more appealing to developers and investors by decreasing their fear of risk and uncertainty. Capital from both public and private sources is allocated into RPS and REOs via the development of a stable and supportive regulatory environment; this leads further to expansion of and innovation in the renewable energy sector (Tovar-Facio et al., 2022).
3. Facilitating industry expansion: renewable power systems (RPS) and energy resources (REOs) are the key elements for the development and growth of renewable energy systems. These measures encourage the growth and enlargement of production by renewable energy companies, manufacturers, service providers, and developers through the emergence of more demand for renewable energy. This supports the application of renewable energy technologies and practices that in turn leads to job creation, economic growth, and local development (Kabeyi and Olanrewaju, 2022).
4. Ensuring market confidence: the long-term market stability is enabled by the commitment to legally enforceable and transparent procurement of renewable energy by RPS and REOs. The regulations force the energy suppliers and the utilities companies to grapple with the set targets within a stipulated period, hence giving stability and predictability to the investors and developers of renewable energy. The cause of this certainty promotes investment in renewable energy initiatives, whereby the shift to a sustainable and resilient energy system is achieved (Soltani et al., 2019).
5. Strengthening energy resilience and security: the carbon retribution, the resilience, and the security are the three key benefits that the RPS and the REOs are boosting through the mechanisms of cutting the fossil fuel dependence and diversifying the energy sources. Unlike with fossil fuels that are normally obtainable, locally situated, and subject to the chances of disruptions or high and low prices, renewable energy sources such as wind, solar, and hydropower are consistently replenishable and hardly ever disrupted. In such a manner, system resilience and reliability in energy are enhanced and the system will be more secure to external shocks, which will result in increased energy security. Finally, both Renewable Portfolio Standards (RPS) and Renewable Energy Obligations (REO) in terms of the investment into renewable energy activities, development of the sector as a

succession of the result, the demand on the market in the future, and stability, independence, and safety of the renewable energy are ensured. Green energy mandates certain utility practices, leading to the low-carbon and sustainable energy future transition polices (Lu et al., 2020).

6.4.3.2 Green Public Procurement (GPP)

The government usually settles for GPP policy acts as a strategic measure to stimulate the establishment of sustainability criteria while documenting products and services.

1. Promoting the demand for renewable energy sources: GPP gives high value to energy efficiency of products and devices as the best way to save energy for government agencies and public services. Governments are capable of creating a remarkable demand for renewable energy technologies through the use of their huge buying power. The augmented demand for clean energy technologies and renewable energy initiatives escalates the innovation process and reduces the costs, contributing to the market growth (Relva et al., 2021).
2. Promoting innovation: government procurement policies (GPP) encourage manufacturers and suppliers to provide innovative energy-saving and renewable energy technologies that meet the sustainability requirements specified by the GPP. This provides incentives for the development of such clean technologies, processes, and products through competition among the suppliers. Therefore, GPP policies accelerate the process of transitioning to a low-pollution economy and spark technological innovation (Guzović et al., 2022).
3. Sustainable market development: GPP policies focus on the creation of sustainable markets with the help of the emissions criteria on energy efficiency and the demand of renewable energy. The effect of this phenomenon indicates that firms in the renewable energy business have an opportunity to expand their market share, attract investors, and expand their business operations. Sustainable markets contribute to economic growth, job creation, and innovation within the renewable energy sector as well as their environmental benefits (Elavarasan et al., 2020).
4. GPP approach through governmental leadership towards sustainability and climate action demonstrates the synchronization of procurement practices with globally applicable environmental and energy targets. Governments show their commitment to the goals of carbon emission reduction, natural resource preservation, and renewable energy acceptance by including sustainability demands in their procurement decisions. This demonstrates the possibility of other organizations following the same path and encourages private sectors to subscribe to similar sustainable procurement procedures (Liu, 2014).
5. GPP measures help to cut carbon footprint by choosing energy-efficient products and using renewable energy for public sector operations. To reduce greenhouse gas emissions and reliance on fossil fuels, government agencies and institutions could install energy-saving technologies and switch to a cleaner energy source. Through this, we achieve the reduction of greenhouse gas emissions while simultaneously improving air quality and public health (Esmaeili Shayan et al., 2022).

In conclusion, Green Public Procurement (GPP) policies have a key role in supporting innovations, creating green markets, providing the leadership in the sustainability field in the government, increasing the demand for the renewable energy, and reducing the carbon footprint of the public sector operations. With sustainability criteria included as part of their procurement procedures, governments can speed up low-carbon and more sustainable future transition through the utilization of their purchasing power (Chu et al., 2024).

6.4.3.3 Research and Development (R&D) Funding

Governmental funding for renewable energy-intensive R&D is not only indispensable but a major form of the help from the government in the matter of R&D progress of renewable energy. The following is a comprehensive analysis of the importance of R&D funding:

1. Accelerating the progress of technology: governments sustain the projects of R&D, so the process of development of renewable energy sources speed up through government intervention. Such allocations allow technicians and scientists to deal with groundbreaking ideas, concepts, theories, and methods, as well as to construct useful solutions and original products in the field of renewable technology by both theoretical and applied research. One of the roles of government is to facilitate transformations in a number of areas, materials science, energy conversion and storage, system integration technologies, and research and development effort through government funding. Through this facilitation there is the continued tuning and optimizing of renewable energy technology (Mlilo et al., 2021).
2. Enabling the mitigation of cost decreases: these funds will be critical to the decrease of the cost of renewable energy technologies through injection of innovation. Through technological advancement, smoothing the process, and utilizing economies of scale, funding for research and development subsidizes reproduction, installation, and operation costs associated with the systems generating renewable energy (Chu et al., 2024). This gives space to renewable forms of energy to supplement the existing sources and, in turn, attracts more of the energy clients and businesses towards its affordability and accessibility. The research and development finance plays the role of helping deployment and dissemination of renewable energy technologies by providing pricing reduction (Lyden et al., 2018).
3. Contributing to technology demonstration initiatives: the allocation of R&D funding is being done with a focus on demonstration projects that will advance the state of the art and show what new renewable technologies can really do when they are reliable, well proven, and applicable. These programs provide a true effort of unpacking the environmental, economic, and technical levels of renewable energy introduction, and so whether the risk will likely be reduced or the stakeholder's confidence increased is an optimistic outcome. Through example situations, technology demonstration initiatives enable the evaluation of the innovative ideas and methods so that successful renewable energy integrations can be expanded on an industry-wide scale (Caineng et al., 2022).
4. Cultivating collaborative alliances: government departments, academic institutions, and industry benefit from the formation of R&D partnerships due to the profit from research funding that leads to the transferring of technologies, multidisciplinary associations, and knowledge exchange among them. Collaborative research projects and development efforts spur inventions in the renewable energy sector and make easier passage through some horrific challenges encountered in this industry by tapping into complementary human and other resources pooled together. These collaborative partnerships act as a mechanism for converting scientific discoveries into actual innovations; this trend further led to the development and mainstreaming of renewable technology giants ready to compete in the marketplace (Jurasz et al., 2020).

Improving international competitiveness: R&D budgets rising lead to the upgrade of renewable energy technology and this makes them competitive on the world market.

With the help of government funding, one can hire the best talent in the area of conducting research, developing new projects, and also increasing the levels of innovation leading to the industrial stand out in the energy sector and taking the advantage of the growing market opportunities. A dynamic energy business ecosystem comes into effect with the afforded R&D funding that allows renewable energy companies, businesses, and entrepreneurs to start, grow, and prosper. The implementation of governmental R&D funding encourages the national fidelity towards the creation and introduction of renewable energy technologies. Such action is not only able to contribute to sustainable economics but also becomes the sign of responsible managerial environmental policy (Soltani et al., 2019).

In simple terms, R&D funding helps to build up technology progress, to lower equipment costs, to allow the meeting of technologies demonstrations, to push collaborative interrelations, and enhance the competitiveness in international markets. Governments can speed up the translation of the renewable energy paradigm into a sustainable, reliable, and environment-fresh future; raise the output as well as the effectiveness of renewable energy technologies; and eradicate technical problems through research and development investments (Jurasz et al., 2020).

6.4.4 Public Awareness Campaigns and Education Programs

From effective educational courses to public campaigns, strategies are able to raise renewable energy consumption and promote a culture of sustainability the same time. A comprehensive analysis of their significance follows:

1. Enhancing awareness: education programs and public awareness drives are undeniably incessant in promoting knowledge of the blessings of renewable energy and also in the need for transitioning to sustainable energy sources (Hou et al., 2023). Such programs aim to publicize information about the environmental, economic, and social value of renewable energy sources to business communities, households, and society generally. These benefits are econometrics, good air, and inhibiting the emission of greenhouse gasses. Consumer awareness campaigns that focus the value of renewable energy on society are among the major drivers of public appreciation and popularity of the efforts in clean energy.
2. Advocates for energy efficiency: educational programs and community awareness activities that prioritize the significance of efficiency measures include renewable energy promotion as well. Through the supply of energy-saving techniques and policies, within their respective socio-economic spheres, these initiatives furnish people and businesses with tools and strategies to reduce their carbon emissions and drop off energy consumption. Alongside green measures like the installations of renewable energy, efficiency measures have a triple purpose: they help to get the best return on investment in clean energy and to generally reduce the consumption of energy (Chapman et al., 2021).
3. Promoting behavior change: education strategy and public informational activities result in encouraging the organizations and individuals to take part in sustainable energy by integrating sustainable energy into their day-to-day lives. The methods can be practical, for example materials or guidance including information with respect to saving energy, waterlessness, and environmentally responsible choice-making. Campaign advocacy urges everybody to become actively involved as we make all necessary changes from the use of non-renewable energy sources to energy efficient appliances, riding public transportation, and installing rooftop solar panels (Lyden et al., 2018).
4. Facilitating informed decision-making: this can be done through the use of public awareness campaigns, public education initiatives, and the creation of an informed citizenry. This will make it possible for people to have the information and means to make responsible choices in terms of energy consumption, wise investments, and endorsement of policies. By way of the spreading of the knowledge about the renewable energy technologies, incentives, and regulations, these undertakings allow consumers to realize the purpose of accepting green energy solutions, for instance installation of solar panels or purchase of the electric vehicles. Sustainable energy technologies are fueled and renewable energy's credibility is developed by providing specific guidance, as in when making selective energy decisions (Ahmad et al., 2022).

5. Facilitating public participation: the programs involving public education and public awareness campaigns help to involve public participation and engagement in renewable energy projects by cultivating platforms that support interactivity, networks, and proactiveness. Such ventures create platforms where all the aforementioned players interact: policymakers—businesses, communities, and individuals—to come up with renewable energy planning and implementation. For this reason, they also consider policy issues regarding the production of green energy. By drawing attention to the principle of mutual interaction through asking for advice, comments, and direct participation, mass events shape the feeling of support and belonging in terms of transition to a new kind of energy. Consequently, there would be the growth of an environment where different policies and community-driven efforts that alleviate renewable energy would be implemented (Huang et al., 2021).
6. Establishing a conducive environment to facilitate policy implementation: transmitting information, constructive behavioral changes, energy conservation, the creation of participation channels, and awareness-building campaigns one can build a friendly environment for the successful gathering and promotion of such energy transition policies and strategies. The strength of public opinion and the energy of advocacy, which are the consequences of awareness campaigns, will encourage policymakers to place renewable energy higher on their priority lists, pass laws that are geared toward advancing it, and set aside funding for programmers that promote renewable sources and that favor projects that are green friendly. The harmoniousness among the public perception as well as the policy goals serves to launch a process of this sort that promotes renewable energy policies' effectiveness as well as longevity and thereby accelerates the move to an appropriate energy landscape (Gatto, 2022).

Basically, for the support for the use of renewable power, the improvement of behavior, the facilitation of well-informed decision-making, and the promotion of public participation as well as the proliferation to enable a conducive environment, immense learning is needed. Awareness campaigns are fundamental in the transition to a sustainable energy future as they provide the information and spark the passion to educate groups of people and the individuals locally.

6.5 CASE STUDY

Low-carbon energy systems require a range of changes within society; however, especially when looking at developing countries, they are in dire need of financial and technical assistance. Explaining the reasons for less development, Adhikari et al. (2018) note that there is no enabling infrastructure or resources available in these areas; they have to depend on outside support (Adhikari et al., 2018). Advanced solutions involving smart technologies and integration of the systems into the energy network are crucial to achieving the goals of ergonomic and sustainable energy networks. This shows that suitable regulation guidelines help in dealing with investment risks and uncertainty, thereby promoting the right long-term investment on low-carbon technologies.

In the study of Streimikiene and Kyriakopoulos (2023), energy poverty intensifies due to the shift towards low-carbon energy, and it is not a straightforward existence. As stated in their literature review, energy poverty raises considerable controversy in both the regional and global contexts (Streimikiene and Kyriakopoulos, 2023). Furthermore, they consider the impact of the COVID-19 pandemic and Russia's aggression against Ukraine on prices in the energy sector and the matter of energy poverty. Therefore, the paper recommends future research to closely focus its investigation to measures to alleviate energy poverty, especially in the residential segment. Overall, Kamali Saraji and Streimikiene (2023) have analyzed challenges in the low-carbon energy transition, elaborated on the basis of the PICOC framework and the PSALSAR protocol, which have been used in the analysis of 123 articles. They are aware of 17 risks in social; economic; environmental; technical;

and institutional contexts and realize that the global accords on climate change have prompted studies in this area (Saraji and Streimikiene, 2023).

Energy transitions and low-carbon development have been examined by Urban (2014) in relation to developing regions such as China and India. In Urban's assessment of the power sector in China and rural Indian household energy requirements by way of energy modeling, the author affirms that such transitions are possible and that they will meaningfully reduce emissions and inform climate change solutions (Urban, 2014). Based on the research carried out by Simsek et al. (2020), the authors provide an energy and environmental model of Chile, estimating the energy demand, supply, and carbon emissions by the year 2030. They asserted that they can prove that enormous energy saving of course comes with impressive emission cuts, which underlined the importance of energy efficiency and renewable energy policies. Relating to the previous point, the map scenarios suggest over 80% of renewable electricity generation by 2030 (Simsek, 2020).

Bugaje et al. (2022) call for the reconsideration of natural gas (NG) in the transition process. As such, they argue that NG can be a centerpiece in the near- to long-term future energy mix, especially considering the political-economic realities within certain contexts and technology development that sees the transition to NG as a "just" transition (Bugaje et al., 2022). Regarding the subject, Murshed et al. (2021) argue for the necessity of a change towards clean energy as a way of mitigating the carbon footprint while stressing the need to shift to non-fossil fuels and hydroelectricity as measures that Bangladesh could undertake as its contribution to global effort in meeting the goals of the Paris Agreement and the United Nations Sustainable Development Goals (Murshed et al., 2021).

Clulow and Reiner (2022) discuss how climate change affects the geopolitical landscape and how different countries plan for the transition to low-carbon energy sources. In their study, it was seen that economic development modulates the impact of democracy as developed countries prefer nuclear energy while the developing countries prefer solar, wind, and hydro energy (Clulow and Reiner, 2022). Regarding reducing energy poverty in the context of climate change mitigation policies, Streimikiene et al. (2021) analyze the outcomes of comparing the impacts of climate change policies on energy poverty in Lithuania and Greece providing a plan of the just transition to low-carbon energy with a focus on energy-vulnerable groups (Streimikiene et al., 2021).

According to Sovacool and Griffiths (2020), an analysis of 28 countries illustrated how cultural factors can obstruct the low-carbon transition predications. They recommend that there is an imperative need to consider culture in bringing about low-carbon technological systems within policy and practice (Sovacool and Griffiths, 2020). In this contextual focus, Kim et al. (2022) examine the distribution of financial budgets toward low-carbon energy transition from emerging to developed nations. They observe that new generation countries can technically minimize greenhouse gas emissions and satisfy energy requirement through shifting energy mix to nuclear and natural gas, although this is risky (Kim et al., 2022).

Energy transitions are important in improving energy supply and demands as well as reducing climate impact on a country's or region's energy subsystem. East Asian economy-specific factors offer a rationale for addressing differing circumstances and commonalities in transitioning energy systems. Based on the work of Bo Li et al. (2020), the region has exhibited some accomplishment towards the renewable energy development and energy efficiency. Nonetheless, the position on coal has rather remained a problem due to major retro infrastructure that already exists in the energy industry even with developments on nuclear energy. There is enormous potential for renewable energy for any part of the region due to its large resource endowment, but it is not effectively supplying the energy needed. Hence, regional energy connectivity is painted as an essential valuable asset that could even more strictly decarbonize the energy sector in East Asia than could the individual nation states (B. Li et al., 2020).

The system change toward low-carbon energy is also associated with industrial decarburization and establishing circular waste bio economy pillars especially in agricultural belts such as Greece. Geographical considerations are imperative for analyzing the probable dynamics of a low-carbon economy platform, as the special focus is placed on decentralized and location-specific options and factors such as location, topography, and spatial heterogeneity (Nikologianni, 2018). In the context of energy substitution dynamics, Agyeman and Lin (2022) examine the pattern of seven selected North African countries and show that technological advance can help decouple fossil energy from economic growth. A technological shift to low-carbon energy sources is clearly reinforced by their discoveries based on Agyeman and Lin (2022) revealing the need to have biased technological change towards cleaner options to advance energy transitions (Agyeman and Lin, 2022). Del Granado et al. (2020) examine scenarios of how the EU can decouple from carbon by the year 2050 by using accessible technologies. They stress the importance of planning energy cooperation levels and decentralization to attain the objective of clean, secure, and efficient energy solutions (Del Granado et al., 2020).

According to Kabeyi and Olanrewaju (2022), there is a structural plan for a green transition through a framework of sustainable indicators intended to meet the objectives of the Paris Agreement. Their work supports the use of technology to increase efficiency in energy systems, reduce and optimize electricity generation, and incorporate renewable resources into the global electricity system to facilitate sustainability transformation. Furthermore, they state that there is an urgent need to establish polices and invest in more infrastructure within the smart grid network to accommodate large consumers of renewables and strengthen grid stability (Kabeyi and Olanrewaju, 2022).

Relva et al. (2021) focus on the difficulties that the developing countries have experienced while attempting to transform their electricity sectors to low-carbon emitting systems. They unearth challenges that include data collection constraints and political instabilities, thus suggesting strategies of energy planning. They highlight the approaches towards these challenges using global frameworks such as the Agenda 21 and Agenda 2030 and go further to call for support from the developed countries and bodies to enable provision of a solution in regard to the assorted problems that surround the developing countries (Relva et al., 2021).

Chapman et al. (2021), studied the worldwide effects of low-carbon energy transition policies for social equity, revealing that the deployment of renewable sources of energy is closely related to enhanced SE scores. Nonetheless, they observe that the advantages of change can be more or less impressive depending on the prior conditions, including the location, income level, and other complementary measures, especially in developing states (Chapman et al., 2021). Fragkos et al. (2021) quantitatively study the low emission scenario for several economies including Australia, Brazil, China, the and EU-28 gross domestic products and find out that low emission scenario towards the world targets are achievable. They highlight the key priority areas of improving the uptake of renewable energy, increasing energy efficiency, and expanding the share of electricity in final energy consumption to achieve emissions reductions while taking into account their country-specific potential and specific advantages (Fragkos et al., 2021).

Increasing renewable share and decarburization efforts: currently, the energy mix in Europe covers various types of energy generations like fossil fuels and renewable energy sources. As suggested by the fresh statistics of the European Environment Agency (EEA), the EU has significantly advanced in its growing use of renewable energy within the EU's energy consumption balance. According to the data obtained for the year 2022, it could be noted that the share of renewable energy sources in the overall total energy consumption in the EU was estimated to be roughly 22%, and the shares of wind, solar, and hydroelectric power sources are slightly dominant.

Figure 6.1 shows the balance of energy from 1990–2022 taken from the EU energy and transport indicator to show energy production, importation, transformation, and consumption in million tons of oil equivalent (Mtoe). This was mainly due to a steady state in domestic energy production in the EU with a slight decrease in productions after the year 2008. On the other hand, imports rose from $52 billion in 1990 to $135 billion in mid-2000 and thereafter remained almost stagnant, implying that continents have increased dependence on imported energy. Import figures are still higher than export figures, which therefore continues to place the EU in the position of a net importer of energy. These include GA and TES, which are constituted by domestic production of energy plus import of

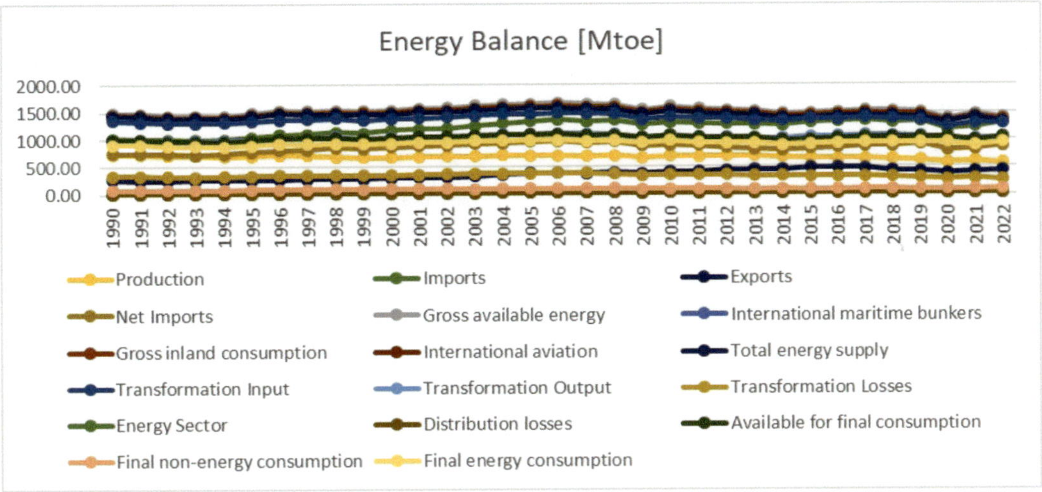

FIGURE 6.1 Energy balance [Mtoe].

energy, and both figures show a steady rise up to the mid-2000s and then remain relatively steady to show the total available energy in the region.

The transformation input and output are still in continual progress for the energy years while the transformation losses are still in a steady condition, signifying that energy conversion practices are facing high efficiency in the EU. The final electricity consumption pattern indicates the constant use of energy through different consumption sectors and FEC rises and falls slightly and then starts rising again until the mid-2000s before becoming stable. They consume only a small proportion of the energy generated through distribution losses implying efficiency in the distribution network across the member states. Moreover, there are some special categories, such as international maritime bunkers and international aviation, that evidence steady energy consumption that demonstrates operating requirements' steadiness. In general, the given chart explains a fairly good and uninterrupted energy system in the EU and balanced and stable trends in production, import, and consumption to provide energy security in the region.

Figure 6.2 shows that the collective energy production in EU countries on an annual basis from 1990 to 2022 with a measurement in million tons of oil equivalent (Mtoe). This comparison of the two datasets shows that there is a common trend of reducing energy production over the analyzed period of 32 years. Energy production in the EU increased during the early 1990s and even reached 800 odd Mtoe during that time period. But there were increases and minima throughout the decade, with production reducing to approximately 700 Mtoe by the end of the nineties.

CO_2 emissions in the EU and energy production during the highs and lows of the 1990s hovered between 675 Mtoe and 725 Mtoe until the beginning of the 2000s when it slightly rose before falling again till 2010 as depicted later. Nonetheless, after the year 2010, the chart shows an increased trend and volatility of reddish mesas and coherent demise, leading to 2022's production drop to nearly 575 Mtoe, which is the lowest figure mentioned in the observed period. This progressive trend observed over the last decade challenges believed notions held in the EU energy industry with regard to ability to sustain the energy production frontier due to resource scarcities, change over to the renewable energy types, and/or policy changes that sought sustainable usage and lower CO_2 emissions. The results of the analysis speak to the importance of a reassessment of the energy policies and investment for reasons concerning the declining trend in domestic energy production observed in the EU.

Based on Figure 6.3, the bar chart shows the energy consumption from renewables in thousand tons of oil-equivalent selected European countries and some of the neighbors of European countries

FIGURE 6.2 Total production over the years.

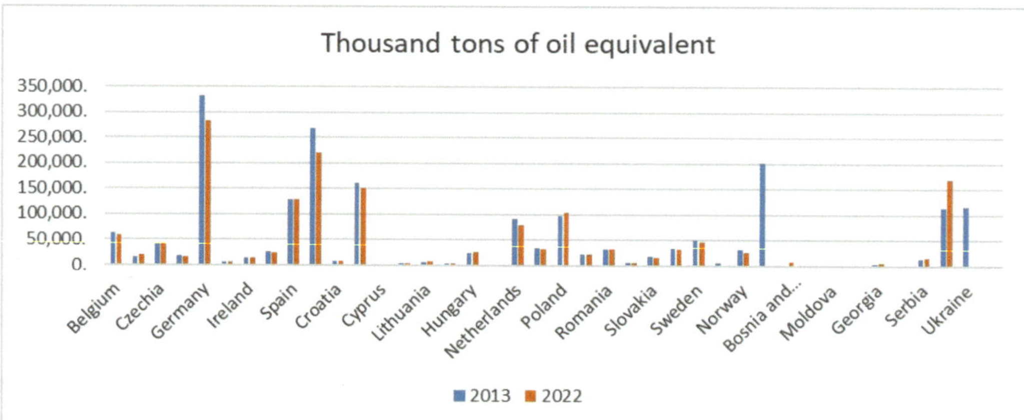

FIGURE 6.3 Thousand tons of oil equivalent.

such Turkey and Ukraine for the years 2013 and 2022. It also shows corresponding changes in the aforementioned measures with a lot of countries having recorded higher consumption in renewable energy during the aforementioned period. Germany, along with the other European countries, consumed more in 2013 with over 300,000,000 tons of oil equivalent but reduced slightly by 2022. Italy also stands out and Spain and the United Kingdom also rank fairly equally high with a considerable amount of renewable energy consumption. Turkey and Ukraine share the most notable improvements in contrast to the rest of the countries while Turkey shows considerable growth in 2022.

This is a trend that can be observed in the chart representing the level of renewable energy consumption, which has been on the rise due to better planning and initiatives that are being taken to build more green energy infrastructure. Some of the countries that had high levels of consumption earlier including Germany, the United Kingdom, and Norway have slightly reduced their consumption levels meaning that there is rational and effective consumption when it comes to energy use

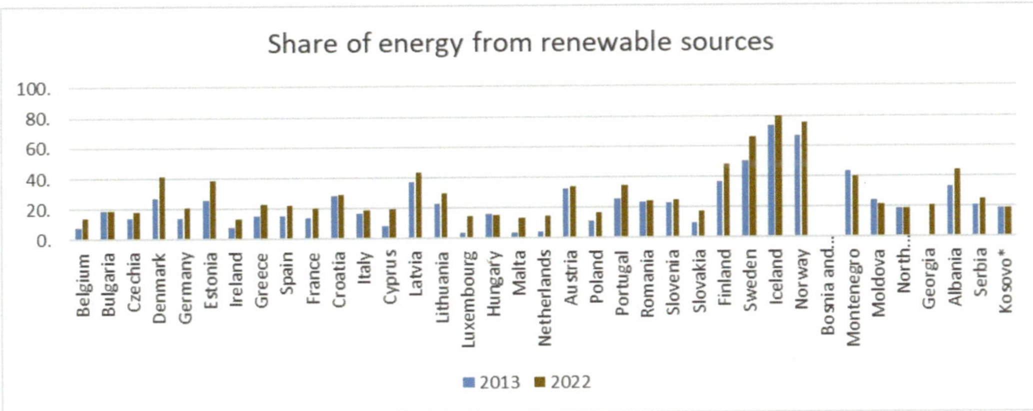

FIGURE 6.4 Share of energy from renewable sources.

or perhaps there are certain changes in their policies. On the other hand, it becomes evident that a number of countries such as Turkey and Ukraine demonstrate such appreciable increase due to the surge in the development of the renewable energy market. This emerging trend reflects a more comprehensive thrust towards the use of sustainable energy and renewable power sources borne out of policy reforms and environmental imperatives towards the development of efficient means of power generation that will check on depletion of fossil resources for energy production and improve energy security in the region.

According to Figure 6.4, the following bar chart presents renewable electricity, expressing as a percentage the share of energy produced from renewable sources in selected European countries and some neighboring countries including Norway and Iceland in both 2013 and 2022. The results suggest that the overall trend throughout most countries is that they increasingly turned to renewable energy sources in the period leading to 2012. The notable performers in renewable energy shares are Iceland, Norway, and Sweden where the renewable power bar has already crossed the 50% mark with Iceland and Norway set to cross the 80% line in the year 2022. These countries have shown a constantly high and long-term interest in renewable energy, proving the efficiency and seriousness of their policies and investments made into the renewable power infrastructure.

Some of these countries show considerable progress rates of renewable electricity share in the range between 2013 and 2022. ESRI showed that, among the EU countries, Denmark, Estonia, and Finland are featuring a marked rise in the share for renewable energy sources. Portugal and Austria also present a significant increase having a relatively high numbers for the share of renewable energy by the year 2022. However, the analyses of the situation by the country still reveals data below some certain shares, which signifies that there are still many problems and further developments are still needed in some countries. This is echoed in the chart that outlines transformations towards cleaner producing processes and focus on environmentally friendly sources of energy, and the success of the countries involved is informed by polices that support the changes across the world.

Figure 6.5 shows that the bar chart gathers the data of the partial contribution of renewable energy to gross final energy consumption for selected European countries and a few neighboring countries including Norway and Iceland in the years 2004, 2013, and a forecasted year, 2022. Most of the countries have displayed an upward trend in mean rank, implying enhanced efforts in the deployment of RE sources in the examined years. On this front, Iceland and Norway remain the two most well-regarded leaders in using renewable energy sources, with figures standing at 73% in 2022. Likewise, Sweden attained a better position of more than 60% in the year 2022, which makes this country one of the best countries. These countries have a continuously high share of actual renewables, which points to high current and stable investment in renewable energy.

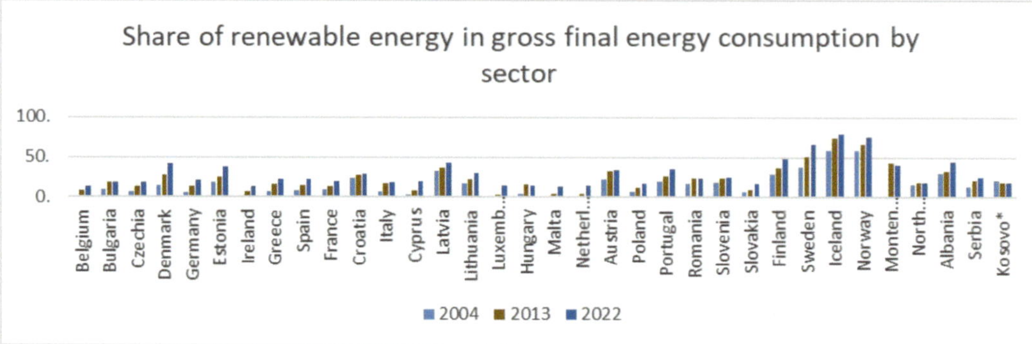

FIGURE 6.5 Share of renewable energy in gross final energy consumption by sector.

It is true that numerous nations have disclosed significant shifts, most importantly increases in the share of renewable energy sources in total final consumption from the level of 2004 to the level of 2022. Denmark showed a remarkable rate of change, while Estonia and Finland also moved significantly, showing good outcomes pointing to the successful application of various policies and investments in place for RE technologies. Both Portugal and Austria also show substantial growth; both countries even reached a high percentage of renewable consumed energy by 2022. This overarching trend of the total chart points out global tendency towards utilizing cleaner energy resources with different levels of effectiveness and achievement among the countries. The data shows that there is an increased willingness by people and companies to use green energy solutions, not only due to environmental impact but also as a result of existing policies.

6.6 CONCLUSION

When looking at the process of achieving global low-carbon energy systems, the centrality of renewable energy cannot be farther from doubt since they provide solutions to global environmental challenges and a way of unleashing great societal and economic benefits. This evolution has its roots deep down in the history of mankind, having evolved from simple wind and water power techniques to the modern sources of power such as the solar panels and wind turbines. The 20th century, characterized by oil shocks, triggered faster development of renewables and research, paving the way for a plethora of renewable energy sources today. This is not a change simply brought about by technology but also by an appreciation of the many faces of value of renewable power first and foremost, the avoidance of global warming and the stabilization of energy supply. Renewable energy sources embody significant benefits such as high accessibility, zero pollution, and declining costs associated with popularity and improvements in technology.

Yet key hurdles like intermittency, system constraints, and regulatory risks stand as daunting obstacles towards mass use. Solving these issues calls for intervention and cooperation from governments, companies, and global communities to promulgate favorable policies, reinforce the electric grid network, and offer fiscal incentives. It is therefore imperative to overcome these barriers for increased use of renewable energy and attaining the 21 Sustainable Development Goals. Internationally, the aspects of renewable electricity include reduced costs of deployment, polices encouraging the utilization of renewable resources, and increased concern for the environment. They are central to helping to establish change towards the use of sustainable energy systems that can withstand competition and fluctuating fuel costs.

Furthermore, spending in renewable energy fosters economic development and employment while enhancing energy availability and utilization, predominantly in the developing world. This dual benefit underlines advantages of renewable energy in the process of creating a new social reality where all stakeholders can benefit and more resources are devoted to ensuring environmental equilibrium on the planet.

Technological innovation also has a strong influence on enhancing renewable energy systems and bringing down the cost of the systems. Nevertheless, more research and development programs are needed all the time to make renewable sources cheaper, more dependable, and easily integrated into the existing energy systems. Hence, financial assistance from the government through the funding programs is crucial in keeping such innovations going and making the renewable energy technologies more capable of providing the required global energy requirements in a more sustainable manner.

Therefore, the changeover to low-carbon systems via renewable energy is perhaps the greatest task facing humanity in the future, socially, economically, and environmentally. Every society has been able to overcome barriers and embrace opportunities towards the achievement of this course through technological developments, creation of conducive polices, and enhanced cooperation with the global community. Adopting renewables reduces climate change effects in addition to relaying and building up the foundation of economic growth and prosperity of the power sector and more so for the future generations. Hence, advancing the energy transition in the form of low-carbon energy systems represents not only a necessity but also a viable plan for a better and stronger energy world.

Thus, detailed comparative overviews of renewable energy policies in various regions may develop useful recommendations for future research on effective ways of improving the pace of renewable energy penetration. Also, studying the social and economic effects of renewable energy transition efforts for local societies and industries will be essential for giving equal mileage on seven continents for sustainable change in the energy sector.

REFERENCES

Adhikari, S., Baral, H., & Nitschke, C. (2018). Adaptation to climate change in Panchase Mountain ecological regions of Nepal. *Environments*, *5*(3), 42.

Agyeman, S. D., & Lin, B. (2022). Nonrenewable and renewable energy substitution, and low–carbon energy transition: Evidence from North African countries. *Renewable Energy*, *194*, 378–395.

Ahmad, T., Madonski, R., Zhang, D., Huang, C., & Mujeeb, A. (2022). Data-driven probabilistic machine learning in sustainable smart energy/smart energy systems: Key developments, challenges, and future research opportunities in the context of smart grid paradigm. *Renewable and Sustainable Energy Reviews*, *160*, 112128.

Albrecht, W., & Steinrücke, M. (2020). Continuous-time scheduling of production, distribution and sales in photovoltaic supply chains with declining prices. *Flexible Services and Manufacturing Journal*, *32*, 629–667.

Al-Falahi, M. D., Jayasinghe, S., & Enshaei, H. (2017). A review on recent size optimization methodologies for standalone solar and wind hybrid renewable energy system. *Energy Conversion and Management*, *143*, 252–274.

Ayaz, M., Namazi, M., ud Din, M. A., Ershath, M. M., & Mansour, A. (2022). Sustainable seawater desalination: Current status, environmental implications and future expectations. *Desalination*, *540*, 116022.

Azarpour, A., Mohammadzadeh, O., Rezaei, N., & Zendehboudi, S. (2022). Current status and future prospects of renewable and sustainable energy in North America: Progress and challenges. *Energy Conversion and Management*, *269*, 115945.

Bandaru, S. H., Becerra, V., Khanna, S., Espargilliere, H., Torres Sevilla, L., Radulovic, J., . . . Khusainov, R. (2021). A general framework for multi-criteria based feasibility studies for solar energy projects: Application to a real-world solar farm. *Energies*, *14*(8), 2204.

Beccali, M., Cellura, M., & Mistretta, M. (2003). Decision-making in energy planning. Application of the Electre method at regional level for the diffusion of renewable energy technology. *Renewable Energy*, *28*(13), 2063–2087.

Black, R., Busby, J., Dabelko, G. D., De Coning, C., Maalim, H., McAllister, C., . . . Barnhoorn, A. (2022). *Environment of Peace: Security in a New Era of Risk*. https://researchprofiles.canberra.edu.au/en/publications/environment-of-peace-security-in-a-new-era-of-risk

Bugaje, A.-A. B., Dioha, M. O., Abraham-Dukuma, M. C., & Wakil, M. (2022). Rethinking the position of natural gas in a low-carbon energy transition. *Energy Research & Social Science*, *90*, 102604.

Caineng, Z., Feng, M., Songqi, P., Minjie, L., Zhang, G., Xiong, B., . . . Zhi, Y. (2022). Earth energy evolution, human development and carbon neutral strategy. *Petroleum Exploration and Development*, *49*(2), 468–488.

Cantoni, R., Caprotti, F., & de Groot, J. (2022). Solar energy at the peri-urban frontier: An energy justice study of urban peripheries from Burkina Faso and South Africa. *Energy Research & Social Science*, *94*, 102884.

Castillo-Calzadilla, T., Cuesta, M., Olivares-Rodriguez, C., Macarulla, A., Legarda, J., & Borges, C. (2022). Is it feasible a massive deployment of low voltage direct current microgrids renewable-based? A technical and social sight. *Renewable and Sustainable Energy Reviews*, *161*, 112198.

Chapman, A., Shigetomi, Y., Ohno, H., McLellan, B., & Shinozaki, A. (2021). Evaluating the global impact of low-carbon energy transitions on social equity. *Environmental Innovation and Societal Transitions*, *40*, 332–347.

Chu, H., Huang, Z., Zhang, Z., Yan, X., Qiu, B., & Xu, N. (2024). Integration of carbon emission reduction policies and technologies: Research progress on carbon capture, utilization and storage technologies. *Separation and Purification Technology*, 127153.

Clemente, D., Rosa-Santos, P., & Taveira-Pinto, F. (2021). On the potential synergies and applications of wave energy converters: A review. *Renewable and Sustainable Energy Reviews*, *135*, 110162.

Clulow, Z., & Reiner, D. M. (2022). Democracy, economic development and low-carbon energy: When and why does democratization promote energy transition? *Sustainability*, *14*(20), 13213.

Davies, C., Chen, W. Y., Sanesi, G., & Lafortezza, R. (2021). The European Union roadmap for implementing nature-based solutions: A review. *Environmental Science & Policy*, *121*, 49–67.

Del Granado, P. C., Resch, G., Holz, F., Welisch, M., Geipel, J., Hartner, M., . . . Bernath, C. (2020). Energy transition pathways to a low-carbon Europe in 2050. *Economics of Energy & Environmental Policy*, *9*(1), 121–136.

Diallo, S., Ouoba, Y., & Gatete, C. (2024). Effect of fossil fuel subsidies on renewable energy transition in sub-Saharan African countries. In *Energy Regulation in Africa: Dynamics, Challenges, and Opportunities* (pp. 429–452). Springer.

Dufo-López, R., Cristóbal-Monreal, I. R., & Yusta, J. M. (2016). Optimisation of PV-wind-diesel-battery stand-alone systems to minimise cost and maximise human development index and job creation. *Renewable Energy*, *94*, 280–293.

Elavarasan, R. M., Shafiullah, G., Padmanaban, S., Kumar, N. M., Annam, A., Vetrichelvan, A. M., . . . Holm-Nielsen, J. B. (2020). A comprehensive review on renewable energy development, challenges, and policies of leading Indian states with an international perspective. *IEEE Access*, *8*, 74432–74457.

Enevoldsen, P., & Jacobson, M. Z. (2021). Data investigation of installed and output power densities of onshore and offshore wind turbines worldwide. *Energy for Sustainable Development*, *60*, 40–51.

Esmaeili Shayan, M., Hayati, M., Najafi, G., & Esmaeili Shayan, S. (2022). The strategy of energy democracy and sustainable development: Policymakers and instruments. *Iranica Journal of Energy & Environment*, *13*(2), 185–201.

Fragkos, P., van Soest, H. L., Schaeffer, R., Reedman, L., Köberle, A. C., Macaluso, N., . . . Qimin, C. (2021). Energy system transitions and low-carbon pathways in Australia, Brazil, Canada, China, EU-28, India, Indonesia, Japan, Republic of Korea, Russia and the United States. *Energy*, *216*, 119385.

Gatto, A. (2022). The energy futures we want: A research and policy agenda for energy transitions. *Energy Research & Social Science*, *89*, 102639.

Gebreslassie, M. G. (2021). Development and manufacturing of solar and wind energy technologies in Ethiopia: Challenges and policy implications. *Renewable Energy*, *168*, 107–118.

Glasson, J., Durning, B., Welch, K., & Olorundami, T. (2022). The local socio-economic impacts of offshore wind farms. *Environmental Impact Assessment Review*, *95*, 106783.

Gungah, A., Emodi, N. V., & Dioha, M. O. (2019). Improving Nigeria's renewable energy policy design: A case study approach. *Energy Policy*, *130*, 89–100.

Guzović, Z., Duic, N., Piacentino, A., Markovska, N., Mathiesen, B. V., & Lund, H. (2022). Recent advances in methods, policies and technologies at sustainable energy systems development. *Energy*, *245*, 123276.

Hassan, Q., Viktor, P., Al-Musawi, T. J., Ali, B. M., Algburi, S., Alzoubi, H. M., . . . Jaszczur, M. (2024). The renewable energy role in the global energy transformations. *Renewable Energy Focus*, *48*, 100545.

Hilpert, S., Günther, S., Kaldemeyer, C., Krien, U., Pleßmann, G., Wiese, F., & Wingenbach, C. (2017). *Addressing Energy System Modelling Challenges: The Contribution of the Open Energy Modelling Framework (OEMOF)*. https://orbit.dtu.dk/en/publications/addressing-energy-system-modelling-challenges-the-contribution-of

Hoang, A. T., & Nguyen, X. P. (2021). Integrating renewable sources into energy system for smart city as a sagacious strategy towards clean and sustainable process. *Journal of Cleaner Production*, *305*, 127161.

Hou, Z.-M., Xiong, Y., Luo, J.-S., Fang, Y.-L., Haris, M., Chen, Q.-J., . . . Huang, L.-C. (2023). International experience of carbon neutrality and prospects of key technologies: Lessons for China. *Petroleum Science*, *20*(2), 893–909.

Huang, M., He, W., Incecik, A., Cichon, A., Królczyk, G., & Li, Z. (2021). Renewable energy storage and sustainable design of hybrid energy powered ships: A case study. *Journal of Energy Storage*, *43*, 103266.

Javid, I., Chauhan, A., Thappa, S., Verma, S., Anand, Y., Sawhney, A., . . . Anand, S. (2021). Futuristic decentralized clean energy networks in view of inclusive-economic growth and sustainable society. *Journal of Cleaner Production*, *309*, 127304.

Jurasz, J., Canales, F., Kies, A., Guezgouz, M., & Beluco, A. (2020). A review on the complementarity of renewable energy sources: Concept, metrics, application and future research directions. *Solar Energy*, *195*, 703–724.

Kabeyi, M. J. B., & Olanrewaju, O. A. (2022). Sustainable energy transition for renewable and low carbon grid electricity generation and supply. *Frontiers in Energy Research*, *9*.

Kalair, A., Abas, N., Saleem, M. S., Kalair, A. R., & Khan, N. (2021). Role of energy storage systems in energy transition from fossil fuels to renewables. *Energy Storage*, *3*(1), e135.

Khare, V., & Bhuiyan, M. A. (2022). Tidal energy-path towards sustainable energy: A technical review. *Cleaner Energy Systems*, *3*, 100041.

Kim, Y. J., Soh, M., & Cho, S.-H. (2022). Identifying optimal financial budget distributions for the low-carbon energy transition between emerging and developed countries. *Applied Energy*, *326*, 119967.

Kumar, A., Sah, B., Singh, A. R., Deng, Y., He, X., Kumar, P., & Bansal, R. C. (2017). A review of multi criteria decision making (MCDM) towards sustainable renewable energy development. *Renewable and Sustainable Energy Reviews*, *69*, 596–609. https://doi.org/10.1016/j.rser.2016.11.191

Kuriqi, A., Pinheiro, A. N., Sordo-Ward, A., Bejarano, M. D., & Garrote, L. (2021). Ecological impacts of run-of-river hydropower plants—Current status and future prospects on the brink of energy transition. *Renewable and Sustainable Energy Reviews*, *142*, 110833.

Li, B., Nian, V., Shi, X., Li, H., & Boey, A. (2020). Perspectives of energy transitions in East and Southeast Asia. *Wiley Interdisciplinary Reviews: Energy and Environment*, *9*(1), e364.

Li, G., & Zhu, W. (2023). Tidal current energy harvesting technologies: A review of current status and life cycle assessment. *Renewable and Sustainable Energy Reviews*, *179*, 113269.

Liu, G. (2014). Development of a general sustainability indicator for renewable energy systems: A review. *Renewable and Sustainable Energy Reviews*, *31*, 611–621.

Lu, Y., Khan, Z. A., Alvarez-Alvarado, M. S., Zhang, Y., Huang, Z., & Imran, M. (2020). A critical review of sustainable energy policies for the promotion of renewable energy sources. *Sustainability*, *12*(12), 5078.

Lyden, A., Pepper, R., & Tuohy, P. G. (2018). A modelling tool selection process for planning of community scale energy systems including storage and demand side management. *Sustainable Cities and Society*, *39*, 674–688.

McPherson, M., Monroe, J., Jurasz, J., Rowe, A., Hendriks, R., Stanislaw, L., . . . Crownshaw, T. (2022). Open-source modelling infrastructure: Building decarbonization capacity in Canada. *Energy Strategy Reviews*, *44*, 100961.

Mirzania, P., Ford, A., Andrews, D., Ofori, G., & Maidment, G. (2019). The impact of policy changes: The opportunities of Community Renewable Energy projects in the UK and the barriers they face. *Energy Policy*, *129*, 1282–1296.

Mlilo, N., Brown, J., & Ahfock, T. (2021). Impact of intermittent renewable energy generation penetration on the power system networks–A review. *Technology and Economics of Smart Grids and Sustainable Energy*, *6*(1), 25.

Montoya, L. T. C., Lain, S., Issa, M., & Ilinca, A. (2021). Renewable energy systems. In *Hybrid Renewable Energy Systems and Microgrids* (pp. 103–177). Elsevier.

Mulder, P., Dalla Longa, F., & Straver, K. (2023). Energy poverty in the Netherlands at the national and local level: A multi-dimensional spatial analysis. *Energy Research & Social Science*, *96*, 102892.

Murshed, M., Ahmed, Z., Alam, M. S., Mahmood, H., Rehman, A., & Dagar, V. (2021). Reinvigorating the role of clean energy transition for achieving a low-carbon economy: Evidence from Bangladesh. *Environmental Science and Pollution Research*, *28*, 67689–67710.

Nikologianni, A. (2018). *The Role of Low Carbon, Spatial Quality and Drawings in Landscape-Based Regional Strategies*. Birmingham City University.

Niu, D., & Wei, Y. (2013). A novel social-environmental-economic dispatch model for thermal/wind power generation and application. *International Journal of Innovative Computing, Information and Control*, *9*(7), 3005–3014.

Ouramdane, O., Elbouchikhi, E., Amirat, Y., & Sedgh Gooya, E. (2021). Optimal sizing and energy management of microgrids with vehicle-to-grid technology: A critical review and future trends. *Energies*, *14*(14), 4166.

Pandey, A., & Asif, M. (2022). Assessment of energy and environmental sustainability in South Asia in the perspective of the Sustainable Development Goals. *Renewable and Sustainable Energy Reviews*, *165*, 112492.

Pfenninger, S., Hawkes, A., & Keirstead, J. (2014). Energy systems modeling for twenty-first century energy challenges. *Renewable and Sustainable Energy Reviews*, *33*, 74–86.

Phillips, T., Chalishazar, V., McJunkin, T., Maharjan, M., Alam, S. S., Mosier, T., & Somani, A. (2020). *A Metric Framework for Evaluating the Resilience Contribution of Hydropower to the Grid*. Paper presented at the 2020 Resilience Week (RWS). https://ieeexplore.ieee.org/document/9241249

Rahman, A., Farrok, O., & Haque, M. M. (2022). Environmental impact of renewable energy source based electrical power plants: Solar, wind, hydroelectric, biomass, geothermal, tidal, ocean, and osmotic. *Renewable and Sustainable Energy Reviews*, *161*, 112279.

Rahman, M. M., Khan, I., Field, D. L., Techato, K., & Alameh, K. (2022). Powering agriculture: Present status, future potential, and challenges of renewable energy applications. *Renewable Energy*, *188*, 731–749.

Rahman, M. M., Paatero, J. V., & Lahdelma, R. (2013). Evaluation of choices for sustainable rural electrification in developing countries: A multicriteria approach. *Energy Policy*, *59*, 589–599.

Relva, S. G., da Silva, V. O., Gimenes, A. L. V., Udaeta, M. E. M., Ashworth, P., & Peyerl, D. (2021). Enhancing developing countries' transition to a low-carbon electricity sector. *Energy*, *220*, 119659.

Rojas-Zerpa, J. C., & Yusta, J. M. (2015). Application of multicriteria decision methods for electric supply planning in rural and remote areas. *Renewable and Sustainable Energy Reviews*, *52*, 557–571.

Rosales-Asensio, E., Diez, D. B., Cabrera, P., & Sarmento, P. (2024). Effectiveness and efficiency of support schemes in promoting renewable energy sources in the Spanish electricity market. *International Journal of Electrical Power & Energy Systems*, *158*, 109926.

Saraji, M. K., & Streimikiene, D. (2023). Challenges to the low carbon energy transition: A systematic literature review and research agenda. *Energy Strategy Reviews*, *49*, 101163.

Sen, S., & Ganguly, S. (2017). Opportunities, barriers and issues with renewable energy development–A discussion. *Renewable and Sustainable Energy Reviews*, *69*, 1170–1181.

Settembre-Blundo, D., González-Sánchez, R., Medina-Salgado, S., & García-Muiña, F. E. (2021). Flexibility and resilience in corporate decision making: A new sustainability-based risk management system in uncertain times. *Global Journal of Flexible Systems Management*, *22*(Suppl 2), 107–132.

Sikiru, S., Oladosu, T. L., Amosa, T. I., Olutoki, J. O., Ansari, M., Abioye, K. J.,. . . Soleimani, H. (2024). Hydrogen-powered horizons: Transformative technologies in clean energy generation, distribution, and storage for sustainable innovation. *International Journal of Hydrogen Energy*, *56*, 1152–1182.

Simsek, Y. (2020). *Long-Term Sustainable Energy Scenarios Modeling and Analysis for Chile: A Roadmap Towards 2030*. https://repositorio.uc.cl/handle/11534/28324

Siwal, S. S., Zhang, Q., Devi, N., Saini, A. K., Saini, V., Pareek, B., . . . Thakur, V. K. (2021). Recovery processes of sustainable energy using different biomass and wastes. *Renewable and Sustainable Energy Reviews*, *150*, 111483.

Soltani, M., Kashkooli, F. M., Dehghani-Sanij, A., Kazemi, A., Bordbar, N., Farshchi, M., . . . Dusseault, M. B. (2019). A comprehensive study of geothermal heating and cooling systems. *Sustainable Cities and Society*, *44*, 793–818.

Sovacool, B. K., Baum, C. M., & Low, S. (2023). Risk–risk governance in a low-carbon future: Exploring institutional, technological, and behavioral tradeoffs in climate geoengineering pathways. *Risk Analysis*, *43*(4), 838–859.

Sovacool, B. K., & Griffiths, S. (2020). The cultural barriers to a low-carbon future: A review of six mobility and energy transitions across 28 countries. *Renewable and Sustainable Energy Reviews*, *119*, 109569.

Sovacool, B. K., Newell, P., Carley, S., & Fanzo, J. (2022). Equity, technological innovation and sustainable behaviour in a low-carbon future. *Nature Human Behaviour*, *6*(3), 326–337.

Strantzali, E., & Aravossis, K. (2016). Decision making in renewable energy investments: A review. *Renewable and Sustainable Energy Reviews*, *55*, 885–898.

Streimikiene, D., Baležentis, T., Volkov, A., Morkūnas, M., Žičkienė, A., & Streimikis, J. (2021). Barriers and drivers of renewable energy penetration in rural areas. *Energies*, *14*(20), 6452.

Streimikiene, D., & Kyriakopoulos, G. L. (2023). Energy poverty and low carbon energy transition. *Energies*, *16*(2), 610.

Streimikiene, D., Kyriakopoulos, G. L., Lekavicius, V., & Siksnelyte-Butkiene, I. (2021). Energy poverty and low carbon just energy transition: Comparative study in Lithuania and Greece. *Social Indicators Research*, *158*(1), 319–371.

Štreimikienė, D., Šliogerienė, J., & Turskis, Z. (2016). Multi-criteria analysis of electricity generation technologies in Lithuania. *Renewable Energy*, *85*, 148–156.

Suwaileh, W., Isaifan, R., Rahighi, R., Bakhshayesh, A., & Ahmed, M. (2023). *Technological Advances in Harnessing Energy from Renewable Sources for Water Production*. https://www.intechopen.com/chapters/86831

Taghizadeh-Hesary, F., & Yoshino, N. (2020). Sustainable solutions for green financing and investment in renewable energy projects. *Energies*, *13*(4), 788.

Tham, R., Morgan, G., Dharmage, S., Marks, G., & Cowie, C. (2020). Scoping review to understand the potential for public health impacts of transitioning to lower carbon emission technologies and policies. *Environmental Research Communications*, *2*. https://doi.org/10.1088/2515-7620/ab9526

Tovar-Facio, J., Cansino-Loeza, B., & Ponce-Ortega, J. M. (2022). Management of renewable energy sources. In *Sustainable Design for Renewable Processes* (pp. 3–31). Elsevier.

Upadhyay, S., & Sharma, M. (2014). A review on configurations, control and sizing methodologies of hybrid energy systems. *Renewable and Sustainable Energy Reviews*, *38*, 47–63.

Urban, F. (2014). *Low carbon transitions for developing countries*. Routledge.

Verbruggen, A., Fischedick, M., Moomaw, W., Weir, T., Nadaï, A., Nilsson, L. J., . . . Sathaye, J. (2010). Renewable energy costs, potentials, barriers: Conceptual issues. *Energy Policy*, *38*(2), 850–861. https://doi.org/10.1016/j.enpol.2009.10.036

Wakiyama, T., & Zusman, E. (2021). The impact of electricity market reform and subnational climate policy on carbon dioxide emissions across the United States: A path analysis. *Renewable and Sustainable Energy Reviews*, *149*, 111337.

Wiese, F., Hilpert, S., Kaldemeyer, C., & Pleßmann, G. (2018). A qualitative evaluation approach for energy system modelling frameworks. *Energy, Sustainability and Society*, *8*, 1–16.

7 Green Jobs

Justas Streimikis

7.1 THE GREEN JOBS CONCEPT

Green jobs emerged as a response to growing environmental concerns and the need for sustainable development practices. While it's challenging to pinpoint a specific origin, the idea gained prominence in the late 20th and early 21st centuries as awareness of environmental issues, such as climate change, pollution, and resource depletion, increased worldwide.

The emergence of the green job concept was caused by (Tammilehto, Nyholm and Jansson, 2022):

1. Environmental movement. The modern environmental movement, which gained momentum in the 1960s and 1970s, played a significant role in raising awareness about the impact of human activities on the environment. Environmental activists and organizations advocated for policies and practices that promote sustainability and conservation, laying the groundwork for the emergence of green jobs.
2. Renewable energy development. The growing interest in renewable energy sources, such as solar, wind, hydroelectric, and geothermal power, spurred the creation of jobs in these emerging industries. The oil crises of the 1970s highlighted the need for alternative energy sources, leading to increased investment in renewable energy technologies and the associated workforce.

Renewable energy development has been a crucial factor in the emergence of green jobs and the transition to a more sustainable economy. The evolution of renewable energy development is presented in Figure 7.1 and Figure 7.2.

Overall, the development of renewable energy has not only contributed to reducing carbon emissions and mitigating climate change but has also created millions of jobs worldwide, spanning a wide range of sectors from manufacturing and construction to research and innovation. As renewable energy continues to expand, the demand for skilled workers in green jobs will likely continue to grow, driving economic development and sustainable prosperity.

3. Sustainable Development Agenda. The concept of sustainable development, as articulated in the 1987 Brundtland Report by the United Nations World Commission on Environment and Development, emphasized the integration of environmental, social, and economic considerations in decision-making. Green jobs became recognized as a key component of sustainable development, contributing to both environmental protection and economic growth.

The Sustainable Development Agenda, as articulated in the United Nations' Sustainable Development Goals (SDGs), provides a comprehensive framework for addressing global challenges while promoting economic prosperity, social inclusion, and environmental sustainability. The concept of green jobs is closely aligned with the Sustainable Development Agenda, particularly Goal 8 (Decent Work and Economic Growth) and Goal 7 (Affordable and Clean Energy). The

DOI: 10.1201/9781003517481-7

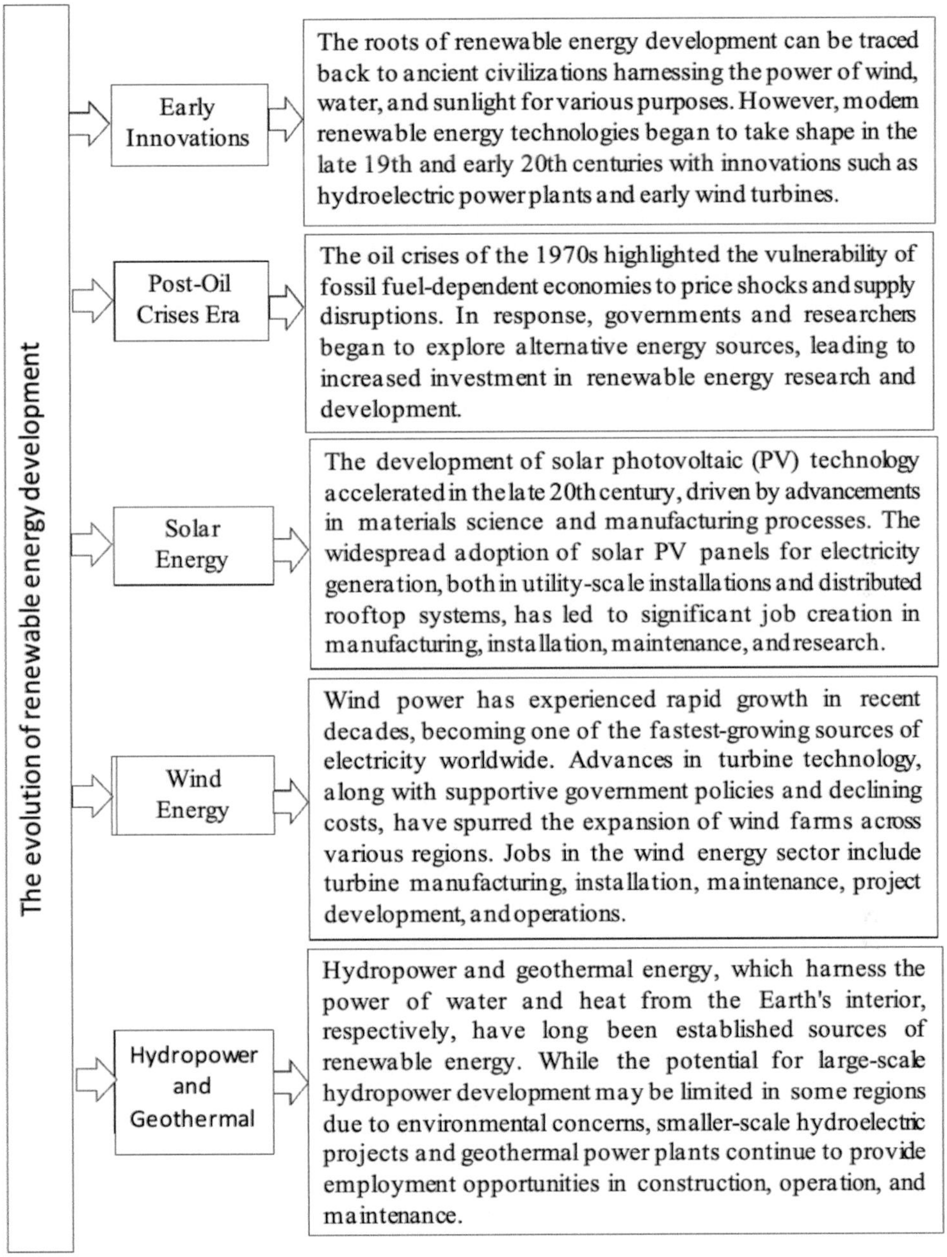

FIGURE 7.1 The evolution of renewable energy development (1).

Source: created by authors according to Li, Luo and Xu (2022).

link between the Sustainable Development Agenda and the concept of green jobs is presented in Figure 7.3.

Overall, the Sustainable Development Agenda provides a holistic framework for integrating environmental, social, and economic considerations into development policies and strategies. Green jobs are essential for achieving sustainable development objectives by promoting economic

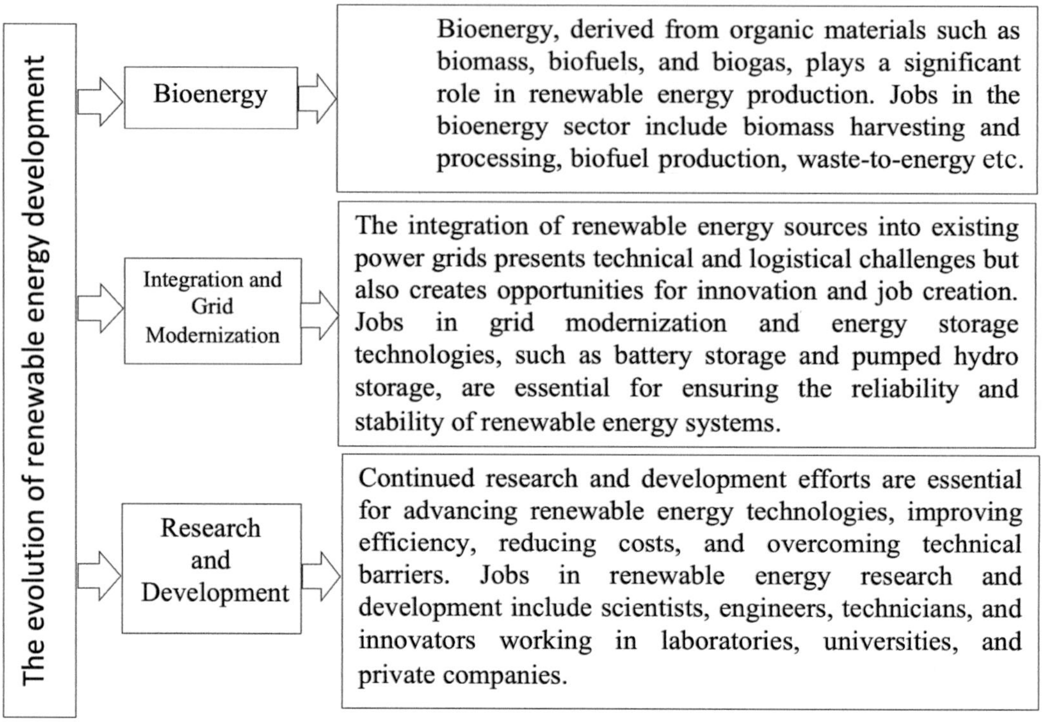

FIGURE 7.2 The evolution of renewable energy development (2).

Source: created by authors according to Su Ali, Lyulyov, Pimonenko and Chen (2023), Li, Luo and Xu (2022).

growth, environmental sustainability, social inclusion, and resilience to global challenges such as climate change and resource depletion. By aligning green job creation efforts with the Sustainable Development Goals, countries can advance progress towards a more prosperous, equitable, and sustainable future for all.

4. Globalization and trade. The globalization of the economy and increased international trade led to discussions about the environmental and social impacts of the global supply chains. Efforts to address issues like carbon emissions, pollution, and labor rights in the context of global trade contributed to the recognition of green jobs as a means to promote sustainable production and consumption patterns.

Globalization has facilitated the international exchange of goods, services, and technologies, allowing countries to access renewable energy technologies and green products from around the world. This increased market access creates opportunities for countries to adopt clean technologies, expand their renewable energy infrastructure, and create green jobs in industries.

Trade policies and agreements can influence the development and adoption of renewable energy technologies and green products. Tariffs, subsidies, and trade barriers can affect the competitiveness of renewable energy technologies and impact the distribution of green jobs across countries. Trade agreements that prioritize environmental sustainability and promote clean technology transfer can facilitate the adoption of renewable energy and support green job creation globally.

Globalization can also create competitive pressures that influence green job creation and environmental policies. Countries may face pressure to lower environmental standards or reduce labor costs to remain competitive in global markets. This can pose challenges for maintaining high

Promoting Decent Work and Economic Growth (Goal 8)

- The Sustainable Development Goal 8 emphasizes the importance of promoting inclusive and sustainable economic growth, full and productive employment, and decent work for all. Green jobs play a crucial role in achieving this goal by providing employment opportunities that contribute to environmental sustainability and support economic development. By investing in renewable energy, energy efficiency, sustainable agriculture, and other green sectors, countries can create jobs while also advancing sustainable development objectives.

Transitioning to Clean Energy (Goal 7)

- Goal 7 of the Sustainable Development Agenda focuses on ensuring access to affordable, reliable, sustainable, and modern energy for all. Green jobs are central to achieving this goal by supporting the transition to clean energy sources such as solar, wind, hydroelectric, and geothermal power. Jobs in renewable energy development, installation, operation, and maintenance contribute to expanding access to clean energy while reducing reliance on fossil fuels and mitigating the impacts of climate change.

Promoting Environmental Sustainability

- Green jobs contribute to environmental sustainability by reducing carbon emissions, minimizing environmental degradation, and promoting resource conservation. By investing in energy efficiency measures, waste management practices, sustainable transportation systems, and ecosystem restoration projects, countries can create employment opportunities while also protecting natural ecosystems and biodiversity.

Addressing Climate Change

- Green jobs play a crucial role in addressing climate change by supporting the transition to a low-carbon economy and reducing greenhouse gas emissions. Jobs in renewable energy, energy efficiency, and sustainable transportation sectors help mitigate climate change impacts while creating economic opportunities and building resilience to climate-related risks.

Fostering Social Inclusion and Equity

- The concept of green jobs promotes social inclusion and equity by providing opportunities for marginalized communities, women, and youth to participate in the green economy. By ensuring equitable access to training, education, and employment opportunities in green sectors, countries can address social inequalities and promote inclusive economic growth.

Building Resilience to Environmental Risks

- Green jobs contribute to building resilience to environmental risks and natural disasters by promoting sustainable land use practices, improving disaster preparedness and response mechanisms, and investing in climate-resilient infrastructure. Jobs in ecosystem restoration, disaster risk reduction, and climate adaptation help communities adapt to changing environmental conditions and reduce vulnerability to environmental hazards.

FIGURE 7.3 The link between the Sustainable Development Agenda and the concept of green jobs.

Source: created by authors according to Tanasie et al. (2022), Kozar and Sulich (2023).

environmental and labor standards while fostering green job growth. However, globalization also provides opportunities for countries to differentiate themselves by investing in sustainable practices, innovation, and green technologies. Globalization and trade influence the development of green jobs by shaping market dynamics, supply chains, technology transfer, and policy frameworks.

5. Policy initiatives. Governments around the world began implementing policies and programs aimed at promoting green jobs and transitioning to a low-carbon economy. This included incentives for renewable energy deployment, energy efficiency

improvements, sustainable transportation initiatives, and green workforce training programs.

6. International agreements. International agreements and frameworks, such as the Kyoto Protocol and the Paris Agreement, underscored the importance of addressing climate change and transitioning to a greener economy. Green jobs became a focal point of discussions about climate mitigation and adaptation strategies at the global level.
7. Advocacy and research. Non-governmental organizations (NGOs), labor unions, research institutions, and advocacy groups played a crucial role in promoting the concept of green jobs and advocating for policies to support their creation. Research studies and reports highlighted the potential economic, social, and environmental benefits of investing in green industries and employment opportunities.

Overall, the origins of the green job concept can be traced back to a combination of environmental activism, renewable energy development, sustainable development principles, policy initiatives, and advocacy efforts aimed at addressing environmental challenges while promoting economic prosperity and social equity.

Green jobs are employment opportunities that contribute to environmental sustainability and conservation while promoting economic growth and social equity. These jobs are typically found in industries and sectors prioritizing reducing carbon emissions, minimizing environmental impact, and fostering sustainable practices. Green jobs encompass various fields, including renewable energy, energy efficiency, waste management, sustainable transportation, conservation, and eco-friendly agriculture. They are crucial in transitioning towards a more sustainable economy and addressing pressing environmental challenges such as climate change, pollution, and resource depletion.

There can be variations in how "green jobs" are defined depending on the context, the organization, or the specific focus of the definition. We can find broad definition that covers a very wide range of problems, and some definitions of green jobs encompass any employment that contributes to environmental sustainability or resource conservation (Bracarense, 2022). This includes jobs in sectors such as renewable energy, energy efficiency, green construction, sustainable transportation, environmental conservation, waste management, and organic agriculture. Expanding on the broad definition of green jobs, we can include a wide array of employment opportunities across diverse industries and sectors that contribute to environmental sustainability and conservation efforts. Some additional examples within this expanded definition are shown in Figure 7.4.

Other narrow definitions may be more specific, focusing only on jobs directly related to renewable energy and energy efficiency sectors. This narrower definition might exclude other environmentally friendly industries like sustainable agriculture or waste management (Unay-Gailhard and Bojnec, 2019). Expanding on the narrow definition of green jobs, which focuses specifically on employment within the renewable energy and energy efficiency sectors, we can provide some examples (Figure 7.5 and Figure 7.6).

Some qualitative definitions emphasize the quality of the jobs, considering factors such as whether they are well paying, offer benefits, provide opportunities for career advancement, and adhere to labor standards. In this sense, green jobs not only contribute to environmental sustainability but also promote social equity and economic prosperity for workers.

Quantitative definitions may quantify green jobs based on specific criteria, such as the percentage of a job's tasks or activities that are environmentally friendly, the amount of carbon emissions reduced or avoided, or the proportion of a company's revenue derived from green products or services (Merino-Saum, Clement and Wyss, 2020). A quantitative definition of green jobs is less commonly used compared to qualitative or policy-based definitions. However, it is still employed

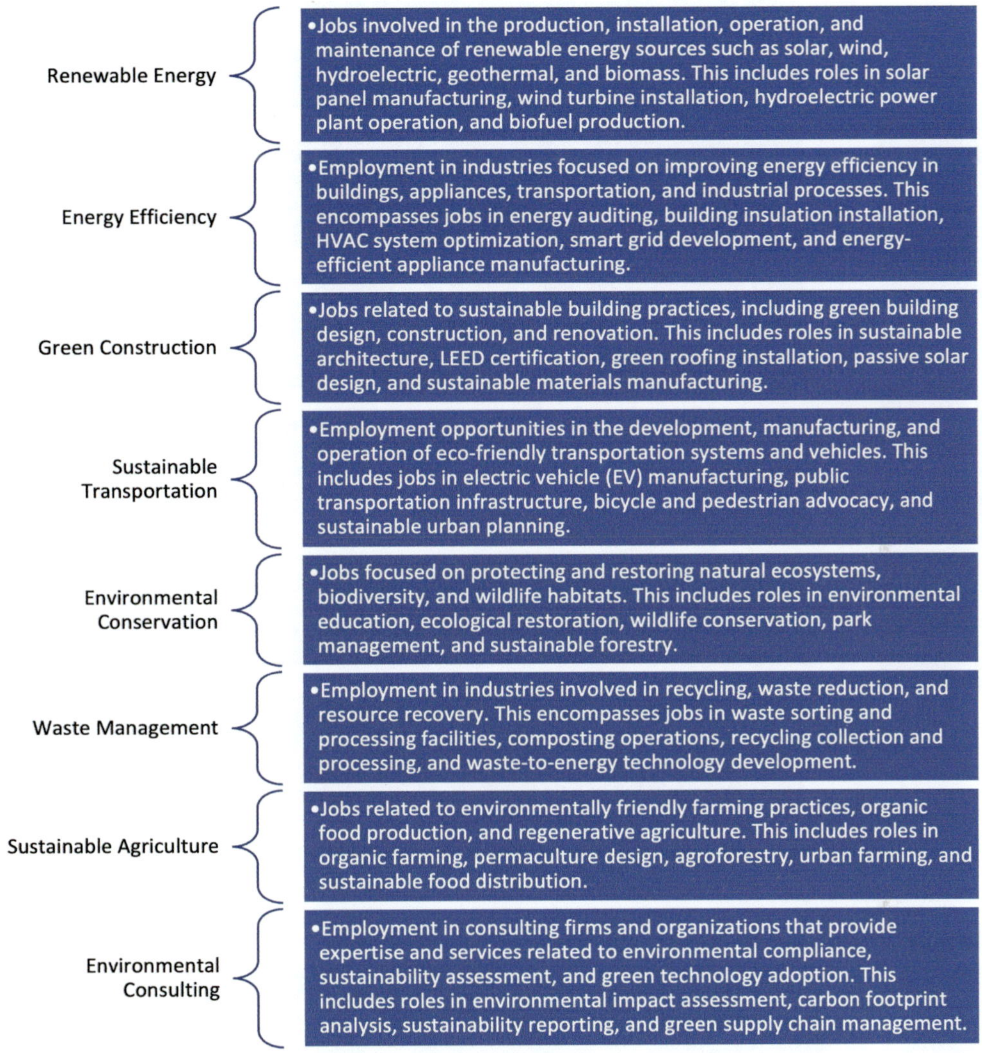

FIGURE 7.4 Green jobs examples according to broad definition.

Source: created by authors.

in certain contexts, particularly in research, policy analysis, and corporate sustainability reporting. Examples of the application of the quantitative definition are given in Figure 7.7.

While quantitative definitions of green jobs provide a more objective and measurable approach to assessing environmental impact and job creation, they can also be challenging to implement due to the complexity of defining and quantifying environmental sustainability in the context of employment. As a result, qualitative and policy-based definitions are often more commonly used in practice, supplemented by quantitative analysis where feasible and appropriate.

Definitions of green jobs may also be shaped by government policies or initiatives aimed at promoting environmentally friendly industries and employment opportunities (Süsser, Ceglarz, Gaschnig and Stavrakas, 2021; Song, Liu, Yuan and Zhao, 2024). These definitions often align with policy goals related to climate change mitigation, renewable energy deployment, or sustainable development.

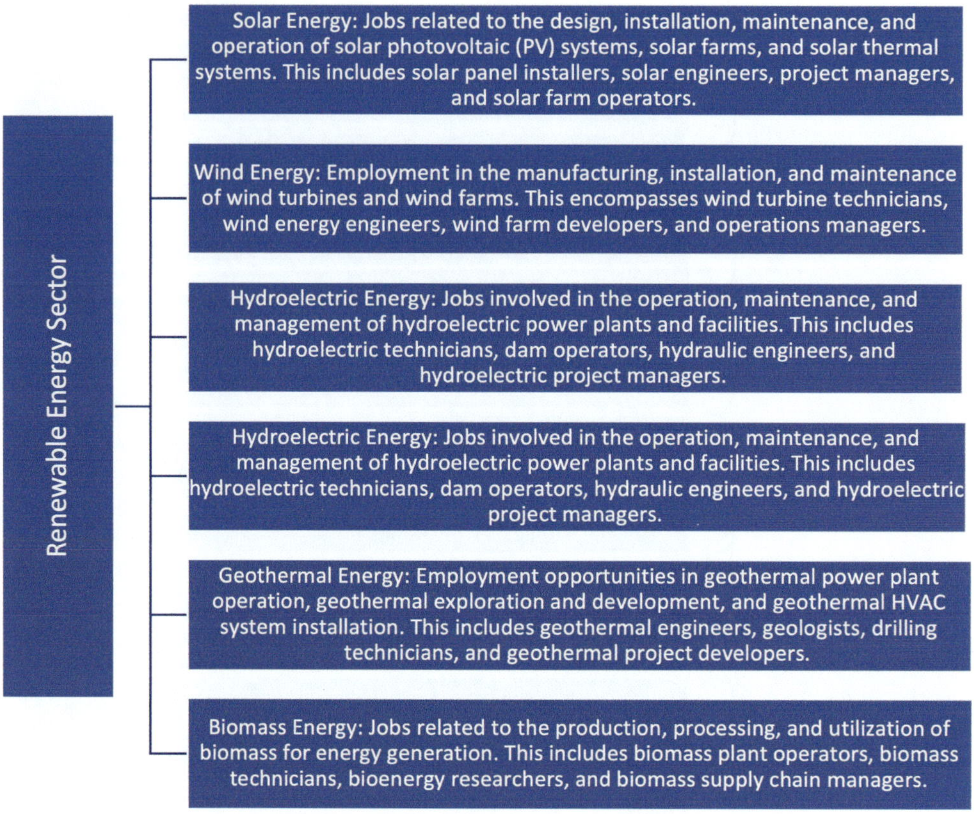

FIGURE 7.5 Green jobs example according to narrow definition in the renewable energy sector.

Source: created by authors.

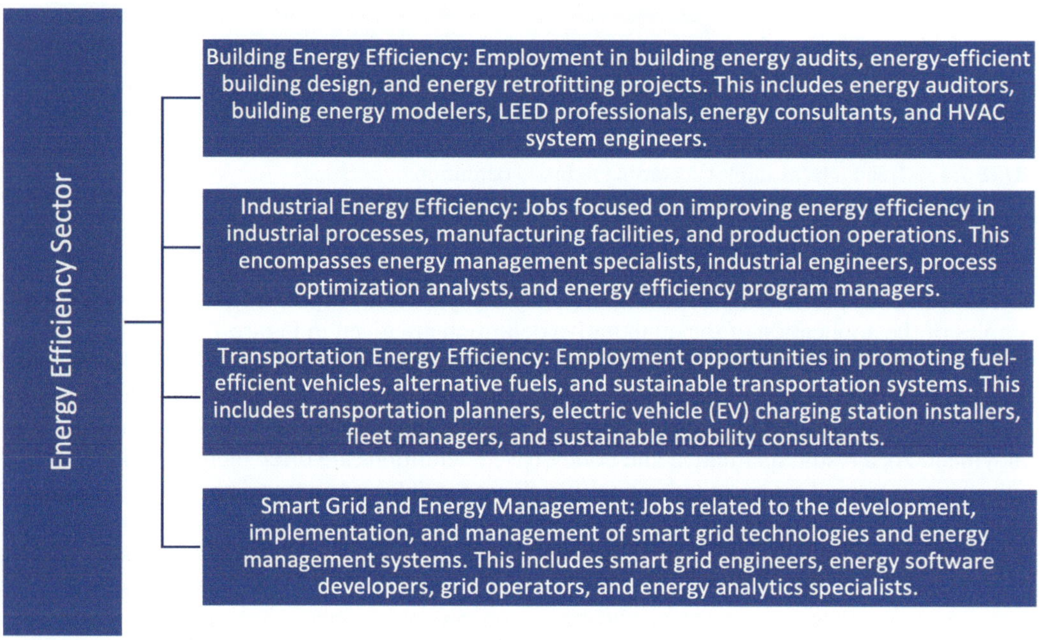

FIGURE 7.6 Green jobs examples according to narrow definition in the energy efficiency sector.

Source: created by authors.

Research Studies

- Researchers studying the green economy or analyzing employment trends may use quantitative definitions to quantify the number of green jobs within specific sectors or regions. This could involve using criteria such as the proportion of a job's tasks related to environmental sustainability or the amount of carbon emissions reduced or avoided by a particular industry.

Policy Evaluation

- Policymakers and government agencies may utilize quantitative definitions to evaluate the effectiveness of green job creation policies or initiatives. By quantifying the number of green jobs created or the environmental benefits achieved as a result of these policies, policymakers can assess their impact and make informed decisions about future initiatives.

Corporate Reporting

- Companies and organizations committed to sustainability may use quantitative definitions to track and report on their green job creation efforts. This could involve quantifying the number of employees working in environmentally friendly sectors or calculating the environmental benefits associated with specific job roles or projects.

International Comparisons

- Quantitative definitions of green jobs may also be used for international comparisons and benchmarking purposes. Organizations such as the International Labour Organization (ILO) or the Organisation for Economic Co-operation and Development (OECD) may develop standardized criteria for quantifying green jobs to facilitate cross-country comparisons and analysis.

FIGURE 7.7 Examples of the application of the quantitative definition of green jobs.

Source: created by authors.

The diversity of definitions underscores the multidimensional nature of the concept of green jobs and the various ways it can be interpreted and applied in different contexts.

The COVID-19 pandemic, the war in Ukraine, and other extreme situations have had both challenges and opportunities for the implementation of the concept of green jobs.

The following challenges can be distinguished (Mentes, 2023; Sulich and Rutkowska, 2020; Dell'Anna, 2021):

1. Economic disruption. The COVID-19 pandemic and other extreme situations have caused economic disruptions, including job losses, business closures, and supply chain disruptions. These challenges have affected the implementation of green jobs by diverting resources away from sustainability initiatives and delaying investments in renewable energy and environmental projects.
2. Policy priorities. During times of crisis, governments may prioritize short-term economic recovery measures over long-term sustainability goals. The focus on immediate economic stabilization and social support programs may lead to a deprioritization of green job creation initiatives and environmental regulations.
3. Funding constraints. Fiscal constraints resulting from the economic impacts of the pandemic and other crises may limit government funding for green job programs and renewable energy projects. Budget cuts and austerity measures can hinder investments in green infrastructure, energy efficiency programs, and sustainable development initiatives.
4. Supply chain disruptions. Global supply chain disruptions caused by the pandemic and geopolitical tensions, such as the war in Ukraine, have affected the availability of critical materials and components needed for renewable energy technologies. Delays in supply chains can hinder the deployment of renewable energy projects and slow down the transition to a low-carbon economy.

The following possibilities are distinguished (Sulich and Rutkowska, 2020; Dell'Anna, 2021):

1. Green recovery. The COVID-19 pandemic has prompted calls for a green recovery that prioritizes investments in sustainable infrastructure, renewable energy, and green jobs. Stimulus packages and economic recovery plans aimed at reviving economies post-pandemic provide an opportunity to integrate green job creation initiatives and advance sustainability goals.
2. Resilience building. Extreme situations like the COVID-19 pandemic and geopolitical conflicts underscore the importance of resilience and sustainability. Investing in green jobs and renewable energy can help build resilience to future crises by diversifying energy sources, reducing dependency on fossil fuels, and creating more resilient and decentralized energy systems.
3. Innovation and adaptation. Crises often catalyze innovation and adaptation in response to emerging challenges. The need to address the health, economic, and environmental impacts of the pandemic and other extreme situations has spurred innovation in areas such as remote work, telemedicine, clean energy technologies, and sustainable agriculture. Green jobs in research, development, and innovation are essential for driving these advancements.
4. Global cooperation. Addressing global challenges such as climate change, pandemics, and geopolitical conflicts requires international cooperation and coordination. Crises can serve as catalysts for enhanced collaboration on green job creation, technology transfer, and capacity building to support Sustainable Development Goals globally.

While the COVID-19 pandemic, the war in Ukraine, and other extreme situations pose challenges to the implementation of the concept of green jobs, they also present opportunities to accelerate efforts toward a more sustainable and resilient future. By leveraging these opportunities and addressing the challenges, governments, businesses, and civil society can advance the transition to a green economy and create a more sustainable and equitable world.

7.2 INDICATORS OF GREEN JOBS

Indicators of green jobs are metrics or measures used to assess the extent to which employment opportunities contribute to environmental sustainability, promote green technologies and practices, and support the transition to a low-carbon economy. These indicators help policymakers, researchers, and organizations track the growth and impact of green jobs across various sectors and industries.

One way to identify green jobs is by classifying industries and sectors based on their environmental impact and contribution to sustainability. This classification may include sectors such as renewable energy (solar, wind, hydroelectric, geothermal), energy efficiency, green construction, sustainable transportation, waste management, and environmental conservation. We can provide some examples of occupations commonly classified as green jobs (Figure 7.8).

Renewable energy technicians professionals are involved in the installation, operation, maintenance, and repair of renewable energy systems, such as solar photovoltaic (PV) panels, wind turbines, hydroelectric generators, geothermal heat pumps, and biomass energy facilities. Renewable energy technicians may specialize in electrical, mechanical, or systems engineering, and they play a crucial role in expanding the deployment of clean energy technologies. Energy auditors assess the energy performance of buildings, industrial facilities, and transportation systems to identify opportunities for energy efficiency improvements. They conduct energy audits, analyze energy consumption data, and recommend measures to reduce energy waste and optimize efficiency. Energy efficiency specialists implement energy-saving measures, such as installing energy-efficient lighting, improving insulation, upgrading HVAC systems, and implementing smart building technologies.

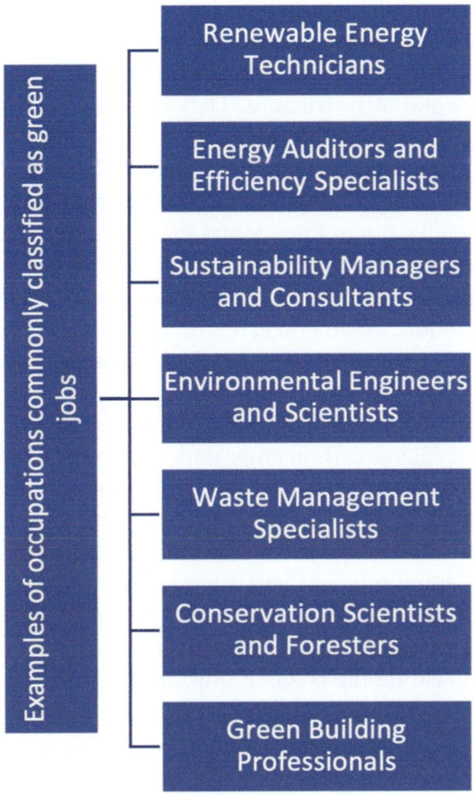

FIGURE 7.8 Some examples of occupations commonly classified as green jobs.

Source: created by authors.

Sustainability managers and consultants work with organizations, businesses, and government agencies to develop and implement sustainability strategies and initiatives. They conduct sustainability assessments, develop sustainability plans, and oversee the implementation of environmentally friendly practices, such as resource conservation, waste reduction, and greenhouse gas emissions reduction. Sustainability managers may also monitor and report on sustainability performance metrics and compliance with environmental regulations. Environmental engineers and scientists are responsible for assessing and mitigating the environmental impact of human activities on ecosystems, air quality, water resources, and soil health. They design and implement environmental remediation projects, develop pollution control technologies, and conduct environmental impact assessments for infrastructure projects. Environmental engineers and scientists may specialize in areas such as air quality management, water resource management, ecological restoration, or hazardous waste management.

Waste management specialists are involved in the collection, recycling, treatment, and disposal of solid waste, hazardous waste, and wastewater. They oversee waste management facilities, such as recycling centers, composting facilities, landfill sites, and wastewater treatment plants. Waste management specialists may also develop and implement waste reduction and recycling programs, educate the public about proper waste disposal practices, and ensure compliance with environmental regulations.

Conservation scientists and foresters work to protect and manage natural ecosystems, wildlife habitats, and forest resources. They conduct research, monitor environmental conditions, and develop conservation strategies to preserve biodiversity, restore degraded ecosystems, and sustainably manage forestlands. Conservation scientists and foresters may also work with landowners, government agencies, and non-profit organizations to implement conservation projects, manage protected areas, and promote sustainable land use practices.

Green building professionals are involved in the design, construction, and operation of environmentally sustainable buildings and infrastructure. They incorporate green building principles, such as energy efficiency, water conservation, indoor air quality, and materials sustainability, into building design and construction practices. Green building professionals may include architects, engineers, construction managers, and building inspectors with specialized training and expertise in green building standards, such as Leadership in Energy and Environmental Design (LEED) certification.

These are just a few examples of occupations classified as green jobs, illustrating the diverse range of roles and professions that contribute to environmental sustainability, conservation, and the transition to a greener economy. As the demand for green jobs continues to grow, new roles are likely to emerge, reflecting advancements in technology, changes in environmental policies, and evolving market trends.

Indicators of green jobs often consider the environmental impact of employment opportunities, such as their contribution to reducing carbon emissions, conserving natural resources, minimizing pollution, and protecting ecosystems. Metrics such as greenhouse gas emissions avoided, energy savings, water conservation, and waste diversion can help quantify the environmental benefits associated with green jobs. These indicators provide valuable insights into the environmental outcomes associated with green job creation initiatives and can inform decision-making, policy development, and resource allocation. Figure 7.9 shows key indicators of the environmental benefits of green jobs.

These indicators provide a comprehensive framework for assessing the environmental benefits of green jobs and can help demonstrate the positive impact of sustainable employment opportunities on environmental sustainability, resource conservation, and ecosystem health. By tracking these indicators, stakeholders can monitor progress, evaluate performance, and identify opportunities for further environmental improvement in green job creation initiatives.

Adopting green technologies and practices within industries and occupations is a key indicator of green jobs. Expanding on the indicator of adopting green technologies and practices within industries and occupations as a key aspect of green jobs involves examining the various ways in which businesses, organizations, and individuals integrate environmentally sustainable practices and technologies into their operations and daily activities (Li, Ahmad, Ayassrah, Irshad and Telba, 2023). Industries and occupations can adopt renewable energy technologies to reduce reliance on fossil fuels and mitigate greenhouse gas emissions. This includes:

- Installing solar photovoltaic (PV) systems.
- Wind turbines.
- Hydroelectric generators.
- Geothermal heating and cooling systems.
- Biomass energy facilities.

Adopting renewable energy technologies not only reduces carbon emissions but also creates jobs in manufacturing, installation, operation, and maintenance.

Improving energy efficiency is another critical aspect of adopting green technologies and practices. Industries and occupations can implement energy-saving measures:

- Upgrading lighting systems.
- Optimizing HVAC (heating, ventilation, and air conditioning) systems.
- Installing energy-efficient appliances and equipment.
- Implementing building automation systems.

Energy efficiency improvements reduce energy consumption, lower operating costs, and create jobs in energy auditing, retrofitting, and building management.

Greenhouse Gas Emissions Reduction

- This indicator measures the amount of greenhouse gas emissions avoided or reduced as a result of green job activities. It quantifies the environmental impact of renewable energy deployment, energy efficiency improvements, sustainable transportation initiatives, and other green technologies and practices. Metrics such as metric tons of CO2 equivalent (CO2e) emissions avoided or reduced can help assess the contribution of green jobs to climate change mitigation efforts.

Energy Savings

- Energy savings indicators quantify the amount of energy saved or conserved through green job activities, such as energy efficiency upgrades, renewable energy installations, and sustainable transportation solutions. This indicator measures the reduction in energy consumption and associated environmental impacts, such as fossil fuel extraction, air pollution, and greenhouse gas emissions. Metrics such as kilowatt-hours (kWh) or British thermal units (BTUs) of energy saved can help assess the effectiveness of energy efficiency measures and renewable energy investments.

Water Conservation

- Water conservation indicators assess the amount of water saved or conserved through green job activities, such as water-efficient technologies, sustainable agriculture practices, and water resource management initiatives. This indicator measures the reduction in water consumption, water waste, and water pollution, contributing to the preservation of freshwater resources and ecosystems. Metrics such as gallons or cubic meters of water saved can help quantify the environmental benefits of water conservation efforts.

Waste Reduction and Recycling

- Waste reduction and recycling indicators measure the amount of waste diverted from landfills and incinerators through green job activities, such as recycling programs, waste reduction initiatives, and sustainable materials management practices. This indicator assesses the environmental impact of waste management activities, such as reducing landfill methane emissions, conserving natural resources, and minimizing pollution. Metrics such as tons or kilograms of waste recycled or diverted can help evaluate the effectiveness of waste reduction and recycling efforts.

Air Quality Improvement

- Air quality improvement indicators assess the reduction in air pollutants, such as particulate matter (PM), nitrogen oxides (NOx), sulfur dioxide (SO2), volatile organic compounds (VOCs), and hazardous air pollutants (HAPs), resulting from green job activities. This indicator measures the positive impact of renewable energy deployment, energy efficiency improvements, and pollution control measures on air quality and public health. Metrics such as pounds or kilograms of air pollutants avoided or reduced can help quantify the environmental benefits of air quality improvement efforts.

Biodiversity Conservation

- Biodiversity conservation indicators assess the preservation and enhancement of biodiversity resulting from green job activities, such as habitat restoration, ecosystem protection, and sustainable land use practices. This indicator measures the positive impact of conservation efforts on ecosystems, species diversity, and ecosystem services. Metrics such as hectares or acres of habitat restored or protected can help evaluate the effectiveness of biodiversity conservation initiatives.

FIGURE 7.9 Key indicators of the environmental benefits of green jobs.

Source: created by authors according to Sulich and Sołoducho-Pelc (2022), Unay-Gailhard and Bojnec (2019).

Adopting clean transportation solutions reduces emissions from vehicles and contributes to air quality improvement. Industries and occupations can:

- Electrify their fleets.
- Use alternative fuels such as biofuels or hydrogen.
- Promote public transportation and carpooling.
- Invest in infrastructure for electric vehicles (EVs) and charging stations. Adopting clean transportation solutions creates jobs in manufacturing electric vehicles, producing alternative fuels, building charging infrastructure, and providing transportation services.

Adopting waste reduction and recycling practices minimizes waste generation, conserves natural resources, and reduces pollution. Industries and occupations can implement:

- Recycling programs.
- Source reduction initiatives.
- Composting systems.
- Waste-to-energy technologies to divert waste from landfills and incinerators.

The economic benefits of green jobs encompass a wide range of positive impacts on individuals, businesses, communities, and economies as a whole. Figure 7.10 shows key indicators of the economic benefits of green jobs.

Job Creation and Employment Opportunities

- Green jobs generate employment opportunities across various sectors and industries, including renewable energy, energy efficiency, waste management, sustainable transportation, green construction, and environmental consulting. By investing in green technologies, infrastructure, and practices, governments and businesses can create new job opportunities, reduce unemployment, and stimulate economic growth.

Economic Growth and Innovation

- Green jobs contribute to economic growth by fostering innovation, entrepreneurship, and technological advancement. Investments in renewable energy, clean technologies, and sustainable practices drive economic activity, attract private investment, and stimulate business development in emerging green industries. Green jobs also promote diversification and resilience in the economy by reducing dependency on fossil fuels and traditional industries.

Cost Savings and Efficiency Improvements

- Green jobs often lead to cost savings and efficiency improvements for businesses and households. Energy efficiency measures, such as building retrofits, energy-efficient appliances, and lighting upgrades, reduce energy consumption and lower utility bills. Renewable energy technologies, such as solar and wind power, offer stable and predictable energy costs, reducing vulnerability to fossil fuel price fluctuations. Adopting green technologies and practices improves resource efficiency, minimizes waste generation, and reduces operating costs for businesses.

Market Growth and Export Opportunities

- Green jobs drive market growth and export opportunities in clean energy, environmental goods, and sustainable products and services. As global demand for renewable energy and environmental solutions continues to rise, countries with a strong green economy can capitalize on export markets and gain a competitive advantage in the global marketplace. Green jobs contribute to trade balance, export revenue, and economic resilience by diversifying export sectors and reducing dependence on imported fossil fuels.

Infrastructure Development and Investment

- Green jobs stimulate infrastructure development and investment in clean energy, transportation, water management, and sustainable urban development projects. Governments and private investors allocate funds for renewable energy installations, public transit systems, smart grid technologies, green buildings, and sustainable infrastructure projects, creating jobs in construction, engineering, manufacturing, and project management. Infrastructure investment in green sectors promotes economic development, enhances quality of life, and attracts investment in communities.

Health and Well-being Benefits

- Green jobs contribute to improved public health and well-being by reducing air and water pollution, mitigating climate change impacts, and enhancing environmental quality. By transitioning to cleaner and healthier energy sources, green jobs reduce healthcare costs associated with air pollution-related illnesses, respiratory diseases, and climate-related disasters. Investing in green infrastructure, such as parks, green spaces, and urban forests, enhances quality of life, promotes physical activity, and improves mental health outcomes.

FIGURE 7.10 Key indicators of the economic benefits of green jobs.

Source: created by authors according to Sulich and Sołoducho-Pelc (2022), Mentes (2023), D'Amato and Korhonen (2021).

Green jobs build resilience to environmental risks and climate change impacts, reducing vulnerability and enhancing adaptive capacity. Investing in climate adaptation measures, such as flood protection infrastructure, coastal defenses, and disaster preparedness programs, creates jobs in disaster risk reduction, emergency response, and community resilience planning. Green jobs contribute to economic stability, risk mitigation, and long-term sustainability by addressing environmental challenges and strengthening resilience.

Overall, green jobs offer a multitude of economic benefits, ranging from job creation and market growth to cost savings and infrastructure development. By harnessing the potential of green technologies, practices, and industries, countries can foster sustainable economic development, enhance competitiveness, and create a more resilient and prosperous future for all.

Green jobs offer a multitude of social benefits, ranging from job creation and skills development to community empowerment and environmental justice. By prioritizing investments in green sectors and sustainable development initiatives, governments, businesses, and civil society can foster inclusive growth, promote social equity, and build resilient and thriving communities. Figure 7.11 shows key indicators of the social benefits of green jobs.

Green jobs contribute to community resilience and adaptation to environmental risks and climate change impacts. By investing in climate-resilient infrastructure, disaster preparedness programs, and ecosystem restoration projects, green jobs initiatives help communities build adaptive capacity, reduce vulnerability, and enhance disaster resilience. Green jobs empower communities to respond effectively to environmental challenges, protect livelihoods, and safeguard natural resources for future generations.

Indicators of green jobs may also consider the policy and regulatory frameworks that support the growth of environmentally sustainable industries and employment opportunities. Policy and regulatory frameworks are essential drivers of the growth of environmentally sustainable industries and employment opportunities. Indicators that consider these frameworks provide insights into the level of government support, regulatory incentives, and policy alignment with sustainability goals. Here are some key indicators related to policy and regulatory frameworks for green jobs (Drago and Gatto, 2022; Sulich and Rutkowska, 2020):

1. Policy alignment with Sustainability Goals. This indicator assesses the extent to which national, regional, and local policies align with environmental sustainability goals, such as reducing carbon emissions, conserving natural resources, and promoting renewable energy deployment. Policies that prioritize sustainability objectives, set targets for greenhouse gas emissions reduction, and promote clean energy transition signal government commitment to fostering green jobs and sustainable development.
2. Renewable energy targets and incentives. This indicator evaluates the presence of renewable energy targets, mandates, and incentives that encourage the adoption of renewable energy technologies and support the growth of green energy industries. Policies such as renewable portfolio standards (RPS), feed-in tariffs (FITs), tax credits, grants, and subsidies for renewable energy generation incentivize investment in solar, wind, hydroelectric, geothermal, and biomass energy projects, creating jobs in the renewable energy sector.
3. Energy efficiency standards and incentives. This indicator examines the existence of energy efficiency standards, regulations, and incentives aimed at improving energy efficiency in buildings, appliances, transportation, and industrial processes. Policies such as building energy codes, appliance efficiency standards, energy labeling programs, and financial incentives for energy efficiency upgrades encourage investment in energy-saving technologies and practices, creating jobs in energy auditing, retrofitting, and building management.
4. Environmental regulations and compliance. This indicator assesses the stringency and enforcement of environmental regulations that govern pollution control, waste management, resource conservation, and environmental protection. Strong regulatory frameworks, such as air quality standards, water quality regulations, waste management regulations, and environmental impact assessment requirements, promote sustainable business practices,

Job Opportunities and Inclusive Growth

- Green jobs provide employment opportunities for individuals from diverse backgrounds, including women, minorities, youth, and disadvantaged communities. By promoting inclusive growth and equal access to employment, green jobs contribute to reducing unemployment, poverty, and social inequality. Investing in green sectors and sustainable industries creates pathways to economic empowerment and social mobility for marginalized groups.

Skills Development and Training

- Green jobs offer opportunities for skills development, training, and lifelong learning in emerging and high-demand sectors such as renewable energy, energy efficiency, sustainable agriculture, and environmental conservation. Training programs, apprenticeships, and vocational education initiatives prepare individuals for green job careers, equip them with relevant skills and competencies, and enhance their employability in the green economy. Green jobs contribute to building a skilled workforce, closing skills gaps, and fostering human capital development.

Community Development and Empowerment

- Green jobs contribute to community development and empowerment by revitalizing local economies, fostering entrepreneurship, and strengthening social cohesion. Investments in green infrastructure, renewable energy projects, and sustainable development initiatives create job opportunities in communities, stimulate business growth, and attract investment. Green jobs empower communities to take ownership of their economic future, promote self-sufficiency, and build resilience to economic shocks.

Health and Well-being Benefits

- Green jobs promote public health and well-being by reducing exposure to environmental hazards, improving air and water quality, and enhancing access to clean and healthy environments. Transitioning to cleaner energy sources, promoting sustainable transportation options, and implementing green infrastructure projects reduce air pollution, mitigate climate change impacts, and enhance quality of life. Green jobs contribute to improved physical health, mental well-being, and overall quality of life for individuals and communities.

Environmental Justice and Equity

- Green jobs advance environmental justice and equity by addressing environmental inequalities and ensuring that the benefits of environmental sustainability are shared equitably among all segments of society. By prioritizing investments in underserved communities, promoting workforce diversity, and engaging with local stakeholders, green jobs initiatives promote social equity and environmental justice. Green jobs empower marginalized communities to participate in decision-making processes, advocate for their rights, and access green job opportunities.

Cultural Preservation and Heritage

- Green jobs support cultural preservation and heritage conservation by promoting sustainable land use practices, protecting natural ecosystems, and preserving cultural landscapes and heritage sites. Indigenous communities, in particular, play a vital role in environmental stewardship and traditional knowledge transmission, contributing to biodiversity conservation and ecosystem resilience. Green jobs initiatives recognize and respect indigenous rights, cultural heritage, and traditional practices, fostering cultural diversity and intergenerational knowledge exchange.

FIGURE 7.11 Key indicators of the social benefits of green jobs.

Source: created by authors according to Sulich and Sołoducho-Pelc (2022).

create demand for environmental compliance services, and generate jobs in environmental consulting, auditing, and compliance management.

5. Green procurement and sustainable supply chains. This indicator evaluates government procurement policies and sustainable supply chain initiatives that prioritize environmentally preferable products, materials, and services. Policies such as green procurement guidelines, eco-labeling programs, and sustainable purchasing requirements encourage public and private sector organizations to adopt green technologies, products, and practices, creating market demand for sustainable goods and services and stimulating job growth in green industries.
6. Skills development and workforce training programs. This indicator looks at government investments in skills development, workforce training, and education programs aimed at preparing individuals for green jobs and emerging sustainability careers. Policies such as workforce development grants, apprenticeship programs, vocational training initiatives, and green job training partnerships support the development of a skilled workforce with the expertise needed to drive innovation and growth in green sectors.
7. Research and innovation funding. This indicator examines government funding for research, development, and innovation in clean technologies, renewable energy, sustainable agriculture, and environmental conservation. Policies such as research grants, innovation funds, technology incubators, and public-private partnerships promote collaboration between government, academia, and industry to advance green technologies, commercialize innovations, and create high-quality green jobs in research, development, and technology deployment.

Interpreting these indicators involves analyzing the presence, effectiveness, and implementation of policy and regulatory frameworks that support the growth of green jobs and sustainable industries. Strong policy alignment with sustainability goals, clear targets and incentives for renewable energy and energy efficiency, robust environmental regulations and enforcement mechanisms, and investments in skills development and innovation signal government commitment to fostering green growth and creating a supportive environment for green job creation. By monitoring these indicators, stakeholders can assess the policy landscape, identify opportunities for policy improvement, and advocate for policies that promote environmental sustainability and inclusive economic development.

The level of investment and financing directed towards green industries and projects is an important indicator of green jobs. This includes public and private investments in renewable energy infrastructure, energy efficiency programs, sustainable transportation initiatives, green buildings, and environmental conservation projects. These indicators provide insights into the availability of financial resources, market demand for green technologies and solutions, and investor confidence in environmentally sustainable businesses and projects. A definition and interpretation of these indicators are shown in Table 7.1.

The level of investment and financing directed towards green industries and projects is one of the important indicators of the growth and development of green jobs. By tracking green investment flows, public sector funding, private sector investment, and international development assistance, stakeholders can assess progress towards sustainability goals, identify investment opportunities, and advocate for policies that promote green finance and investment.

Tracking labor market trends, such as job growth rates, employment levels, wages, and skills demand, within green industries and occupations provides insights into the dynamics of green job creation and workforce development. This includes monitoring job creation in emerging green sectors, skill gaps, and training needs for green jobs. Indicators for monitoring the dynamics of workforce development are represented in Table 7.2.

By monitoring these key labor market trends and workforce development indicators, policymakers, employers, educators, and workforce development professionals can assess the dynamics of

TABLE 7.1
Indicators of the Growth and Development of Green Jobs

Indicator	Definition	Interpretation
Green investment flows	This indicator refers to the amount of capital invested in green industries, projects, and initiatives aimed at promoting environmental sustainability and reducing carbon emissions. Green investment flows encompass various types of funding sources, including public and private investment, venture capital, impact investment, green bonds, and project financing. Tracking green investment flows provides a measure of market demand for green technologies, products, and services, as well as investor confidence in the long-term viability and profitability of green businesses and projects.	Increasing green investment flows indicates growing interest and support for environmentally sustainable businesses and projects, driving innovation, job creation, and economic growth in green sectors. High levels of green investment signal a shift towards a low-carbon economy and the mainstreaming of sustainability principles in financial decision-making. Conversely, declining green investment flows may indicate market uncertainty, regulatory barriers, or investor hesitation, highlighting the need for policy interventions to stimulate green finance and investment.
Public sector funding and incentives	This indicator refers to government funding, subsidies, grants, and incentives directed towards green industries, research and development (R&D) initiatives, and sustainable infrastructure projects. Public sector funding plays a crucial role in catalyzing private sector investment, supporting early-stage technologies, and addressing market barriers to green innovation and adoption. Government incentives such as tax credits, grants, loan guarantees, and feed-in tariffs encourage investment in renewable energy, energy efficiency, clean transportation, and environmental conservation.	Adequate public sector funding and incentives signal government commitment to fostering green growth, creating a favorable investment climate for green industries, and accelerating the transition to a low-carbon economy. Effective use of public funds can leverage private sector investment, stimulate job creation, and drive innovation in green technologies and solutions. However, insufficient or inconsistent public sector funding may hinder market development, limit access to capital for green projects, and slow progress towards sustainability goals.
Private sector investment and corporate sustainability initiatives	This indicator refers to private sector investment, corporate sustainability initiatives, and responsible investment practices aimed at promoting environmental sustainability and social responsibility. Private sector investors, including institutional investors, asset managers, and corporate entities, allocate capital to green industries, sustainable businesses, and impact-focused ventures that demonstrate positive environmental and social outcomes. Corporate sustainability initiatives encompass efforts to integrate environmental, social, and governance (ESG) criteria into business strategies, operations, and investment decisions.	Increasing private sector investment and corporate sustainability initiatives reflect growing awareness of environmental risks, climate change impacts, and the business case for sustainability. Private sector investors and companies recognize the potential for competitive advantage, brand differentiation, and long-term value creation through green investments and sustainable practices. Alignment between corporate sustainability goals and investor preferences signals market demand for environmentally sustainable products, services, and business models, driving innovation and job creation in green sectors.
International development assistance and climate finance	This indicator refers to international development assistance, climate finance, and multilateral funding mechanisms aimed at supporting climate change mitigation, adaptation, and sustainable development initiatives in developing countries. International donors, development banks, and climate funds provide financial support for green projects, capacity-building programs, and policy reforms that promote environmental sustainability, renewable energy deployment, and resilience-building efforts in vulnerable communities.	International development assistance and climate finance play a critical role in mobilizing resources, leveraging co-financing, and supporting green job creation in developing countries. By providing financial support for climate-smart investments and sustainable development projects, international donors contribute to poverty reduction, economic growth, and environmental resilience in vulnerable regions. Effective coordination and alignment of international funding efforts with national development priorities can maximize the impact of climate finance and accelerate progress towards Sustainable Development Goals.

Source: created by authors according to Belmonte-Ureña Plaza-Úbeda, Vazquez-Brust and Yakovleva (2021), D'Alessandro, Cieplinski, Distefano and Dittmer (2020), Fang, Liu and Putra (2022).

TABLE 7.2
Indicators for Monitoring the Dynamics of Workforce Development

Indicator	Monitoring area	
Job growth rates	Tracking job growth rates within green industries and occupations provides insights into the pace of expansion and demand for green jobs.	This involves analyzing data on the number of new job openings, job creation trends, and industry growth projections. Job growth rates can be monitored through labor market reports, industry surveys, and government employment data. Comparing job growth rates between green sectors and traditional industries helps assess the relative importance of green jobs in driving economic growth and job creation.
Employment levels	Monitoring employment levels within green industries and occupations involves tracking the total number of employed individuals, as well as changes in employment over time.	This includes analyzing data on workforce participation rates, employment-to-population ratios, and sectoral employment shares. Employment levels can be monitored through labor force surveys, establishment surveys, and administrative data sources. Identifying shifts in employment patterns and trends in green job creation helps gauge the overall health and resilience of green sectors and the labor market.
Wages and compensation	Monitoring wages and compensation within green industries and occupations provides insights into labor market dynamics, workforce incentives, and skill premiums.	This involves analyzing data on average wages, wage growth rates, and compensation packages for green jobs compared to non-green jobs. Wage and compensation data can be obtained from labor market surveys, employer reports, and government wage statistics. Assessing wage differentials between green and non-green sectors helps evaluate the attractiveness of green jobs, skill requirements, and labor market competitiveness.
Skills demand and job requirements	Monitoring skills demand and job requirements within green industries and occupations involves identifying the specific skills, qualifications, and competencies needed to succeed in green jobs.	This includes analyzing job postings, skill requirements, and employer preferences for green job candidates. Skills demand can be assessed through job vacancy data, employer surveys, and skills gap analyses. Identifying emerging skills trends, training needs, and skill shortages helps inform workforce development strategies and education programs tailored to the needs of green industries.
Training and education programs	Monitoring training and education programs within green industries and occupations involves tracking enrollment levels, completion rates, and outcomes of green-focused training initiatives.	This includes assessing the availability and accessibility of training programs, certifications, and credentials relevant to green jobs. Training and education data can be obtained from educational institutions, training providers, and workforce development agencies. Evaluating the effectiveness of training programs, job placement rates, and skills acquisition helps ensure alignment with labor market needs and demand for green skills.
Occupational Mobility and Career Pathways	Monitoring occupational mobility and career pathways within green industries and occupations involves tracking transitions between occupations, career advancement opportunities, and workforce mobility patterns.	This includes analyzing data on job turnover rates, promotion rates, and occupational shifts within and across green sectors. Occupational mobility data can be obtained from longitudinal surveys, employment records, and career-tracking studies. Identifying barriers to career advancement, skills mismatches, and opportunities for upward mobility helps support career development and talent retention in green industries.

Source: created by authors according to Burnett and Lisk (2021).

green job creation, identify emerging opportunities and challenges, and design targeted interventions to support workforce development and skills training in green industries and occupations. This proactive approach helps ensure that the workforce is equipped with the skills and competencies needed to meet the demands of a rapidly evolving green economy.

Indicators of green jobs may also consider social equity and inclusion aspects, such as the distribution of green employment opportunities across demographic groups, income levels, and geographic regions. Ensuring equitable access to green jobs and addressing barriers to entry for underrepresented communities is essential for promoting inclusive growth and environmental justice. Monitoring social equity and inclusion aspects within green jobs involves examining the distribution of employment opportunities, access to training and advancement, and participation rates across demographic groups, income levels, and geographic regions. Some key indicators (Sulich and Rutkowska, 2020):

1. The demographic representation indicator assesses the representation of different demographic groups within green industries and occupations, including gender, race, ethnicity, age, and disability status. It involves analyzing workforce demographics to identify disparities and inequities in employment outcomes and opportunities:
   - Gender—representation of men and women in green jobs, including leadership positions, technical roles, and traditionally male-dominated or female-dominated occupations.
   - Race and ethnicity—participation of racial and ethnic minorities in green industries and occupations, addressing barriers to entry, workplace discrimination, and opportunities for advancement.
   - Age—the employment of younger and older workers in green jobs, addressing age-related barriers to entry, training needs, and opportunities for lifelong learning and career development.
   - Disability status—the inclusion of individuals with disabilities in green employment, ensuring accessibility, accommodation, and support for equal participation in the workforce.
2. Income distribution examines the distribution of green employment opportunities across income levels and socioeconomic groups. It involves assessing the extent to which green jobs provide pathways to economic mobility, reduce income inequality, and support living wage standards:
   - Living wage jobs—identification of green jobs that offer competitive wages, benefits, and job security, ensuring that workers earn a living wage that meets basic needs and supports financial stability.
   - Economic mobility—evaluation of the role of green jobs in facilitating upward economic mobility and reducing poverty rates, particularly for low-income communities and marginalized groups.
   - Income disparities—addressing income disparities within green industries and occupations, ensuring equitable access to higher-paying roles, career advancement opportunities, and financial rewards for all workers.
3. Geographic distribution examines the distribution of green employment opportunities across geographic regions, including urban, suburban, and rural areas. It involves assessing regional disparities in job availability, accessibility, and infrastructure support for green industries:
   - Urban-rural divide—evaluation the concentration of green jobs in urban centers versus rural communities, addressing disparities in access to employment opportunities, transportation, and support services.
   - Regional development—supporting regional development initiatives that promote green job creation, infrastructure investment, and economic diversification in underserved or economically distressed areas.
   - Rural revitalization—investing in rural revitalization efforts that leverage natural resources, renewable energy, and sustainable agriculture to create employment opportunities, support local economies, and enhance the quality of life in rural communities.

4. Access to training and advancement assesses access to training, education, and career advancement opportunities within green industries and occupations. It involves evaluating barriers to entry, skill development needs, and support for lifelong learning and professional growth:
   - Training programs—expanding access to green job training programs, apprenticeships, and vocational education initiatives for individuals from underrepresented and disadvantaged backgrounds.
   - Career pathways—creating clear career pathways and advancement opportunities within green industries, providing support for skills development, credentialing, and career navigation.
   - Mentorship and support services—offering mentorship programs, career counseling, and support services to help individuals overcome barriers to entry, navigate career transitions, and achieve success in green occupations.

By expanding these indicators to consider social equity and inclusion aspects, stakeholders can assess the extent to which green jobs contribute to equitable economic growth, social mobility, and community development. Addressing disparities and promoting diversity within green industries and occupations helps build a more inclusive and resilient workforce, fostering sustainable development and prosperity for all.

Overall, indicators of green jobs provide a comprehensive framework for assessing the environmental, economic, and social dimensions of employment opportunities that contribute to sustainability and support the transition to a green economy. By tracking these indicators, stakeholders can monitor progress, identify opportunities for intervention, and promote policies and initiatives that foster the growth of green jobs.

7.3 DRIVERS OF AND BARRIERS TO GREEN JOBS

Drivers to green jobs are factors that influence the growth, demand, and development of employment opportunities in environmentally sustainable industries and occupations. These drivers shape the trajectory of green job creation and contribute to the transition towards a more sustainable economy.

The first of the drivers to green jobs is the environmental sustainability imperatives (ESI). Environmental sustainability imperatives refer to the pressing needs and challenges related to the preservation, conservation, and responsible management of natural resources, ecosystems, and the environment. These imperatives arise from the recognition of the finite nature of Earth's resources, the interconnectedness of human activities with ecological systems, and the urgency to address environmental degradation and mitigate the impacts of climate change (Starostka-Patyk, Bajdor and Białas, 2024; Li and Rabeeu, 2024). Increasing awareness of environmental challenges, such as climate change, resource depletion, and pollution, drives demand for green technologies, products, and services. Concerns about environmental degradation and the need for sustainable development incentivize businesses, governments, and consumers to adopt green practices, invest in clean technologies, and create employment opportunities in green industries. An expanded definition of environmental sustainability imperatives is shown in Table 7.3.

Addressing these environmental sustainability imperatives requires coordinated action, collaboration, and commitment from governments, businesses, civil society organizations, and individuals to achieve a more sustainable and resilient future for people and the planet. By recognizing the interconnectedness of environmental, social, and economic systems, stakeholders can work together to implement solutions that balance ecological integrity, social equity, and economic prosperity.

The next driver to green jobs is policy and regulatory support. Policy and regulatory support refers to the framework of laws, regulations, incentives, and measures implemented by governments to promote environmentally sustainable practices, industries, and technologies (Figure 7.12). These policies aim to create an enabling environment for the transition to a green economy; drive

TABLE 7.3
Environmental Sustainability Imperatives

	Environmental sustainability imperatives	Explanation
ESI1	Climate change mitigation	The imperative to mitigate climate change involves reducing greenhouse gas emissions, transitioning to renewable energy sources, and implementing measures to adapt to the impacts of global warming. Climate change imperatives seek to limit temperature rise to well below 2 degrees Celsius above pre-industrial levels and pursue efforts to limit it to 1.5 degrees Celsius, as outlined in the Paris Agreement. Urgent action is needed to reduce carbon emissions from fossil fuel combustion, deforestation, industrial processes, and land use change to mitigate the adverse effects of climate change on ecosystems, communities, and economies.
ESI2	Biodiversity conservation	The imperative to conserve biodiversity involves protecting and preserving the variety of life on Earth, including species, ecosystems, and genetic diversity. Biodiversity imperatives aim to prevent species extinction, safeguard habitats, and maintain ecosystem resilience in the face of environmental pressures such as habitat loss, pollution, invasive species, and climate change. Conservation efforts focus on establishing protected areas, restoring degraded ecosystems, and promoting sustainable land use practices to preserve biodiversity and ecosystem services essential for human well-being.
ESI3	Resource efficiency and circular economy	The imperative to promote resource efficiency and adopt a circular economy involves optimizing resource use, minimizing waste generation, and promoting the reuse, recycling, and recovery of materials and resources throughout their lifecycle. Imperatives for resource efficiency aim to decouple economic growth from resource consumption, reduce environmental impacts, and enhance resource productivity while fostering sustainable production and consumption patterns. Transitioning to a circular economy involves redesigning products, processes, and systems to minimize waste generation, conserve resources, and promote sustainable resource management.
ESI4	Water conservation and management	The imperative to conserve and manage water resources involves ensuring equitable access to clean and safe water, protecting aquatic ecosystems, and promoting efficient water use in agriculture, industry, and urban areas. Water conservation imperatives address water scarcity, pollution, and degradation of freshwater ecosystems, emphasizing the sustainable management of water resources, watershed protection, and integrated water management approaches to meet the needs of current and future generations.
ESI5	Land use planning and sustainable agriculture	The imperative to promote sustainable land use planning and agriculture involves balancing competing demands for land while preserving natural habitats, soil fertility, and ecosystem services. Imperatives for sustainable land use and agriculture aim to promote land conservation, minimize land degradation, and enhance agricultural productivity in an environmentally sustainable manner. Sustainable agriculture practices focus on soil conservation, water management, biodiversity conservation, and agroecological approaches to ensure food security, livelihoods, and ecosystem health.
ESI6	Waste reduction and pollution prevention	The imperative to reduce waste generation and prevent pollution involves minimizing the release of pollutants into the environment, reducing the production of hazardous materials, and promoting waste reduction, reuse, and recycling. Imperatives for waste reduction and pollution prevention aim to minimize environmental contamination, protect human health, and conserve natural resources by adopting pollution prevention measures, implementing waste management strategies, and promoting cleaner production processes across industries.
ESI7	Environmental justice and equity	The imperative to promote environmental justice and equity involves ensuring fair and equitable distribution of environmental benefits and burdens, addressing environmental disparities, and empowering marginalized communities to participate in environmental decision-making processes. Environmental justice imperatives seek to address environmental racism, socioeconomic inequalities, and disproportionate exposure to environmental hazards and pollution faced by marginalized and vulnerable populations. Efforts to promote environmental justice involve advocating for inclusive policies, community engagement, and participatory approaches to environmental governance.

Source: created by authors according to Singh, Jyoti and Kumar (2021), Liu, Brown and Casazza (2017).

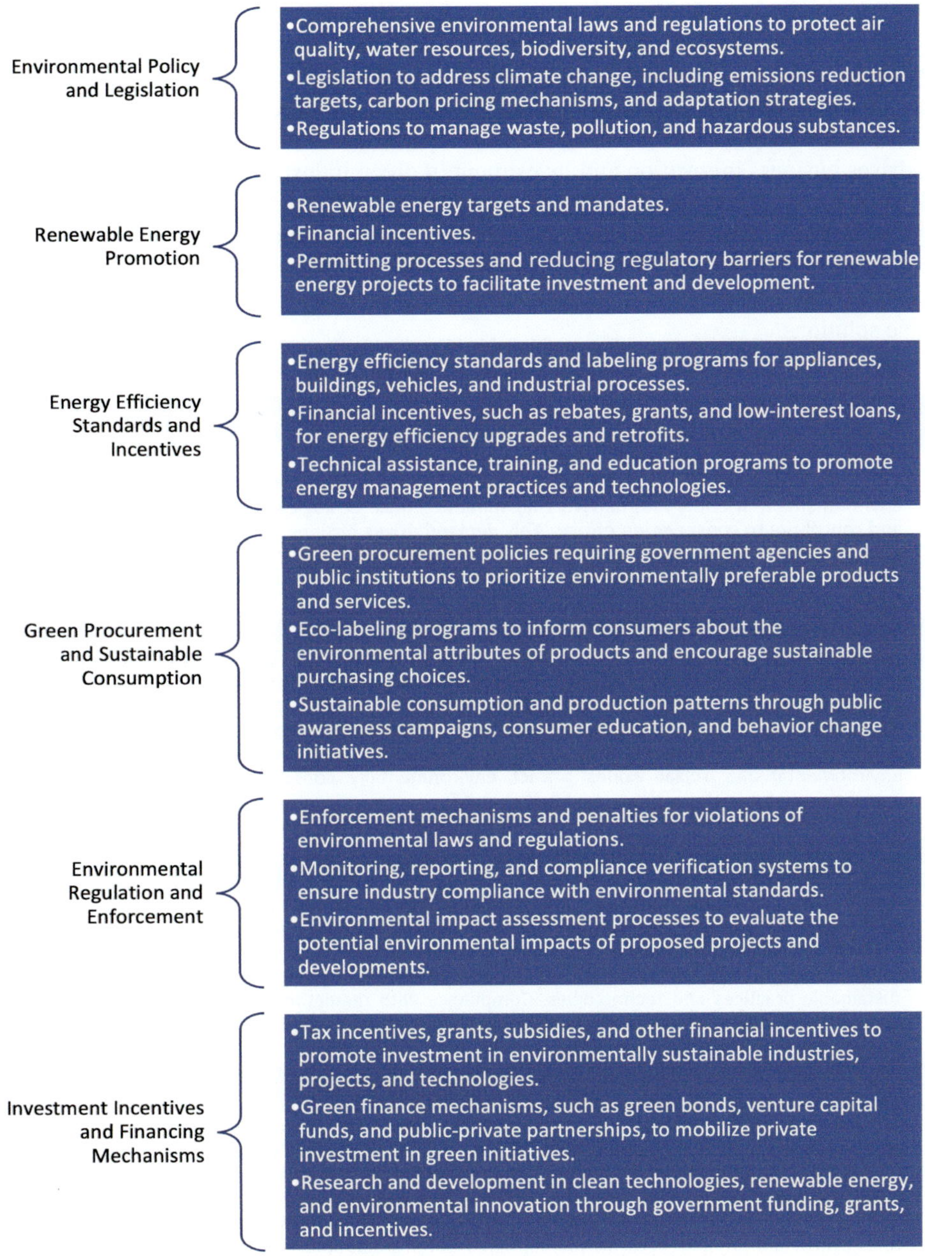

FIGURE 7.12 Policy and regulatory support system.

Source: created by authors according to Dell'Anna (2021), Green and Gambhir (2020).

investment in clean energy, conservation, and sustainable development; and address environmental challenges such as climate change, pollution, and resource depletion.

Supportive policy and regulatory frameworks play a critical role in driving green job creation by providing incentives, mandates, and support mechanisms for environmentally sustainable practices and industries. Policies such as renewable energy targets, energy efficiency standards, carbon pricing mechanisms, and green procurement policies create market demand, stimulate investment, and promote innovation in green sectors.

Market demand and consumer preferences play a significant role in driving the growth of green jobs by influencing business decisions, shaping industry trends, and driving innovation in environmentally sustainable products and services. How market demand and consumer preferences act as drivers of green jobs are shown in Figure 7.13. Growing market demand for sustainable products and services, coupled with shifting consumer preferences towards environmentally friendly options, drives business innovation and investment in green industries. Companies respond to consumer demand for eco-friendly products, energy-efficient technologies, and sustainable solutions by creating new business models, products, and services, leading to job creation in green value chains.

Overall, market demand and consumer preferences act as powerful drivers to green jobs by shaping industry practices, influencing business strategies, and driving investment in environmentally sustainable solutions.

Demand for Sustainable Products and Services

- Growing awareness of environmental issues, climate change, and resource scarcity has led to increased consumer demand for sustainable products and services. Consumers are seeking environmentally friendly alternatives that minimize their ecological footprint, reduce pollution, and contribute to conservation efforts. This demand drives businesses to develop and offer green products and services, creating opportunities for green job growth in areas such as renewable energy, energy efficiency, sustainable transportation, organic agriculture, and eco-friendly consumer goods.

Preference for Ethical and Responsible Brands

- Consumers are increasingly choosing to support brands and companies that demonstrate social and environmental responsibility. Ethical considerations, such as corporate social responsibility (CSR), environmental stewardship, and ethical sourcing practices, influence purchasing decisions and brand loyalty. Businesses that prioritize sustainability and demonstrate a commitment to environmental and social values attract environmentally conscious consumers and gain a competitive edge in the market, driving demand for green products and services and creating jobs in sustainable industries.

Regulatory Compliance and Market Access

- Government regulations, policies, and standards related to environmental protection, energy efficiency, and product stewardship influence market dynamics and drive demand for green solutions. Regulatory requirements, such as emissions standards, energy efficiency mandates, and product labeling regulations, create market incentives for businesses to adopt environmentally sustainable practices and technologies. Compliance with regulatory requirements and access to green markets often require companies to invest in green technologies, innovation, and workforce training, generating demand for skilled workers and green jobs.

Cost Savings and Economic Benefits

- In many cases, environmentally sustainable practices and technologies offer cost savings, efficiency gains, and economic benefits for businesses and consumers. Energy-efficient products, renewable energy systems, waste reduction measures, and sustainable business practices can lower operating costs, increase productivity, and enhance competitiveness in the market. Businesses that invest in green technologies and sustainability initiatives can achieve long-term cost savings and financial returns, driving demand for green solutions and creating jobs in green industries and sectors.

Innovation and Technological Advancements

- Market demand for green products and services stimulates innovation and technological advancements in clean technologies, renewable energy, and sustainable solutions. Businesses respond to consumer preferences and market trends by investing in research, development, and innovation to create new products, processes, and technologies that are environmentally friendly, energy-efficient, and socially responsible. Technological innovation drives job creation in research and development, engineering, manufacturing, and other specialized fields, contributing to the growth of green industries and the green economy.

FIGURE 7.13 Market demand and consumer preferences act as drivers of green jobs.

Source: created by authors according to 2022; Dell'Anna (2021); Yamane and Kaneko (2021), Zhang and Yousaf (2020).

One more driver to green jobs is investment and financing opportunities. Increasing investment and financing opportunities in green industries and projects, including venture capital, impact investment, green bonds, and public-private partnerships, provide capital for innovation, infrastructure development, and business expansion in environmentally sustainable sectors. Access to finance facilitates the scaling up of green initiatives, accelerates market penetration, and stimulates job creation in green value chains. Elements of the investment and financing opportunities as drivers of green jobs are shown in Figure 7.14.

FIGURE 7.14 Elements of the investment and financing opportunities as drivers of green jobs.

Source: created by authors according to Dell'Anna (2021), Ning, Cherian, Sial, Álvarez-Otero and Kour (2023), Sulich and Rutkowska (2020).

Green investment funds and impact investors focus on generating positive environmental and social outcomes alongside financial returns by investing in environmentally sustainable businesses and projects. These funds target investments in renewable energy, clean technology, sustainable agriculture, and conservation initiatives that address pressing environmental challenges and contribute to Sustainable Development Goals. Impact investors provide patient capital, technical assistance, and support for capacity-building efforts, enabling green enterprises to scale up operations, create jobs, and deliver measurable environmental and social benefits.

Overall, investment and financing opportunities are essential drivers to green jobs, providing the financial resources, incentives, and support needed to accelerate the transition towards a low-carbon, resource-efficient economy. By mobilizing capital for green initiatives, fostering innovation, and creating market incentives for sustainable investments, investment and financing opportunities play a pivotal role in unlocking the economic potential of the green economy and driving job creation in environmentally sustainable industries and occupations.

Investments in workforce development, skills training, and education programs tailored to the needs of green industries and occupations help build a skilled workforce equipped with the expertise needed to drive innovation, support technological adoption, and meet the demands of a transitioning economy (Wang and Wang, 2021). Training programs, apprenticeships, and vocational education initiatives prepare individuals for green careers and enhance employability in emerging green sectors.

Global efforts to address environmental challenges, such as the Paris Agreement on climate change, the Sustainable Development Goals (SDGs), and international environmental agreements, create opportunities for cooperation, knowledge sharing, and capacity building in green technology transfer, sustainable development, and green job creation (Adamowicz, 2022). Collaboration among countries, organizations, and stakeholders promotes the exchange of best practices, fosters innovation, and catalyzes green growth at the global level.

By understanding and leveraging these drivers to green jobs, stakeholders can harness the potential of environmentally sustainable industries and occupations to drive economic growth, promote social inclusion, and advance environmental sustainability goals. Aligning policies, investments, and strategies with key drivers to green jobs helps accelerate the transition towards a low-carbon, resource-efficient economy while creating opportunities for job creation, innovation, and prosperity.

While green jobs offer numerous benefits, including environmental sustainability, economic growth, and social well-being, they also face several barriers that hinder their widespread adoption and implementation. These barriers can vary depending on factors such as geographical location, economic conditions, policy frameworks, and technological readiness. Here are some common barriers to green jobs (Chiappinelli, Gerres, Neuhoff et al., 2021; Gholami, Bachok, Saman and Streimikiene, 2020; Sulich and Sołoducho-Pelc, 2022; Chien, Ngo, Hsu, Chau and Iram, 2021):

1. Cost and Economic Viability. One of the primary barriers to green jobs is the perceived or actual higher upfront costs associated with environmentally sustainable practices, technologies, and industries. Investments in renewable energy, energy efficiency, sustainable transportation, and green infrastructure often require significant capital expenditure, which can deter businesses and individuals, particularly in regions where conventional, fossil fuel-based alternatives are cheaper in the short term. Lack of access to affordable financing, limited financial incentives, and uncertainty about long-term returns on investment can further impede the economic viability of green initiatives.

Cost and economic viability encompass various factors (Table 7.4) that influence the affordability, profitability, and financial feasibility of adopting green technologies, practices, and industries.

By considering these aspects of cost and economic viability, stakeholders can assess the financial implications, risks, and opportunities associated with green investments, identify barriers to adoption, and develop strategies to overcome challenges and promote the economic viability of green jobs and industries.

TABLE 7.4
Cost and Economic Viability Factors in the Context of Green Jobs

Factor	Definition	Interpretation
Upfront investment costs	Cost and economic viability often revolve around the initial capital investment required to adopt green technologies and practices.	Upfront investment costs include expenditures on equipment, infrastructure, technology deployment, and implementation of environmentally sustainable measures. For example, installing solar panels, retrofitting buildings for energy efficiency, transitioning to electric vehicles, or implementing sustainable manufacturing processes may require substantial upfront investments, which can be perceived as barriers to adoption, particularly for small businesses or individuals with limited financial resources.
Operational and lifecycle costs	In addition to upfront investment costs, cost and economic viability considerations extend to operational and lifecycle costs associated with maintaining and operating green technologies and systems.	These costs include ongoing expenses for maintenance, repairs, upgrades, and replacement of equipment and infrastructure over their lifespan. Understanding and accurately assessing operational and lifecycle costs are essential for evaluating the long-term financial viability and return on investment of green initiatives, as they impact the total cost of ownership and economic feasibility of sustainable solutions compared to conventional alternatives.
Return on investment (ROI) and payback period	Cost and economic viability are often evaluated in terms of return on investment (ROI) and payback period, which measure the financial returns and time required to recoup initial investment costs.	Calculating ROI involves comparing the benefits or savings generated by green investments, such as reduced energy bills, operational efficiencies, productivity gains, or environmental benefits, to the upfront investment costs. A shorter payback period indicates faster recoupment of investment and higher economic viability, while a longer payback period may pose challenges to financial feasibility and hinder investment attractiveness.
Total cost of ownership (TCO)	Cost and economic viability considerations extend beyond upfront and operational costs to encompass the total cost of ownership (TCO) over the entire lifecycle of green technologies and solutions.	TCO analysis takes into account all direct and indirect costs associated with acquiring, deploying, operating, maintaining, and disposing of assets or systems, including energy consumption, labor costs, financing expenses, environmental externalities, and end-of-life disposal costs. Assessing TCO provides a comprehensive understanding of the economic implications and cost-effectiveness of green investments compared to conventional alternatives, helping stakeholders make informed decisions about resource allocation and investment priorities.
Financial incentives and support mechanisms	Cost and economic viability can be influenced by the availability of financial incentives, subsidies, grants, tax credits, and other support mechanisms that reduce the financial barriers and improve the economic attractiveness of green investments.	Government incentives, such as renewable energy feed-in tariffs, energy efficiency rebates, green building certifications, or low-interest loans for sustainable projects, can offset upfront costs, enhance ROI, and accelerate the adoption of environmentally sustainable practices and technologies by making them more financially competitive and attractive to businesses and individuals.
Economies of scale and technological advancements	Cost and economic viability considerations are influenced by economies of scale, technological advancements, and innovation in green technologies and industries.	As production volumes increase and technologies mature, economies of scale may lead to cost reductions, economies in production, and efficiencies in resource utilization, making green solutions more affordable and economically viable over time. Technological advancements, such as improvements in efficiency, performance, and cost-effectiveness of renewable energy systems, energy storage technologies, or sustainable materials, contribute to lowering costs and improving the economic competitiveness of green alternatives compared to conventional options.
Externalities and non-market considerations	Cost and economic viability assessments also take into account externalities and non-market considerations associated with environmental impacts, social benefits, and public health outcomes.	Externalities, such as carbon emissions, air pollution, water contamination, and ecosystem degradation, impose hidden costs and environmental risks that may not be reflected in market prices but have significant economic implications and societal consequences. Accounting for externalities and non-market considerations through tools such as cost-benefit analysis, environmental valuation, and lifecycle assessment helps integrate environmental and social factors into economic decision-making and promotes more sustainable and equitable outcomes.

Source: created by authors according to Zhao, Patwary, Qayyum, Alharthi and Bashir (2022), Dell'Anna (2021).

2. Lack of Policy Support and Regulatory Uncertainty. Inadequate policy support and regulatory uncertainty present significant barriers to the growth of green jobs. The absence of supportive policy frameworks, such as renewable energy targets, carbon pricing mechanisms, energy efficiency standards, and green procurement policies, can hinder market development, investment, and innovation in green industries. Regulatory inconsistencies, changing policy priorities, and political instability create uncertainty for businesses, investors, and entrepreneurs, undermining confidence in green initiatives and deterring long-term planning and investment.
3. Technology and infrastructure constraints. Technological and infrastructure constraints pose challenges to the widespread adoption of green technologies and practices. Limited access to advanced technologies, inadequate infrastructure, and underdeveloped supply chains can hinder the deployment and scalability of renewable energy systems, energy-efficient technologies, and sustainable solutions. Insufficient research and development, skills gaps, and lack of technical expertise further impede technological innovation and inhibit the growth of green industries and job opportunities.

Technology and infrastructure constraints as barriers to green jobs (Figure 7.15) refers to the challenges and limitations arising from inadequate technological capabilities, underdeveloped infrastructure, and insufficient support systems that hinder the adoption, deployment, or scalability of environmentally sustainable practices, technologies, or industries, thereby impeding the creation or expansion of green employment opportunities.

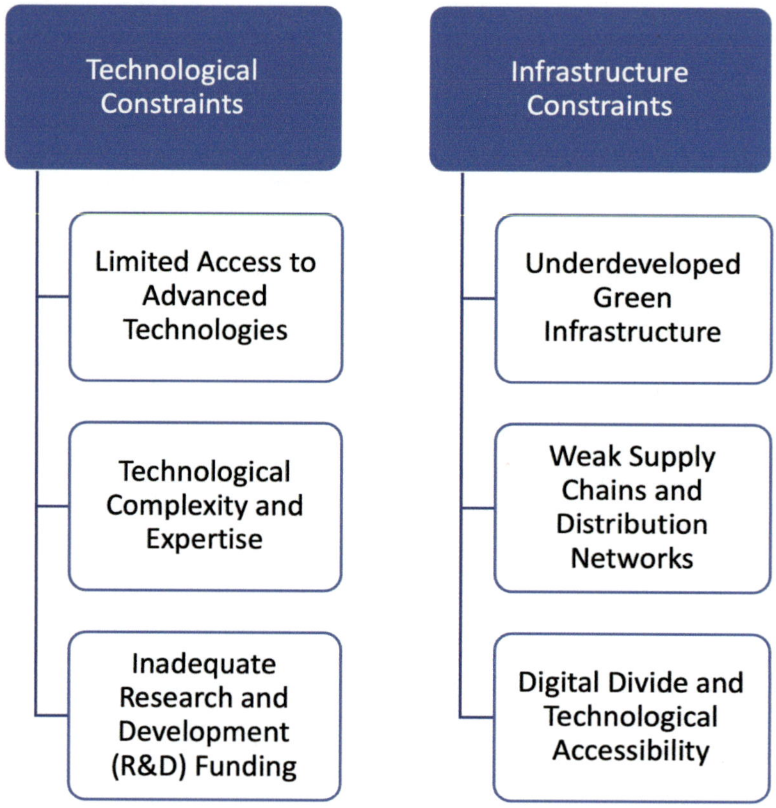

FIGURE 7.15 Technology and infrastructure constraints as barriers to green jobs.

Source: created by authors.

Technology constraints may arise from limited access to advanced or state-of-the-art green technologies, innovations, or solutions that are essential for implementing environmentally sustainable practices or processes. Lack of access to cutting-edge technologies, research and development (R&D) facilities, or intellectual property rights may hinder businesses or organizations from adopting or developing innovative solutions to address environmental challenges and create green jobs. High technical complexity, specialized skills requirements, or lack of qualified professionals in fields such as renewable energy, energy efficiency, waste management, or green manufacturing may pose barriers to workforce participation, skill development, or technology transfer in green industries.

Infrastructure constraints may arise from underdeveloped or insufficient green infrastructure, including renewable energy generation capacity, energy-efficient buildings, sustainable transportation systems, waste management facilities, or water and sanitation networks. Inadequate infrastructure investment, lack of planning, or outdated infrastructure systems may limit the availability or accessibility of green services, technologies, or amenities required to support sustainable living, economic activities, or job creation.

Infrastructure constraints can stem from weak supply chains, logistics, or distribution networks that impede the delivery, installation, or maintenance of green technologies or products to end users. Fragmented supply chains, logistical bottlenecks, or inadequate transportation networks may increase costs, lead times, or inefficiencies in sourcing, distributing, or servicing green solutions, thereby hindering market access and adoption of environmentally sustainable practices.

In summary, technology and infrastructure constraints as barriers to green jobs encompass challenges related to inadequate technological capabilities, underdeveloped infrastructure, and digital inequalities that hinder the adoption, deployment, or scalability of environmentally sustainable practices, technologies, or industries, thereby limiting the potential for green job creation and economic growth. Addressing these barriers requires investment in research, development, and innovation, upgrading infrastructure systems, enhancing technological accessibility, and promoting inclusive green growth strategies to unlock the full potential of green jobs and foster sustainable development.

4. Market barriers and consumer preferences. Market barriers, such as lack of consumer awareness, low demand for green products and services, and preferences for conventional alternatives, pose challenges to the expansion of green jobs. Consumer behavior, cultural norms, and perceptions of green products may influence purchasing decisions, limiting market uptake and inhibiting business growth in environmentally sustainable sectors. Negative perceptions about the quality, performance, and affordability of green products and services can deter adoption and hinder market penetration, particularly in price-sensitive markets. Market barriers and consumer preferences present challenges to the widespread adoption and growth of green jobs by influencing purchasing decisions, market dynamics, and business strategies. Elements of market barriers and consumer preferences in the context of green jobs are shown in Table 7.5.

By addressing market barriers and understanding consumer preferences, businesses, policymakers, and stakeholders can overcome challenges to the growth of green jobs and industries, promote market acceptance of environmentally sustainable solutions, and accelerate the transition to a more sustainable and resilient economy. Strategies such as consumer education, market segmentation, product innovation, pricing strategies, and policy interventions can help unlock market potential, drive demand for green products and services, and create new opportunities for green job creation and economic growth.

5. Skills and workforce development challenges. Skills and workforce development challenges represent significant barriers to green job creation and recruitment. Green jobs often require specialized skills, training, and expertise in areas such as renewable energy, energy

TABLE 7.5
Elements of Market Barriers and Consumer Preferences in the Context of Green Jobs.

Element of market barriers and consumer preferences	Definition	Interpretation
Lack of consumer awareness	Market barriers often stem from a lack of consumer awareness and understanding of green products, services, and technologies.	Many consumers may be unfamiliar with the environmental benefits, performance characteristics, and cost savings associated with green alternatives compared to conventional products. Limited awareness of green options may result in low demand, slow market uptake, and challenges in reaching target audiences with green messaging and education initiatives.
Perception of higher costs	Consumer preferences may be influenced by the perception that green products and services are more expensive or less affordable than conventional alternatives.	While green solutions may offer long-term cost savings and environmental benefits, upfront investment costs or premium pricing may deter price-sensitive consumers from choosing green options, particularly in markets where affordability and budget constraints are primary considerations. Overcoming perceptions of higher costs through pricing strategies, financial incentives, and value proposition communication is essential to expanding market demand for green products and services.
Quality and performance concerns	Market barriers may arise from consumer concerns about the quality, reliability, and performance of green products and services compared to conventional counterparts.	Skepticism about the durability, efficiency, and effectiveness of green technologies or sustainable solutions may deter consumers from making purchasing decisions, particularly if they perceive green alternatives as inferior or less capable of meeting their needs or expectations. Addressing quality and performance concerns through product certification, standards compliance, and performance guarantees is essential for building consumer trust and confidence in green options.
Limited availability and accessibility	Market barriers can be attributed to the limited availability and accessibility of green products and services in retail outlets, distribution channels, and geographic regions.	Accessibility challenges, such as inadequate market penetration, limited product availability, or lack of distribution networks, may restrict consumer access to green options, particularly in underserved or remote areas where green infrastructure and supply chains are less developed. Improving accessibility through expanded distribution channels, online platforms, and partnerships with retailers and suppliers can enhance market reach and facilitate consumer access to green alternatives.
Preference for familiar brands and technologies	Consumer preferences may favor familiar brands, products, and technologies over new or unfamiliar green alternatives, particularly if consumers perceive established brands as trustworthy, reliable, or culturally resonant.	Brand loyalty, habit formation, and emotional attachments to familiar products may influence purchasing decisions, even if green alternatives offer comparable or superior performance and environmental benefits. Overcoming brand loyalty and promoting behavior change through marketing campaigns, product differentiation, and consumer education efforts can help shift consumer preferences towards green options and foster market acceptance of sustainable alternatives.
Cultural and social norms	Market barriers may arise from cultural, social, and behavioral norms that shape consumer attitudes, values, and purchasing behaviors.	Cultural preferences, lifestyle choices, and social influences may impact consumer perceptions of green products and sustainability, influencing adoption rates and market acceptance of environmentally friendly options. Addressing cultural and social barriers through targeted messaging, community engagement, and culturally tailored marketing strategies can help align green solutions with consumer values, preferences, and aspirations, driving demand and market growth.
Regulatory and policy environment	Market barriers can be influenced by the regulatory and policy environment, including government regulations, standards, and incentives that shape market dynamics and consumer choices.	Inconsistent or conflicting regulations, lack of enforcement, or absence of supportive policies and incentives may hinder market development, discourage investment, and limit consumer confidence in green options. Clear and consistent regulatory frameworks, along with supportive policies such as tax incentives, subsidies, and green procurement requirements, can create market incentives, stimulate demand, and drive the adoption of green products and services.

Source: created by authors according to Śmiglak-Krajewska and Wojciechowska-Solis (2021), Chien, Ngo, Hsu, Chau and Iram (2021).

efficiency, environmental management, and sustainable practices. However, skills shortages, mismatches between job requirements and available talent, and lack of formal education and training programs tailored to green industries can hinder workforce participation, limit job mobility, and impede the growth of green employment opportunities.

6. Institutional and organizational barriers. Institutional and organizational barriers, such as bureaucratic inefficiencies, institutional inertia, and resistance to change, can impede the implementation of green initiatives and the adoption of environmentally sustainable practices. Lack of coordination between government agencies, insufficient interagency cooperation, and siloed decision-making processes can slow down policy implementation, delay project approvals, and increase administrative burdens for businesses and stakeholders involved in green projects. Organizational culture, management practices, and stakeholder engagement strategies may also influence the pace and success of green transitions within companies and institutions.

Institutional and organizational barriers as barriers to green jobs (Table 7.6) refer to the challenges and obstacles arising from bureaucratic structures, organizational practices, and decision-making processes within institutions and businesses that hinder the creation, adoption, or growth of environmentally sustainable employment opportunities.

TABLE 7.6
Institutional and Organizational Barriers as Barriers to Green Jobs.

Institutional and organizational barriers		The cause of the barriers	Examples
Institutional barriers	Bureaucratic inefficiencies	Institutional barriers may arise from bureaucratic inefficiencies, rigid procedures, and administrative hurdles within government agencies, regulatory bodies, or public institutions responsible for overseeing environmental regulations, permitting processes, or funding allocation for green projects.	Cumbersome administrative procedures, lengthy approval timelines, and bureaucratic red tape can delay project implementation, increase compliance costs, and discourage investment in green initiatives.
	Regulatory fragmentation and complexity	Institutional barriers may stem from regulatory fragmentation, overlapping jurisdictions, or inconsistent enforcement of environmental laws and regulations across different government agencies or levels of governance.	Regulatory complexity, ambiguous standards, or conflicting requirements may create compliance challenges for businesses, leading to regulatory uncertainty, legal risks, or delays in obtaining permits or approvals for green projects.
	Lack of coordination and collaboration	Institutional barriers can result from a lack of coordination, collaboration, or communication between government agencies, departments, or stakeholders involved in green initiatives.	Fragmented governance structures, siloed decision-making processes, or turf battles between agencies may hinder interagency cooperation, impede information sharing, and slow down progress on cross-cutting environmental issues or initiatives.
	Policy inertia and resistance to change	Institutional barriers may arise from policy inertia, resistance to change, or entrenched interests within government institutions or regulatory bodies that are resistant to adopting new approaches, technologies, or practices to promote environmental sustainability.	Institutional inertia, bureaucratic resistance, or opposition from vested interests may impede policy innovation, limit regulatory flexibility, or delay the adoption of progressive measures to support green job creation or industry transformation.

(Continued)

TABLE 7.6 *(Continued)*
Institutional and Organizational Barriers as Barriers to Green Jobs.

Organizational barriers	Resistance to innovation and risk aversion	Organizational barriers may stem from resistance to innovation, risk aversion, or cultural norms within businesses, corporations, or organizations that inhibit the adoption of environmentally sustainable practices or technologies.	Organizational cultures that prioritize short-term profits, maintain status quo, or resist change may be reluctant to invest in green initiatives, implement sustainability measures, or embrace disruptive technologies that could transform operations or business models.
	Lack of strategic alignment and vision	Organizational barriers can result from a lack of strategic alignment, vision, or leadership commitment to sustainability goals within businesses or institutions.	Absence of clear sustainability objectives, performance metrics, or accountability mechanisms may hinder integration of environmental considerations into organizational decision-making processes, strategic planning, or corporate governance practices.
	Inadequate resources and capacity	Organizational barriers may arise from inadequate resources, capacity constraints, or limited expertise within businesses or institutions to implement green initiatives, develop sustainable products, or adopt eco-friendly practices.	Lack of dedicated funding, technical skills, or specialized knowledge may impede organizational efforts to invest in green technologies, conduct environmental assessments, or comply with sustainability standards.
	Resistance to change and stakeholder engagement	Organizational barriers can stem from resistance to change, lack of employee engagement, or insufficient stakeholder involvement in decision-making processes related to sustainability initiatives.	Resistance from employees, management, or shareholders to embrace sustainability goals, participate in green projects, or align business practices with environmental objectives may hinder organizational transformation efforts and impede progress toward green job creation or industry sustainability.

Source: created by authors according to Kiefer, Del Río González and Carrillo-Hermosilla (2019).

Institutional and organizational barriers to green jobs encompass challenges related to bureaucratic inefficiencies, regulatory fragmentation, policy inertia, resistance to innovation, and organizational culture within institutions and businesses that hinder the adoption, implementation, or scaling up of environmentally sustainable practices, technologies, or industries. Overcoming these barriers requires fostering collaboration, enhancing coordination, promoting leadership commitment, and cultivating a culture of innovation and sustainability within institutions and organizations to drive progress toward a green economy and create opportunities for sustainable employment.

7. Socioeconomic and equity considerations. Socioeconomic disparities, inequitable access to resources, and social exclusion can exacerbate barriers to green jobs, particularly for marginalized and vulnerable populations. Economic inequalities, lack of access to education and training opportunities, and discrimination based on gender, race, ethnicity, or socioeconomic status can limit participation in green industries and perpetuate disparities in employment outcomes. Addressing equity considerations, promoting social inclusion, and ensuring equal access to green employment opportunities are essential for building a more inclusive and equitable green economy.

When considering social and economic aspects as barriers to green jobs, we're examining how factors related to society and the economy can hinder the creation, growth, or adoption of environmentally sustainable employment opportunities.

Social Aspects as Barriers to Green Jobs (Sulich and Sołoducho-Pelc, 2022; Kodua, Xiao, Adjei, Asante and Ofosu, 2022):

- Social aspects as barriers to green jobs encompass various social dynamics, behaviors, and cultural norms that may impede the transition to a green economy or hinder the creation of green employment opportunities. These barriers can include:
  - Resistance to change. Societal resistance to adopting new technologies, practices, or industries due to cultural norms, traditions, or perceptions.
  - Lack of awareness and education. Limited understanding or awareness of environmental issues, sustainability concepts, or green job opportunities within the workforce or society.
  - Skills mismatch and training gaps. Inadequate skills, training, or education among the workforce to meet the demands of green industries or occupations leads to mismatches between job requirements and available talent.
  - Social equity and inclusion. Inequitable access to green employment opportunities or disparities in workforce participation due to factors such as gender, race, ethnicity, socioeconomic status, or geographic location.
  - Community resistance or nimbyism. Opposition from local communities to green projects or initiatives, such as renewable energy installations or waste management facilities, due to concerns about property values, environmental impacts, or perceived risks.

Economic Aspects as Barriers to Green Jobs (Tumpa, Ali, Rahman and Paul, 2019; García-Quevedo, Jové-Llopis and Solà, 2020):

- Economic aspects as barriers to green jobs encompass various economic factors, market dynamics, and structural challenges that may hinder the growth or viability of green industries or employment opportunities. These barriers can include:
  - High initial costs and investment barriers. Upfront investment costs or financial barriers associated with adopting green technologies, practices, or infrastructure may deter businesses or individuals from pursuing green initiatives.
  - Market barriers and consumer preferences. Limited market demand or consumer preferences for green products or services compared to conventional alternatives lead to challenges in scaling up green businesses or attracting investment.
  - Regulatory uncertainty and policy constraints. Inconsistent or unclear regulatory frameworks, lack of supportive policies, or absence of financial incentives may hinder business investment in green projects or limit market development.
  - Economic disruptions and transition challenges. Economic disruptions or structural changes resulting from the transition to a green economy, such as job displacement in traditional industries or challenges in retraining workers for green occupations.
  - Access to financing and capital. Limited access to financing, credit, or investment capital for green businesses, startups, or projects, particularly in regions or sectors with underdeveloped financial markets or investor skepticism about green investments.

In summary, social and economic aspects can act as barriers to green jobs by impeding the adoption of environmentally sustainable practices, technologies, or industries due to factors such as societal attitudes, cultural norms, market dynamics, financial constraints, or regulatory challenges. Overcoming these barriers requires concerted efforts to address social equity, promote awareness and education, incentivize green investment, and create supportive policy frameworks that foster the growth of green industries and employment opportunities.

By addressing these barriers through targeted policies, investments, capacity-building initiatives, and stakeholder engagement efforts, governments, businesses, and civil society organizations can overcome challenges and unlock the full potential of green jobs to drive sustainable development, economic growth, and environmental resilience.

7.4 POLICIES AND MEASURES TO PROMOTE GREEN EMPLOYMENT

Policies to promote green employment are essential for fostering a sustainable economy and creating jobs that contribute to environmental protection, resource efficiency, and low-carbon growth.

There are several types of policies that governments and institutions can implement to promote green employment. First is regulatory policies:

- Environmental regulations (see Table 7.7).
- Energy efficiency standards (see Table 7.8).
- Renewable energy mandates (see Table 7.9).

Regulatory policies set the rules and standards that guide the behavior of businesses, industries, and individuals toward more sustainable practices.

These environmental regulations not only help protect the environment and public health but also stimulate economic growth by creating green jobs and promoting innovation in sustainable technologies and practices.

TABLE 7.7
Examples of Environmental Regulations That can Help Promote Green Jobs by Encouraging Sustainable Practices and Technologies

Regulatory policy	**Explanation**	
Environmental regulations	Implementing stringent environmental regulations, such as emissions standards, waste management rules, and pollution controls, can drive businesses to adopt greener practices and technologies, creating demand for green jobs in compliance and monitoring.	
Examples of environmental regulations	**Overview**	**Impact**
Clean Air Act (United States)	The Clean Air Act regulates air emissions from stationary and mobile sources to ensure that air quality meets health-based standards.	It requires industries to adopt pollution control technologies, leading to job creation in the fields of environmental engineering, air quality monitoring, and the manufacturing of emission control equipment.
European Union Emissions Trading System (EU ETS)	The EU ETS is a cap-and-trade system that limits the total amount of greenhouse gases that can be emitted by covered entities, allowing them to buy and sell emission allowances.	This system incentivizes companies to reduce emissions and invest in low-carbon technologies, creating jobs in renewable energy, carbon management, and energy efficiency services.
Renewable Energy Directive (RED II) (European Union)	The Renewable Energy Directive sets binding targets for renewable energy consumption within the EU, aiming for a 32% share of renewable energy by 2030.	It drives investments in renewable energy projects such as wind, solar, and bioenergy, resulting in job creation in the installation, maintenance, and operation of renewable energy systems.

TABLE 7.7 *(Continued)*

Examples of Environmental Regulations That can Help Promote Green Jobs by Encouraging Sustainable Practices and Technologies

Examples of environmental regulations	Overview	Impact
Endangered Species Act (United States)	The Endangered Species Act aims to protect and recover imperiled species and the ecosystems upon which they depend.	It creates jobs in conservation biology, wildlife management, and environmental consultancy, as industries must assess and mitigate impacts on protected species and habitats.
Waste Electrical and Electronic Equipment (WEEE) Directive (European Union)	The WEEE Directive sets collection, recycling, and recovery targets for electrical and electronic equipment to reduce waste and encourage resource efficiency.	It fosters job growth in the recycling and waste management sectors, as companies develop systems to collect and process electronic waste.
National Environmental Policy Act (NEPA) (United States)	NEPA requires federal agencies to assess the environmental impacts of their proposed actions and consider alternatives before making decisions.	This regulation promotes job creation in environmental impact assessment, planning, and consulting, as well as in the implementation of mitigation measures.
Fuel Economy Standards (Various Countries)	Fuel economy standards set requirements for the average fuel efficiency of new vehicles, reducing greenhouse gas emissions and reliance on fossil fuels.	These standards drive innovation in the automotive industry, leading to jobs in the design, manufacture, and servicing of fuel-efficient and electric vehicles.
Extended Producer Responsibility (EPR) Regulations (Various Countries)	EPR regulations hold producers responsible for the entire lifecycle of their products, including take-back, recycling, and disposal.	These regulations create jobs in product design, recycling, and waste management industries, as companies work to meet EPR requirements.
Climate Action Plans (Various Countries and Cities)	Climate action plans outline strategies and policies to reduce greenhouse gas emissions and enhance climate resilience at national, regional, or local levels.	Implementing these plans creates jobs in sectors such as renewable energy, energy efficiency, public transportation, and sustainable urban development.

Source: created by authors.

TABLE 7.8

Some Examples of Energy Efficiency Standards That Help Reduce Energy Consumption, Lower Greenhouse Gas Emissions, and Create Green Jobs

Regulatory policy	Explanation
Energy efficiency standards	Establishing minimum energy performance standards for buildings, appliances, and industrial equipment encourages the adoption of energy-efficient technologies, generating jobs in retrofitting, auditing, and manufacturing of efficient products.

Examples of environmental regulations	Overview	Impact
Energy Star Program (United States)	The Energy Star program is a voluntary labeling initiative that identifies and promotes energy-efficient products and buildings.	Energy Star standards cover a wide range of products, including appliances, electronics, lighting, and heating and cooling systems, as well as commercial and residential buildings. The program drives demand for energy-efficient products and services, creating jobs in manufacturing, installation, and maintenance.

(*Continued*)

TABLE 7.8 ***(Continued)***
Some Examples of Energy Efficiency Standards That Help Reduce Energy Consumption, Lower Greenhouse Gas Emissions, and Create Green Jobs

Examples of environmental regulations	Overview	Impact
Building Energy Codes (Various Countries)	Building energy codes set minimum energy efficiency requirements for new and renovated buildings.	These codes create jobs in architecture, construction, engineering, and building inspections as professionals work to meet and enforce energy efficiency standards.
Minimum Energy Performance Standards (MEPS) (Various Countries)	MEPS are regulations that set minimum energy efficiency levels for appliances and equipment.	MEPS encourage manufacturers to produce energy-efficient products, leading to job creation in design, testing, and compliance verification.
Corporate Average Fuel Economy (CAFE) Standards (United States)	CAFE standards regulate the average fuel economy of new cars and light trucks sold in the US	These standards incentivize automotive manufacturers to develop and produce more fuel-efficient vehicles, creating jobs in automotive engineering, research and development, and manufacturing of advanced vehicle technologies.
Japan Top Runner Program	The Top Runner Program sets energy efficiency standards based on the best-performing products in each category, requiring other manufacturers to meet or exceed this benchmark.	This program covers a wide range of products, including vehicles, appliances, and industrial equipment. It drives innovation and job creation in product development, testing, and certification.
California Title 24 Building Energy Efficiency Standards	Title 24 sets energy efficiency requirements for new residential and nonresidential buildings in California.	These standards promote energy-efficient building design and construction practices, creating jobs in architecture, construction, energy auditing, and compliance verification.
Energy Conservation Building Code (ECBC) (India)	The ECBC sets energy efficiency standards for commercial buildings in India, focusing on aspects such as building envelope, HVAC systems, lighting, and electrical systems.	The ECBC promotes energy-efficient building design and construction practices, creating jobs in architecture, engineering, construction, and building energy management.
China's Energy Efficiency Standards for Appliances	China has implemented mandatory energy efficiency standards and labeling for a wide range of appliances, including air conditioners, refrigerators, and washing machines.	These standards drive the production and adoption of energy-efficient appliances, creating jobs in manufacturing, product testing, and energy auditing.
Australia's National Construction Code (NCC)	The NCC includes energy efficiency requirements for residential and commercial buildings, focusing on aspects such as insulation, windows, and HVAC systems.	The NCC promotes energy-efficient building practices, creating jobs in construction, architecture, engineering, and building inspection.

Source: created by authors.

These energy efficiency standards not only reduce energy consumption and greenhouse gas emissions but also stimulate economic growth by creating jobs in various sectors, including manufacturing, construction, engineering, and compliance verification.

These renewable energy mandates are critical for accelerating the transition to a sustainable energy system, reducing carbon emissions, and creating green jobs across various sectors of the economy.

TABLE 7.9
Some Notable Examples of Renewable Energy Mandates

Regulatory policy	Explanation	
Renewable energy mandates	Policies such as renewable portfolio standards (RPS) require a certain percentage of energy to come from renewable sources, stimulating investment in renewable energy projects and associated employment opportunities.	
Examples of environmental regulations	**Overview**	**Impact**
Renewable Portfolio Standards (RPS)	RPS are state-level policies that mandate that a specific percentage of electricity sold by utilities must come from renewable sources.	These standards drive investment in renewable energy projects, such as wind and solar farms, and create jobs in construction, operation, and maintenance of renewable energy facilities.
China's Renewable Energy Law	This law mandates that a certain percentage of electricity consumed must come from renewable sources.	The law has driven massive investments in wind, solar, hydro, and biomass energy projects, creating numerous jobs in these sectors and supporting industries.
Germany's Renewable Energy Sources Act (EEG)	The EEG provides a framework for promoting renewable energy through feed-in tariffs and other incentives.	The act has led to significant growth in renewable energy installations, particularly in wind and solar, creating jobs in manufacturing, installation, and maintenance of renewable energy systems.
Brazil's Proinfa Program	Proinfa (Program of Incentives for Alternative Electricity Sources) aims to diversify Brazil's energy mix by promoting wind, biomass, and small hydroelectric projects.	Proinfa has contributed to the development of Brazil's renewable energy sector, generating jobs in the construction and operation of renewable energy projects.
Australia's Renewable Energy Target (RET)	The RET requires that a specific amount of Australia's electricity comes from renewable sources.	The RET has driven significant investment in wind and solar projects, creating jobs in project development, construction, and ongoing operations.
South Africa's Renewable Energy Independent Power Producer Procurement Programme (REIPPPP)	REIPPPP is a competitive bidding process to procure renewable energy from independent power producers.	The program has attracted billions in investment and created thousands of jobs in construction, engineering, and operations of renewable energy projects.
Japan's Renewable Energy Act	The act supports the introduction of feed-in tariffs and other measures to promote renewable energy.	The act has driven investment in solar, wind, and geothermal energy, creating jobs in these sectors and related industries.
European Union Renewable Energy Directive	The directive sets binding targets for renewable energy across the EU.	The directive fosters job creation in renewable energy sectors such as wind, solar, biomass, and hydroelectric power, as well as in supporting industries like manufacturing and engineering.

Source: created by authors.

The next policies that promote green employment are fiscal policies:

- Subsidies and grants (see Figure 7.16 and Figure 7.17).
- Tax incentives (see Table 7.10).
- Green public procurement (see Figure 7.18).

Investment Tax Credit (ITC) (United States)

- The ITC provides a tax credit for investing in renewable energy projects.
- The credit is currently 26% of the installation cost for solar, wind, and other renewable energy systems. It has significantly boosted the solar industry by reducing upfront costs for both residential and commercial installations, leading to job creation in solar installation, manufacturing, and maintenance.

Production Tax Credit (PTC) (United States)

- The PTC provides a per-kilowatt-hour tax credit for electricity generated by qualified renewable energy facilities.
- This credit has been crucial in supporting the development of wind energy projects by providing financial incentives based on the amount of electricity produced, thus promoting investment and job growth in the wind energy sector.

Feed-in Tariffs (FITs) (Various Countries)

- FITs guarantee a fixed payment for renewable energy producers for the electricity they generate and feed into the grid.
- Germany: The Renewable Energy Sources Act (EEG) introduced FITs, significantly increasing renewable energy capacity and creating jobs in the wind, solar, and biomass sectors.
- Japan: Japan's FIT scheme, introduced after the Fukushima disaster, incentivizes solar, wind, and other renewables, leading to a rapid expansion of solar power installations and associated job growth.
- FITs provide long-term financial security for investors in renewable energy projects, driving the development of renewable energy industries and creating employment opportunities.

Renewable Energy Grants (Australia)

- Various grant programs support renewable energy projects, research, and development.
- Australian Renewable Energy Agency (ARENA): Provides funding for innovative renewable energy projects and research to increase the supply and lower the cost of renewable energy.
- Clean Energy Finance Corporation (CEFC): Offers finance and investment for renewable energy, energy efficiency, and low-emission technologies.
- These grants have helped fund numerous renewable energy projects, fostering innovation and job creation in Australia's renewable energy sector.

Green Homes Grant (United Kingdom)

- The Green Homes Grant provides vouchers to homeowners for energy efficiency improvements.
- Homeowners can receive up to £5,000 (or £10,000 for low-income households) for insulation, heat pumps, and other energy-saving measures, boosting jobs in the construction and home improvement sectors.

Subsidies for Electric Vehicles (EVs) (Various Countries)

- Governments provide subsidies to reduce the purchase cost of electric vehicles.
- Norway: Offers significant tax exemptions and subsidies for EV purchases, making EVs affordable and leading to high adoption rates.
- China: Provides subsidies for EV purchases and supports domestic EV manufacturing, creating jobs in the automotive and battery production industries.
- These subsidies promote the adoption of cleaner transportation options and stimulate job growth in the manufacturing, sales, and maintenance of electric vehicles.

FIGURE 7.16 Examples of subsidies and grants that promote green employment (1).

Source: created by authors.

Renewable Energy Subsidies (India)

- •The government provides various subsidies to support the deployment of renewable energy technologies.
- •Solar Subsidy Scheme: Offers capital subsidies for residential and commercial solar installations.
- •Wind Energy Subsidies: Provides incentives for wind power projects, including accelerated depreciation and generation-based incentives.
- •These subsidies have driven significant growth in India's solar and wind energy sectors, creating numerous jobs in installation, operations, and manufacturing.

Eco-friendly Appliance Rebates (Various Countries)

- •Rebates are offered to consumers who purchase energy-efficient appliances.
- •United States: Many states offer rebates for Energy Star-rated appliances like refrigerators, washing machines, and water heaters.
- •Australia: Rebates for energy-efficient appliances help reduce household energy consumption and encourage the use of eco-friendly products.
- •These rebates promote energy conservation and create jobs in the manufacturing and retail sectors of energy-efficient appliances.

Agricultural Grants for Renewable Energy (United States)

- •The Rural Energy for America Program (REAP) provides grants and loans to agricultural producers and rural small businesses for renewable energy systems and energy efficiency improvements.
- •This program helps rural communities reduce energy costs and supports job creation in renewable energy installation and maintenance in rural areas.

Energy Efficiency Grants (European Union)

- •The EU offers various grants and funding programs to support energy efficiency projects.
- •Horizon 2020: Provides funding for research and innovation in energy efficiency technologies and projects.
- •European Structural and Investment Funds (ESIF): Supports energy efficiency improvements in buildings and industries across EU member states.
- •These grants help improve energy efficiency, reduce energy costs, and create jobs in energy auditing, retrofitting, and building renovations.

FIGURE 7.17 Examples of subsidies and grants that promote green employment (2).

Source: created by authors.

TABLE 7.10
Examples of Tax Incentives

Fiscal policy	**Explanation**	
Tax incentives	Tax incentives are financial mechanisms used by governments to encourage investments in renewable energy, energy efficiency, and other environmentally sustainable practices. Offering tax credits, deductions, or exemptions for investments in green technologies, research and development, or environmentally friendly practices can incentivize businesses and individuals to pursue sustainable projects, fostering green job growth.	
Examples of tax incentives	**Overview**	**Impact**
Investment Tax Credit (ITC) (United States)	Provides a tax credit for investments in renewable energy projects.	It significantly reduces the upfront cost of renewable energy projects, encouraging investment and job creation in the solar and wind industries.

(*Continued*)

TABLE 7.10 *(Continued)*
Examples of Tax Incentives

Examples of tax incentives	Overview	Impact
Production Tax Credit (PTC) (United States)	Provides a per-kilowatt-hour tax credit for electricity generated by renewable energy facilities.	This incentive has driven substantial growth in the wind energy sector, promoting investments and creating jobs in wind farm development and operation.
Accelerated Depreciation Benefits (India)	Allows businesses to depreciate renewable energy assets at an accelerated rate.	This reduces the taxable income of companies investing in renewable energy, encouraging more businesses to adopt sustainable energy solutions and creating jobs in the renewable energy sector.
Enhanced Capital Allowances (ECAs) (United Kingdom)	Allows businesses to claim 100% first-year tax relief on investments in energy-saving equipment. ECAs cover a wide range of energy-efficient equipment, including heating, lighting, and insulation systems.	This incentive encourages businesses to invest in energy-efficient technologies, reducing their energy costs and carbon footprint while creating jobs in the manufacturing and installation of such technologies.
Renewable Energy Certificates (RECs) Tax Exemption (Australia)	Provides tax exemptions for income derived from the sale of RECs. Businesses and individuals generating renewable energy can earn RECs, which can be sold to offset the cost of renewable energy projects.	This exemption incentivizes the generation and sale of renewable energy, promoting investment in renewable energy projects and supporting job growth in the sector.
Green Investment Tax Allowance (GITA) (Malaysia)	Offers tax allowances for investments in green technology projects and services.	GITA encourages businesses to invest in green technologies, reducing their tax burden and promoting job creation in the green technology sector.
Energy Efficient Commercial Buildings Deduction (United States)	Provides a tax deduction for energy efficiency improvements in commercial buildings.	This deduction encourages the retrofitting of commercial buildings to improve energy efficiency, creating jobs in construction, engineering, and energy auditing.
Feed-in Tariff (FIT) Income Tax Exemption (Germany)	Exempts income earned from feed-in tariffs for renewable energy from taxation. Individuals and businesses that generate renewable energy and receive payments through FIT schemes can enjoy tax-exempt income.	This exemption promotes the adoption of renewable energy systems, leading to increased investment and job creation in renewable energy generation.
Tax Credits for Electric Vehicles (EVs) (Various Countries)	Provides tax credits for the purchase of electric vehicles.	These credits make EVs more affordable, encouraging their adoption and supporting jobs in the EV manufacturing and infrastructure sectors.
Renewable Heat Incentive (RHI) (United Kingdom)	Provides payments to individuals and businesses that use renewable energy to heat their buildings. Participants receive quarterly payments for seven years based on the amount of renewable heat generated by systems such as heat pumps, biomass boilers, and solar thermal panels.	The RHI encourages the use of renewable heating technologies, reducing carbon emissions and creating jobs in the installation and maintenance of renewable heating systems.

Source: created by authors.

FIGURE 7.18 Examples of Green Public Procurement.

Source: created by authors.

Fiscal policies use government spending and taxation to influence economic activities and promote green employment.

These subsidies and grants play a crucial role in reducing the financial barriers to adopting renewable energy and energy-efficient technologies, driving market growth, and creating a wide range of green jobs.

Tax incentives are financial mechanisms used by governments to encourage investments in renewable energy, energy efficiency, and other environmentally sustainable practices.

These tax incentives play a crucial role in lowering the financial barriers to adopting renewable energy and energy-efficient technologies, promoting sustainable practices, and creating a wide range of green jobs across various sectors.

Government policies that prioritize the purchase of environmentally friendly goods and services can stimulate demand for green products and services, supporting job creation in sustainable manufacturing and service sectors.

General definition of *Green Public Procurement* (GPP): Green Public Procurement (GPP) is a process whereby public authorities seek to procure goods, services, and works with a reduced environmental impact throughout their life cycle when compared to goods, services, and works with the same primary function that would otherwise be procured (Bryngemark, Söderholm and Thörn, 2023; Krieger and Zipperer, 2022). According to the United Nations Environment Programme (UNEP), Green Public Procurement is "the approach by which public authorities integrate environmental criteria into all stages of their procurement process, thus encouraging the spread of environmentally friendly goods and services." The European Commission defines GPP as "a process whereby public authorities seek to procure goods, services, and works with a reduced environmental impact throughout their life cycle compared to goods, services, and works with the same primary function that would otherwise be procured." Some examples of Green Public Procurement are shown in Figure 7.18.

Impact of Green Public Procurement:

- Environmental benefits. GPP reduces greenhouse gas emissions, conserves natural resources, minimizes waste, and promotes biodiversity by prioritizing products and services with lower environmental impacts.
- Economic benefits. It can lead to cost savings over the product lifecycle, stimulate green innovation and markets, and create jobs in the green economy.
- Social benefits. GPP can improve public health by reducing pollution and exposure to hazardous substances, promoting fair labor practices, and enhancing the overall quality of life in communities.

By adopting GPP, public authorities can leverage their purchasing power to drive market demand for sustainable products and services, leading to broader environmental and societal benefits.

The next policies that promote green employment are *labor market policies*. Labor market policies aim to support workforce development and ensure that workers have the skills needed for green jobs. Some of the important policies are:

- Training and education programs. Training and education programs are crucial for equipping the workforce with the skills needed to support and expand green industries. These programs help workers transition to green jobs, enhance their qualifications, and ensure that businesses have access to a skilled labor pool capable of meeting the demands of a sustainable economy. Investing in vocational training, education, and apprenticeship programs focused on green skills can equip workers with the knowledge and capabilities needed for employment in emerging green industries such as renewable energy, energy efficiency, and sustainable construction. Some examples and a wider explanation of training and education programs that promote green employment are shown in Table 7.11.

TABLE 7.11
Examples and a Wider Explanation of Training and Education Programs that Promote Green Employment

Example	Description	Impact
	Vocational training and certification programs	
Renewable energy training programs	Programs that offer specialized training in the installation, maintenance, and operation of renewable energy systems such as solar panels, wind turbines, and geothermal systems.	These programs provide certifications that validate workers' expertise, making them more attractive to employers in the renewable energy sector and ensuring that installations are performed to high standards.
	Higher education degrees and courses	
Environmental Engineering and Sustainable Design degrees	Universities and colleges offer undergraduate and graduate degrees focused on environmental engineering, sustainable design, and green technologies.	Graduates from these programs are equipped with the theoretical knowledge and practical skills needed to design and implement sustainable solutions in various industries, from construction to manufacturing.
	Apprenticeships and on-the-job training	
Green construction apprenticeships	Apprenticeship programs in green construction provide hands-on training in sustainable building practices, energy-efficient construction techniques, and the use of eco-friendly materials.	Apprentices gain valuable experience and certifications, making them well-prepared for careers in green building and retrofitting projects, which are essential for reducing the environmental impact of the construction industry.
	Government-sponsored training initiatives	
Workforce Investment Act (WIA) Green Job Training Programs (United States)	Under the WIA, various programs are funded to provide training for workers in green industries, focusing on skills needed for renewable energy, energy efficiency, and environmental protection jobs.	These programs help unemployed or underemployed individuals gain the skills necessary to enter the growing green job market, thereby reducing unemployment and supporting economic growth.
	Industry partnership programs	
Solar Energy Training Network (SETN) (United States)	SETN is a collaborative effort among industry, academia, and government to provide high-quality solar workforce training.	The network supports the development of a skilled solar workforce, ensuring that the industry has access to qualified workers, which helps drive the expansion of solar energy projects.
	International training and certification	
The Global Wind Organization (GWO) Training Standards	The GWO sets international standards for training in the wind industry, ensuring that workers have the necessary skills and safety knowledge.	Standardized training and certification improve safety and efficiency in the wind industry globally, promoting the growth of wind energy projects and employment.
	Job retraining programs for displaced workers	
Just Transition Fund (Various Countries)	Programs aimed at retraining workers from declining industries, such as coal mining, to prepare them for employment in renewable energy and other green sectors.	These programs help mitigate the economic impact of the transition to a low-carbon economy by providing new career opportunities for displaced workers.

Source: created by authors.

These programs equip workers with the technical and professional skills needed in green industries, enhancing their employability and career prospects (skills development). By fostering a skilled workforce, these programs support the expansion of green industries, contributing to economic development and job creation (economic growth). Training and education programs promote the adoption of sustainable practices, reducing environmental impacts and supporting the transition to a low-carbon economy (environmental benefits). These programs can provide opportunities for underrepresented groups, promoting social inclusion and equity in the green job market (social equity).

By investing in training and education programs, governments and organizations can ensure that the workforce is prepared to meet the demands of a sustainable economy, driving growth in green employment and supporting the broader goals of environmental sustainability and economic resilience.

- Job transition support. Job transition support programs are designed to help workers move from declining or traditional industries to emerging green sectors. These programs provide the necessary resources, training, and assistance to facilitate a smooth transition and ensure that workers have the skills needed for new, sustainable job opportunities. Providing support for workers transitioning from declining industries to green sectors through retraining programs, career counseling, and job placement services can help mitigate the social impacts of economic transitions and promote green employment. Some examples and a wider explanation of job transition support programs are shown in Table 7.12 and Table 7.13.

TABLE 7.12
Some Examples of Job Transition Support Programs

Overview	Details	Impact
	Just transition fund (European Union)	
A fund aimed at supporting regions and workers affected by the transition to a low-carbon economy.	The Just Transition Fund provides financial resources for reskilling programs, job search assistance, and economic diversification in regions heavily dependent on fossil fuels.	This helps workers in coal mining and other high-emission industries transition to jobs in renewable energy, energy efficiency, and other sustainable sectors, ensuring they are not left behind as economies shift towards greener practices.
	The Workforce Innovation and Opportunity Act (WIOA) (United States)	
A federal program that supports job training and employment services for displaced workers.	WIOA offers funding for job training programs in various sectors, including green industries, and it provides career counseling, job placement services, and support for earning credentials.	This program helps workers who have lost their jobs due to economic changes or industry shifts gain new skills and find employment in emerging green sectors such as renewable energy and energy efficiency.
	Green Skills Agreement (Australia)	
An initiative to integrate green skills into vocational education and training (VET) programs.	The Green Skills Agreement focuses on updating training packages and courses to include sustainability competencies, ensuring that workers in traditional industries acquire the skills needed for green jobs.	This helps workers from sectors like manufacturing and construction transition to roles in green building, renewable energy, and sustainable practices, promoting a skilled and adaptable workforce.
	Retraining programs for coal workers (Germany)	
Programs designed to support coal workers transitioning to renewable energy jobs.	These programs provide training in renewable energy technologies, energy efficiency, and environmental management, along with financial support during the transition period.	By offering targeted retraining and financial assistance, these programs ensure that coal workers can smoothly transition to new careers in the growing renewable energy sector.

TABLE 7.12 *(Continued)*
Some Examples of Job Transition Support Programs

Overview	Details	Impact
Environmental Workforce Development and Job Training (EWDJT) Grants (United States)		
EPA grants aimed at providing job training in environmental fields.	EWDJT grants fund programs that train unemployed and underemployed residents in brownfields remediation, environmental health and safety, and sustainable practices.	These grants create job opportunities in environmental cleanup and sustainable practices, helping workers transition from traditional industries to green jobs.
Transition training programs (Canada)		
Initiatives to support workers transitioning from traditional energy sectors to renewable energy and energy efficiency roles.	Programs include training in solar panel installation, wind turbine maintenance, and energy auditing, as well as financial support during the training period.	These programs facilitate the shift from fossil fuel-based jobs to renewable energy careers, supporting economic diversification and job creation in sustainable industries.
SkillsFuture (Singapore)		
A national initiative to promote lifelong learning and skills development, including green skills.	SkillsFuture offers credits and subsidies for training courses in green technologies, sustainable practices, and environmental management.	This initiative helps workers across various sectors acquire the skills needed for green jobs, promoting continuous learning and adaptability in a rapidly changing economy.
Green Economy Transition (GET) Program (International)		
Programs run by organizations like the European Bank for Reconstruction and Development (EBRD) to support job transitions in green sectors.	GET programs provide funding and training for workers transitioning to jobs in renewable energy, energy efficiency, and sustainable resource management.	These programs support the global shift towards a green economy by facilitating job transitions and promoting sustainable practices across different regions and industries.

Source: created by authors.

TABLE 7.13
Wider Explanation of Job Transition Support Programs

Description	Impact
Skill development and retraining	
Job transition programs often include comprehensive retraining initiatives that equip workers with the skills needed for new green jobs. This can involve technical training in renewable energy technologies, environmental management, and sustainable practices.	These programs ensure that workers have the necessary competencies to succeed in green industries, reducing unemployment and supporting the growth of a skilled green workforce.
Financial assistance	
Financial support during the transition period is crucial for workers moving from declining industries to green sectors. This can include unemployment benefits, stipends during training, and relocation assistance.	Financial assistance helps mitigate the economic impact on workers and their families, providing stability while they acquire new skills and seek green employment opportunities.

(*Continued*)

TABLE 7.13 ***(Continued)***
Wider Explanation of Job Transition Support Programs

Description	Impact
Career counseling and job placement	
Many job transition programs offer career counseling, job search assistance, and job placement services. These services help workers identify suitable green job opportunities and navigate the job market.	Career counseling and job placement services increase the likelihood of successful transitions by providing personalized support and connecting workers with employers in green industries.
Partnerships with industry	
Effective job transition programs often involve partnerships with green industry employers, educational institutions, and training providers. These partnerships ensure that training programs are aligned with industry needs and that workers have access to real-world job opportunities.	Collaboration with industry partners enhances the relevance and effectiveness of training programs, leading to higher employment rates and better job matches in green sectors.
Community and regional support	
Regional and community-based programs address the specific needs of workers in areas heavily impacted by economic transitions, such as former coal mining regions.	Tailored support for communities ensures that local economies can adapt and thrive in the green transition, fostering regional economic development and social stability.

Source: created by authors.

By providing comprehensive support for workers transitioning to green jobs, these programs help ensure that the shift towards a sustainable economy is inclusive and equitable. They reduce the risk of unemployment, promote skills development, and create a workforce capable of driving green innovation and growth.

- Occupational standards and certifications. Occupational standards and certifications establish clear criteria for the skills and knowledge required in green jobs. They ensure that workers are adequately trained and that employers can trust the qualifications of their employees. These standards and certifications are crucial for maintaining high-quality work, ensuring safety, and promoting the growth of green industries. Developing and promoting industry standards and certifications for green jobs can enhance the credibility and recognition of green skills and occupations, encouraging more individuals to pursue careers in these fields. Some examples and a broader explanation of the occupational standards and certifications are shown in Table 7.14 and Table 7.15.

By establishing clear standards and offering certifications, these programs ensure that the workforce is adequately prepared to meet the demands of green industries. They promote quality, safety, and continuous improvement, supporting the overall growth and sustainability of the green economy.

The next policies that promote green employment are innovation and research policies. Policies that promote innovation and research can drive the development and commercialization of new green technologies and solutions.

- **Research and development (R&D) funding** is crucial for driving innovation in green technologies and sustainable practices. By investing in R&D, governments, private companies, and international organizations can develop new solutions to environmental challenges, improve existing technologies, and create new green jobs. Some examples are taken in Table 7.16.

TABLE 7.14
Examples of Occupational Standards and Certifications

Overview	Details	Impact
Leadership in Energy and Environmental Design (LEED) certification		
Administered by the US Green Building Council (USGBC), LEED certification is a globally recognized symbol of sustainability achievement.	LEED certification provides a framework for healthy, efficient, and cost-saving green buildings. Professionals can become LEED Accredited Professionals (LEED AP) to demonstrate expertise in green building principles.	This certification ensures that buildings are designed and constructed to high environmental standards, promoting energy efficiency, water conservation, and sustainable building materials.
Building Research Establishment Environmental Assessment Method (BREEAM)		
A leading sustainability assessment method for master planning projects, infrastructure, and buildings.	BREEAM sets standards for best practices in sustainable building design, construction, and operation. Professionals can gain BREEAM accreditation to certify their ability to implement these standards.	BREEAM certification encourages the creation of sustainable environments that enhance the well-being of people and contribute to the protection of natural resources.
Certified Energy Manager (CEM)		
Offered by the Association of Energy Engineers (AEE), the CEM certification is a recognized credential for energy management professionals.	The CEM program covers topics such as energy auditing, energy efficiency, renewable energy, and energy procurement. Certification demonstrates expertise in optimizing energy use and implementing sustainable energy solutions.	CEM certification helps ensure that energy managers are qualified to improve energy performance, reduce costs, and enhance sustainability in various industries.
National Australian Built Environment Rating System (NABERS)		
A performance-based rating system for buildings, measuring their environmental impact.	NABERS ratings cover energy efficiency, water usage, indoor environment quality, and waste management. Professionals can become accredited to assess and improve building performance.	NABERS certification drives improvements in building performance, reducing environmental impacts and operational costs while promoting sustainable practices.
Global Wind Organisation (GWO) Certification		
GWO sets industry-wide training standards for the wind energy sector.	GWO certification includes training in areas such as working at heights, first aid, fire awareness, and manual handling. These certifications ensure that workers have the skills and knowledge to operate safely in the wind energy industry.	GWO certification enhances worker safety, improves operational efficiency, and supports the growth of the wind energy sector by ensuring a skilled and competent workforce.
ISO 14001 Environmental Management Certification		
An international standard for environmental management systems (EMS).	ISO 14001 provides a framework for organizations to manage their environmental responsibilities. Certification demonstrates a commitment to reducing environmental impact and compliance with regulations.	ISO 14001 certification helps organizations improve their environmental performance, enhance sustainability, and gain a competitive advantage by demonstrating environmental responsibility.
Solar Energy International (SEI) Solar Professional Certification		
SEI offers certifications for various roles in the solar energy industry.	Certifications include Solar Electric Design and Installation, Solar Business and Technical Sales, and Solar Water Heating. These programs provide comprehensive training in solar technologies and best practices.	SEI certifications ensure that solar energy professionals are equipped with the skills needed to design, install, and maintain solar energy systems, supporting the growth and reliability of the solar industry.
Certified Sustainable Development Professional (CSDP)		
Offered by the Association of Energy Engineers (AEE), this certification focuses on sustainable development practices.	The CSDP program covers topics such as sustainable building design, renewable energy, water conservation, and environmental policy. Certification demonstrates expertise in implementing sustainable development projects.	CSDP certification promotes the integration of sustainable practices in various projects and industries, contributing to environmental protection and resource conservation.

Source: created by authors.

TABLE 7.15
Broader Explanation of the Occupational Standards and Certifications

Ensuring quality and consistency	
Occupational standards and certifications provide a benchmark for the skills and knowledge required in green jobs. They ensure that all workers meet a minimum standard of competency, leading to higher quality and consistency in work.	This reliability fosters trust in green industries, encouraging investment and growth. For example, LEED-certified professionals are trusted to deliver high-performance buildings that meet rigorous sustainability criteria.
Enhancing safety and compliance	
Certifications often include training on safety protocols and regulatory compliance, ensuring that workers can perform their jobs safely and within legal requirements.	For instance, GWO certification in the wind energy sector ensures that technicians are trained to work safely at heights and handle emergencies, reducing workplace accidents and improving overall safety.
Supporting career development	
Certifications provide a clear pathway for career advancement by validating specialized skills and knowledge. This helps workers gain recognition and opens up new job opportunities.	Professionals with certifications such as CEM or ISO 14001 are often more competitive in the job market, can command higher salaries, and have greater opportunities for career progression.
Driving industry growth	
By setting high standards for skills and performance, certifications help drive the adoption of best practices and innovative technologies in green industries.	This leads to improved efficiency, reduced environmental impact, and accelerated growth of green sectors. For example, widespread adoption of BREEAM and LEED standards has significantly advanced sustainable building practices worldwide.
Fostering global standards and recognition	
International certifications and standards facilitate the global exchange of skills and knowledge, allowing workers to operate in different regions and industries.	Certifications like ISO 14001 and GWO are recognized globally, enabling professionals to work across borders and contribute to international projects, thereby promoting global sustainability efforts.
Encouraging lifelong learning	
Certifications often require ongoing education and recertification, encouraging professionals to stay updated with the latest advancements and trends in their field.	This commitment to continuous learning helps workers remain relevant in a rapidly evolving job market and ensures that green industries benefit from the latest innovations and best practices.

TABLE 7.16
Examples of Research and Development (R&D) Funding Initiatives

Overview	Details	Impact
	Horizon Europe (European Union)	
Horizon Europe is the EU's key funding program for research and innovation, with a budget of €95.5 billion for 2021–2027	The program supports projects in various fields, including climate action, clean energy, and circular economy. It aims to enhance the EU's scientific and technological bases and boost innovation capacity.	Horizon Europe funds projects that develop cutting-edge green technologies and solutions, fostering innovation and creating high-quality green jobs across the EU.
	Advanced Research Projects Agency-Energy (ARPA-E) (United States)	
ARPA-E is a US government agency that funds high-potential, high-impact energy technologies that are too early for private-sector investment.	ARPA-E focuses on innovative energy projects that can significantly improve efficiency, reduce emissions, and enhance energy security. Grants are awarded to researchers, startups, and companies working on breakthrough technologies.	By supporting transformative energy projects, ARPA-E helps develop new technologies that can lead to the creation of green jobs and promote sustainable energy solutions.

TABLE 7.16 ***(Continued)***
Examples of Research and Development (R&D) Funding Initiatives

Overview	Details	Impact
	Mission Innovation (global initiative)	
Mission Innovation is a global initiative involving 24 countries and the European Union, aimed at accelerating global clean energy innovation.	Member countries commit to doubling their clean energy R&D investments over five years and collaborating on various innovation challenges. Projects focus on renewable energy, smart grids, energy storage, and other green technologies.	Mission Innovation drives international collaboration and significant funding in clean energy R&D, fostering innovation and creating green jobs worldwide.
	Green Climate Fund (GCF)	
The GCF is a financial mechanism under the UNFCCC that supports climate adaptation and mitigation projects in developing countries.	The fund provides grants and loans for projects that promote low-emission and climate-resilient development. It includes a focus on technology development and transfer.	By funding climate projects, the GCF helps developing countries implement green technologies, build capacity, and create sustainable job opportunities.

A broader explanation of research and development (R&D) funding initiatives is shown in Table 7.17.

By investing in R&D funding, governments and organizations can drive the innovation needed to develop sustainable technologies and practices. This not only addresses pressing environmental challenges but also fosters the growth of green industries and creates a wide range of job opportunities, contributing to a more sustainable and prosperous future.

- **Innovation clusters and hubs** are critical drivers of economic growth and technological advancement. By bringing together businesses, research institutions, and other stakeholders, these clusters create ecosystems that foster collaboration, innovation, and commercialization of new technologies.

Innovation clusters and hubs are geographical concentrations of interconnected companies, research institutions, and other organizations focused on driving innovation and technological advancements in specific industries. These clusters foster collaboration and knowledge exchange among businesses, academia, and government, creating an ecosystem conducive to innovation. Innovation hubs often provide startups and small businesses with access to funding, mentoring, and networking opportunities, accelerating their growth and development. These clusters attract top talent from around the world, offering a vibrant environment for creativity and entrepreneurial activities. Some examples are shown in Table 7.18.

By co-locating resources and expertise, innovation clusters reduce barriers to innovation, such as high costs and fragmented knowledge. Governments and private investors often support these clusters through incentives, grants, and infrastructure development to spur regional economic growth. Successful innovation clusters can transform local economies, generating high-quality jobs and fostering a culture of continuous improvement and competitiveness. The collaborative nature of innovation clusters enables rapid prototyping and commercialization of new products and services, driving progress in industries like green technology and renewable energy. Wider explanation and impact are shown in Table 7.19.

TABLE 7.17
A Broader Explanation of Research and Development (R&D) Funding Initiatives

Description	Impact
Driving technological innovation	
R&D funding enables the development of new technologies and the improvement of existing ones, addressing critical environmental challenges. This includes advancements in renewable energy, energy efficiency, sustainable agriculture, and waste management.	Innovation in green technologies can lead to more efficient and cost-effective solutions, making sustainable practices more accessible and widespread. This, in turn, drives the growth of green industries and creates new job opportunities.
Supporting startups and small businesses	
R&D funding often targets startups and small businesses, which are key drivers of innovation. These companies can develop and commercialize new technologies that larger firms might overlook.	By providing financial support to innovative startups, R&D funding helps bring new green products and services to market, fostering entrepreneurship and creating high-quality jobs in emerging industries.
Enhancing Competitiveness	
Investments in R&D can enhance the competitiveness of a country's green industries on a global scale. By staying at the forefront of technological advancements, countries can attract investment and create export opportunities.	A competitive green economy can generate significant economic benefits, including job creation, increased productivity, and trade surpluses in green technologies and services.
Facilitating collaboration	
R&D funding programs often encourage collaboration among academia, industry, and government. This multidisciplinary approach leverages diverse expertise and resources to address complex environmental challenges.	Collaborative projects can lead to more comprehensive and effective solutions, accelerating the development and deployment of green technologies and practices.
Promoting sustainable economic growth	
By investing in green R&D, governments can stimulate sustainable economic growth. Green technologies can reduce dependency on fossil fuels, lower greenhouse gas emissions, and promote resource efficiency.	Sustainable economic growth can create resilient economies that are less vulnerable to environmental and economic shocks, providing stable and long-term employment opportunities.
Meeting climate goals	
R&D funding is essential for developing the technologies needed to meet national and international climate targets, such as those set by the Paris Agreement.	Achieving climate goals requires significant advancements in renewable energy, energy storage, carbon capture, and other green technologies. R&D funding supports these advancements, helping to mitigate climate change and its impacts.

TABLE 7.18
Examples of Innovation Clusters and Hubs

Overview	Details	Impact
	Silicon Valley (California, USA)	
Silicon Valley is perhaps the most famous innovation cluster globally, known for its concentration of high-tech companies, startups, and venture capital.	The cluster primarily focuses on technology and digital innovation, including software development, hardware manufacturing, and internet services.	Silicon Valley has produced some of the world's largest tech companies, such as Apple, Google, and Facebook. It attracts top talent and significant investment, driving global technological advancements.
	Boston-Cambridge (Massachusetts, USA)	
The Boston-Cambridge area is a leading cluster for biotechnology, pharmaceuticals, and medical research.	This cluster leverages the presence of world-renowned universities like MIT and Harvard, focusing on biotech innovation, healthcare, and life sciences.	Companies like Biogen and Moderna are based here, and the cluster has been instrumental in developing breakthrough medical treatments and vaccines.

TABLE 7.18 *(Continued)*
Examples of Innovation Clusters and Hubs

Overview	Details	Impact
	Research Triangle Park (North Carolina, USA)	
Research Triangle Park (RTP) is one of the largest research parks in the world, located between the cities of Raleigh, Durham, and Chapel Hill.	RTP focuses on research and development across various sectors, including information technology, environmental sciences, and biotechnology.	RTP hosts over 300 companies and has created thousands of high-skilled jobs, contributing significantly to the regional economy.
	Clean Energy Corridor (Denmark)	
Denmark's Clean Energy Corridor is a prominent cluster focusing on renewable energy technologies, especially wind energy.	The cluster includes companies, research institutions, and governmental bodies working on wind power, smart grids, and energy efficiency.	Denmark is a global leader in wind energy technology, and this cluster has driven significant advancements and exports in the renewable energy sector.
Tech City (London, UK)		
Tech City, also known as Silicon Roundabout, is a technology cluster located in East London	The cluster focuses on digital technology, fintech, and creative industries.	Tech City has attracted numerous startups and global tech giants, contributing to London's reputation as a leading tech hub in Europe.

TABLE 7.19
Wider Explanation and Impact of Innovation Clusters and Hubs

Description	Impact
Collaboration and knowledge exchange	
Innovation clusters promote close collaboration among companies, universities, and research institutions. This proximity facilitates the exchange of ideas, knowledge, and best practices.	Enhanced collaboration leads to faster innovation cycles, more robust research outputs, and the development of cutting-edge technologies.
Access to funding and resources	
Clusters often attract venture capitalists, angel investors, and government funding, providing critical financial resources for startups and established companies.	Access to funding enables businesses to scale their operations, invest in R&D, and bring new products to market more quickly.
Talent attraction and retention	
Innovation clusters draw top talent from around the world due to their dynamic and stimulating environments.	A high concentration of skilled professionals fosters a competitive and innovative culture, driving continuous advancements in technology and business practices.
Economic growth and job creation	
By concentrating high-tech industries and research institutions, clusters generate significant economic activity and create numerous high-quality jobs.	This economic growth benefits local and regional economies, improving living standards and increasing GDP.
Accelerated commercialization	
Clusters provide a supportive ecosystem for startups, including incubators, accelerators, and mentorship programs.	These resources help startups commercialize their innovations faster, leading to rapid market entry and growth.

(*Continued*)

TABLE 7.19 ***(Continued)***
Wider Explanation and Impact of Innovation Clusters and Hubs

Description	Impact
Enhanced global competitiveness	
Innovation clusters enhance the global competitiveness of their regions by fostering industries that lead to technological advancements.	Regions with strong innovation clusters often become global leaders in specific sectors, attracting international business and investment.
Infrastructure development	
Governments and private investors often invest in infrastructure to support innovation clusters, including advanced research facilities, co-working spaces, and transportation networks.	Improved infrastructure supports the efficient functioning of businesses and research institutions, enhancing the overall productivity of the cluster.
Cross-disciplinary innovation	
Clusters often encourage interdisciplinary collaboration, bringing together diverse fields such as technology, biology, and environmental science.	This interdisciplinary approach leads to innovative solutions that address complex challenges, such as climate change and sustainable development.
Community and network building	
Innovation clusters foster a sense of community among businesses, researchers, and entrepreneurs, creating strong professional networks.	These networks facilitate partnerships, knowledge sharing, and collaborative projects, enhancing the overall innovation capacity of the cluster.

Innovation clusters and hubs are vital for fostering technological advancements, driving economic growth, and creating high-quality jobs. By concentrating resources, talent, and infrastructure in specific regions, they create dynamic ecosystems that promote sustained innovation and competitiveness.

- **Intellectual property rights (IPR) protection**. Strengthening IPR protection for green innovations can encourage investment in R&D by ensuring that inventors and businesses can reap the benefits of their innovations, promoting job creation in research-intensive green industries

Intellectual Property Rights (IPR) protection is essential for fostering innovation and supporting the growth of green jobs. IPR includes patents, trademarks, copyrights, and trade secrets that protect the creations and inventions of individuals and businesses. Effective IPR protection encourages investment in research and development (R&D) by ensuring that innovators can reap the benefits of their inventions. Some examples of the impact of IPR protection on innovation and green employment are shown in Table 7.20.

Intellectual Property Rights (IPR) protection is vital for fostering a robust environment for innovation, especially in green technologies and sustainable practices. By ensuring that inventors and companies can secure and benefit from their innovations, IPR protection encourages investment, attracts talent, and drives the growth of green industries. This not only advances environmental goals but also creates high-quality jobs, contributing to a more sustainable and prosperous economy.

TABLE 7.20
Examples of the Impact of IPR Protection on Innovation and Green Employment

Overview	Details	Impact
	Patent Systems	
Patents protect new inventions by granting the inventor exclusive rights to use, sell, and license the invention for a certain period, typically 20 years.	Patents cover a wide range of innovations, including new technologies, processes, and products that contribute to environmental sustainability.	Effective patent systems encourage companies to invest in green technologies, knowing their inventions will be protected and can be commercially exploited.
	Trademarks	
Trademarks protect brand names, logos, and other identifiers that distinguish goods and services.	Trademarks ensure that companies can build and protect their brand identity, which is crucial for marketing green products and services.	Strong trademark protection helps green companies establish market presence and consumer trust, promoting sustainable products.
	Copyrights	
Copyrights protect original works of authorship, such as literature, music, and software.	In the context of green innovation, copyrights can protect software for energy management, educational materials on sustainability, and more.	Copyright protection encourages the creation and dissemination of information and tools that support environmental sustainability.
	Trade Secrets	
Trade secrets protect confidential business information that provides a competitive edge.	Trade secrets can include proprietary processes, formulas, and designs that contribute to green technology and practices.	Protecting trade secrets ensures that companies can maintain their competitive advantage, encouraging investment in green R&D.

7.5 CASE STUDY

EUROSTAT develops several indicators to measure the expansion of green jobs in the European Union. The main indicator is employment in the environmental goods and service sector, which provides information on the employment state in the environmental sector. This important indicator is calculated using Eurostat statistics on employment and European environmental goods and service sector (EGSS) accounts. The data on employment in the environmental goods and service sector in European Union member states and the EU-27 average in 2014 and recent years are given in Figure 7.19.

According to Figure 7.19, the percentage of people employed in the environmental goods and services sectors has increased in all EU member states from 2014 to 2020, except for Finland and Romania. In Finland, the proportion of total employment in green jobs decreased by 3%, and in Romania by 15%. The three countries having the largest growth in green employment in the EU were: Bulgaria (98%), Luxembourg (87%), and Ireland (79%).

The distribution of green jobs varies among European Union member states due to many reasons linked to barriers and drivers of green employment. In Estonia, over 50% of environmental employment in 2020 was in the management of energy and water resources. Similar patterns can be observed in Finland, Luxembourg, and Sweden. Meanwhile, in Belgium and Croatia, the majority of green employment was in waste and wastewater management, accounting for 78% and 73% respectively. Estonia and Finland had the highest shares of green employment as a percentage of

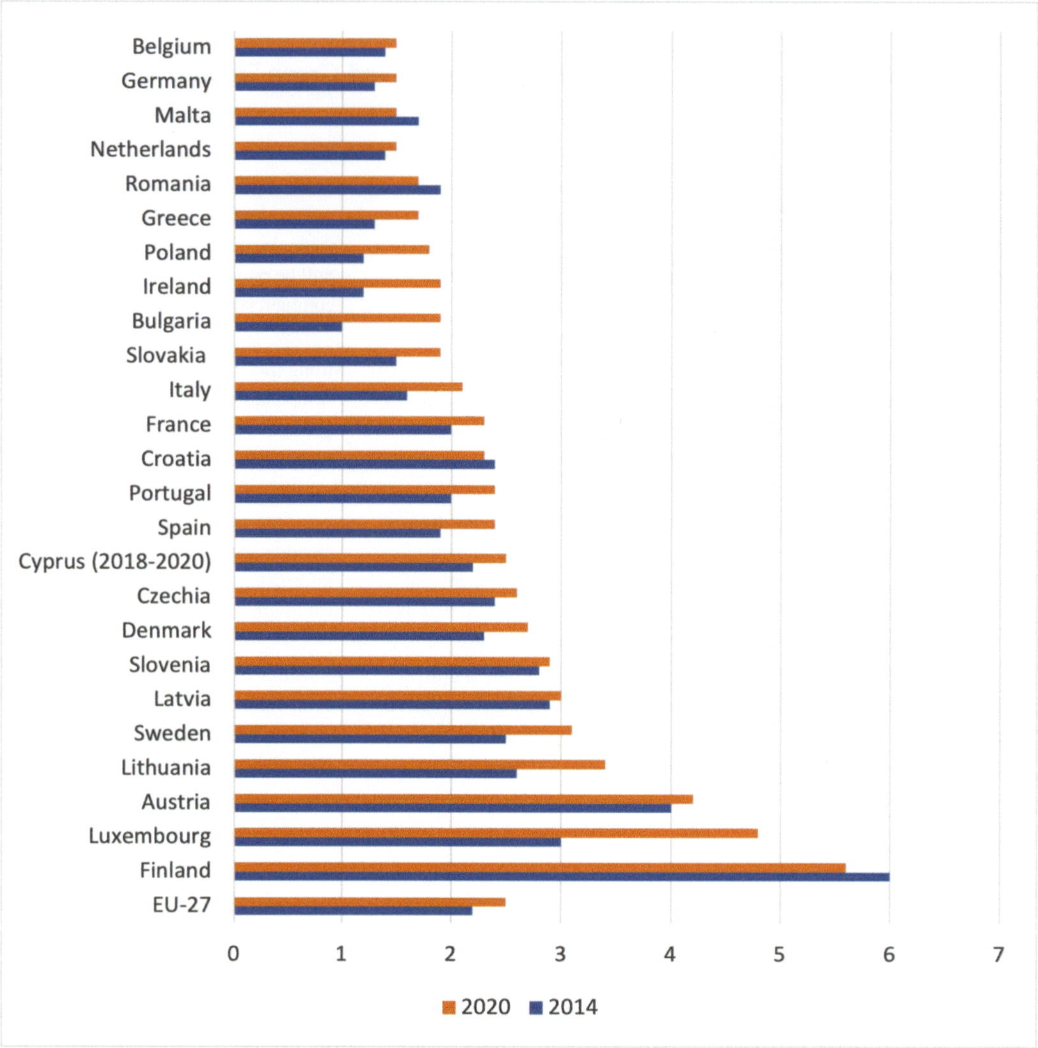

FIGURE 7.19 Employment in the environmental goods and service sector in European Union member states and EU-27 average in 2014 and 2020, %.

Source: European Environment Agency, 2024.

total employment, both exceeding 5% in 2020. Although Finland's share decreased after 2014, it still remained the highest among the member states in 2020. Conversely, the Netherlands, Malta, Germany, and Belgium had the lowest shares of green employment, all below 1.5% of total employment in 2020.

There is available data from EUROSTAT on individuals employed in circular economy sectors, calculated as a percentage of total full-time equivalent (FTE) employment. This measure indicates the proportion of individuals employed in recycling, repair, reuse, rental, and leasing sectors compared to the total employed population in the country. Table 7.21 provides the employment trends in circular economy sectors for EU member states.

The employment numbers in the circular economy sector can also be considered as green jobs, as they represent employment in recycling, repair and reuse, rental, and leasing sectors. These sectors provide services that reduce pollution and save resources. The leading countries in terms of

TABLE 7.21
Employment in Circular Economy Sectors of EU Member States and EU-27 Average, % of Total Employment—Numerator in Full-time Equivalent (FTE)

	2012	2013	2014	2015	2016	2017	2018	2019	2020	2021
EU-27	2.0	2.0	2.1	2.1	2.1	2.1	2.2	2.1	2.2	2.2
Belgium	1.4	1.4	1.3	1.4	1.3	1.3	1.3	1.3	1.3	1.3
Bulgaria	1.5	1.4	1.4	1.4	1.5	1.5	1.5	1.5	1.5	1.5
Czechia	2.4	2.3	2.2	2.3	2.3	2.3	2.2	2.3	2.3	2.3
Denmark	1.2	1.2	1.2	1.2	1.2	1.2	1.2	1.2	1.2	1.2
Germany	1.5	1.4	1.5	1.5	1.6	1.6	1.7	1.6	1.7	1.7
Estonia	1.8	1.9	2	2	1.9	1.9	2.1	2.0	2.2	2.2
Ireland	1.2	1.3	1.3	1.3	1.3	1.4	1.4	1.4	1.5	1.4
Greece	1.4	1.3	1.5	1.3	1.5	1.4	1.5	1.5	1.3	1.3
Spain	1.9	2	2	2.1	2.1	2.1	2.2	2.2	2.3	2.3
France	1.8	1.8	1.8	1.7	1.7	1.9	1.8	1.8	1.8	1.8
Croatia	2.7	2.8	2.9	2.9	2.8	2.9	3.1	3.5	3	3.1
Italy	2.3	2.4	2.4	2.5	2.5	2.4	2.3	2.5	2.5	2.4
Cyprus	1.4	1.5	1.5	1.7	1.8	1.8	1.9	1.9	1.9	2.0
Latvia	2.4	2.4	2.6	2.6	2.7	2.6	2.5	2.4	2.7	2.8
Lithuania	2.4	2.6	2.6	2.6	2.6	2.6	2.6	2.7	2.8	2.8
Luxembourg	0.5	0.5	0.5	0.5	0.5	0.5	0.5	0.4	0.4	0.4
Hungary	2.8	2.5	2.2	2.4	2.5	2.5	2.6	2.3	2.3	2.3
Malta	2.5	2.4	2.3	2.2	2.3	2.1	2.1	2.0	1.9	1.9
Netherlands	1.1	1.1	1.1	1.1	1.1	1.1	1.1	1.1	1.1	1.1
Austria	1.1	1.2	1.1	1.1	1.1	1.1	1.1	1.1	1.1	1.1
Poland	2.5	2.6	2.5	2.6	2.6	2.6	2.6	2.7	2.7	2.7
Portugal	1.8	1.8	1.8	1.8	1.8	1.9	2	1.8	1.8	1.8
Romania	1.0	1.0	1.0	1.0	1.0	1.0	1.0	1.0	1.0	1.2
Slovenia	1.6	1.6	1.6	1.6	1.6	1.6	1.7	2.0	1.5	1.5
Slovakia	2.0	2.0	1.9	2.0	2.0	2.0	2.0	2.0	2.1	2.2
Finland	1.7	1.7	1.7	1.7	1.6	2	2.3	1.8	1.6	1.5
Sweden	1.6	1.6	1.6	1.6	1.7	1.7	1.6	1.6	1.7	1.7

Source: EC (2024).

employment in the circular economy are Croatia, Poland, Hungary, Lithuania, and Latvia. On the other hand, Austria, the Netherlands, and Romania are among the worst-performing countries in this regard.

The European Commission has started monitoring the renewable energy progress of each EU member state in 2021, in accordance with the Renewable Energy Directive 2018/2001, also known as the RED II directive. The *EU Observer* (EC, 2022) has provided information on the number of jobs available in the renewable energy sector for 2020 and 2021.

The renewable energy jobs in EU member states in 2020 and 2021 are given in Figure 7.20.

According to 2022 data from the EC, there are approximately 1.5 million people working in the renewable energy sector in the EU. This includes both direct and indirect jobs associated with renewable energy. Between 2020 and 2021, the number of green jobs increased by 12%, resulting in the creation of 156,700 jobs. In 18 EU member states, the number of green energy jobs either

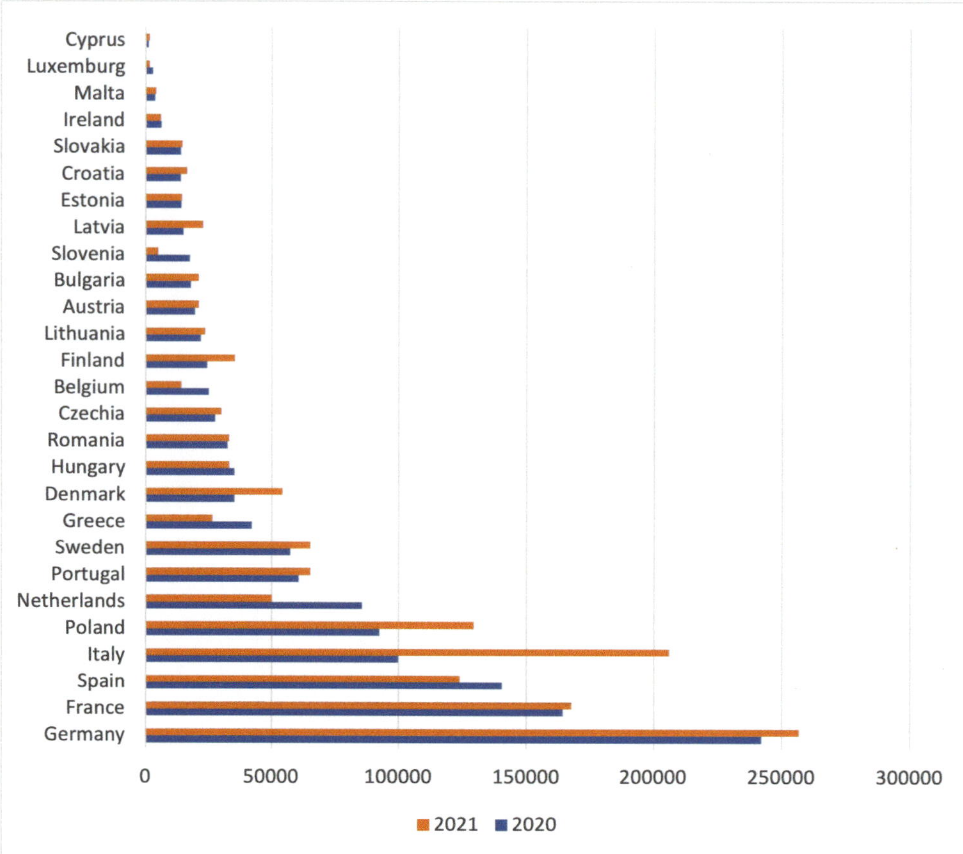

FIGURE 7.20 Employment in the renewable sector of EU member states.

Source: EC (2022),

increased or remained stable during this period. The top five countries with the largest number of jobs in the EU renewable energy sector are as follows:

1. Germany: 256,800 jobs, accounting for 17% of all EU renewable employment.
2. Italy: 206,100 jobs, representing 14% of all EU renewable jobs.
3. France: 167,800 jobs, making up 11% of total renewable jobs in the EU.
4. Poland: 129,300 jobs, constituting 9% of the overall EU renewable jobs.
5. Spain: 124,000 renewable jobs, accounting for 8% of the total EU jobs in the renewable energy sector.

In 2021, Italy is expected to experience the highest growth in green jobs. Poland saw a significant increase with the creation of 36,700 new renewable energy jobs, marking a 40% increase from 2020. Denmark witnessed the emergence of 19,000 new renewable energy jobs in 2021, reflecting a 54% increase since 2020. Meanwhile, Austria created 10,500 new green jobs, representing a 53% growth in renewable energy employment. On the other hand, Spain experienced a decrease of 16,500 renewable energy jobs, leading to a 12% decline in the renewable energy sector. Greece and Slovenia also saw declines, with Greece experiencing a decrease of 15,700 jobs (a 37% decrease) and Slovenia with a decrease of 12,500 jobs (a 71% decrease) in renewable energy employment during the 2020–2021 period.

During the 2020–2021 period, the renewable energy sector saw the largest increase in employment in the solid biomass sector, with 70,800 new jobs created, representing a 25% growth. This was followed by the heat pumps and solar PV sectors, which created 58,500 and 57,400 new jobs, respectively. Employment in biofuels, hydropower, and geothermal sectors increased, while a decline was observed in the biogas and wind technologies sectors.

7.6 CONCLUSIONS

1. The concept of green jobs is essential in the transition towards a sustainable and resilient economy. By encompassing a broad range of industries and occupations, green jobs not only help mitigate environmental issues but also drive economic growth and social development. The evolution of green jobs from a niche concern to a mainstream economic driver highlights their importance in addressing global challenges such as climate change, resource depletion, and environmental degradation. As we continue to innovate and adopt sustainable practices, the demand for green jobs will grow, underscoring their critical role in shaping a sustainable future.
2. Indicators of green jobs provide valuable insights into the progress and impact of sustainable employment initiatives. By tracking the adoption of green technologies, economic and social benefits, policy and regulatory frameworks, and environmental outcomes, stakeholders can gauge the effectiveness of green job programs and policies. These indicators not only measure the current state of green employment but also guide future efforts to enhance sustainability and inclusivity in the workforce. A comprehensive understanding of these indicators ensures that green job initiatives contribute meaningfully to environmental protection, economic resilience, and social equity.
3. The development of green jobs is influenced by a complex interplay of drivers and barriers. Environmental sustainability imperatives, policy and regulatory support, market demand, and investment opportunities drive the growth of green employment. However, challenges such as cost and economic viability, market barriers, social and economic aspects, institutional and organizational constraints, and technology and infrastructure limitations can impede progress. Addressing these barriers through strategic interventions and supportive policies is crucial for unlocking the full potential of green jobs. By overcoming these challenges, we can accelerate the transition to a sustainable economy and create a more inclusive and prosperous society.
4. Effective policies and measures are vital for promoting green employment and driving the transition to a sustainable economy. Environmental regulations, energy efficiency standards, renewable energy mandates, subsidies, grants, tax incentives, and green public procurement create a supportive framework for green industries. Labor market policies, including training and education programs, job transition support, and occupational standards and certifications, ensure that the workforce is prepared for green jobs. Innovation and research policies, such as funding for R&D, innovation clusters, and intellectual property rights protection, foster technological advancements and market growth. Together, these policies and measures create an enabling environment for sustainable economic development, job creation, and environmental protection.
5. A case study on green jobs assessment in EU member states showed that the best results in green job expansion can be noticed in Estonia and Finland, where the shares of green employment in total employment make up more than 5% of all jobs. At the same time, the lowest shares of green employment in total employment, less than 1.5%, were reported for the Netherlands, Malta, Germany, and Belgium. The variations in green job statistics among European Union member states indicate the need for tailored policies and measures to promote and maximize green jobs in the EU while addressing barriers.

REFERENCES

Adamowicz, M. (2022). Green deal, green growth and green economy as a means of support for attaining the sustainable development goals. *Sustainability*. *14*(10): 5901. https://doi.org/10.3390/su14105901

Belmonte-Ureña, L. J., Plaza-Úbeda, J. A., Vazquez-Brust, D. A., & Yakovleva, N. (2021). Circular economy, degrowth and green growth as pathways for research on sustainable development goals: A global analysis and future agenda. *Ecological Economics*. 185. https://doi.org/10.1016/j.ecolecon.2021.107050

Bracarense, N. (2022). Green jobs: Sustainable path for environmental conservation and socio-economic stability and inclusion. *Review of Political Economy*. https://doi.org/10.1080/09538259.2022.2041311

Bryngemark, E., Söderholm, P., & Thörn, M. (2023). The adoption of green public procurement practices: Analytical challenges and empirical illustration on Swedish municipalities. *Ecological Economics*. 204(A). https://doi.org/10.1016/j.ecolecon.2022.107655

Burnett, J. R., & Lisk, T. C. (2021). The future of employee engagement: Real-time monitoring and digital tools for engaging a workforce. In *Routledge eBooks* (pp. 117–128). https://doi.org/10.4324/9781003142492-9

Chiappinelli, O., Gerres, T., Neuhoff, K., et al. (2021). A green COVID-19 recovery of the EU basic materials sector: identifying potentials, barriers and policy solutions. *Climate Policy*. 21(10): 1328–1346. https://doi.org/10.1080/14693062.2021.1922340

Chien, F., Ngo, Q. T., Hsu, C. C., Chau, K. Y., & Iram, R. (2021). Assessing the mechanism of barriers towards green finance and public spending in small and medium enterprises from developed countries. *Environmental Science and Pollution Research*. 28: 60495–60510. https://doi.org/10.1007/s11356-021-14907-1

D'Alessandro, S., Cieplinski, A., Distefano, T., & Dittmer, K. (2020). Feasible alternatives to green growth. *Nature Sustainability*. 3: 329–335. https://doi.org/10.1038/s41893-020-0484-y

D'Amato, D., & Korhonen, J. (2021). Integrating the green economy, circular economy and bioeconomy in a strategic sustainability framework. *Ecological Economics*. https://doi.org/10.1016/j.ecolecon.2021.107143

Dell'Anna, F. (2021). Green jobs and energy efficiency as strategies for economic growth and the reduction of environmental impacts. *Energy Policy*. 149. https://doi.org/10.1016/j.enpol.2020.112031 https://www.sciencedirect.com/science/article/pii/S0301421520307424

Drago, C., & Gatto, A. (2022). Policy, regulation effectiveness, and sustainability in the energy sector: A worldwide interval-based composite indicator. *Energy Policy*. https://doi.org/10.1016/j.enpol.2022.112889

European Commission (2022). *The State of Renewable Energies in Europe, Edition 2022*. 21st EurObserv'ER Report. https://www.eurobserv-er.org/21st-annual-overview-barometer/ (Accessed on 28 January 2024).

European Commission (2024). *EUROSTAT database. Employment performance monitor*. https://ec.europa.eu/eurostat/data/database (Accessed on 28 January 2024).

European Environment Agency (2024). *Employment in the Environmental Goods and Services Sector by EU Member States, 2014 and 2020*. https://www.eea.europa.eu/data-and-maps/daviz/employment-in-the-environmental-goods#tab-chart_1 (Accessed on 28 January 2024)

Fang, W., Liu, Z., & Putra, A. R. S. (2022). Role of research and development in green economic growth through renewable energy development: Empirical evidence from South Asia. *Renewable Energy*. 194: 1142–1152. https://doi.org/10.1016/j.renene.2022.04.125

García-Quevedo, J., Jové-Llopis, E., & Solà, J. (2020). Barriers to the circular economy in European small and medium-sized firms. *Business Strategy and the Environment*. 29(6): 2450–2464. https://doi.org/10.1002/bse.2513

Gholami, H., Bachok, M. F., Saman, M. Z. M., & Streimikiene, D. (2020). An ISM approach for the barrier analysis in implementing green campus operations: Towards higher education sustainability. *Sustainability*. 12(1): 363. https://doi.org/10.3390/su12010363

Green, F., & Gambhir, A. (2020). Transitional assistance policies for just, equitable and smooth low-carbon transitions: Who, what and how? *Climate Policy*. 20(8): 902–921. https://doi.org/10.1080/14693062.2019.1657379

Kiefer, C. P., Del Río González, P., & Carrillo-Hermosilla, J. (2019). Drivers and barriers of eco-innovation types for sustainable transitions: A quantitative perspective. *Business Strategy and the Environment*. https://doi.org/10.1002/bse.2246

Kodua, L. T., Xiao, Y., Adjei, N. O., Asante, D., & Ofosu, B. O. (2022). Barriers to green human resources management (GHRM) implementation in developing countries: Evidence from Ghana. *Journal of Cleaner Production*. 340. https://doi.org/10.1016/j.jclepro.2022.130671

Kozar, Ł., & Sulich, A. (2023). Green jobs: Bibliometric review. *International Journal of Environmental Research and Public Health*. https://doi.org/10.3390/ijerph20042886

Krieger, B., & Zipperer, V. (2022). Does green public procurement trigger environmental innovations? *Research Policy*. 51(6). https://doi.org/10.1016/j.respol.2022.104516

Li, C., Ahmad, S. F., Ayassrah, A. Y., Irshad, M., & Telba, A. A. (2023). Green production and green technology for sustainability: The mediating role of waste reduction and energy use. *Heliyon*. 9(12). https://www.cell.com/heliyon/fulltext/S2405-8440(23)09704-9

Li, M., Luo, Y., & Xu, X. (2022). Mapping the scientific structure and evolution of renewable energy for sustainable development. *Environmental Science and Pollution Research*. https://doi.org/10.1007/s11356-022-20361-4

Li, M., & Rabeeu, A. (2024). How corporate social responsibility motivation drives customer extra-role behavior and green purchase intentions: The role of ethical corporate identity. *Sustainability*. 16: 5611. https://doi.org/10.3390/su16135611

Liu, G., Brown, M. T., & Casazza, M. (2017). Enhancing the sustainability narrative through a deeper understanding of sustainable development indicators. *Sustainability*. 9(6): 1078. https://doi.org/10.3390/su9061078

Mentes, M. (2023). Sustainable development economy and the development of green economy in the European Union. *Energy, Sustainability and Society*. 13(32). https://doi.org/10.1186/s13705-023-00410-7

Merino-Saum, A., Clement, J., & Wyss, R. (2020). Unpacking the Green economy concept: A quantitative analysis of 140 definitions. *Journal of Cleaner Production*. 242: 118549. https://doi.org/10.1016/j.jclepro.2019.118549

Ning, Y., Cherian, J., Sial, M. S., Álvarez-Otero, S., & Kour, R. (2023). Green bond as a new determinant of sustainable green financing, energy efficiency investment, and economic growth: A global perspective. *Environmental Science and Pollution Research*. 30: 61324–61339. https://doi.org/10.1007/s11356-021-18454-7

Singh, A. K., Jyoti, B., & Kumar, S. (2021). Assessment of global sustainable development, environmental sustainability, economic development and social development index in selected economies. *International Journal of Sustainable Development*. 16(1): 123–138. https://doi.org/10.18280/ijsdp.160113

Śmiglak-Krajewska, M., & Wojciechowska-Solis, J. (2021). Consumer versus organic products in the COVID-19 pandemic: Opportunities and barriers to market development. *Energies*. 14(17): 5566. https://doi.org/10.3390/en14175566

Song, C., Liu, Z., Yuan, M., & Zhao, C. (2024). From text to effectiveness: Quantifying green industrial policies in China. *Journal of Cleaner Production*. 446: 141445. https://doi.org/10.1016/j.jclepro.2024.141445

Starostka-Patyk, M., Bajdor, P., & Białas, J. (2024). Green logistics performance index as a benchmarking tool for EU countries' environmental sustainability. *Ecological Indicators*. https://doi.org/10.1016/j.ecolind.2023.111396

Su, H., Ali, F., Lyulyov, O., Pimonenko, T., & Chen, Y. (2023). Renewable energy development efficiency: Spatial dynamic evolution and influencing factors. *Natural Resources Forum*. https://doi.org/10.1111/1477-8947.12368

Sulich, A., & Rutkowska, M. (2020). Green jobs, definitional issues, and the employment of young people: An analysis of three European Union countries. *Journal of Environmental Management*. https://www.sciencedirect.com/science/article/pii/S0301479720302498

Sulich, A., & Sołoducho-Pelc, L. (2022). The circular economy and the Green Jobs creation. *Environmental Science and Pollution Research*. 29: 14231–14247. https://doi.org/10.1007/s11356-021-16562-y

Süsser, D., Ceglarz, A., Gaschnig, H., & Stavrakas, V. (2021). Model-based policymaking or policy-based modelling? How energy models and energy policy interact. *Energy Research & Social Science*. https://www.sciencedirect.com/science/article/pii/S2214629621000773

Tammilehto, M., Nyholm, M., & Jansson, J. (2022). Green jobs—A literature review. *International Journal of Environmental Research and Public Health*. 19(13): 7998. https://doi.org/10.3390/ijerph19137998

Tanasie, A., Nastase, L. L., Vochita, L., Manda, A. M., Boțoteanu, G. I., & Sitnikov, C. S. (2022). Green economy—Green jobs in the context of sustainable development. *Sustainability*. https://doi.org/10.3390/su14084796

Tumpa, T. J., Ali, S. M., Rahman, M. H., & Paul, S. K. (2019). Barriers to green supply chain management: An emerging economy context. *Journal of Cleaner Production*. 236. https://doi.org/10.1016/j.jclepro.2019.117617

Unay-Gailhard, İ., & Bojnec, Š. (2019). The impact of green economy measures on rural employment: Green jobs in farms. *Journal of Cleaner Production*. 208: 541–551. https://doi.org/10.1016/j.jclepro.2018.10.160

Wang, X., & Wang, Q. (2021). Research on the impact of green finance on the upgrading of China's regional industrial structure from the perspective of sustainable development. *Resources Policy*. 74. https://doi.org/10.1016/j.resourpol.2021.102436

Yamane, T., & Kaneko, S. (2021). Is the younger generation a driving force toward achieving the sustainable development goals? Survey experiments. *Journal of Cleaner Production*. 292. https://doi.org/10.1016/j.jclepro.2021.125932

Zhang, X., & Yousaf, H. M. A. U. (2020). Green supply chain coordination considering government intervention, green investment, and customer green preferences in the petroleum industry. *Journal of Cleaner Production*. https://doi.org/10.1016/j.jclepro.2019.118984

Zhao, J., Patwary, A. K., Qayyum, A., Alharthi, M., & Bashir, F. (2022). The determinants of renewable energy sources for the fueling of green and sustainable economy. *Energy*. 238(C). https://doi.org/10.1016/j.energy.2021.122029

8 Green Investments

Karolis Andriuskevicius

8.1 THE GREEN INVESTMENTS CONCEPT

Green investment may be defined as an endeavor involving investments directed towards projects or businesses dedicated to conserving natural resources, promoting alternative energy generation, safeguarding air and water quality, and embracing other eco-friendly practices. It can be viewed as a socially responsible investment as it emphasizes adherence to environmental regulations, social responsibility, and substantial benefits. Aligning with the views of Porter and Van der Linde (1995), green investment encompasses all endeavors aimed at enhancing environmental well-being. This perspective is supported by Xu et al. (2017), who propose that green investment plays a role in reducing reliance on harmful energy sources like coal and fossil fuels. Investments that enhance overall manufacturing process efficiency are considered green investments, extending beyond energy efficiency and renewable energy to include waste processing and recycling, water purification, pollution control in industries, biodiversity preservation, and climate change mitigation. Chițimiea et al. (2021) carried out bibliometric and systematic analyses for the period 1990–2020 and then analyzed a sample of 444 articles on the topic of green, sustainable, and responsible investments. Authors conclude that not only are companies concerned about environmental issues and efficient use of resources, but businesses interests in green investments have also dramatically increased since 2015.

The financial industry is currently experiencing a significant transformation to align with the global trend towards a more sustainable and environmentally conscious way of life and business practices. Striking a balance between economic advancement and environmental conservation has highlighted green finance as a vital strategy for addressing environmental challenges. The growing emphasis on green investments can be seen as a direct result of various key advancements in policy (e.g., the Paris Agreement on Climate Change, the UN Sustainable Development Goals of Agenda 2030, the implementation of the European Union Green Deal Investment Plan), regulations (including the recent European Union Taxonomy Regulation, the EU's updated Non-Financial Reporting Directive, and the UK Government's Green Finance Strategy post-Brexit), and socio-economic factors (such as the influence of movements like Fridays for Future and the COVID-19 pandemic).

There is no one single definition of the term green investment. In contrast, definition of green investment is broad and reflects dynamic nature of its developments. Table 8.1 summarizes different definitions.

Tran et al. (2020) defines the concept of green investment as involving leveraging environmentally conscious funds from both public and private sectors to support initiatives like: a) the creation of environmental goods and services, such as water treatment, ecosystem preservation, and landscape conservation and b) the prevention, mitigation, or compensation for environmental or climate harm, such as energy conservation and adoption of renewable energy sources.

Green investments, sometimes referred to as environmental investments, specifically denote social investments directed towards environmental enhancement. Various types of green investments exist, ranging from individual environmental donations to socially responsible investments made by enterprises. Green investments pertain to investments aimed at reducing greenhouse gas and air pollutant emissions while minimizing the impact on non-energy goods production and consumption. Similarly, Martin and Moser (2016) define green investments as a specific form of

DOI: 10.1201/9781003517481-8

TABLE 8.1
Definitions of Green Investment

Author	Definition of green investment
Lindenberg (2014)	Green investment comprises three dimensions, encompassing: a) funding public and private green projects, b) supporting governmental environmental initiatives, and c) involving the financial sector.
Inderst et al. (2012)	The concept of "green investment" carries diverse interpretations and extents. It may stand alone, form part of a broader investment strategy, or intersect with other methodologies such as socially responsible investing (SRI), environmental, social, and governance investing (ESG), sustainable investing, long-term investing, or related principles.
OECD (2017)	Green investments serve as a strategy to attain economic growth concurrently with minimizing environmental damage.
EBF (2017)	Green investments cover environmental and climate change considerations, but they extend beyond these specific elements.
G20 Green Finance Study Group (2021)	Green investments offer environmental advantages within the broader framework of sustainable environmental development.
European Commission (2021)	Green finance falls under the umbrella of sustainable finance, with a specific emphasis on the "environmental" aspect. This includes goals like addressing climate change through mitigation and adaptation, sustainably managing and safeguarding water and marine resources, conserving biodiversity, reducing pollution, and advancing the circular economy.
Rizzello, A. (2022)	Green investments form part of a sustainable financial system that facilitates the distribution of capital towards policies, organizations, investments, and activities that deliver direct or indirect environmental benefits. These benefits are aimed at reducing environmental harm compared to the current situation or generating positive environmental impacts.
Ding et al. (2023)	Investing in green initiatives involves allocating funds to decrease greenhouse gas emissions and air pollutants generated by companies, while maintaining their usual levels of production and product consumption.

corporate social responsibility that aids in carbon emissions reduction. Drawing from practical contexts and existing literature, Du et al. (2019) see green investments being predominantly utilized in three key areas: investments in urban infrastructure, pollution treatment, and environmental facility enhancements.

Green investments involve allocating financial resources to projects or companies that prioritize sustainable practices, eco-friendly technologies, and the preservation of natural resources. This concept entails investing in initiatives that have beneficial environmental effects or support sustainability. Green investments cover various sectors such as renewable energy, energy efficiency, sustainable agriculture, clean technology, waste management, and green infrastructure.

Magalhães (2021) recognizes that the term "green investment" became prominent around 2010 as a primary response to ecological crises. Definitions of green investments are often unclear, implied, or lacking. For instance, Campiglio (2016) defines green investments as: "investments in all productive sectors that help to improve the environmental sustainability of the economic system." Golka et al. (2024) define green investments as transforming the economy by transition to net-zero carbon emission. Magalhães, N. (2021) introduced the term "Green Investment Paradigm" (GIP), which constitutes "the framework advocating the need for significant green investments to achieve the energy transition."

To reinforce the global response to climate change, sustainable development, and poverty eradication, the significance of low-carbon investments cannot be overstated. McCollum et al. (2018) emphasized the urgent need for a significant shift in investment trends towards green technology projects. The United Nations (UN) highlighted an annual demand exceeding $1.5 trillion for green

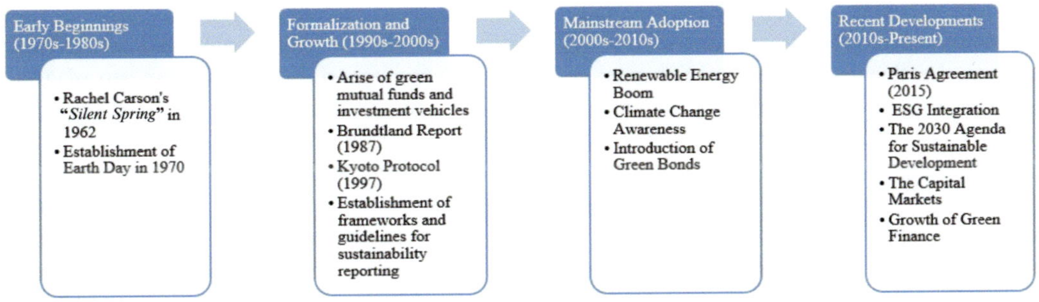

FIGURE 8.1 Evolution and adoption of green investment concept.

Source: author's own.

investments to meet the criteria outlined in the Paris Agreement. The recent challenges posed by the COVID-19 pandemic in sourcing funds for clean energy and green technology underscore the importance of green financing, as emphasized in recent studies (Zhan and Santos-Paulino, 2021). Madaleno et al. (2022) assert that establishing diverse green finance sources is essential for ensuring sustainable economic growth. They argue that green finance facilitates the provision of resources for environmental stewardship, the funding of green technologies, and the promotion of renewable energy production, leading to sustainable development.

The concept of green investment originated from corporate social responsibility (CSR). Since 1924, CSR has emphasized that companies are accountable not only for financial gains but also for fostering societal and economic development in harmony with the natural environment. This viewpoint gave rise to socially responsible investments, which evolved into green investments over time. While the roots of green investment can be traced back to the broader sustainability and environmental protection movement, its significant growth has occurred only in recent decades. Figure 8.1 provides a concise overview of the evolution of the green investment concept.

During the beginning (1970s–1980s) stage the concept of socially responsible investing began to take shape, where investors considered environmental and social factors alongside financial returns. Origins of socially responsible investing are circumstanced by the increase of environmental awareness. Specifically, the 1970s saw growing awareness of environmental issues, spurred by events like the publication of Rachel Carson's *Silent Spring* in 1962 and the establishment of Earth Day in 1970. Rachel Carson's groundbreaking book *Silent Spring* played a significant role in raising awareness about the negative environmental impacts of indiscriminate pesticide use, particularly DDT. The book highlighted the devastating effects of pesticides on bird populations and the broader ecosystem, sparking public concern and leading to the modern environmental movement. While *Silent Spring* specifically focused on the dangers of pesticides, its broader message about the need to protect the environment and promote sustainability has had a lasting impact on environmental awareness and activism. This increased environmental consciousness has influenced various aspects of society, including the world of finance and investment. *Silent Spring* is seen as a catalyst for greater consideration of environmental factors in investment decisions. The establishment of Earth Day on April 22, 1970 is another significant milestone in the history of the environmental movement. Earth Day was founded as a day of environmental education, awareness, and activism, aimed at highlighting environmental issues and promoting sustainability. The event helped catalyze public support for environmental protection and conservation efforts. Earth Day played a crucial role in raising public awareness about environmental issues and the need for responsible environmental stewardship. This increased awareness has had a ripple effect on various sectors, including the financial industry. Following Earth Day and the growing environmental movement, there has been a greater emphasis on integrating environmental considerations into investment decisions. Investors have become more conscious of the environmental impact of their investments and are increasingly seeking opportunities to support sustainable and eco-friendly initiatives. Furthermore, the establishment of Earth Day has contributed to the growth of green investment by fostering a broader understanding of environmental issues and encouraging individuals and organizations to align their investment practices with principles of sustainability and responsible stewardship of the planet.

Formalization and growth (1990s–2000s) of the green investment concept is circumstanced by several trends and events. First, in the 1990s, dedicated green mutual funds and investment vehicles started to emerge, focusing specifically on companies with strong environmental practices. Second, the international treaty of Kyoto Protocol (1997) aimed at addressing global climate change by setting binding targets for greenhouse gas emissions reduction. The treaty outlined specific commitments for countries to limit or reduce their emissions of carbon dioxide and other greenhouse gases, with the overall goal of mitigating climate change impacts. The Kyoto Protocol had significant implications for green investment in several ways:

1. Encouraging investment in clean energy: the Kyoto Protocol provided incentives for investing in renewable energy sources, energy-efficient technologies, and other clean energy endeavors. Nations that ratified the agreement were prompted to move towards low-carbon energy systems, resulting in heightened investments in green technologies and sustainable energy undertakings.
2. Carbon trading and offset initiatives: through mechanisms like emissions trading and clean development projects, the Kyoto Protocol introduced avenues for countries and businesses to trade emission allowances or participate in emission reduction projects in developing nations. This created fresh opportunities for green investments in carbon trading and offsetting actions.
3. Fostering corporate sustainability: compliance with the Kyoto Protocol compelled companies to evaluate and diminish their carbon footprints, fostering a greater emphasis on corporate sustainability and environmental stewardship. Consequently, this spurred investments in sustainable approaches, technologies, and eco-friendly supply chains.
4. Influence on financial markets: the Kyoto Protocol and subsequent climate agreements have impacted financial markets by heightening awareness of climate-related risks and the imperative to transition towards a low-carbon economy. This shift has brought about the introduction of green financial instruments like green bonds, climate funds, and other financial products that bolster environmentally sustainable projects.

Overall, the Kyoto Protocol has played a crucial role in shaping the landscape of green investment by creating a framework that encourages investment in sustainable and climate-friendly initiatives, while also promoting the transition towards a more environmentally responsible global economy.

Third, companies began to report on their environmental impact more regularly, providing data that investors could use to make informed decisions in 1990s–2000s. In the 1990s, there was a growing emphasis on environmental sustainability, leading to a rise in environmental reporting by companies. Environmental reports focused on issues such as energy consumption, greenhouse gas emissions, waste management, and pollution control. The late 1990s and early 2000s saw the establishment of various frameworks and guidelines for sustainability reporting. One of the most notable initiatives was the Global Reporting Initiative (GRI), which introduced standardized guidelines for reporting on economic, environmental, and social performance.

Green investment as a concept was adopted into the mainstream in the 2000s–2010s. It is shaped by the renewable energy boom, climate change awareness, and introduction of green bonds. The renewable energy boom influenced green investment in several ways. First, the growth of the renewable energy sector created new investment opportunities in technologies such as solar, wind, hydro, and geothermal power. As these technologies became more economically viable, more investors started focusing on renewable energy projects. Second, the renewable energy boom highlighted the potential for sustainable investments in clean energy. Investors started looking for opportunities that not only offered financial returns but also contributed to environmental sustainability. Third, government policies and incentives aimed at promoting renewable energy, such as feed-in tariffs and tax credits, made green investments more attractive. These supportive policies helped drive investments in renewable energy projects. Fourth, the increasing awareness of climate change risks and the need to transition to a low-carbon economy made investments in renewable energy more appealing. Diversifying investment portfolios with green assets helped mitigate risks associated with traditional investments tied to fossil fuels. Finally, the renewable energy boom spurred innovation in the green finance sector, leading to the development of new financial instruments and investment products tailored to renewable energy projects. This innovation made it easier for investors to participate in green investments. Climate change awareness has played a vital role in shaping the mainstream adoption of green investment practices by influencing environmental considerations, regulatory changes, risk perception, public pressure, and the long-term outlook of investors. This increased awareness has driven the shift towards sustainable and environmentally friendly investment opportunities. Introduction of green bonds has been instrumental in shaping the mainstream adoption of green investment practices by providing dedicated funding for sustainable projects, enhancing market transparency and credibility, driving mainstream acceptance of green investments, mitigating risks, and fostering innovation in sustainable finance. Green bonds have been a key catalyst in accelerating the transition towards a more environmentally conscious and sustainable investment landscape.

The Paris Agreement (2015), ESG integration, and the growth of green finance are three major recent developments that reflect a growing interest in incorporating environmental and social considerations into investment strategies for both financial and ethical reasons. The Paris Agreement acts as a catalyst for the growth of green finance by providing a roadmap for climate action, setting regulatory standards, sending market signals, and driving innovation towards a more sustainable future. This, in turn, influences and shapes the concept of green investment by aligning it with environmental goals and long-term sustainability. In September 2015, global leaders approved the 2030 Agenda for Sustainable Development, which was supplemented in January 2016 by the establishment of the 17 Sustainable Development Goals (SDGs). These goals are a series of objectives aimed at uniting nations in eradicating poverty, addressing disparities, and combating climate change. ESG integration shapes the concept of green investment by enhancing risk assessment, promoting long-term value creation, fostering stakeholder engagement and transparency, and responding to market demand for sustainable investments. By incorporating ESG factors into investment strategies, investors can align their financial goals with environmental and social considerations, contributing to a more sustainable and responsible investment landscape. The growth of green finance plays a pivotal role in shaping the concept of green investment by increasing capital flow towards sustainable projects, promoting innovative financial instruments, enhancing risk management practices, aligning with policies for sustainability, and fostering the development of vibrant green investment markets.

Having looked into historical events and developments, the question arises of how the green investment concept may evolve in the future. Beisenbina et al. (2023) performed an extensive bibliometric analysis and reviewed over 1,000 studies on sustainable investment. Authors identified current trends, gaps, and future research topics. Results of their work suggest that over 40% of the publications correspond to analysis of classical investment tools with a gradual evolution towards the concept of "green" instruments. Current trends towards sustainable and green investments reflect several global instruments and incentives. These incentives, agreements, and instruments are

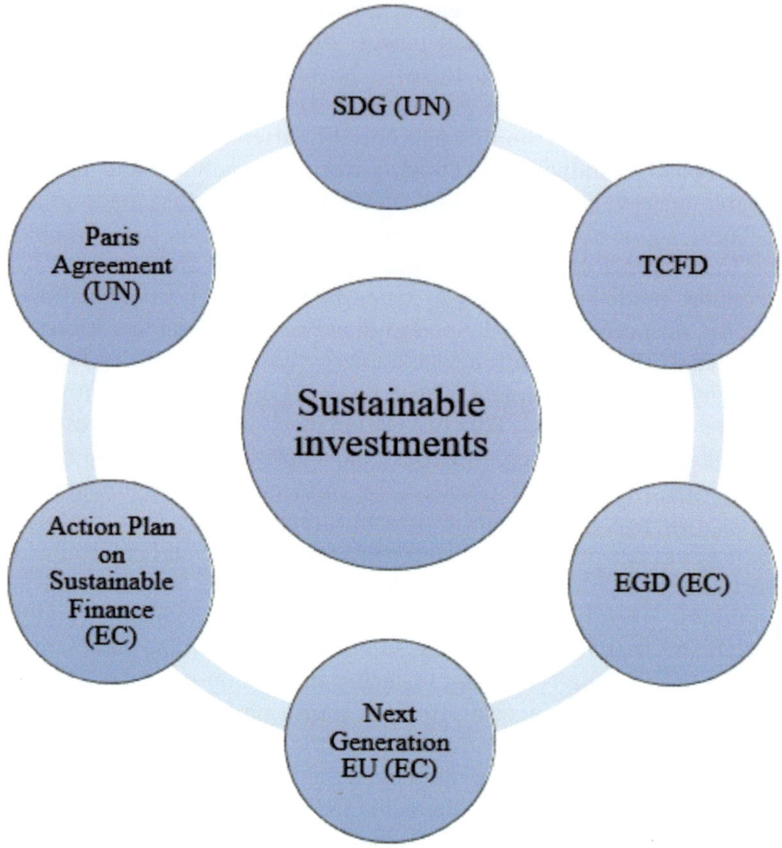

FIGURE 8.2 Global incentives, agreements, and instruments driving the financial system towards sustainable investments.

Source: author's own.

presented in Figure 8.2 and establish the framework towards a sustainable development, promoting sustainable investment.

1. Paris Agreement (Agreement, 2015). The Paris Agreement plays a pivotal role in promoting sustainable investment by providing a global framework for addressing climate change. By committing countries to reduce greenhouse gas emissions, it creates certainty and signals to investors that the transition to a low-carbon economy is inevitable. This certainty encourages investment in renewable energy, energy efficiency, and other sustainable technologies, driving innovation and economic growth while mitigating climate risks. Additionally, the agreement fosters transparency and accountability, enhancing investor confidence in sustainable projects and ensuring that investments contribute to long-term climate goals.
2. Sustainable Development Goals (SDGs, United Nations, 2015). The SDGs provide a comprehensive roadmap for global development, covering social, economic, and environmental dimensions. They are crucial for promoting sustainable investment because they offer a clear framework for identifying investment opportunities that contribute to sustainable development. Investing in projects aligned with the SDGs can generate positive impacts on issues such as poverty alleviation, gender equality, clean energy, and climate action, while also delivering financial returns. Furthermore, the SDGs create alignment and coherence

across various stakeholders, including governments, businesses, and investors, facilitating collaboration and mobilizing resources towards common objectives. Incorporating the SDGs into investment strategies helps investors assess the sustainability performance of companies and projects, manage risks, and seize opportunities in emerging markets and industries. Overall, the SDGs serve as a guiding principle for promoting responsible and impactful investment that contributes to a more sustainable and equitable world.

3. Task Force on Climate-related Financial Disclosures (Board, 2017). The Task Force on Climate-related Financial Disclosures (TCFD) is instrumental in promoting sustainable investment by enhancing transparency and accountability regarding climate-related risks and opportunities. By providing a standardized framework for companies to disclose climate-related financial information, the TCFD enables investors to make more informed decisions. Investors increasingly recognize that climate change poses significant risks to their portfolios, including physical risks from extreme weather events and transition risks associated with policy changes and shifts in market preferences. The TCFD helps investors assess and manage these risks by encouraging companies to disclose information on their climate-related governance, strategy, risk management, and metrics. Moreover, by disclosing their climate-related performance in line with TCFD recommendations, companies can demonstrate their resilience to climate risks, improve their access to capital, and attract investment from sustainable investors who prioritize environmental considerations. This incentivizes companies to integrate climate considerations into their business strategies and operations, driving the transition to a low-carbon economy. Overall, the TCFD fosters greater transparency, consistency, and comparability in climate-related financial disclosures, which supports the integration of climate factors into investment decision-making and promotes sustainable investment practices.
4. The Action Plan on Sustainable Finance (European Commission, 2018a) by the European Commission is crucial for promoting sustainable and green investment for several reasons. First, it sets clear standards and guidelines for financial institutions, encouraging them to integrate environmental, social, and governance (ESG) factors into their investment decisions. This helps redirect capital towards sustainable projects and away from activities that harm the environment. Second, the action plan aims to enhance transparency and disclosure requirements, allowing investors to make informed decisions about where to allocate their funds based on sustainability criteria. This transparency fosters trust and confidence in the market for sustainable investments. Additionally, the action plan includes measures to mobilize private capital towards sustainable projects, complementing public financing efforts. By providing incentives and removing barriers for investors to engage in sustainable finance, the plan helps scale up investments in renewable energy, energy efficiency, sustainable infrastructure, and other green sectors. Overall, the action plan on sustainable finance plays a crucial role in accelerating the transition to a low-carbon, sustainable economy by aligning financial flows with the objectives of the Paris Agreement and the European Green Deal.
5. The European Green Deal (European Commission, 2019) is instrumental in driving sustainable and green investment in several ways. First, it provides a comprehensive framework and ambitious targets for achieving climate neutrality by 2050. This long-term vision signals to investors and businesses that the transition to a green economy is not only desirable but inevitable, creating certainty and reducing investment risks in sustainable projects. Second, the Green Deal includes a wide range of policies and initiatives aimed at decarbonizing key sectors such as energy, transportation, agriculture, and buildings. These measures create new market opportunities for clean technologies and innovative solutions, attracting investment into renewable energy, electric vehicles, sustainable agriculture, energy-efficient buildings, and more. Furthermore, the Green Deal emphasizes the importance of sustainable finance as a key enabler of the transition to a green economy. It aligns financial regulations

and incentives with sustainability objectives, encouraging financial institutions to incorporate environmental, social, and governance (ESG) factors into their investment decisions. Moreover, the Green Deal fosters international cooperation and partnerships to address global environmental challenges, such as climate change and biodiversity loss. By leveraging diplomatic efforts and financial support, the European Union can mobilize resources for sustainable development projects worldwide, facilitating green investment on a global scale. Overall, the European Green Deal serves as a catalyst for sustainable and green investment by providing a clear roadmap, policy incentives, and international leadership to accelerate the transition to a more sustainable and resilient economy.

6. The Next Generation EU initiative (European Commission, 2020) holds significant importance in propelling sustainable and environmentally friendly investments for various reasons. To begin with, it channels a notable portion of its funding towards facilitating the green transition, committing to allocating 30% of the overall budget for climate-centric expenditures. This substantial financial dedication serves as a compelling driver for member states to prioritize investments in renewable energy, energy efficiency, sustainable transport, and other green industries. Additionally, Next Generation EU is designed to bolster the objectives outlined in the European Green Deal by backing projects that support climate action, adaptation, and the preservation of biodiversity. Through the financing of ventures such as renewable energy infrastructure, eco-friendly mobility solutions, and conservation initiatives, Next Generation EU expedites the shift towards a low-carbon and resilient economic landscape. Moreover, the initiative integrates sustainability benchmarks into its funding criteria, mandating that member states ensure projects funded through the recovery and resilience facility align with environmental sustainability targets. This ensures that investments contribute towards enduring sustainability and resilience, steering clear of perpetuating harmful environmental practices. Furthermore, Next Generation EU drives innovation and technological advancements in green sectors by endorsing research, development, and innovation initiatives. By investing in clean technologies and sustainable innovation, the program enhances competitiveness and progress in the green economy, thereby fostering fresh prospects for sustainable investments and employment opportunities. Overall, Next Generation EU emerges as a pivotal force in catalyzing sustainable and environmentally conscious investments by furnishing financial support, policy incentives, and regulatory structures to facilitate the transition towards a more sustainable, inclusive, and resilient European economy.

These incentives are agreements and norms also affect the financial system, whose activity should be oriented towards a sustainable development, promoting sustainable investment.

Thee green investment concept is shaped by an ecosystem that consists of a minimum of four key components: the capital demand, the capital supply, the intermediaries facilitating connections between the two, and the broader operational context of all stakeholders (Figure 8.3).

On the *supply side*, financial entities actively engage in investment decisions, utilizing a variety of financial instruments deployed through diverse mechanisms. The supply side of the green investment ecosystem consists of entities that produce environmentally focused capital and the mechanisms through which these funds are channeled. The primary category includes:

- Asset owners, including public sector/government investments, environmentally focused trusts and foundations, corporations, high net worth individuals (HNWIs), retail investors, and asset management firms.
- Asset management firms encompass distributors of green funds, family offices, institutional investors, and financial institutions.

Within these groups, banks, financial institutions, and funds play a growing role in the green finance network. This is due to the establishment and improvement of regulatory structures for

Demand			Supply	
Buyers of green benefits	*Organizations motivated by sustainability*	Main forms of investment	*Channel of green investments*	*Sources of green investments*
Public sector procurement and commissioning	Nonprofit generating green driven income	Secured/unsecured loans	Green investment banks	Public sector green investments
Public sector green debt issuers	Environmentally motivated social enterprises	Green bonds	Green banking	Green minded trusts and foundations
Foundations and NGOs	Hybrid organizations (profit with environmental purpose)	Environmental impact bonds	Development finance institutions	Corporates
Green minded consumers	Corporations adopting green business models	Equity and quasi equity	Green investment fund managers	Institutional investors, banks
Green minded corporations	For profit organizations in green business	Grants	Crowdfunding platforms	Green fund wholesalers, family offices

FIGURE 8.3 Ecosystem of green investment concept.

Source: Rizzello, 2022.

green investments, notably in the EU area, encouraging their participation in the sustainable financial system (Hafner et al., 2020). Consequently, the responsibilities of these financial bodies are quickly broadening, especially concerning the issuance of green bonds.

The channels through which green capital flows into the market are diverse. In addition to the growth and progress of green finance within the banking sector, the importance of various financial intermediaries is noteworthy, including:

- Green investment banks: public or semi-public entities that utilize limited public funds to attract increased private investment in environmentally sustainable activities, particularly in renewable energy projects.
- Development finance institutions (Fuchs et al., 2021): these organizations play a vital role in redirecting financial flows towards low-emission and climate-resilient development pathways to achieve environmental and climate goals, primarily focusing on direct project financing.
- Alternative banks, also referred to as ethical or social banks: these financial institutions prioritize creating environmental value for a wide range of stakeholders over maximizing profits for shareholders.

Asset management is rapidly expanding as a key sector within the green financial industry. Typically, asset managers specialize in providing advisory or discretionary management services for investors, involving comprehensive financial analysis, asset and stock selection, plan execution, and regular monitoring and reporting of investment activities. Important sectors in green asset management include mutual funds, pension funds, and private client assets.

Conversely, a variety of subcategories exist within green investment fund management, like fossil-free fund managers or climate fund managers, depending on whether their operational focus lies within specific green sectors. Green investment funds carefully select companies that adhere to environmentally responsible practices.

In recent times, crowdfunding platforms have experienced substantial growth within the green finance arena. Crowdfunding entails engaging numerous individual investors (referred to as the "crowd") through specialized websites known as crowdfunding platforms to finance projects, particularly those related to green finance that emphasize environmentally conscious initiatives.

On the *demand side*, various organizations are structured around capital generation, aiming to achieve both financial returns and environmental benefits in response to the green investment imperatives set by diverse sustainability-focused entities. These stakeholders typically operate within a broader framework of enabling conditions that translate green investment needs into concrete demands. From the demand perspective of a green investment ecosystem, there are two primary categories: those seeking environmental outcomes and environmentally motivated organizations. The first group includes public sector entities pursuing environmental objectives through procurement and environmental program implementation. Governments have recently aligned their green goals with the financial markets' need for green capital by issuing sovereign green bonds, as highlighted by Schmittmann and Chua (2021). Other entities involved in delivering environmental results include green-focused foundations and international non-governmental organizations active in this field. Another significant aspect of demand for green outcomes includes environmentally conscious consumers and sustainable corporations.

8.2 INDICATORS

Daniek (2020) explains that an indicator is a tool used to describe a specific condition. Indicators provide insights into both past and current states of a system, helping to identify important trends that can reveal relationships within the system. Depending on the issue being examined and the quality and availability of data, indicators can provide information in either quantitative or qualitative form. Quantitative indicators, which are measurable, allow for consistent and widespread comparisons between different phenomena. Consequently, qualitative indicators are often translated into quantitative formats, such as rankings or percentages.

When dealing with complex phenomena, it is often necessary to combine multiple indicators. Proxy indicators are used when a phenomenon cannot be directly measured, allowing for a more accurate description. Before using indicators, they should undergo a thorough evaluation, taking into account key aspects as mentioned by UNEP (2014):

- Policy relevance: guaranteeing that the indicator focuses on issues that are in the public interest and are relevant to policymaking.
- Analytical robustness: requiring that the indicator is based on the most up-to-date scientific knowledge to ensure its accuracy.
- Measurability: the ability to accurately represent a particular reality in a timely and appropriate manner, while also allowing for comparisons across different countries or regions and over time.

The term "green investment indicator" typically refers to a measure or metric used to assess the environmental sustainability or impact of investment decisions. It can encompass various factors such as carbon footprint, renewable energy usage, resource efficiency, and adherence to environmental regulations. Investors and organizations often use these indicators to evaluate the environmental performance and sustainability of their investments, portfolios, or projects. Green investment indicators have been developed by various organizations, financial institutions, industry associations, and regulatory bodies to help investors assess the environmental and sustainability performance of companies, projects, and financial products.

Becchetti et al. (2022) recognize that steering towards a sustainable economy requires robust environmental and social measurements. Sustainable finance involves the consideration of environmental, social, and governance (ESG) factors in evaluating financial investments. Within sustainable finance lies green finance, which specifically focuses on the environmental aspect. This includes goals such as mitigating and adapting to climate change, managing water and marine resources sustainably, preserving biodiversity, reducing pollution, and advocating for circular economy principles (European Commission, 2021).

The United Nations Environment Program (UNEP) defines green finance as increasing financial resources from the public, private, and non-profit sectors to achieve Sustainable Development Goals, aligning with the targets set for 2030 in the Sustainable Development Goals (SDGs) framework. Despite various definitions, the primary aim of green and sustainable finance remains facilitating the shift towards a sustainable economy by directing financial investments away from unsustainable practices towards those that support green and sustainable growth patterns. There is an urgent need for investments and innovations to meet environmental demands. Green finance is expected to provide a clear roadmap with consistent indicators to guide the transition and monitor progress.

According to the G20 (2017), promoting green finance not only tackles climate change but also stimulates economic growth and job creation. However, challenges exist at both microeconomic and macroeconomic levels. Microeconomic barriers include information disparities leading to extra costs and credit limitations, mismatched timeframes between long-term green investments and short-term financial horizons, as well as skills gaps. Macroeconomic obstacles range from general economic issues like exchange rate fluctuations and inflation to more specific concerns such as limited fiscal capacities in developing countries and differing regulatory and political risks across nations. To address these challenges, four main areas are identified: promoting sustainable finance standards, improving transparency among stakeholders, strengthening sustainable investment markets, and assisting developing nations in green financing.

These interventions have the potential to unlock essential resources across all economic tiers. The core objective is to make sustainable investments financially viable for market participants while preventing or minimizing the risks associated with greenwashing—false perceptions or communications portraying activities as sustainable. Greenwashing is a significant hindrance for all involved parties (issuers, investors, and intermediaries) and underscores the necessity for stringent and reliable standards. Notably, the European Union has been at the forefront of regulating these issues, collaborating with international entities to ensure the provision of reliable, comparable, and verifiable information.

In 2016, the European Commission (EC) formed a High-Level Expert Group on Sustainable Finance, stressing the importance of defining "green" and "sustainable" finance early on in their efforts. Subsequently, the European Commission released a plan for financing sustainable growth. One of the initial objectives outlined by the EC was the development of a "Taxonomy," a framework categorizing sustainable economic activities according to widely accepted scientific criteria. This responsibility was assigned to a Group of Technical Experts (TEG), leading to the creation of Regulation (EU) 2020/852, commonly referred to as the "Taxonomy Regulation" or simply "Taxonomy." This regulation benefited financial institutions offering sustainable finance products and businesses required to disclose non-financial statements.

The EU Taxonomy defines six environmental goals: climate change mitigation, climate change adaptation, sustainable water and marine resource management, transition to a circular economy, prevention and control of pollution, and protection and regeneration of biodiversity and ecosystems.

TABLE 8.2
Green Investment Indicators from the EU Taxonomy Perspective

Area	KPI (unit)
Climate change mitigation	Net emission of GHGs (kg CO_2, eq)
Climate change adaptation	Climate change vulnerability proxy (dimensionless)
Sustainable use and protection of water and marine resources	Water scarcity footprint (m3eq.)
Shift towards a circular economy	Utilization of fossil fuels and non-renewable biomass (MJ)
	Utilization of primary minerals (kg)
	Production of non-recyclable waste (kg)
Pollution mitigation and management	Release of particulate matter (linked to disease occurrence)
	Formation of photochemical ozone (kg NMVOCeq.)
	Acid deposition (mol H + eq.)
	Freshwater eutrophication (kg Peq.)
Conservation and rehabilitation of biodiversity and ecosystems	Land use for human activities (m2a)
	Net deforestation rate (m2)

Source: Becchetti et al., 2022.

Building on these objectives, Becchetti et al. (2022) present several Key Performance Indicators (KPIs) (Table 8.2).

The OECD's framework for measurement includes 26 indicators designed to capture key aspects of green growth and monitor advancements in this direction. The most recent updated and expanded set of green growth indicators introduced by the OECD was unveiled in 2017. Following the approach of previous iterations, the OECD classifies these indicators into four interconnected groups:

- *Environmental and resource productivity of the economy*—indicators demonstrating the efficiency with which various economic activities utilize energy and other environmental resources (e.g., carbon productivity).
- *Natural asset base*—indicators tracking the sustainable utilization of natural capital (e.g., availability of renewable natural resource stocks).
- *Environmental dimension of quality of life*—indicators showcasing the influence of environmental conditions and risks on societal well-being (e.g., natural disasters).
- *Economic opportunities and policy responses*—indicators evaluating the economic opportunities linked with green growth (e.g., innovation policy).

In the OECD's set of indicators, there are primary indicators, along with their related components or enhancements, and proxy indicators utilized when the primary indicators are not presently measurable. The primary indicators are highlighted and numbered in the table, while secondary components or supplements are also numbered, and proxy indicators are marked with bullets for improved clarity (as depicted in Table 8.3).

The United Nations Environment Programme (UNEP) has introduced a set of metrics for evaluating the green economy and environmentally friendly investments. UNEP's strategy centers on using green economy indicators to guide a holistic approach to policymaking. Moreover, UNEP underscores the significance of initiating transformations in national economic policies through the promotion of clean technology investments, enhancement of ecosystem services, and emphasis on environmental preservation.

UNEP highlights that although certain challenges have a global scope, their resolutions must start at the national level. Because each country faces issues influenced by local dynamics, investment

TABLE 8.3
Review of Green Growth Indicators by OECD

The socio-economic context and characteristics of growth	
Economic growth, productivity, and competitiveness	Economic growth and structure • Expansion and composition of GDP. • Disposable income. Efficiency and trade: • Workforce productivity. • Comprehensive productivity. Trade-weighted unit labor expenses: • Proportionate significance of trade: (exports + imports)/GDP. Inflation and commodity prices • Consumer Price Index. • Prices of food; crude oil; minerals, ores, and metals.
Labor markets, education and income	Job market Workforce participation Rate of unemployment Socio-demographic patterns • Population growth, structure, and density. • Life expectancy: healthy years at birth. • Economic inequality: GINI index. • Education level: quality and availability of education.
Economic efficiency in terms of environmental and resource utilization	
Carbon and energy productivity	1. CO_2 efficiency. 1.1. CO_2 emissions efficiency based on production (GDP per energy-related CO_2 emissions). 1.2. CO_2 emissions efficiency based on demand (real income per energy-related CO_2 emissions). 2. Energy efficiency. 2.1. Energy efficiency (GDP per unit of total primary energy supply). 2.2. Energy intensity across sectors (manufacturing, transportation, households, services). 2.3. Percentage of renewable energy (in total primary energy Supply and electricity generation).
Resource productivity	3. Material efficiency (excluding energy). 3.1. Demand-based material efficiency (an inclusive metric; original measurements in physical quantities). Real income per material units incorporated in final demand, material composition. 3.2. Production-based (domestic) material efficiency (GDP per material units consumed, material composition). • Biological materials (food, other biomass). • Inorganic materials (metallic minerals, industrial minerals). 3.3. Waste generation rates and recovery ratios (by sector, per GDP or value added unit, per person). 3.4. Nutrient flows and balances (N, P). Nutrient balances in agriculture (N, P). Per agricultural land area and variation in agricultural production. 4. Water efficiency. Economic value generated per unit of water used, categorized by sector (e.g., for agriculture: irrigation water per irrigated hectare).
Multifactor productivity	5. Environmentally enhanced comprehensive productivity (based on monetary units).
Natural asset base	
Natural resources stock	6. Natural resources index. A comprehensive measure presented in monetary units.

(Continued)

TABLE 8.3 ***(Continued)***
Review of Green Growth Indicators by OECD

The socio-economic context and characteristics of growth	
Sustainable reserves	7. Freshwater assets. Accessible replenishable resources (groundwater, surface water) and associated withdrawal rates (national, regional). 8. Woodland assets. Forest extent and capacity; alterations in stock levels over time. 9. Aquatic resources. Percentage of fish supplies within sustainable biological boundaries (worldwide).
Non-sustainable reserves	10. Global stocks or reserves of specified minerals (to be determined): metallic minerals, industrial minerals, fossil fuels, critical raw materials, and corresponding extraction rates.
Variety of life and ecological systems	11. Land assets. Changes in land cover, transitioning from natural to artificial states. Land utilization: status and alterations.

Source: Daniek, 2020

strategies and incentives need customization to suit unique political, economic, and institutional frameworks. This tailoring is crucial for shaping policies designed to boost social well-being and fairness, in alignment with local socio-economic and environmental settings. Implementing monitoring indicators that trace policy advancement in line with green economy principles requires the development of a dedicated framework for their execution, as illustrated in the table provided (Table 8.4).

Besides OECD and the UN, other entities that have also played a role in developing green investment indicators include:

1. United Nations Principles for Responsible Investment (UN PRI): UN PRI is a global initiative dedicated to understanding the impact of sustainability on investors and fostering the growth of sustainable financial markets. They have established criteria and metrics for responsible investing that encompass environmental considerations.
2. Global Reporting Initiative (GRI): GRI is an autonomous international organization that advocates for sustainability reporting and has crafted a comprehensive set of standards for reporting on environmental, social, and governance (ESG) matters, incorporating indicators pertinent to eco-conscious investments.
3. Climate Disclosure Standards Board (CDSB): CDSB concentrates on financial disclosure related to climate aspects to aid companies and investors in addressing climate opportunities and risks. They have devised standards and metrics for disclosing climate-based information in financial reports.
4. Carbon Disclosure Project (CDP): CDP administers a global data platform for investors, companies, cities, states, and regions to manage their environmental footprints. They gather environmental data and offer businesses a platform to divulge their climate-related details, encompassing vital indicators concerning carbon emissions and climate vulnerabilities.
5. Institutional Investors Group on Climate Change (IIGCC): IIGCC represents a European consortium of institutional investors focusing on climate change and provides recommendations on assimilating climate factors into investment strategies. They have contributed to formulating measures for evaluating climate hazards and prospects in investments.
6. Green Bond Principles (GBP): GBP is a set of voluntary guidelines created by the International Capital Market Association (ICMA) to uphold trust and transparency in the green bond sector. These principles outline suggestions on fund usage, project assessment, and reporting, serving as benchmarks for green investments.

TABLE 8.4
Review of Green Growth Indicators by UNEP

Environmental	
Climate change	Carbon dioxide emissions (tons per year)
	Proportion of renewable energy in electricity generation (%)
	Per capita energy consumption (British thermal units per person)
Management of ecosystems	Forest area (hectares)
	Percentage of water scarcity
	Area conserved on land and sea (hectares)
Efficient resource utilization	Energy efficiency (British thermal units per USD)
	Material efficiency (tons per USD)
	Water efficiency (cubic meters per USD)
	Carbon dioxide efficiency (tons per USD)
Management of chemicals and waste	Percentage of waste collected
	Percentage of waste recycled or reused
	Quantity of waste generated (tons per year) or area used for landfills (hectares)
Regulations	
Green investment	Research and development investment (as a percentage of GDP)
	Environmental Goods and Services Sector (EGSS) expenditure (in USD per year)
Environmental fiscal restructuring	Subsidies for fossil fuels, water, and fisheries (in USD or percentage)
	Taxation on fossil fuels (in USD or percentage)
	Incentives for renewable energy (in USD or percentage)
Pricing external factors and assessing ecosystem services	Carbon pricing (in USD per ton)
	Value of ecosystem services (e.g., water provision)
Eco-conscious procurement	Expenditure on sustainable procurement (USD per year and as a percentage)
	Carbon dioxide and material productivity of government operations (tons per USD)
Training for environmentally sustainable jobs	Training expenditure (in USD per year and as a percentage of GDP)
	Number of individuals trained (persons per year)
Quality of Life and Fairness	
Job opportunities	Construction sector employment (persons, %)
	Personnel engaged in operation and management (persons, %)
	Annual income generated (USD)
	GINI coefficient
Performance of the Environmental Goods and Services Sector (EGSS)	Value added annually (USD)
	Employment opportunities
Natural and human resources	Value of natural resources (USD)
	Net yearly value added or removal (USD)
	Literacy rate (%)
Resource accessibility	Proportion with access to modern energy (%)
	Proportion with access to clean water (%)
	Proportion with access to sanitation (%)
	Proportion with access to healthcare (%)
Well-being	Level of harmful chemicals in drinking water (grams/liter)
	Individuals hospitalized due to air pollution (individuals)
	Road traffic fatalities per 100,000 residents (transport-related)

Source: Daniek, 2020

These organizations and initiatives have played a role in establishing frameworks, standards, and indicators to guide investors in evaluating the environmental impact and sustainability performance of their investments. By utilizing these green investment indicators, investors can make more informed decisions that support environmental goals and contribute to a more sustainable financial landscape. Some common green investment indicators include:

1. The carbon footprint is a crucial green investment indicator used to quantify the amount of greenhouse gas emissions, primarily carbon dioxide, produced directly and indirectly by an organization, project, or investment. It is a key metric for assessing the environmental impact of activities and investments.

   Carbon footprint measurement involves calculating the total greenhouse gas emissions resulting from all stages of a product's lifecycle or an organization's operations, including raw material extraction, manufacturing, transportation, product use, and disposal. This indicator helps investors understand and evaluate the environmental impact of their investments and make informed decisions to support more sustainable practices.

   Monitoring and reducing the carbon footprint is essential for green investing as it indicates a commitment to reducing overall emissions and mitigating climate change. By considering carbon footprint as a green investment indicator, investors can prioritize environmentally responsible investments that contribute to building a more sustainable future.
2. Renewable energy usage is a fundamental green investment indicator that assesses the proportion of energy derived from renewable sources in a company, project, or investment portfolio. It serves as a key metric for evaluating the sustainability and environmental impact of energy consumption.

   Tracking renewable energy usage involves monitoring the percentage of energy generated from renewable sources such as solar, wind, hydroelectric, geothermal, and biomass. This indicator reflects a commitment to reducing reliance on fossil fuels, minimizing carbon emissions, and supporting the transition to cleaner energy alternatives.

   By prioritizing investments with high renewable energy usage, investors can directly contribute to reducing greenhouse gas emissions and promoting sustainability. Companies and projects with a significant percentage of renewable energy usage demonstrate their environmental responsibility and commitment to combating climate change, making them attractive targets for green investment initiatives.
3. Water usage efficiency as a green investment indicator measures how effectively water resources are utilized in a company, project, or investment. It plays a critical role in assessing environmental sustainability and responsible water management practices.

   Water usage efficiency involves evaluating the amount of water consumed relative to the output or production level. This indicator helps determine how efficiently water is used in various processes, operations, or products, highlighting opportunities for conservation and reducing water waste.

   By monitoring and improving Water Usage Efficiency, investors can support investments that prioritize responsible water stewardship, conserve water resources, and minimize environmental impact. Companies and projects that demonstrate high water efficiency not only contribute to sustainability goals but also enhance operational resilience and reduce overall water-related risks. Water usage efficiency is a key metric in green investing that promotes water conservation and sustainable resource management.
4. Waste management performance is a crucial green investment indicator that evaluates the efficiency and effectiveness of waste reduction, recycling, and disposal practices within a company, project, or investment portfolio. It plays a significant role in assessing environmental responsibility and sustainability efforts.

   Monitoring waste management performance involves assessing various aspects of waste generation, handling, treatment, and disposal to minimize environmental impact. This indicator measures the percentage of waste diverted from landfills through recycling, composting, or other sustainable waste management practices.

By focusing on waste management performance as a green investment indicator, investors can identify opportunities to support investments that prioritize waste reduction, recycling, and environmentally friendly waste disposal methods. Companies and projects that exhibit strong waste management practices not only contribute to environmental conservation but also demonstrate a commitment to sustainable operations and resource efficiency. Waste management performance is a key metric in green investing that promotes responsible waste management and supports a circular economy approach.

5. Social impact metrics are a set of indicators used in green investing to measure and evaluate the social impact of investments on communities, employees, stakeholders, and society at large. These metrics focus on assessing how investments contribute to positive social outcomes and align with sustainability goals beyond environmental considerations.

   Social impact metrics within green investing can encompass a wide range of factors, including but not limited to:

   5.1 Community engagement: evaluating how investments interact with and benefit local communities, including job creation, infrastructure development, and social programs.
   5.2 Labor practices: assessing investment practices related to fair labor standards, employee well-being, diversity, equity, and inclusion in the workplace.
   5.3 Philanthropic initiatives: tracking investments in charitable activities, community development projects, and initiatives that support social causes.
   5.4 Human rights: examining how investments respect and uphold human rights principles within their operations and value chain.
   5.5 Stakeholder engagement: evaluating relationships and communication with stakeholders, feedback mechanisms, and responsiveness to stakeholder concerns.

   By incorporating social impact metrics into green investment strategies, investors can ensure that their investments not only deliver financial returns but also create positive social benefits and contribute to sustainable development. Companies and projects that demonstrate strong social impact practices are more likely to attract socially responsible investors and effectively address broader societal challenges while fostering long-term sustainability and positive change.

6. Expenditure on green innovation serves as a crucial metric for green investments, concentrating on monitoring funds allocated towards researching, developing, and implementing eco-friendly technologies, products, and methodologies. This metric plays a crucial role in evaluating a company's dedication to innovative solutions for sustainability, energy conservation, and minimized environmental footprint.

   Key aspects of green innovation spending as a green investment indicator include:

   6.1 Research and development (R&D): gauging the extent of investment in R&D endeavors directed towards creating novel green technologies, processes, and remedies for environmental and sustainability issues.
   6.2 Technology incorporation: reviewing the implementation and assimilation of green technologies, such as renewable energy systems, energy-efficient machinery, and sustainable materials, within operations.
   6.3 Product development: examining investments in the creation and enhancement of environmentally friendly products, services, and solutions that promote ecological sustainability.
   6.4 Efficiency improvements: monitoring efforts to enhance operational efficiency, reduce resource consumption, and minimize waste generation through innovative practices and technologies.

By considering green innovation spending as a green investment indicator, investors can identify companies and projects that prioritize innovation for sustainable development and environmental stewardship. Investments in green innovation not only contribute to mitigating environmental impact but also drive business growth, competitiveness, and resilience in a rapidly changing market focused on sustainability. Companies that allocate resources to green innovation initiatives are positioned to lead in sustainability practices, drive positive environmental change, and create long-term value for investors. Incorporating green innovation spending as a green investment indicator promotes a focus on innovation-driven sustainability and supports investments in cutting-edge solutions for a greener future.

Overall, green investment indicators help investors, organizations, and policymakers evaluate the environmental impact and sustainability of their investments, contributing to a more environmentally conscious and sustainable economy.

8.3 DRIVERS AND BARRIERS

Inderst et al. (2012) acknowledge that motives behind green investments differ (see Table 8.5). Recognizing the variety of motivations among green investors is essential as it shapes their interpretation and definition of "green investment." These motivations are diverse, with different investors

TABLE 8.5
Motivations for green investment

Financial factors	Non-financial factors	Public image	Adherence to regulations and legal responsibilities
Common return factors: anticipated gains from green entities or investments	Ecological	Investor and investee company reputations	National laws and regulations (such as socially responsible investment policies and ESG disclosure requirements)
Common risk factors: instability, potential losses, value at-risk (VaR), default risks, etc.	Scientific	Influence from political figures, media, NGOs, etc.	International conventions (e.g. UN Global Compact)
Common diversification factors aiming for lower correlation of green investments with other assets	Ethical, religious	"Intangible asset," e.g., community investing	Voluntary industry codes and principles (e.g. UN PRI, Carbon Disclosure Project (CDP), Global Reporting Initiative (GRI))
Long-term risk consideration • Non-standard risk criteria, (e.g. integration of tail-risk or black swan events, reduction of catastrophic risks by reducing long-term carbon emission)	Political, social	Marketing tool	Disclosure regulation
Internalization of (negative and positive) externalities: • Via taxes and subsidies. • Via collective action of investor groups.	Other "norm-based"		Good governance codes for institutional investors and companies; corporate social responsibility (CSR)
	"Double bottom-line" or "triple bottom-line"		Part of fiduciary obligations

Source: Inderst et al., 2012.

placing varying importance on priorities, risk and return preferences, and motivation hierarchies. The ranking of these motivations depends on factors such as legal environment, institutional structure, and the preferences of decision-makers.

For most companies and financial institutions, financial considerations typically take precedence. Conversely, a foundation or charity may prioritize explicit green policies or ethical objectives. There has been an ongoing debate regarding what "fiduciary duty" entails in terms of responsible investing. Traditionally, there have been arguments suggesting that an SRI approach might conflict with the fiduciary responsibility of maximizing financial returns or ensuring financial security for beneficiaries like pension holders or life insurance policyholders. Hence, it was advocated that "green investments" should, at a minimum, match the financial performance of non-green investments.

The UNEP Finance Initiative (UNEP FI) highlights varying interpretations of pension fund fiduciary duties across regions and the acceptance of integrating ESG factors. Some experts propose that fiduciaries should not only consider ESG factors but also play a proactive role in addressing climate change concerns.

Chițimiea et al. (2021) conducted bibliometric and systematic analyses to pinpoint the factors influencing decisions regarding green investments. Their research demonstrates that the drivers and incentives for embracing green investments are broad and varied (refer to Figure 8.4).

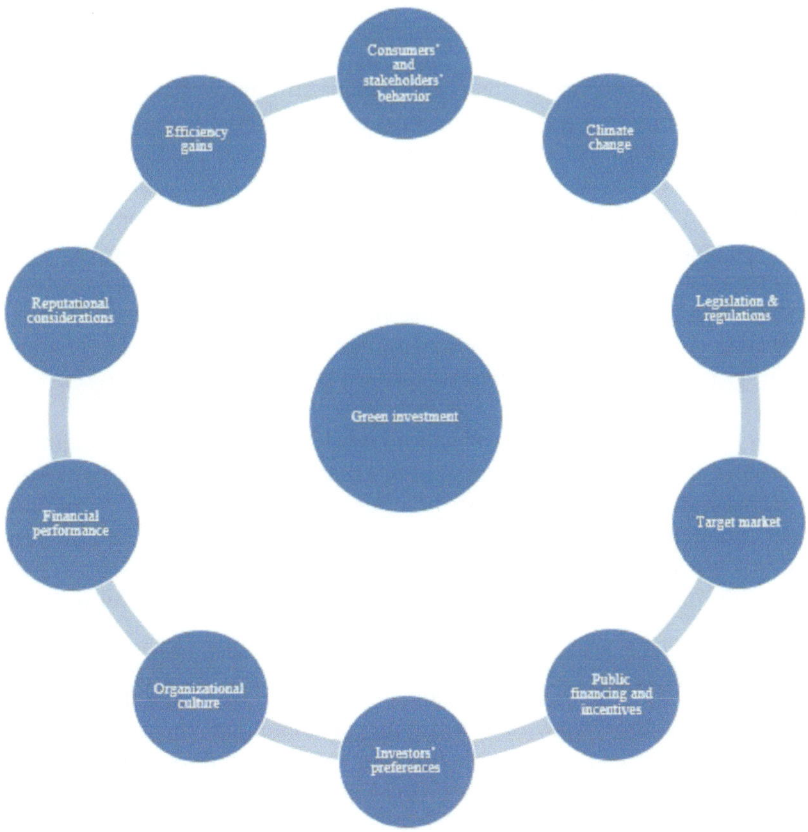

FIGURE 8.4 Drivers and motivation for the adoption of green investments.

Source: Chițimiea et al., 2021.

Consumer and stakeholder behavior: the behavior of consumers and stakeholders is evolving. Customers are becoming more knowledgeable and mindful of the environmental impacts of the products they buy. Despite increasing pressure for green investments from various stakeholders—such as consumers, investors, shareholders, and NGOs—companies remain hesitant to invest in green technologies due to the higher costs and risks involved.

Climate change: green investments are closely linked with actions to combat or adapt to climate change. Sustainable environmental technologies are crucial in reducing pollution, including CO_2 emissions and fuel usage, while ensuring compliance with environmental regulations. The environmental effects have prompted companies to introduce innovative green projects to minimize pollution.

Legislation and regulations: Environmental regulatory policies are pivotal in raising environmental standards. Various studies have explored how regulatory pressure fosters green innovations and investments. Understanding how this pressure influences organizations to improve their green investment strategies is essential. Legislative mandates directly impact companies, sometimes resulting in increased operational or financial expenses, like taxes and levies.

Target market: markets respond positively to green investments, indicating that companies are aligning with significant trends and pursuing profit opportunities. Adapting to market shifts is crucial for effectively implementing green investments. However, there's a persistent risk that resultant services or products may not meet market expectations or lack competitiveness in pricing, potentially leading to financial setbacks for companies.

Public financing and incentives: public financial investments (PFIs) have the flexibility to utilize both traditional and innovative approaches to link green initiatives with funding, improving capital accessibility, reducing risks, promoting cooperation, empowering market players, and influencing overall market dynamics. Companies that maintain long-term relationships with banks are more likely to pursue green investment strategies. Conversely, having multiple credit relationships may hinder a firm's environmental investment initiatives. Companies facing credit constraints and high indebtedness often encounter greater challenges in pursuing green investments.

Investor preferences: environmental impact is a key focus for investors, compelling companies to address it to improve their environmental practices. Presently, investors seem to be at a crossroads between traditional, financially driven investments and sustainable options. Financing green investments entails specific risks, driving investors to seek greater investment security through reliable indicators. There is an increasing demand for indicator-based tools that incorporate sustainability principles; however, these initiatives may lack substantial involvement from scientists.

Organizational culture: entrepreneurs often find motivation to embrace sustainability through their personal beliefs. To enhance an organization's sustainability efforts, there should be internal and vertical integration of sustainability practices across departments, strategies, and divisions. Managers' financial motivations serve as crucial indicators of companies' market value and sustainable investment endeavors. Internal factors like managerial mindset, supply chain relationships, and digital capabilities minimally affect value creation within the green supply chain.

Financial performance: there is a clear correlation between financial success and sustainability, attributed to high performance across various metrics like Carhart's alpha, Sharpe ratio, net return, and cost efficiency. By employing diligent management and strategic planning, companies can utilize tools to boost performance, promoting sustainability without sacrificing financial outcomes. This strategy not only enhances profitability but also increases the likelihood of long-term survival.

Reputational considerations: mutual funds have the potential to influence companies to develop eco-friendly technologies and enhance their corporate social responsibility (CSR) standards. This influence becomes especially significant when peer firms in the investment network emphasize CSR practices. As companies strengthen their commitment to sustainability, their brand equity typically grows. Companies that invest resources in reducing their environmental impact and complying with environmental regulations often create more financial value in the medium to long term.

Efficiency gains: sustainability has been a prominent concern for companies over many years as they strive to secure future competitive advantages through cost efficiencies, meeting consumer demands, managing risks, leveraging tax benefits, and optimizing resource usage in competitive markets. By collaborating on value creation, businesses can navigate towards investing in green technology and sharing the resulting benefits among partners.

Duong and Trang (2019) discovered that five key drivers accounted for over 60% of the transition towards eco-friendly business practices. Market competition emerged as the primary motivation, followed by fuel scarcity, new government regulations, smart technology, and finally, knowledge and innovation. In contrast, three factors—low-carbon emissions (37%), green culture (22%), and green finance (5%)—had a lesser impact on the investment decision-making process.

Du et al. (2019) undertook a thorough evaluation to identify the catalysts behind green investments, examining the direct and spillover effects of political, economic, and environmental aspects through a hybrid analysis. Their results revealed that economic and environmental factors wielded a more significant influence compared to political factors. Despite this, the researchers noted a gradual strengthening of the government's role in promoting green investments and improving environmental conditions in Chinese provinces between 2003–2016, particularly through the introduction and enforcement of environmental regulations.

Prior studies highlight two primary reasons why companies engage in green investments. On one hand, green initiatives may compete for resources crucial for a company's day-to-day operations, affecting production and sales, potentially impacting financial performance. For instance, considerable funding is required for acquiring eco-friendly equipment and developing green technologies, which carry uncertain returns, elevating investment risks. Additionally, green technology investments often yield societal benefits, while companies shoulder all costs, creating a disincentive for firms to adopt new technologies with public advantages. Given the environmental aspect as a public good, negative environmental impacts from economic activities can only be remedied through governmental environmental policies. Hence, companies often view green investments as a reactive step to comply with these regulations.

Conversely, engaging in green investments demonstrates a proactive approach that assists in cultivating a positive social image, enhancing environmental responsibility practices, and gaining competitive edges. This proactive stance towards green investments aligns with signal theory, where taking proactive green actions signals a company's social responsibility and eco-friendly stance to external stakeholders. By making green investments, companies can send positive messages to consumers, build trust, and contribute to improved firm performance, albeit with a delayed effect. Proactive green investments, particularly those focusing on pollution control at the production source, are more likely to yield additional benefits, leading to both environmental and economic advantages.

While the aforementioned drivers motivate companies to prioritize sustainability and green investments, several barriers can hinder and diminish progress in adopting green practices. These are presented in Figure 8.5.

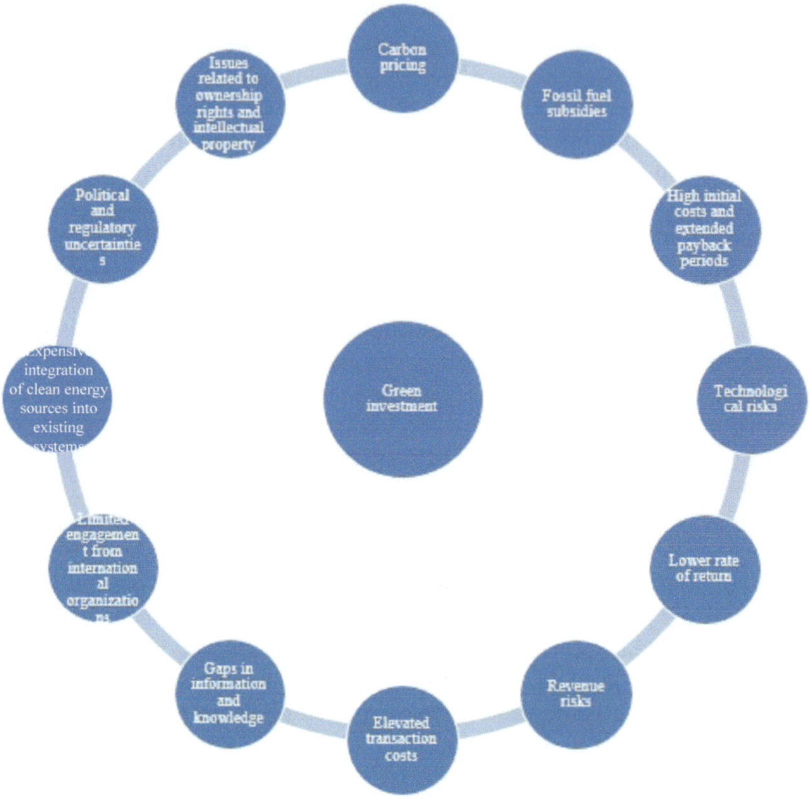

FIGURE 8.5 Barriers for the adoption of green investments.

Source: author's own.

According to a World Bank study, the primary challenges for green investment include:

1. Carbon pricing: the attractiveness of green investments to investors hinges on the stability, predictability, and reliability of carbon prices. The UN Secretary-General's High-level Advisory Group on Climate Change Financing (AGF 2010) suggests that a carbon emissions price of $25 per ton could potentially boost annual investments to $100 billion, catering to the needs of developing nations.
2. Fossil fuel subsidies: distorted pricing and resource allocation caused by underpriced fossil fuels discourage green investment by making traditional technologies appear more economically viable.
3. High initial costs and extended payback periods: while certain green investments justify their expenses from a financial standpoint, many do not. This necessitates higher return rates for green investments, sometimes yielding significantly more than conventional investments. Therefore, countries may need to support the profitability of green investments through price mechanisms or initiatives like green certificates.
4. Technological risks: implementing new technologies entails inherent uncertainties and risks related to viability and functionality. Securing financing and developing new technologies can be challenging without public funding or venture capital backing, as financial investors are often hesitant to fund projects until the risk-return dynamic stabilizes.
5. Revenue risks: given the uncertainties and income variability associated with new technologies, investors and financial institutions exercise caution when considering funding.

> The unpredictability of returns may lead to challenges in obtaining funding, especially for projects with lower credit ratings that require collateral to mitigate risks and meet financial obligations.

Further, a lower rate of return and a much higher risk compared to fossil fuel projects as two major barriers may be added.

In addition to the primary challenges mentioned earlier, the green investments of businesses are also impacted, albeit to a lesser degree, by various other factors such as elevated transaction costs, gaps in information and knowledge, disparities in beliefs, limited engagement from international organizations, the necessity of replacing aging assets, the expensive integration of clean energy sources into existing systems, political and regulatory uncertainties, issues related to ownership rights and intellectual property, as well as inadequacies in domestic financial instruments.

Addressing these barriers requires collaborative efforts from policymakers, investors, businesses, and civil society to create a conducive environment for green investment and accelerate the transition to a sustainable economy.

Du et al. (2019) examined the debate in previous literature regarding the relationship between economic growth and environmental pollution. Additionally, authors demonstrated that environmental policies can influence the progression of green investments under certain circumstances. Consequently, they proposed assessing the determinants of green investments from a triangle perspective encompassing political, economic, and environmental considerations (Figure 8.6).

Political factors exhibit a strong correlation with green investment, portraying the government's commitment to environmental improvements. These factors typically yield dual impacts. First, political involvement leads to increased government investments in green initiatives and

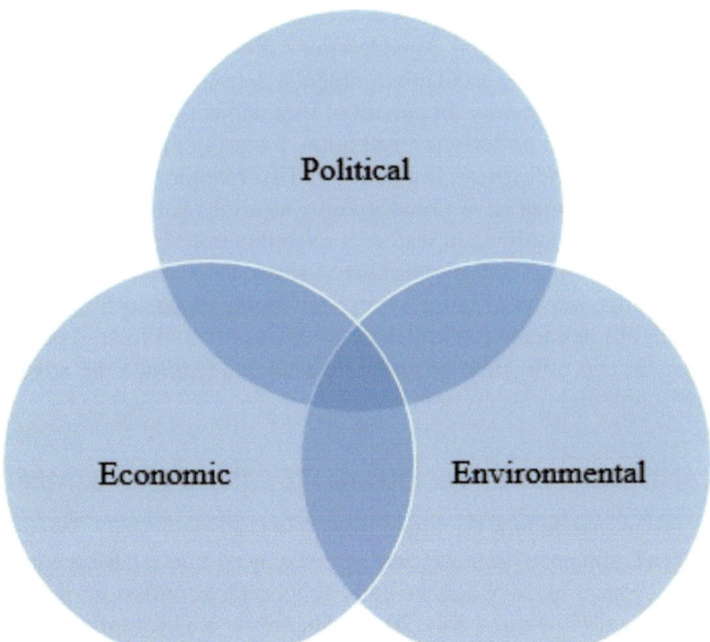

FIGURE 8.6 Triangle of the determinants of green investments.

Source: Du et al., 2019.

enhanced promotion of pro-environmental behaviors. Research by Iyer et al. (2015) reveals that policies facilitate the adoption of low-carbon technologies, thus reducing emissions abatement costs. Similarly, Fullerton and Kim (2008) note that stringent environmental policies positively influence environmental quality. Conversely, escalating carbon tax rates may dampen firms' research and development (R&D) investments in non-carbon energy technologies. Environmental policies serve as a guiding force and intervention tool for both economic growth and environmental enhancement (Drake et al., 2016). For instance, these policies encompass various measures such as green subsidies, consumer rebates, environmental taxes, and pollution fines to incentivize pro-environmental practices and regulate activities detrimental to the environment. Gadenne et al. (2009) highlight that regulations foster public environmental awareness and encourage organizations to modify their business processes and environmental strategies. Research by Krass et al. (2013) assesses the impact of regulatory incentives and finds that environmental taxes influence firms' decisions regarding the adoption of green technologies. In the realm of political factors, environmental policy may demonstrate diverse political orientations, with many studies focusing primarily on regulatory aspects (Gadenne et al., 2009; Drake et al., 2016).

Economic factors. Economic drivers significantly influence green investment initiatives. With a growing emphasis on environmental conservation, there is an increased demand for eco-friendly business approaches (Gadenne et al., 2009). Corporate social responsibility (CSR) investments in environmental sustainability can notably boost a company's stock price. Direct and indirect economic impacts and performance associated with green investments shall be evaluated. Shahbaz et al. (2013) explore the relationship among CO_2 emissions, financial development, energy consumption, and economic growth, noting that financial development leads to a reduction in CO_2 emissions. Particularly within the realm of economic factors, GDP and GDP per capita are undeniably critical in driving green investment (Zhang et al., 2018; Drake et al., 2016). Additionally, fixed asset investments should be considered an economic influencer as they are closely tied to urban infrastructure investments. Wang et al. (2017) observe that fixed asset investments notably contribute to increases in carbon emissions.

Environmental factors. Environmental considerations are paramount in the drive towards promoting green investments, aiming to curtail environmental deterioration, avert ecological crises, and progress towards a low-carbon economy. In pursuit of these objectives, the Chinese government has dedicated substantial resources to pollution remediation, fostering the growth of green technologies (Zhao et al., 2019) and green industries (Liu et al., 2019). Encouragingly, advancements in purification and recycling technologies have transformed waste into valuable resources. For instance, membrane bioreactors are being utilized in water reuse applications for wastewater treatment, while chemical scrubbers offer a cost-effective alternative to oxidative catalysts in waste gas treatment systems. Despite the positive impact of current green technologies on curbing environmental degradation, it remains crucial to delineate the influence of environmental factors on green investments. The effective deployment and utilization of green investments facilitate the advancement of green technologies and environmental preservation.

8.4 POLICIES AND MEASURES TO PROMOTE GREEN INVESTMENTS

Significant investments are needed for the green transition (Liu et al., 2022; Wu et al., 2024). These investments can affect a company's financial performance by competing for resources and introducing income uncertainty. However, they also bring benefits such as emissions reduction and enhancing the natural environment (Chen and Ma, 2021). Therefore, it is essential to find ways to guide and encourage companies towards making wise green investments for environmental improvement (Yang et al., 2023). Governments worldwide have recognized environmental challenges and have introduced reforms that merge economic growth with environmental protection, with green finance regulations playing a crucial role (Lee and Lee, 2022).

Green investment policies refer to governmental regulations, incentives, and frameworks designed to promote investment in environmentally sustainable projects and initiatives. These policies aim to address climate change, reduce greenhouse gas emissions, conserve natural resources, and promote the transition to a low-carbon economy. Given the complex and swiftly advancing landscape of the green investments, there is a need for various harmonization efforts to aid all stakeholders in constructing a green financial system capable of directing funds and investments towards eco-friendly endeavors and projects. Over the past two decades, several initiatives have been formulated to enable the financial sector to bolster the financial aspect of the green transition. As per the Green Finance Platform (2024), globally, more than 784 policy and regulatory measures have been implemented, reflecting a 317% rise since 2015.

The relationship among environmental regulations, policies, and green investments can be understood by drawing on theoretical frameworks such as institutional theory, environmental stress theory, and Porter's win-win hypothesis (Figure 8.7). *Institutional theory* posits that organizations operate as institutions influenced by societal norms and the environment, which significantly shape their strategies beyond typical market forces. Companies are attentive to social norms, routines, regulations, and environmental performance, with increasing pollution intensifying institutional pressure on businesses to prioritize environmental sustainability. Consequently, heightened environmental regulations may appear to discourage investments; however, exploring green innovations could mitigate this negative impact on corporate investment. Luo and Qu (2023) utilized institutional theory to establish a theoretical connection between environmental corporate social responsibility, corporate image, pro-environmental behavior, and the environmental performance of businesses. In contrast, *environmental stress theory* elucidates the relationship between these core variables more effectively. Governments implement various environmental regulations

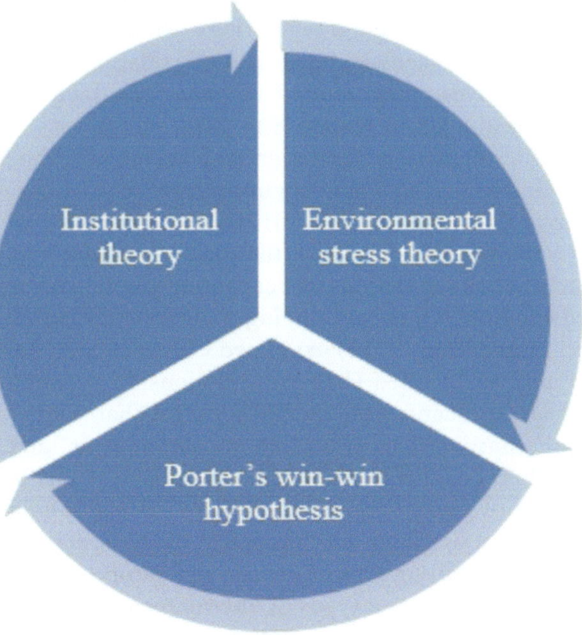

FIGURE 8.7 Theoretical framework surrounding environmental regulations, policies, and green investments.

Source: author's own.

like environmental taxation, fines, and penalties to uphold environmental sustainability. These regulations intensify pressure on enterprises and potentially displace investment activities, as environmental taxation limits funds available for new investment endeavors. Yet, the pressure from increased environmental regulations can be alleviated through the exploration of green innovation initiatives (Zhao and Sun, 2016). In 1995, Porter and Linde introduced the "Porter Hypothesis," suggesting that investments made by companies in environmental protection could stimulate innovation and adoption of cleaner technologies. This, in turn, could help offset environmental costs, leading to a mutually beneficial outcome where both environmental sustainability and economic profitability are achieved. Porter's win-win theory asserts that engaging in more green innovation activities can lead to financial gains and ensure sustainable growth. Green innovation boosts product quality, enhances investment returns, and incentivizes corporate managers to increase investments. Furthermore, green innovation plays a crucial role in guaranteeing environmental sustainability.

The recent 2021 United Nations Climate Change Conference has enhanced the comprehension of climate-related risks and environmental degradation among policymakers and scholars (Ren et al., 2022; Zhu and Lee, 2022). This increasing public awareness is compelling firms to minimize their environmental footprint and partake in environmental preservation, possibly through boosting their green investments (Berrone et al., 2013). Existing literature indicates that governmental policies and regulations play a pivotal role in propelling companies towards green investments (Zhang et al., 2022). Governments worldwide have formulated various environmental policies to incentivize such initiatives (Clemens et al., 2008; Garrone et al., 2018; Irfan et al., 2022). International green investments standards and policies are set by governments, trailed by both public and private financial institutions. Additionally, non-financial and non-governmental entities and associations also do play an active role in this domain. Main standard policies on green investments are described in Table 8.6.

When examining the connection between environmental regulations and corporate green investments, it is important to note (Huang and Lei, 2021) that existing studies haven't come to a consistent conclusion. This divide can generally be categorized into three perspectives. The first viewpoint suggests that environmental regulations can drive corporate green investments. For instance, Leiter et al. (2011) discovered a positive correlation between environmental regulations and corporate green investments in the European manufacturing industry, where firms tend to increase green investments in response to legal obligations. The second perspective proposes that environmental regulations hinder green investments. Van Soest (2005), through a mathematical model, indicated that stringent environmental regulations could actually decrease green investments. Sanyal (2007) and Gans (2012) also suggested that environmental regulations negatively affect such investments. Last, the third viewpoint suggests a non-linear relationship between environmental regulations and corporate green investments. Tang et al. (2013) discovered a U-shaped relationship by developing a comprehensive environmental regulation index. On the contrary, Li and Tian (2016) examined Chinese listed companies in heavily polluting industries and found an inverted U-shaped relationship between environmental regulations and green investments. Huang and Lei (2021) state that while most researchers agree that environmental regulations will have a notable impact on firms' green investment behavior, there is inconsistency in their conclusions regarding the nature of this impact. One contributing factor is the varying choice of proxy variables for environmental regulations. However, a more critical reason is that existing literature has generally treated environmental regulations as a broad, unified concept without delving into different regulatory types. This oversight neglects the potential varied effects that different types of environmental regulations could have on corporate green investments, resulting in overly generalized conclusions lacking policy specificity. Several studies have highlighted the significant disparities in the effects of different environmental regulations (Ren et al., 2018). Thus, there is a crucial need to categorize environmental

TABLE 8.6
Summary of Policies on Green Investments

Typology	Name of issuer (main list)	Standard/framework outputs (main list)
Governments and national/ sovranational/ international entities and organizations	United Nations UNEP-FI United Nations Principles of Responsible Investment (UNPRI) United Nations Development Program (UNDP) OECD European Union People's Bank of China French Ministry of the Ecology	United Nations Global Compact Principles Partnership for Action in Green Economy UNPRI Private Equity Action on Climate Change UNDP-SDG Impact Practice Standards for Private Equity Funds UNDP-OECD—Impact Standards for Financing Sustainable Development OECD—Framework for SDG Aligned Finance EU Sustainable Finance Taxonomy Chinese Green Bond Catalogue Greenfin Label—France
Public and private finance institutions/ organizations	Asian Development Bank (ADB) International Finance Corporation (IFC) Climate Bonds Initiatives (CBI) Green Climate Fund (GCF) Multilateral Development Banks (MDB) World Bank European Investment Bank	ADB Safeguard Policy Statement CBI standards and certification scheme GCF Investment Framework MDB Common Principles for Climate Mitigation World Bank Environmental and Social Framework Environmental and Social Standards
Non-financial and non-governmental institutions	Global Reporting Initiative (GRI) Loan Market Association (LMA) Equator Principles International Integrated Reporting Council (IIRC) Financial Accounting Standards Board (FASB) International Sustainability Standards Board (ISSB) Sustainability Accounting Standard Board (SASB) CiCERO Task Force for Climate Related Financial Disclosure (TCFD) European Financial Reporting Advisory Group (EFRAG)	GRI Standards LMA Green Loan Principles Equator Principles (I, II, III, and IV) IIRC Integrated Reporting framework FASB Non Financial Reporting Standards SASB Investment and Commercial Banking Standard CiCERO Shades of Green TCFD—Climate Related Financial Disclosures EFRAG Sustainability Reporting Standards Roadmap
Associations and networks	Global Investor Coalition Partnership for Carbon Accounting Financials (PCAF) International Capital Market Association (ICMA) Global Impact Investing Network World Business Council for Sustainable Development (WBCSD) ASEAN Capital Market Forum	The Low Carbon Registry PCAF—Global GHG Accounting and Reporting Standard for the Financial Industry ICMA Green Bond Principles; ICMA Sustainability Bond Guidelines ICMA Harmonized Framework for Impact Reporting WBCSD Framework for Portfolio Sustainability Assessments ASEAN Green Bond Standards

Source: Inderst et al., 2012.

regulations and conduct comparative analyses on how different types of regulations impact corporate green investments.

Huang and Lei (2021) categorize environmental regulation and corporate green investment together with their typical tools into three categories (Figure 8.8).

Command-control environmental regulation is a form of mandatory regulation where the government directly oversees and manages firms' green practices by establishing environmental laws and regulations. Firms are required to adhere to these regulations regarding emissions and technical standards, with non-compliance resulting in penalties. When the intensity of such regulation is low, the cost of green investments to meet the requirements is lower than the penalties, as it is comparatively easier to fulfill the regulations. As per the factor endowment hypothesis, firms tend to proactively increase green investments to meet these regulatory demands (Leiter et al., 2011). However, as the strictness of command-control environmental regulation intensifies, firms may struggle to adjust to their specific circumstances due to the regulations' lack of consideration for firm heterogeneity. This leads to situations where the costs of meeting these regulations outweigh the penalties, prompting firms to opt for sanctions over increased green investments for economic reasons.

In contrast, market-based environmental regulation employs mechanisms like charges, subsidies, and market incentives to encourage firms to independently evaluate costs, revenues, and determine their levels of green investments and emissions. This approach includes tools such as environmental taxes, emission trading schemes, and subsidies. Many studies suggest that market-based environmental regulation is cost-effective, efficient, flexible, and easier to implement compared to command-control counterparts (Jung et al., 1996; Montero, 2002; Requate and Unold, 2003). By allowing firms to make choices based on economic benefits, market-based regulations create economic incentives for firms to invest in green initiatives, innovate in pollution control technologies, and reduce emission costs.

Public-participation environmental regulation aims to stimulate firms to increase green investments through public opinion and social pressures. Public demand for environmental quality exerts social pressure on firms to fulfill their green responsibilities. Additionally, studies indicate that firms demonstrating environmental responsibility can enhance their social image and gain competitive advantages in the market (Miles and Covin, 2000; Mohr and Webb, 2005). Some research also suggests that public attention to environmental issues, regardless of the country's development

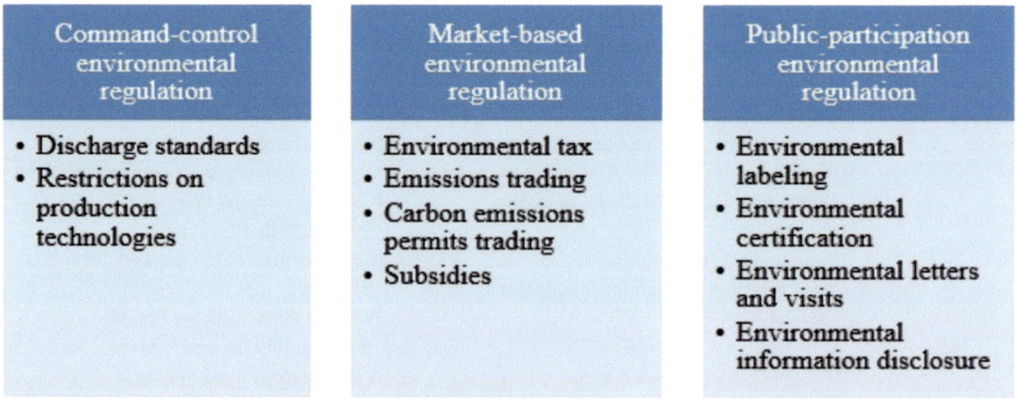

FIGURE 8.8 Theoretical framework surrounding environmental regulations, policies, and green investments.

Source: author's own.

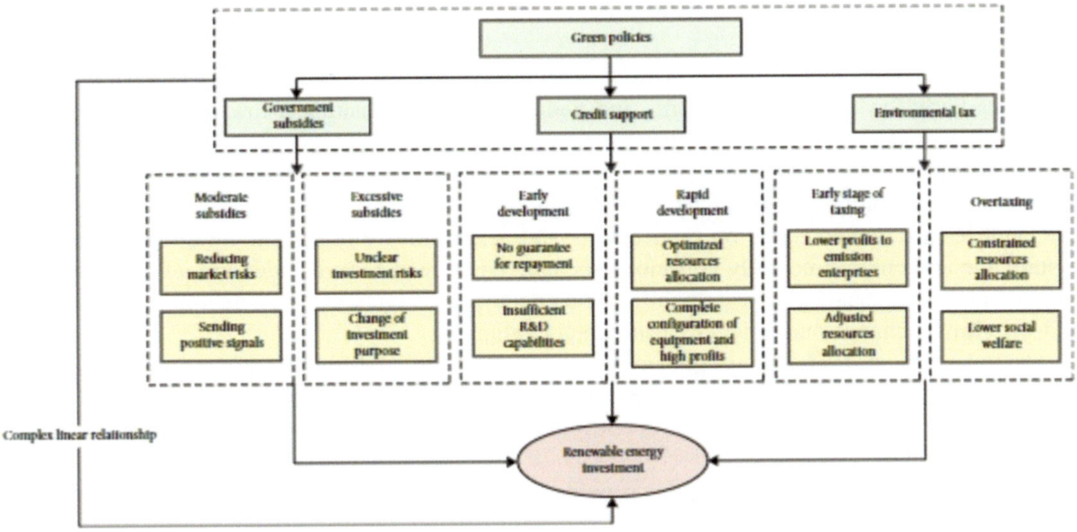

FIGURE 8.9 Theoretical framework on relationship among the green policy environment, green investments, and the influencing mechanisms.

status, can significantly enhance the environmental performance of polluting firms (Langpap and Shimshack, 2010).

The complex green policy environment, encompassing coordinated energy planning and investments, is made up of various elements. Key channels such as government subsidies, credit support, and environmental taxes play a significant role in impacting green investments. Figure 8.9 illustrates the relationship among the green policy environment, green investments, and the influencing mechanisms.

8.5 CASE STUDY

Finding accurate and up-to-date data on global green investment statistics is challenging due to several reasons:

- Lack of standardization: green investments encompass a wide range of sectors and can vary in definitions and criteria across different regions and organizations. This lack of standardization makes it difficult to aggregate data consistently.
- Fragmented reporting: data on green investments may be reported by various sources, including governments, financial institutions, and non-profit organizations. This fragmented reporting can lead to inconsistencies and gaps in the data.
- Rapidly evolving landscape: the field of green finance is constantly evolving, with new technologies, policies, and investment opportunities emerging regularly. Keeping track of these changes and updating data accordingly can be a complex task.
- Limited transparency: some green investments may not be publicly disclosed or reported due to confidentiality concerns or competitive reasons, making it challenging to obtain a complete picture of global green investment activities.

Considering the aforementioned points, a case study on the Green Future Index developed by MIT with an analysis of green financial policy outputs in OECD countries is performed. This case provides a comprehensive view of how countries are progressing in their green investment efforts.

The Green Future Index, created by MIT Technology Review Insights, ranks 76 countries and territories based on their advancements and dedication to establishing a low-carbon future. This index evaluates the extent to which these regions are transitioning their economies towards clean energy, sustainable industries, agriculture, and communities by investing in renewables, fostering innovation, and promoting green finance (European Commission, 2024).

Countries are ranked based on their efforts and achievements in transitioning towards a more sustainable and environmentally conscious society. The Green Future Index aims to raise awareness about the importance of sustainability and encourage countries to take meaningful actions to address climate change and other environmental challenges.

The index provides valuable insights into the green progress of different nations and helps policymakers, businesses, and communities make informed decisions to support a more sustainable future for our planet. The Green Future Index is relevant to measuring green investment in several ways:

1. Benchmarking progress: the index provides a benchmark for countries to assess their performance in promoting sustainability and transitioning towards a green future. By ranking countries based on various environmental factors, including green investments, the index highlights where progress is being made and where there is room for improvement.
2. Highlighting best practices: countries that perform well in the Green Future Index are likely making significant investments in green technologies, renewable energy, sustainable infrastructure, and other environmentally friendly initiatives. Studying the approaches and strategies of top-ranking countries can offer insights into effective green investment practices.
3. Encouraging investment: the index can serve as a motivation for countries to increase their green investments in order to improve their ranking and contribute to global sustainability efforts. It can encourage governments, businesses, and investors to allocate resources towards projects that have a positive impact on the environment.
4. Informing policy decisions: Policymakers can use the insights from the Green Future Index to guide their decision-making processes related to environmental policies, regulations, and incentives. Understanding how green investments contribute to a country's overall sustainability goals can help shape effective policies that support further investment in green technologies and initiatives.

Most recent is the Green Future Index 2023 edition, which comprises 23 indicators selected from distinct sets of country-level data. According to MIT Technology Review Insights (2023), the Green Future Index 2023 represents the third annual comparative assessment of 76 nations and territories (which collectively account for approximately 90% of global GDP) regarding their capacities to foster a sustainable, low-carbon future within their economies and societies. The index was formulated through thorough primary and secondary research processes. The secondary research involved ongoing scrutiny of numerous articles, research reports released by climate-focused organizations and NGOs, government policy documents, and scientific literature. Primary research entailed close to 20 in-depth interviews with a diverse group of global climate change experts, including climate scientists, researchers, technologists, government officials, NGO advocates, and clean technology innovators and investors.

The Green Future Index 2023 comprises five fundamental pillars (Figure 8.10):

Pillar 1: *carbon emission*. This pillar evaluates countries' effectiveness in reducing carbon dioxide emissions both overall and within key sectors. It includes indicators such as:

- Total carbon dioxide emissions in 2020 relative to GDP in millions of tons.
- Average annual change in carbon dioxide emissions between 2015 and 2020, both overall and within the industry, transportation, and agriculture sectors.

FIGURE 8.10 Theoretical framework on relationship among the green policy environment, green investments, and the influencing mechanisms.

Pillar 2: *energy transition.* This pillar focuses on the growth rate and contribution of renewable and clean energy generation in each country. Indicators in this pillar cover:

- Increase in renewable energy production in gigawatt-hours between 2015 and 2020.
- Percentage of energy from renewable sources in final energy consumption in 2019.
- Growth of nuclear energy production in gigawatt-hours between 2015 and 2020.
- Percentage of nuclear-generated energy in final energy consumption in 2019.

Pillar 3: *green society.* This pillar assesses efforts by government, industry, and society to promote environmentally friendly practices. It includes indicators such as:

- Number of LEED-certified green buildings in 2020 per million urban population.
- Proportion of recycled solid waste in total waste managed.
- Net change in forestation between 2015 and 2020, combining natural forest growth and planned afforestation projects.
- Quantity of electric passenger vehicles per million urban population in 2021.

Pillar 4: *clean innovation.* This pillar evaluates the innovation environment for advancing a low-carbon future, including metrics like:

- Growth in green intellectual property indicated by the rise in sustainable technology patents relative to GDP between 2017 and 2021.
- Investment in cross-border clean energy and food technology as a percentage of GDP between 2016 and 2020.
- Number of food tech startups per million urban population.

Pillar 5: *climate policy.* This pillar measures the ambition and effectiveness of climate policies, encompassing areas such as carbon financing, sustainable agriculture, and the utilization of pandemic recovery funds to achieve green economic restoration. It includes indicators such as:

- Evaluation of policy action aligned with the Paris Agreement and nationally determined contributions (NDCs).
- Assessment of frameworks supporting carbon capture and sequestration (CCS) efforts.
- Implementation of financial incentives like carbon taxes and carbon bond markets to encourage emission reduction.
- Evaluation of sustainable agriculture policies and their implementation effectiveness.
- Analysis of the impact of COVID-19 recovery stimulus packages on decarbonization efforts, specifically focusing on energy transition initiatives versus conventional fossil fuel projects.

Together, these pillars offer a comprehensive assessment of each country's commitment to a sustainable future in two main aspects: their current progress in meeting carbon reduction targets and engaging in climate-friendly societal practices and the aspirations they need to fulfill for a carbon-neutral economy. The initial four progress pillars make up 60% of the index weight, focusing on evaluating advancements made. The fifth pillar, regarding climate policy, evaluates the degree to which investments and policy initiatives support green infrastructure projects and legislation frameworks. These elements collectively drive the establishment and maintenance of a country's sustainable future, with this pillar carrying a 40% weighting in the index.

According to the report, in 2023 Iceland, Finland, and Norway are the countries most prepared for a low-carbon future. Other countries making up the top five of the Green Future Index 2023 report are Denmark and Sweden (Figure 8.11). On the other side of the scale are Bangladesh, Qatar, Zambia, Algeria, and Iran, which are least prepared for a low-carbon future.

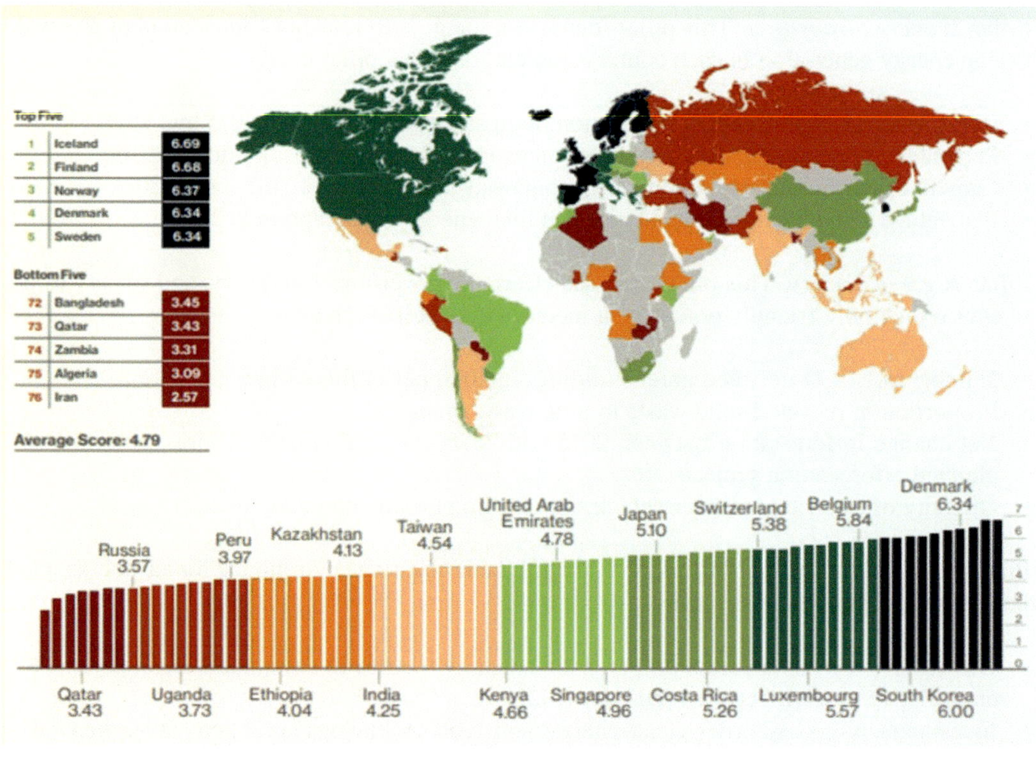

FIGURE 8.11 The Green Future Index 2023 rankings world map.

Source: MIT Technology Review Insights, 2023.

Green Leaders — The 20 countries making the greatest progress and commitment toward building a low-carbon future.

Greening Middle — The 20 countries that are making progress or commitment toward building a green future.

Climate Laggards — The 20 countries that are making slow and uneven progress or commitment toward building a green future.

Climate Abstainers — The 16 countries that will be left behind in the green future through their lack of progress and commitment toward developing a modern, clean, and innovative economy.

Group	Rank 2023		Rank 2022	Country	Score
Green Leaders	1	—	1	Iceland	6.69
Green Leaders	2	↑	6	Finland	6.68
Green Leaders	3	↑	5	Norway	6.37
Green Leaders	4	↓	2	Denmark	6.34
Green Leaders	5	↑	9	Sweden	6.34
Green Leaders	6	↓	3	Netherlands	6.22
Green Leaders	7	↓	4	United Kingdom	6.12
Green Leaders	8	↑	10	South Korea	6.00
Green Leaders	9	↓	7	France	5.99
Green Leaders	10	↑	13	Spain	5.92
Green Leaders	11	↓	8	Germany	5.92
Green Leaders	12	↓	11	Belgium	5.84
Green Leaders	13	↑	17	Italy	5.70
Green Leaders	14	↑	15	Canada	5.69
Green Leaders	15	↓	12	Ireland	5.69
Green Leaders	16	↑	28	Luxembourg	5.57
Green Leaders	17	↑	22	Greece	5.56
Green Leaders	18	—	18	Portugal	5.52
Green Leaders	19	↑	21	United States	5.39
Green Leaders	20	↓	14	Switzerland	5.38
Greening Middle	21	↓	16	Poland	5.38
Greening Middle	22	↑	23	Austria	5.37
Greening Middle	23	↑	25	Bulgaria	5.31
Greening Middle	24	↓	20	Costa Rica	5.26
Greening Middle	25	↑	31	South Africa	5.23
Greening Middle	26	↑	38	Uruguay	5.13
Greening Middle	27	↓	26	China	5.12
Greening Middle	28	↓	19	Japan	5.10
Greening Middle	29	↓	24	Hungary	5.09
Greening Middle	30	↑	33	Chile	5.08
Greening Middle	31	↓	27	Czech Republic	4.97
Greening Middle	32	↓	29	Singapore	4.96
Greening Middle	33	↑	45	Hong Kong, China	4.89
Greening Middle	34	↓	32	Colombia	4.83
Greening Middle	35	↑	46	Slovakia	4.81
Greening Middle	36	↑	41	United Arab Emirates	4.78
Greening Middle	37	↓	36	Morocco	4.73
Greening Middle	38	↓	34	Brazil	4.70
Greening Middle	39	↑	47	Romania	4.66
Greening Middle	40	—	40	Kenya	4.66
Climate Laggards	41	↓	39	New Zealand	4.57
Climate Laggards	42	↑	52	Australia	4.56
Climate Laggards	43	↓	30	Israel	4.56
Climate Laggards	44	↑	53	Philippines	4.56
Climate Laggards	45	↑	54	Mexico	4.54
Climate Laggards	46	↓	37	Taiwan	4.54
Climate Laggards	47	↑	61	Ukraine	4.38
Climate Laggards	48	↑	68	Argentina	4.31
Climate Laggards	49	↑	70	Indonesia	4.29
Climate Laggards	50	↓	42	India	4.25
Climate Laggards	51	↓	43	Nigeria	4.18
Climate Laggards	52	↑	63	Ecuador	4.14
Climate Laggards	53	↑	56	Vietnam	4.13
Climate Laggards	54	↓	49	Kazakhstan	4.13
Climate Laggards	55	↓	48	Thailand	4.12
Climate Laggards	56	↓	51	Saudi Arabia	4.11
Climate Laggards	57	↑	58	Kuwait	4.05
Climate Laggards	58	↓	35	Ethiopia	4.04
Climate Laggards	59	↓	50	Angola	4.01
Climate Laggards	60	↓	59	Egypt	3.99
Climate Abstainers	61	↑	66	Dominican Republic	3.98
Climate Abstainers	62	↑	67	Peru	3.97
Climate Abstainers	63	↑	69	Turkey	3.83
Climate Abstainers	64	↓	44	Cameroon	3.82
Climate Abstainers	65	↑	71	Ghana	3.80
Climate Abstainers	66	↓	57	Uganda	3.73
Climate Abstainers	67	↓	55	Pakistan	3.72
Climate Abstainers	68	↓	65	Malaysia	3.67
Climate Abstainers	69	↑	72	Guatemala	3.57
Climate Abstainers	70	↓	64	Russia	3.57
Climate Abstainers	71	↑	74	Paraguay	3.55
Climate Abstainers	72	↓	62	Bangladesh	3.45
Climate Abstainers	73	—	73	Qatar	3.43
Climate Abstainers	74	↓	60	Zambia	3.31
Climate Abstainers	75	—	75	Algeria	3.09
Climate Abstainers	76	—	76	Iran	2.57

FIGURE 8.12 The Green Future Index country rankings, 2023.

Source: MIT Technology Review Insights, 2023.

When looking into The Green Future Index country rankings (Figure 8.12) several trends are observed.

1. In the Green Future Index 2023 rankings, there is more stability than progress noted. Among the top 20 scorers, known as the Green Leader group, all but three were part of this category in the 2022 rankings.
2. The Green Leaders maintained their emphasis on low-carbon practices, including ongoing efforts to decrease emissions, boost renewable energy usage, enhance sustainability in transportation and utilities through green policies, and targeted investments. However, around half of them experienced a decline in scores, notably with Iceland, the top-ranked country, seeing a significant drop. While carbon reduction efforts are intensifying and policies are strengthening, initial outcomes show a slight decline in overall performance.
3. Within the Greening Middle segment, comprising countries ranked 21st to 40th in the Green Future Index, there is notable movement and considerable shifts in rankings. These nations are actively translating sustainable policy frameworks into practical initiatives.

This group includes various emerging economies, particularly those with robust export sectors that can align sustainable policies with economic incentives. For instance, Uruguay and South Africa climbed 12 and six positions, respectively, in the 2023 rankings. Uruguay, with a large share of emissions linked to its agriculture and forestry sectors, is pursuing decarbonization efforts, like implementing a carbon-neutral certification process for its beef exports to Europe. South Africa introduced a Green Finance Taxonomy in April 2022, setting standards for sustainable economic activities to attract foreign investment.

4. Despite notable attempts to integrate economic and sustainable development, it is evident that emerging economies struggle to match the performance of wealthier counterparts in the Green Future Index rankings. Analyzing each country's ranking alongside its GDP per capita highlights a stark reality: prosperity significantly influences the capacity to shape a sustainable, low-carbon future. The index indicates a strong correlation between high green society scores and national GDP numbers at 76% and high climate policy scores and GDP at 64%.
5. Various factors contribute to countries lagging behind, including excessive reliance on fossil fuel production or natural resource extraction, which can hinder the implementation of decarbonization initiatives.
6. The lowest quadrant in the Green Future Index rankings comprises the Climate Abstainers, characterized by a lack of substantial policy changes or advancements in decarbonization practices. This group encompasses nations like Russia in 70th place and Iran at the bottom. While some countries, such as Peru (62nd) or Malaysia (68th), display notable sustainability aspirations, the overall picture for these countries positioned at the bottom (many observing a decrease in scores from 2022) is concerning. Notable improvement was absent across all pillars. The burden of limited income levels is particularly challenging for most nations in this category. For instance, countries like flood-prone Pakistan allocate scant resources to combat climate change-induced disasters.

Addressing the challenge of mitigating severe climate change necessitates significant investments in low-carbon technologies. As a result, the reorientation of financial resources stands as a critical objective of the Paris Agreement, prompting countries to implement measures that influence financial institutions in this direction. Steffen (2021) found that the level of transparency regarding such policies is limited, making international comparisons challenging. In response to this issue, Steffen (2021) conducted a comparative evaluation of green financial policy outcomes in OECD nations spanning from 2001 to 2019. This analysis was based on a newly assembled catalogue of 136 policies from 29 countries and the European Union.

The financial sector, comprising firms and institutions offering financial services, plays a crucial role in advanced and developing economies. Policy experts have persistently focused on ensuring the sector's functionality through financial policies, typically defined as measures related to regulating, supervising, and overseeing financial and payment systems, including markets and institutions. These policies aim to enhance financial stability, create market efficiency, and safeguard client assets and consumer interests (IMF, 2000; OECD, 2008). The significance of the financial sector in climate change mitigation arises not only from the necessity to reallocate capital in alignment with fundamental economic transitions but also due to the characteristics of many low-carbon technologies. For instance, renewable energy solutions typically require more capital investment compared to traditional high-carbon alternatives (Egli et al., 2018), resulting in substantial funding requirements for transitioning to low-carbon practices (Polzin, 2017; Paroussos et al., 2020). In recent times, policymakers have introduced financial policies with a specific focus not only on enhancing overall financial market efficiency but also on "greening" the financial sector. Steffen (2021) defines green policies as "public policies that target the financial sector to align financial flows with the objectives set forth in the Paris Agreement." The comparative assessment of green financial policy outcomes in OECD countries has yielded the following findings:

1. Prior to 2009, there was a scarcity of policies in place (see Figure 8.16). The early policies largely relied on either governmental authority or treasury resources (specifically for green loan guarantees). From 2010 to 2014, the quantity of policies rose, driven by the emergence of new policies leveraging governmental authority and organizational structures. During this period, many countries introduced climate-related disclosure mandates, while several opted to establish dedicated green state investment banks. The significant expansion of green financial policies commenced in 2015, coinciding with the year of the Paris Agreement, which underscored the reorientation of financial flows as a key long-term climate policy objective alongside mitigation and adaptation efforts. The traction gained in addressing the financial sector began to intensify during the Agreement negotiations (Ministry of Finance and Public Accounts 2015), with policy initiatives rapidly gaining momentum thereafter. This growth is evidenced by a compound annual growth rate of around 24%, compared to 16% up to 2014. Between 2015 and 2019, the number of policies more than doubled, with numerous fresh policies leveraging government authority and organizational frameworks and increasingly incorporating the input of public actors. Notably, the number of policies primarily dependent on treasury funds remained limited, reflecting the norm that climate policies using taxation or subsidies tend to target emitting sectors directly rather than the financial sector (Figure 8.13).

2. Various policy instruments were introduced as part of the financial policies (Table 8.7).

Regarding their mode of information dissemination, governments utilize it both as a means of gathering data (e.g., through surveys involving financial stakeholders and expert panels) and as a mechanism for influencing outcomes (e.g., by offering voluntary government endorsements for sustainable financial products—a traditional environmental policy instrument applied to the financial sector). Government studies and expert panels are commonly employed instruments in emerging policy domains. Within the realm of green finance, these tools predominantly emerged post-Paris Agreement and have since found widespread use. While information policies typically suggest a minimal degree of market interference, they can serve as an initial phase before implementing more rigorous measures informed by these assessments or panels.

Government authority features prominently in various policy instruments, often directed at enhancing the transparency of climate-related data provided by financial stakeholders to enable the

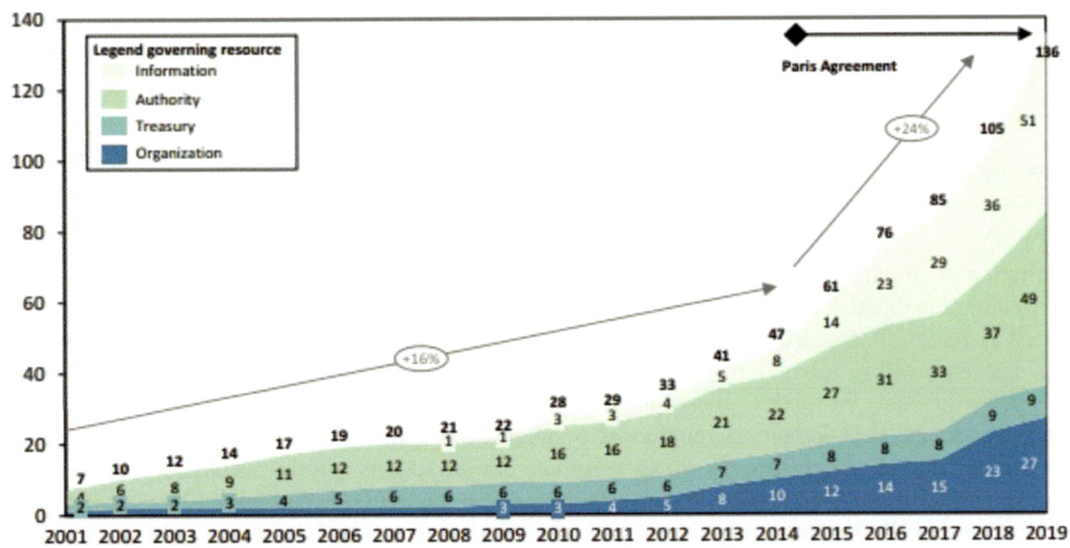

FIGURE 8.13 Trends of green financial policies over time.

Source: Steffen, 2021.

TABLE 8.7
Summary of Green Financial Policy Instruments Based on Rationales and Policy Instruments of 136 Policies in OECD Countries.

Governing resource	Policy rationale	Policy instruments that are part of policy inventory
Information	Collect information and generate knowledge	Government study/survey/consultation Government-convened expert panel/research initiative/exchange forum
	Provide general guidance and specific tools related to green finance	Government label for green finance products/actors Voluntary guidelines/recommendations related to green finance Public green finance strategy/program
Authority	Provide regulatory framework for green finance products	Guidelines for green bonds Guidelines for other green finance products/indices Special legal entities/setups for green finance
	Improve disclosure of climate-related information from companies in general	Disclosure requirements concerning climate/ESG for corporations Regulatory guidance on disclosure concerning climate/ESG for corporations
	Improve disclosure of climate-related information from fund managers and other financial actors	Disclosure requirement of whether and how climate/ESG considered in portfolio Disclosure requirement on climate/ESG impact of portfolio
	Mandate/promote climate-related risk management by financial actor	Specific regulations on climate risk management
Treasury	Using public funds to establish new green financial products	Subsidy for private green bond issuance Subsidies for innovative green loan offerings
	Using public funds to de-risk green financing	Loan guarantees for immature technologies/solutions
Organization	Leverage general financial market activity of public authorities	Investment policy of central bank Investment policy of public funds (e.g. government pension funds) Sovereign green bond issuance
	Engage in dedicated financial market activity specifically for low-carbon goals	Green programs at an existing state investment bank Establishment of a dedicated green state investment bank Establishment of other financial actors (e.g. accelerator, public–private fund)

Source: Steffen, 2021.

integration of climate considerations into investment decisions. However, there is debate surrounding whether heightened transparency can genuinely alter investor behavior in a substantial manner (Ameli et al., 2020). Disclosure requirements for publicly listed companies, encompassing climate considerations within broader ESG (environmental, social, and corporate governance) reporting obligations in many instances, constitute the most prevalent policy instrument. With regards to funds and financial stakeholders, numerous policies exist, although many are relatively lenient, merely necessitating disclosure on whether climate considerations inform portfolio choices at all. Another group of policies employs regulatory authority—not across broad asset categories in general but specifically targeting sustainable financial products—to prevent "greenwashing" wherein products may appear to incorporate climate considerations superficially but fail to do so in practice. Beyond facilitating the growth of a marketplace for sustainable financial products, these policies also prioritize consumer protection, a core objective of financial regulation. Last, since 2016, certain policies have begun directly regulating the climate-related risk management practices of financial entities.

Government funds are utilized for offering loan guarantees to reduce the risk associated with low-carbon financing, although such mechanisms are deployed in only a select few nations. Additionally, a handful of policies extend subsidies to support the creation of low-carbon financial products, with many of these policies having been in effect for an extended period. Notably, no instances of green financial policies involving taxation have been documented. In the aftermath of the 2008–2009 financial downturn, the notion of expanding financial activity taxation, including for climate considerations, has been a topic of significant discourse (Shackelford et al., 2010). However, up until 2019, no policies related to taxing the green financial sector had been introduced.

Conversely, several countries employ government structures for green finance initiatives. On one front, governments leverage financial market operations that public entities must perform in any case and adapt these actions to align with climate objectives. This includes aspects such as aligning the investments made by central banks or government pension funds with climate-conscious investment policies and issuing segments of sovereign bonds known as "green bonds" (Tolliver et al., 2019), which have witnessed substantial adoption since 2018. On the other front, governments embark on fresh, specialized financial market operations to provide green financial services not currently offered by private entities, to cultivate expertise and set industry standards, thereby nurturing a green financial sector within their nations. In some scenarios, existing state-owned investment banks are repurposed for this cause. In other instances, new entities known as "green state investment banks" are established. In essence, the overall landscape reveals the implementation of diverse green financial policy instruments over the past decade, drawing upon a broad spectrum of governmental resources available.

Although the number of green financial policies has increased greatly since the Paris Agreement, there is wide variation among countries. Figure 8.14 shows the policy density by country.

On average, OECD countries (excluding EU institutions) have implemented 3.3 policies by the conclusion of 2019, with the range varying from 0 to 18 policies. France boasts the highest concentration with 18 policies, trailed by the United Kingdom with 12 policies, along with several other nations having more than five policies (such as Australia, Canada, Germany, Italy, Japan, the Netherlands, and the United States). In contrast, several predominantly Eastern European countries—including the Czech Republic, Estonia, Hungary, the Slovak Republic, and Slovenia—have yet to establish any green financial policies. At the supra-national level, notable policy initiatives have unfolded within the European Union framework, largely aligning with the EU's Sustainable Finance Action Plan introduced in March 2018. By the close of 2019, the number of policies at the EU level rivals that of the most prolific individual country (France), contributing to an increased regulatory landscape across all EU member states. However, numerous national policies exceed EU stipulations, rendering direct summation of policy densities inconclusive.

This case illustrates a notable uptick in policy output concerning green financial policies post-Paris Agreement, indicating concerted efforts by OECD countries to channel financial resources towards sustainability. However, akin to other climate policy domains, significant variations exist in the ambitions of different nations. The examination underscores the continued expansion of the green finance sector and green policy initiatives, offering promising avenues for funding sustainable development. Looking ahead, policymakers, regulators, and other stakeholders must confront three primary challenges:

First, there are spillover effects stemming from national and regional regulatory frameworks, with repercussions for companies globally. These effects often manifest through interconnected global supply chains and investment networks, where major corporations and financial institutions increasingly mandate their suppliers or investee companies to disclose their sustainability practices.

Second, there is a need to integrate sustainable finance principles within national strategies for sustainable development. These strategies are typically aligned with overarching national

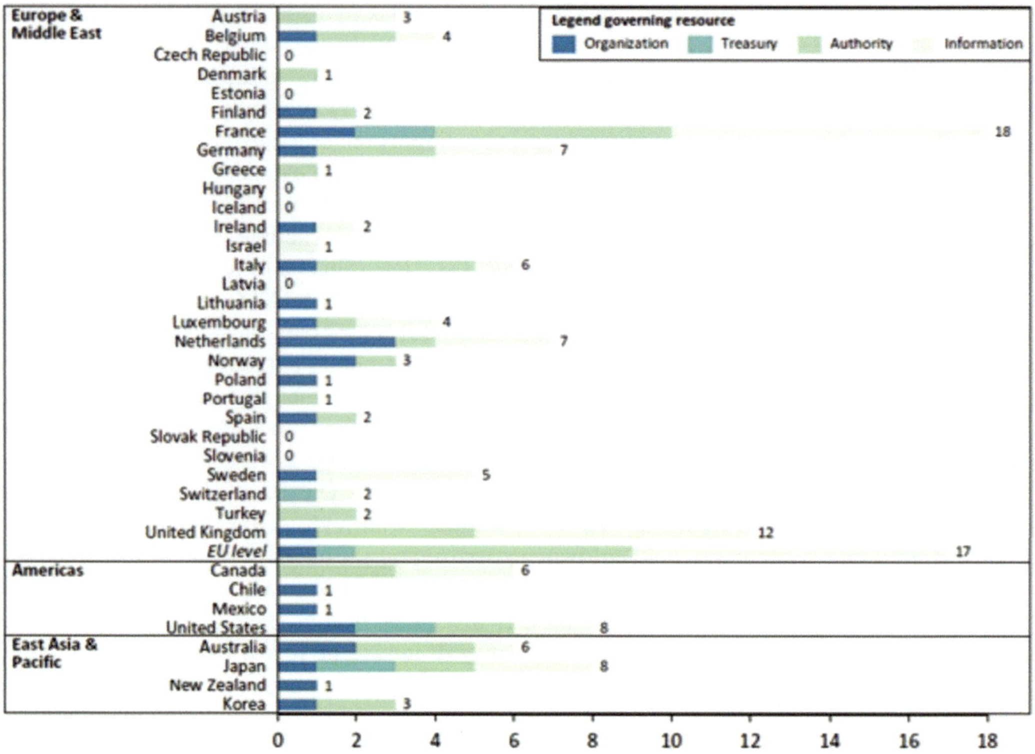

FIGURE 8.14 Cumulative number of green financial policies.

Source: Steffen, 2021.

development objectives, in harmony with national targets outlined in the 2030 Agenda for Sustainable Development and the Paris Agreement. By incorporating sustainable finance frameworks, these strategies can set out policy goals, key priorities, and strategic areas for action to guide and catalyze domestic efforts in bolstering sustainable finance.

Last, it is imperative that policymaking in sustainable finance adopts an impact-oriented approach, prioritizing the effectiveness of policies. Placing a premium on the impact and outcomes of sustainable finance measures is crucial, particularly amidst concerns regarding potential resistance to sustainable investments.

The signals from markets can wield considerable influence, steering and shaping a forthcoming economy that is environmentally conscious, socially just, and transparently governed. Addressing policy hurdles and implementation challenges, including questions of policy coherence and spillover effects, will be paramount in realizing the advantages of sustainable finance in advancing the objectives of the 2030 Agenda for Sustainable Development.

8.6 CONCLUSIONS

This chapter of the book focused on green investment, financial activities that support environmentally sustainable projects or companies. The field of green investment has significantly advanced and is seen as a way to not only make a profit but also to contribute to a more sustainable future for the planet. Several key takeaway are as follows:

1. Green investments involve allocating financial resources to projects or companies that prioritize sustainable practices, eco-friendly technologies, and the preservation of natural resources. This concept entails investing in initiatives that have beneficial environmental effects or support sustainability. Green investments cover various sectors such as renewable energy, energy efficiency, sustainable agriculture, clean technology, waste management, and green infrastructure.
2. The growing emphasis on green investments can be seen as a direct result of various key advancements in policy (e.g., the Paris Agreement on Climate Change, the UN Sustainable Development Goals of Agenda 2030, the implementation of the European Union Green Deal Investment Plan), regulations (including the recent European Union Taxonomy Regulation, the EU's updated Non-Financial Reporting Directive, and the UK Government's Green Finance Strategy post-Brexit), and socio-economic factors (such as the influence of movements like Fridays for Future and the COVID-19 pandemic).
3. The ecosystem of green investment concept consists of a minimum of four key components: the capital demand, the capital supply, the intermediaries facilitating connections between the two, and the broader operational context of all stakeholders.
4. The term "green investment indicator" typically refers to a measure or metric used to assess the environmental sustainability or impact of investment decisions. Green investment indicators shall be assessed from the perspectives of policy relevance, analytical robustness, and measurability.
5. Green investment indicators from the EU Taxonomy perspective cover climate change mitigation, climate change adaptation, sustainable use and protection of water and marine resources, shift towards a circular economy, pollution mitigation and management, and conservation and rehabilitation of biodiversity and ecosystems.
6. The green growth indicators by OECD cover environmental and resource productivity of the economy, natural asset base, environmental dimension of quality of life, and economic opportunities and policy responses.
7. Besides OECD and UN, United Nations Principles for Responsible Investment (UN PRI), Global Reporting Initiative (GRI), Climate Disclosure Standards Board (CDSB), Carbon Disclosure Project (CDP), Institutional Investors Group on Climate Change (IIGCC), and Green Bond Principles (GBP) have and are developing green investment indicators.
8. Motives from green investments are diverse, with different investors placing varying importance on priorities, risk and return preferences, and motivation hierarchies. In general, driving forces reflect consumer and stakeholder behavior, climate change, legislation and regulations, target market, public financing and incentives, investor preferences, organizational culture, financial performance, reputational considerations, and potential efficiency gains.
9. Barriers can hinder and diminish progress in adopting green practices. Among others, barriers include carbon pricing, fossil fuel subsidies, high initial costs and extended payback periods, technological risks, revenue risks, lower rate of return and a much higher risk compared to fossil fuel projects, elevated transaction costs, gaps in information and knowledge, disparities in beliefs, limited engagement from international organizations, the necessity of replacing aging assets, the expensive integration of clean energy sources into existing systems, political and regulatory uncertainties, issues related to ownership rights and intellectual property, as well as inadequacies in domestic and international financial instruments.
10. Over the past two decades, several initiatives have been formulated to enable the financial sector to advance the financial aspect of the green transition (Heaps and Guyatt, 2017). As per the Green Finance Platform (2024), globally, more than 784 policy and regulatory measures have been implemented to support green transition.

11. The relationship among environmental regulations, policies, and green investments can be understood by drawing on theoretical frameworks such as institutional theory, environmental stress theory, and Porter's win-win hypothesis.
12. The regulatory framework for green investments in Europe primarily involves the European Union's initiatives such as the EU Sustainable Finance Action Plan. This plan includes regulations like the EU Taxonomy Regulation, which sets criteria for determining environmentally sustainable economic activities and the Sustainable Finance Disclosure Regulation (SFDR), which requires financial market participants to disclose sustainability-related information in their investment processes. Additionally, the European Green Deal aims to make Europe the first climate-neutral continent by 2050 and includes various legislative measures to promote green investments and sustainable finance practices across the EU.
13. The case study on the Green Future Index developed by MIT with an analysis of green financial policy outputs in OECD countries has shown:
    - 13.1 In 2023 Iceland, Finland, were Norway are the countries most prepared for a low-carbon future. On the other side of the scale are Bangladesh, Qatar, Zambia, Algeria, and Iran, which were least prepared for a low-carbon future.
    - 13.2 The financial sector, comprising firms and institutions offering financial services, plays a crucial role in advanced and developing economies.
    - 13.3 The significance of the financial sector in climate change mitigation arises not only from the necessity to reallocate capital in alignment with fundamental economic transitions but also due to the characteristics of many low-carbon technologies.

REFERENCES

Primary Sources

Ameli, N., Drummond, P., Bisaro, A., Grubb, M., & Chenet, H. (2020). Climate finance and disclosure for institutional investors: Why transparency is not enough. *Climatic Change*, *160*(4), 565–589.

Becchetti, L., Cordella, M., & Morone, P. (2022). Measuring investments progress in ecological transition: The Green Investment Financial Tool (GIFT) approach. *Journal of Cleaner Production*, *357*, 131915.

Beisenbina, M., Fabregat-Aibar, L., Barberà-Mariné, M. G., & Sorrosal-Forradellas, M. T. (2023). The burgeoning field of sustainable investment: Past, present and future. *Sustainable Development*, *31*(2), 649–667.

Berrone, P., Fosfuri, A., Gelabert, L., & Gomez-Mejia, L. R. (2013). Necessity as the mother of 'green' inventions: Institutional pressures and environmental innovations. *Strategic Management Journal*, *34*(8), 891–909.

Campiglio, E. (2016). Beyond carbon pricing: The role of banking and monetary policy in financing the transition to a low-carbon economy. *Ecological Economics*, *121*, 220–230. https://doi.org/10.1016/j.ecolecon.2015.03.020

Chen, Y., & Ma, Y. (2021). Does green investment improve energy firm performance?. *Energy Policy*, *153*, 112252.

Chițimiea, A., Minciu, M., Manta, A. M., Ciocoiu, C. N., & Veith, C. (2021). The drivers of green investment: A bibliometric and systematic review. *Sustainability*, *13*(6), 3507. https://doi.org/10.3390/su13063507

Clemens, B., Bamford, C. E., & Douglas, T. J. (2008). Choosing strategic responses to address emerging environmental regulations: Size, perceived influence and uncertainty. *Business Strategy and the Environment*, *17*(8), 493–511.

Daniek, K. (2020). Green economy indicators as a method of monitoring development in the economic, social and environmental dimensions. *Nierówności społeczne a wzrost gospodarczy*, *62*, 150–173.

Ding, Q., Huang, J., & Chen, J. (2023). Does digital finance matter for corporate green investment? Evidence from heavily polluting industries in China. *Energy Economics*, *117*, 106476.

Drake, D. F., Kleindorfer, P. R., & Van Wassenhove, L. N. (2016). Technology choice and capacity portfolios under emissions regulation. *Production and Operations Management*, *25*(6), 1006–1025.

Du, H. S., Zhan, B., Xu, J., & Yang, X. (2019). The influencing mechanism of multi-factors on green investments: A hybrid analysis. *Journal of Cleaner Production*, *239*, 117977.

Duong, N. T. H., & Trang, T. N. L. (2019). Determinants of green banking implementation in emerging country: Evidence from Vietnam banks. *European Journal of Business and Management*, *11*(15).

Egli, F., Steffen, B., & Schmidt, T. S. (2018). A dynamic analysis of financing conditions for renewable energy technologies. *Nature Energy*, *3*(12), 1084–1092.

Fullerton, D., & Kim, S. R. (2008). Environmental investment and policy with distortionary taxes, and endogenous growth. *Journal of Environmental Economics and Management*, *56*(2), 141–154.

Gadenne, D. L., Kennedy, J., & McKeiver, C. (2009). An empirical study of environmental awareness and practices in SMEs. *Journal of Business Ethics*, *84*, 45–63.

Gans, J. S. (2012). Innovation and climate change policy. *American Economic Journal: Economic Policy*, *4*(4), 125–145.

Garrone, P., Grilli, L., & Mrkajic, B. (2018). The role of institutional pressures in the introduction of energy-efficiency innovations. *Business Strategy and the Environment*, *27*(8), 1245–1257.

Golka, P., Murau, S., & Thie, J. E. (2024). Towards a public sustainable finance paradigm for the green transition. *Finance and Society*, *10*(1), 38–50.

Hafner, S., Jones, A., Anger-Kraavi, A., & Pohl, J. (2020). Closing the green finance gap–A systems perspective. *Environmental Innovation and Societal Transitions*, *34*, 26–60.

Huang, L., & Lei, Z. (2021). How environmental regulation affect corporate green investment: Evidence from China. *Journal of Cleaner Production*, *279*, 123560.

Inderst, G., Kaminker, C., & Stewart, F. (2012). Defining and measuring green investments: Implications for institutional investors' asset allocations. *OECD Working Papers on Finance, Insurance and Private Pensions*, No. 24, OECD Publishing. http://doi.org/10.1787/5k9312twnn44-en

Irfan, M., Razzaq, A., Sharif, A., & Yang, X. (2022). Influence mechanism between green finance and green innovation: Exploring regional policy intervention effects in China. *Technological Forecasting and Social Change*, *182*, 121882.

Iyer, G. C., Clarke, L. E., Edmonds, J. A., Hultman, N. E., & McJeon, H. C. (2015). Long-term payoffs of near-term low-carbon deployment policies. *Energy Policy*, *86*, 493–505.

Krass, D., Nedorezov, T., & Ovchinnikov, A. (2013). Environmental taxes and the choice of green technology. *Production and Operations Management*, *22*(5), 1035–1055.

Langpap, C., & Shimshack, J. P. (2010). Private citizen suits and public enforcement: Substitutes or complements?. *Journal of Environmental Economics and Management*, *59*(3), 235–249.

Leiter, A. M., Parolini, A., & Winner, H. (2011). Environmental regulation and investment: Evidence from European industry data. *Ecological Economics*, *70*(4), 759–770.

Li, Q., & Tian, S. S. (2016). Can environmental regulation promote corporate environmental investment—considering the influence of market competition. *Journal of Beijing Institute of Technology (Social Sciences Edition)*, *18*(4), 1–8.

Lindenberg, N. (2014). Definition of green finance. *DIE mimeo, 2014*. https://ssrn.com/abstract=2446496

Liu, L., Zhao, Z., Zhang, M., & Zhou, D. (2022). Green investment efficiency in the Chinese energy sector: Overinvestment or underinvestment?. *Energy Policy*, *160*, 112694.

Liu, R., Wang, D., Zhang, L., & Zhang, L. (2019). Can green financial development promote regional ecological efficiency? A case study of China. *Natural Hazards*, *95*, 325–341.

Luo, H., & Qu, X. (2023). Impact of environmental CSR on firm's environmental performance, mediating role of corporate image and pro-environmental behavior. *Current Psychology*, *42*(36), 32255–32269.

Madaleno, M., Dogan, E., & Taskin, D. (2022). A step forward on sustainability: The nexus of environmental responsibility, green technology, clean energy and green finance. *Energy Economics*, *109*, 105945.

Magalhães, N. (2021). The green investment paradigm: Another headlong rush. *Ecological Economics*, *190*, 107209.

Martin, P. R., & Moser, D. V. (2016). Managers' green investment disclosures and investors' reaction. *Journal of Accounting and Economics*, *61*(1), 239–254.

McCollum, D. L., Zhou, W., Bertram, C., de Boer, H.-S., Bosetti, V., Busch, S., Després, J., et al. (2018). Energy investment needs for fulfilling the Paris agreement and achieving the sustainable development goals. *Nature Energy*, *3*(7), 589–599. https://doi.org/10.1038/s41560-018-0179-z

Miles, M. P., & Covin, J. G. (2000). Environmental marketing: A source of reputational, competitive, and financial advantage. *Journal of Business Ethics*, *23*, 299–311.

Mohr, L. A., & Webb, D. J. (2005). The effects of corporate social responsibility and price on consumer responses. *Journal of Consumer Affairs*, *39*(1), 121–147.

Paroussos, L., Fragkiadakis, K., & Fragkos, P. (2020). Macro-economic analysis of green growth policies: The role of finance and technical progress in Italian green growth. *Climatic Change*, *160*(4), 591–608.

Polzin, F. (2017). Mobilizing private finance for low-carbon innovation–A systematic review of barriers and solutions. *Renewable and Sustainable Energy Reviews*, *77*, 525–535.

Porter, M. E., & Van der Linde, C. (1995). Toward a new conception of the environment-competitiveness relationship. *Journal of Economic Perspectives*, *9*(4), 97–118.
Ren, S., Li, X., Yuan, B., Li, D., & Chen, X. (2018). The effects of three types of environmental regulation on eco-efficiency: A cross-region analysis in China. *Journal of Cleaner Production*, *173*, 245–255.
Ren, X., Li, Y., Wen, F., & Lu, Z. (2022). The interrelationship between the carbon market and the green bonds market: Evidence from wavelet quantile-on-quantile method. *Technological Forecasting and Social Change*, *179*, 121611.
Requate, T., & Unold, W. (2003). Environmental policy incentives to adopt advanced abatement technology: Will the true ranking please stand up?. *European Economic Review*, *47*(1), 125–146.
Rizzello, A. (2022). *Green investing: Changing paradigms and future directions*. Springer Nature.
Sanyal, P. (2007). The effect of deregulation on environmental research by electric utilities. *Journal of Regulatory Economics*, *31*, 335–353.
Schmittmann, J., & Chua, H. T. (2021). *How green are green debt issuers?*. International Monetary Fund.
Shackelford, D. A., Shaviro, D. N., & Slemrod, J. (2010). Taxation and the financial sector. *National Tax Journal*, *63*(4), 781–806.
Steffen, B. (2021). A comparative analysis of green financial policy output in OECD countries. *Environmental Research Letters*, *16*(7), 074031.
Tang, G. P., Li, L. H., & Wu, D. J. (2013). Environmental regulation, industry attributes and corporate environmental investment. *Accounting Research*, *6*, 83–89.
Tolliver, C., Keeley, A. R., & Managi, S. (2019). Green bonds for the Paris agreement and sustainable development goals. *Environmental Research Letters*, *14*(6), 064009.
Tran, T., Do, H., Vu, T., & Do, N. (2020). The factors affecting green investment for sustainable development. *Decision Science Letters*, *9*(3), 365–386.
Van Soest, D. P. (2005). The impact of environmental policy instruments on the timing of adoption of energy-saving technologies. *Resource and Energy Economics*, *27*(3), 235–247.
Wang, C., Wang, F., Zhang, X., Yang, Y., Su, Y., Ye, Y., & Zhang, H. (2017). Examining the driving factors of energy related carbon emissions using the extended STIRPAT model based on IPAT identity in Xinjiang. *Renewable and Sustainable Energy Reviews*, *67*, 51–61.
Wu, S., Wei, T., Qu, Y., Xue, R., Wang, H., & Shan, Y. (2024). How does global value chain embeddedness affect environmental pollution? Evidence from Chinese enterprises. *Journal of Cleaner Production*, *434*, 140232.
Xu, Q., Lei, Y., Ge, J., & Ma, X. (2017). Did investment become green in China? Evidence from a sectoral panel analysis from 2003 to 2012. *Journal of Cleaner Production*, *156*, 500–506.
Yang, Y., Xue, R., Zhang, X., Cheng, Y., & Shan, Y. (2023). Can the marketization of urban land transfer improve energy efficiency?. *Journal of Environmental Management*, *329*, 117126.
Zhan, J. X., & Santos-Paulino, A. U. (2021). Investing in the sustainable development goals: Mobilization, channeling, and impact. *Journal of International Business Policy*, *4*, 166–183. https://doi.org/10.1057/s42214-020-00093-3
Zhang, D., Zhao, Z., & Lau, C. K. M. (2022). Sovereign ESG and corporate investment: New insights from the United Kingdom. *Technological Forecasting and Social Change*, *183*, 121899.
Zhang, G., Zhang, N., & Liao, W. (2018). How do population and land urbanization affect CO2 emissions under gravity center change? A spatial econometric analysis. *Journal of Cleaner Production*, *202*, 510–523.
Zhao, X., & Sun, B. (2016). The influence of Chinese environmental regulation on corporation innovation and competitiveness. *Journal of Cleaner Production*, *112*, 1528–1536.
Zhao, Y., Cui, Z., Wu, L., & Gao, W. (2019). The green behavioral effect of clean coal technology on China's power generation industry. *Science of the Total Environment*, *675*, 286–294.
Zhu, C., & Lee, C. C. (2022). The effects of low-carbon pilot policy on technological innovation: Evidence from prefecture-level data in China. *Technological Forecasting and Social Change*, *183*, 121955.

Secondary (Derivative) Sources

Agreement, P. (2015). United nations. *United Nations treaty collect*, 1–27.
Board, F. S. (2017). Task force on climate-related financial disclosures. *Final Report: Recommendations of the Task Force on Climate-Related Financial Disclosures.*
European Banking Federation (EBF). (2017). *Towards a green finance frame-work*. https://www.ebf.eu/wp-content/uploads/2017/09/Geen-finance-complete.pdf

European Commission. (2018a). The European central bank, the European economic and social committee and the committee of the regions on the action plan: Financing sustainable growth. *The European Council, the Council. COM/2018/097 Final. Communication from the Commission to the European Parliament.* https://eur-lex.europa.eu/legal-content/EN/TXT/?uri=CELEX%3A52018DC0097

European Commission. (2019). COM(2019) 640 final. Communication from the commission to the European parliament the European economic and social committee and the committee of the regions on the European green deal. *The European Council, the Council.* https://eur-lex.europa.eu/legal-content/EN/TXT/?uri=CELEX:52019DC0640.

European Commission. (2020). Europe's moment: Repair and prepare for the next generation. *Press Release*, 27, 2020.

European Commission. (2021). *Overview of sustainable finance*. https://ec.europa.eu/info/business-economy-euro/banking-and-finance/sustainable-finance/overvi ew-sustainable-finance it.

European Commission. (2024). https://composite-indicators.jrc.ec.europa.eu/explorer/explorer/indices/gfi/green-future-index

Fuchs, S., Kachi, A., Sidner, L., & Westphal, M. (2021). Aligning financial intermediary investments with the Paris agreement. *World Resources Institute*. https://doi.org/10.46830/wriwp.20.00037

G20. (2017). *Fostering sustainable global growth through green finance—what role for the G20?*, 12, April. https://www.g20-insights.org/wp-content/uploads/2017/04/Climate_Green-Finance_V2.pdf

G20Green Finance Study Group. (2021). https://g20sfwg.org/wp-content/uploads/2021/10/G20-SFWG-Synthesis-Report.pdf

Green Finance Platform. (2024). https://www.greenfinanceplatform.org/financial-measures/browse

Heaps, T., & Guyatt, D. (2017). A review of international financial standards as they relate to sustainable development. *UN Environment Inquiry*. https://wedocs.unep.org/bitstream/handle/20.500.11822/16806/Review_International_Financial_Standards_Sustainable_Development.pdf?sequence=1&amp%3BisAllowed=

IMF. (2000). Supporting document to the code of good practices on transparency in monetary and financial policies—approved by the IMF executive board on July 24. Washington DC.

Jung, C., Krutilla, K., & Boyd, R. (1996). Incentives for advanced pollution abatement technology at the industry level: An evaluation of policy alternatives. *Journal of Environmental Economics and Management*, *30*(1), 95–111.

Lee, C. C., & Lee, C. C. (2022). How does green finance affect green total factor productivity? Evidence from China. *Energy Economics*, *107*, 105863.

MIT. (2023). Technology review insights. *The Green Future Index*. https://mittrinsights.s3.amazonaws.com/GFI23report.pdf

Montero, J. P. (2002). Permits, standards, and technology innovation. *Journal of Environmental Economics and Management*, *44*(1), 23–44.

OECD. (2008). *OECD Glossary of Statistical Terms*, OECD Publishing, Paris, https://doi.org/10.1787/9789264055087-en

OECD. (2017). *Green growth indicators 2017*. Paris: OECD Publishing. http://doi.org/10.1787/9789264268586-en

Shahbaz, M., Solarin, S. A., Mahmood, H., & Arouri, M. (2013). Does financial development reduce CO2 emissions in Malaysian economy? A time series analysis. *Economic Modelling*, *35*, 145–152.

United Nations. (2015). Transforming our world by 2030: A new agenda for global action Zero. *Draft of the Outcome Document for the UN Summit to Adopt the Post-2015 Development Agenda*. New York: United Nations.

United Nations Environment Programme. (2014). *Using indicators for green economy policymaking*. https://www.unep.org/resources/report/using-indicators-green-economy-policymaking

Index

A

access to justice, 162–164
advertising, 142, 164, 186, 193
agricultural productivity, 253, 294
agroecological, 294
AHP, 101, 110
air conditioning, 8, 284
Algeria, 364, 372
analytic, 10, 101
Andriuskevicius, K., viii, xiii, 333
animal waste, 240
atmosphere, 2, 8, 143, 158, 160, 180, 237, 253
Australia, 51, 155, 264, 270, 308–309, 312, 316, 369
Austria, 52, 76, 82, 90, 131–135, 180, 186, 267–268, 327–328
automated switches, 10
automation systems, 284

B

Bathaei, A., vii, xiii, 235
Belgium, 82, 90, 93, 131–133, 135, 325–327, 329
Benefits and Features, vii, 242
Benefits of Eergy Efficiency, vi, 1
biodiversity conservation, 46, 294
bioenergy, 91, 240, 306
biomass, 1, 22, 25, 53, 66–67, 146, 235–236, 240, 243, 246–247, 252–253, 256, 258, 272, 282, 284, 287, 309, 312, 329, 344–345, 348
biomass energy, 236, 240, 253, 282, 284, 287, 309
biotechnology, 322–323
Brazil, 51, 154, 187, 244, 264, 270, 309
BREEAM, 319–320
Brexit, 333, 371
Building and Construction Practices, 8–9
Bulgaria, 63–64, 82, 85–86, 132–135, 325, 327
business models, 25, 33, 290, 296, 304
Butkus, M., vi, xiii, 191

C

campaigns, vii, 26, 143, 160, 181, 214, 254, 256–257, 261–262, 302
Canada, 51, 171, 189, 270–271, 317, 369
capabilities, 21, 24, 35, 59, 69, 91, 118, 122, 130, 142, 158–159, 163, 216, 220, 243, 300–301, 314, 352
capable, 12, 158, 162, 211, 259, 269, 302, 314, 318, 357
capacity building, 21, 254, 282, 298
capital constrains, 21
carbon budget, 13
carbon capture, 8, 10, 18, 26, 36, 43, 49, 52, 240, 270, 322, 364
carbon dioxide, 2, 8, 25, 195, 197, 204, 240, 247, 273, 336, 347–348, 362
carbon footprint, 32, 128, 239, 242, 259, 263, 312, 336, 343, 348
carbon neutrality, 12–13, 15, 18, 25, 32, 51, 54, 271
carbon pricing, v, 16–17, 26, 28, 42–43, 49, 51, 295, 300, 347, 354, 371–372
carbon pricing mechanisms, v, 17, 26, 42–43, 49, 295, 300
carbon sequestration, 12
carbon tax, 17, 26, 43, 208, 212, 254, 356, 364
carbon-neutral, i, ix–x, 243, 366
case study, i, v–x, 78, 88, 91–92, 94, 97–115, 122–124, 137, 187–188, 220, 230, 262, 270–271, 325, 329, 361, 372–373
CCS, 8, 18, 26, 43, 185, 192, 232, 364
CDSB, 346, 371
CEDAW, 144, 163–164, 185
CEM, 319–320
certification, 6, 8, 51, 91, 284, 291, 299, 302, 308, 315, 318–320, 329, 359, 366
Certified Energy Manager, 319
changing conditions, 125
characteristics, 20, 83, 119, 136, 143, 165, 196, 199, 213, 216, 218, 242, 250, 302, 345–346, 366, 372
Chile, 94, 244, 263, 272
China, 51, 53–54, 94, 97, 101–103, 106–107, 109–110, 114–115, 137–138, 151, 155, 164, 169, 186, 243, 263–264, 270–271, 308–309, 331–332, 359, 372–375
Chinese, 51, 353, 356, 358–359, 373–374
CHP, 5, 7
circular business, 33
circular economy, 17, 25–26, 33, 294, 320, 326–327, 330–331, 334, 343–344, 349, 371
civil society, 17, 24–25, 33, 46, 162–163, 169, 185, 217, 254, 282, 287, 293, 306, 355
Clean Energy Corridor, 323
clean energy infrastructure, 21, 42
clean energy project, 17, 36, 244
clean energy transition, 51, 272, 287
clean technology, v, 24, 28, 43–45, 259, 276, 289, 293, 298, 334, 339–340, 344, 362, 371
climate action, vi, 12–14, 17, 21, 26, 35, 43, 48–49, 51, 53, 82–83, 139–140, 144, 147–149, 168–169, 184, 186, 259, 307, 320, 337–338, 340
Climate Change Act, 13, 50
climate change adaptation, 147, 184, 188, 253, 343–344, 371
Climate Change Committee, 13, 50
climate change concerns, 27–28, 33, 49, 351
climate change impacts, 26, 48, 287, 290, 336
climate change mitigation, i, ix, 13, 17, 27, 52–53, 72, 77, 88, 90, 118, 146–147, 149, 168, 215, 232, 237, 247, 250–251, 253, 263, 279, 333, 343–344, 366, 371–372
Climate Disclosure Standards Board, 346, 371
climate goals, 8, 16–17, 43, 48, 322, 338, 341
climate mitigation, 34, 36, 42, 46, 48, 147, 167, 169, 220, 254, 278, 290
climate policies, 13, 24, 26, 48, 50–51, 139, 145, 148–149, 168–169, 364, 367
climate resilience, 20, 25–26, 35–36, 43, 145, 147, 307
climate-resilient, 14, 21, 26, 287, 321, 341
CO2, 8, 19, 25–26, 49, 94, 102, 112, 137, 191–192, 195, 198, 201, 204–205, 215, 221, 232, 245, 247, 265, 344–345, 352, 356, 374–375
coal generator, 1

cogeneration, 5, 7
collaborative efforts, 8, 17, 355
collaborative research, 6, 36, 254, 260
combustion, 2, 14, 227, 243, 252, 294
common barriers, 298
communication, 9, 50, 53, 89–90, 124–125, 169, 251, 273, 302–303, 343, 349, 375
communication technology, 169
community engagement, 294, 302, 349
community resilience, 147–148, 167–168, 287
comparative assessment, vi, 57, 68–69, 79, 122–125, 362, 366
comparison, 19, 52–53, 76, 82, 85, 89, 94, 110, 122, 125, 137, 153, 200–201, 203–206, 224, 226, 229, 231–232, 265, 342, 366
compatibility, 245
competitive advantage, 208, 225, 250, 290, 319, 325, 353, 360, 374
competitive edge, 147, 209, 325, 353
competitive prices, 248
compliance, 6, 26–27, 43, 77, 129, 179–180, 214, 257, 283, 287, 289, 302–303, 306, 308, 319–320, 336, 352, 360
composting systems, 286
comprehensive evaluation, 218
concepts differences, v, 59
conceptualization, 115
Conferences of the Parties, 36
conservation efforts, 278, 294, 335
constrained reason, 196
construction practices, 8–9, 284, 308
consumer awareness, 33–35, 49, 261, 301–302
consumer awareness, 33–35, 49, 261, 301–302
consumer demand, 296, 353
continuous improvement, 8, 318, 321
contribute to the transition, 21, 293
control apparatus, 10
control systems, 5, 7
conventional energy, 3, 146, 224, 227, 237, 248, 250, 252, 256–257
conventional power plants, 2
conventions, ix, 140–143, 145, 148, 150, 156–158, 165–167, 170–171, 180, 183, 350
cooling system, 65, 77–78, 195, 226, 272, 284, 307
cooperation, 21, 24–25, 33–34, 36, 43, 48, 65, 120, 122, 129, 244, 254, 264, 268–269, 282, 298, 303, 340, 352
cooperatives, 56
coordinated efforts, 21, 24, 29, 167
coordination, 24–25, 46, 49, 282, 290, 303–304, 332
COP, 36
cost reduction, 30, 192, 209–210, 237, 248, 255–256, 299
cost-effectiveness, 95, 196, 198, 205, 213, 217–218, 231, 238, 299
country differences, 76
country-specific factor, 174
COVID-19, 62, 64, 95, 120, 153–154, 158, 178–179, 184–185, 187, 189–190, 262, 281–282, 330, 333, 335, 364, 371
CRC, iii–iv, 141, 144
creating jobs, 3, 239, 287, 306–309, 312
critical materials, 281
criticism, 4
Croatia, 99, 132–135, 137, 325, 327
crop residue, 66, 240
CSR, 50, 335, 350, 353, 356, 373
cultural and normative indicators, vi, 164
cultural element, 152, 157, 162
cultural factors, 27, 142, 155, 211, 263
cultural shifts, 36, 49
customer, 64–65, 79, 92, 193–196, 202–203, 212–214, 216, 219–220, 251, 331–332, 352
cybersecurity, 11, 17
cybersecurity measure, 11
Cyprus, 82, 86, 131–135, 327
Czech Republic, 74, 92, 132–135, 369

D

damage, 24, 57, 240, 334
data analytics, 10
data collection, 27, 67, 79, 89, 264
DEA, 137–138
decarbonize, 15, 263
declaration, 164
definition, i, ix, 5, 57–58, 60, 62–68, 70–71, 73–74, 92, 95–96, 200–201, 203–204, 278–281, 289–290, 293, 299, 302, 314, 331, 333–334, 343, 350, 361, 373
deforestation, 68, 240, 247, 294, 344
demand and supply, 127, 216
Denmark, v, 14–16, 50, 52, 82, 132–135, 244, 267–268, 323, 327–328, 364
dependence, 12, 24, 30, 36, 46, 95, 102, 111, 118, 121, 133, 206, 210–211, 222, 226, 248, 258, 264
depletion, 95, 108, 116–117, 127, 239, 267, 274, 276, 278, 293, 295, 329
detrimental, 78, 141–144, 146, 152, 164–165, 252, 356
detrimental stereotypes, 142
developability, 117
developing countries, 21–24, 34–35, 50, 58, 60, 90, 94, 160, 184, 186, 189, 231, 251–252, 262–264, 272–273, 290, 321, 330, 343
development objective, 24, 151, 275, 370
diesel, 270
different countries, vii, 36, 55, 70–71, 97, 115, 122, 134, 200–201, 206, 208, 263, 342
digital engagement, 171
digital gender divide, 169–170
dimensions of sustainability, 25, 124, 136
disadvantages, 3–4, 50, 52, 71, 156, 158, 171, 237, 239, 242
disaster risk, 239, 287
discount rate, 196
discrepancies, 142, 145, 150–151, 155–157, 171, 179–180
distribution of employment, 157, 292
distributional energy justice, 56, 76
diversification, 2, 46, 95, 98–99, 103–105, 109–111, 121–122, 127, 244, 252, 292, 316–317, 350
diversity initiative, 142
drivers of energy poverty, v, 65, 75–77
droughts, 144, 167, 253

E

Eastern Europe, 63
Eastern European, 71, 74–75, 79, 369
ECEI, 203–204
eco-friendly, 241, 252, 278, 296, 304, 315, 333–334, 336, 340, 349, 353, 356–357, 371

ecological footprint, 8
economic benefits, 119, 130, 195, 241, 268, 286–287, 314, 322, 360
economic challenges, 3
economic considerations, 202, 205, 274–275, 299
economic development priorities, 24
economic efficiency, 194, 198, 213, 345
economic impact, 24, 186, 270, 281, 315, 317, 356
economic indicator, vi, 154, 160, 199, 205, 222
economic indicators, vi, 154, 160, 199, 205, 222
economic productivity, 154, 198
economic sustainability, 139
economic viability, 4, 238, 298–299, 329
economics, xi, xiii, 50–52, 90–94, 137, 184, 186, 191, 196–198, 205–206, 208, 211–212, 231–234, 261, 270–271, 330, 372–375
ecosystem health, 284, 294
ecosystem resilience, 46, 294
ecosystem services, 294, 344, 347
ecosystems, 2, 25, 33, 118, 166–167, 239, 241–243, 283–284, 293–294, 307, 321, 324, 343–344, 347, 371
Education Programs, vii, 261, 289, 291, 298, 314–316, 329
EEE, 203
Effective Communication, 125
Efficacy of Energy Efficiency, vii, 217
Efficiency Deficits, vii, 218
efficiency gain, 209, 353, 371
Efficiency Initiatives, vii, 196, 212, 217
Efficiency of Energy, vi–vii, 5, 68, 129, 194, 199, 213
efficiency outcome, 197
Efficiency Policies, vii, 40, 50, 202, 215–217, 219, 230
efficiency trends, 199, 202
EKC, 206–207, 212, 224, 229–230
electric vehicle, 10–11, 14, 18, 26, 31, 33, 43, 45–46, 261, 285, 299, 307, 312, 339
electricity grid, 9
electrification, 8, 14, 18, 66, 90, 94, 99, 245, 253, 272
emerging opportunities, 292
emerging trends, i, 184
emission reduction, v, 13, 16–17, 19–21, 29, 43–44, 48–49, 53, 64, 78, 259, 270, 336, 364
emission reduction targets, v, 13, 16, 29, 43–44, 48–49
emissions profile, 20
emissions reduction, 12–13, 16, 20–21, 25–26, 36, 43, 51, 264, 283, 287, 334, 336, 356
emphasis, 21, 59, 74, 96, 113, 116, 123, 143–144, 147, 171, 195, 216, 229, 245, 333–336, 344, 356, 365, 371
Employee training, 6
employment opportunities, 146, 148, 156, 173, 195, 241, 253, 278–279, 282, 284, 287, 292–293, 300, 303–306, 309, 322, 340, 347
empower, 43, 141, 144, 161, 163–164, 168, 170, 172, 180, 237, 287
Energy Access-Consumption Matrix, 68
energy affordability, 55, 58–60, 62, 65, 67, 70–71, 73–74, 76, 79, 91, 98–99, 116, 118, 120, 127, 133
Energy audits, 5, 7, 77–78, 90, 121, 282
Energy Conservation, vi, 5, 121, 192, 194, 197–198, 210, 212, 215, 220, 229, 262, 308, 333, 349
energy cost-efficiency, 203
energy costs, ix, 8, 59–60, 62–63, 65, 70–72, 78, 82–85, 127–128, 192–194, 203–204, 206–207, 214, 228–230, 273, 312
energy demand, 3–4, 12, 53, 72, 79, 83, 95, 98, 121, 191, 194, 196, 200, 206–207, 209, 212, 215–216, 232, 238, 247, 251–252, 263
energy deployment, 25, 36, 49, 243, 245, 277, 279, 287, 290
energy efficiency concept, vi, 191
energy efficiency deficits, vii, 218
energy efficiency gap, vi, 93, 195–196, 210, 231–232
energy efficiency in consumption, vi, 193, 198
energy efficiency indicator, 203
energy efficiency indicators, vii, 198–202, 204–205, 216, 231–233
energy efficiency indices, vii, 200
energy efficiency losses, vi, 195
energy efficiency standards, v, 8, 40–41, 49, 287, 295, 300, 306–308, 329
energy expenses, x, 68, 80, 105, 194, 196, 207
energy facilities, 226, 242, 249, 251, 257, 282, 284, 309
energy flows, 10
Energy Grants, vii, 257
energy grids, 122
energy industry, 24, 97, 109, 117, 213, 241, 263, 265, 287, 319
energy inequalities, v, 55–57
energy input, 67, 191, 197, 200, 226–227
energy inputs, 66–68, 128, 191, 198, 224
energy intensity, vi, 5, 26, 99, 101, 191, 193, 197–202, 205–213, 218–221, 224–234, 345
energy management system, 204, 214
energy market, 54, 86, 95, 115, 119, 122–123, 129–130, 137, 214, 246, 267, 296
energy need, 58, 61–62, 66–67, 71–72, 75, 77, 80, 85, 110, 117, 125–126, 133, 195, 209, 240, 247, 252
energy penetration, ix, 269, 273
energy poverty, i, v, ix, 22, 46, 54–85, 87–95, 115, 119, 127–128, 133, 146, 262–263, 271, 273
energy poverty indicators, v, 62, 65, 78–83, 85, 88–90, 93
energy poverty risk, ix, 58, 74
energy production, 31, 99, 101–102, 112, 117–118, 222, 236, 238, 241, 255, 264–265, 267, 335, 363
energy program, 13, 147, 243, 315
energy requirements, 2, 61, 68, 201, 206, 211, 219, 229, 263, 269, 308
energy savings, 5, 7–8, 65, 77, 192–193, 196–197, 210, 284
energy security, i, vi, ix–x, 1–2, 27, 30, 33, 36, 40, 49, 70, 93, 95–125, 127–131, 133–138, 194, 198, 209–210, 213, 219, 237, 248, 251–254, 258, 265, 267
energy security challenges, vi, 121–122
energy security concept, vi, 95–96, 105, 117, 124
energy security indicators, vi, 98, 100, 102, 104, 106, 108, 110, 112, 114, 117, 136
energy share, 26, 267
energy shortage, 57, 73
energy storage, ix, 9–12, 18, 30, 43, 46, 53, 78, 119, 121–122, 130, 228, 237, 244, 249–250, 271, 299, 322
energy subsidies, 24, 192, 208, 210, 226, 241, 248–261, 269–270, 274, 276, 281–282, 284, 286–287, 316–317, 320, 323
energy technology, 93, 148, 178, 207–210, 232, 235, 242, 249, 252, 260, 269, 323
energy transformation, i, iii, ix, 51, 199, 201, 270
energy vulnerability, v, 57–58, 62–63, 70, 74–76, 78, 92
energy waste, 5–8, 119, 219, 282
energy-efficient lighting, 6, 282

energy-intensive, 25, 202, 206–209, 224–226, 228–229, 260
engineering, 3, 49, 52, 54, 148, 156–157, 172, 178, 181, 191, 197–198, 204–205, 231, 282, 306
enterprise, x, 54, 148, 158–159, 193, 208, 212, 215, 225, 227, 233, 241–242, 298, 330, 333, 358, 374
entrepreneurship, 147, 159–160, 169–170, 177, 182, 184, 186, 189, 254, 322
environmental and sustainability indicators, vi, 165
environmental benefits, 17, 195, 239, 259, 278, 284–285, 294, 299, 302, 314, 316, 334, 342
environmental burden, 2
environmental challenges, 168, 268, 278, 287, 293, 295, 298, 301, 318, 321–322, 333, 340, 356, 362
environmental consideration, 96, 113, 118, 304, 335, 337, 339, 346, 349, 355–356
environmental criteria, 314
environmental degradation, 166, 293, 329, 356, 358
environmental deterioration, 165, 356
environmental footprint, 346, 349, 358
environmental impacts, 46, 68, 96, 119, 218, 245, 251, 282, 294, 299, 305, 307, 314, 316, 319, 330, 334–335, 352–353
environmental improvement, 284, 355–356
environmental justice, 48, 91, 287, 292, 294
environmental Kuznets curve, 206–207, 212, 229–231
environmental objective, 120, 304, 342
environmental protection, 49, 54, 101, 110, 167, 195, 247, 274, 287, 306, 315, 329, 335, 356, 358
environmental quality, 51, 356, 360
environmental stewardship, 12, 32, 36, 166, 168–169, 335–336, 350
Environmental Workforce Development and Job Training, 317
environmentally conscious, 333, 337, 340, 342, 350, 362, 370
environmentally friendly industry, 278–279
environmentally friendly practices, 283, 311, 363
EPA, 195, 317
Equal Pay Act, 156
EROI, 1
ESG, 290, 334, 337, 339–340, 343, 346, 350–351, 368, 374
ESR, 13, 233
essential skills, 160
Estonia, 82, 86, 131–135, 267–268, 325, 327, 329, 369
ethics, 373
EU Countries, i, v–vi, ix–x, 59, 62, 82–83, 86, 105, 114–115, 122–123, 133–134, 173, 185, 220, 225, 265, 267, 331
EU Energy Poverty, v, 78–79, 88, 91
EU member state, i, 50, 62, 74, 88, 114, 124–125, 130–131, 133–134, 325–330, 369
EU-27, 325–327
EU-28, 51, 264, 270
Europe, 53, 62–64, 70, 74–75, 90, 92–93, 137, 161, 165, 171, 232, 244, 264, 270, 320, 323, 330, 366, 372, 375
European Commission, 59, 61, 70, 78–82, 89, 91, 93, 201, 314, 327, 330, 339–340, 343, 362, 375
European Green Deal, 339–340, 372, 375
evaluating the efficiency, vii, 204
evaluation of energy efficiency, 205
EWDJT, 317
executing energy, vii, 216
expertise, 22, 24, 36, 167, 215, 217, 225, 254, 284, 289, 298, 300–301, 304, 315, 319, 321–322, 369
export opportunities, 322
Expression of Gender Equality, vi, 144

F

fair trade, 33
FDI, 210–212, 220–223, 225–226, 229–230
feasibility, 52, 182, 235, 239, 242, 269, 298–299
Feed-in Tariffs, vii, 39, 248, 254–255, 287, 290, 299, 309, 312, 337
feedback, 27, 53, 127, 195–197, 219–220, 248, 349
female participation, 173–174, 180–181, 188
fertilizer use, 12
financial barriers, 256, 305, 313–314
financial constraints, 61, 76, 305
financial institution, 213, 249, 339–341, 343, 351, 354, 358, 361, 366, 369
financial performance, 178, 188, 351–353, 356, 371
financial resources, v, 21, 24, 35, 67, 140, 144, 147–148, 157, 160, 177, 257, 289, 298–299, 316, 334, 343, 366, 369, 371
financial viability, 299
Finland, 75, 132–135, 267–268, 325–327, 329, 364, 372
fiscal, 74, 78, 90, 222, 268, 281, 309, 311, 313, 343, 347
FiTs, vii, 255, 287
flexibility, 10–11, 43, 58, 115–116, 125, 127, 240, 255–256, 272, 303, 352
flexibility issues, 116
flexible generation, 1
floods, 144, 167, 253
food intake, 65, 71
foreign direct investment, 53, 206, 210–212, 225–226, 229–231
foreign policy, 115
forest resources, 283
forestry, 12, 240, 366
Formulating and Executing Energy, vii, 216
fossil energy, 248, 264
fossil fuel extraction, 247
fossil fuel resource, 96
fossil fuel subsidies, 24, 250, 270, 354, 371
fossil resource, 267
France, v, 12–13, 51–52, 74, 82, 90, 92–93, 132–135, 327–328, 359, 369
fundamental energy needs, 72
future generations, 27, 48, 124, 182, 198, 269, 287, 294
Future Research, vi, 110, 113, 124, 134, 262, 269, 271, 337

G

gas price, 81
gasoline, 197, 231
GBP, 346, 371
Gender Equality, vi, x, 139–173, 175–189, 338
Gender Equality Concept, vi
gender inclusion, 159
Gender Inequality Index, 172, 176
gender stereotype, 139, 142, 144, 158, 161, 164–165, 170, 181, 186, 189
gender vulnerabilities, 148
gender wage gap, 156, 160, 180, 186, 188
gender-based discrimination, 139–141, 144, 158, 164, 187

gender-responsive, 148–150, 168–169, 178–179, 181–182, 185, 187
gender-responsive budgeting, 179, 187
genetic diversity, 294
geographic distribution, 104, 292
geopolitical relation, 116, 119, 125, 129, 135
geopolitical tensions, 116, 119, 281
geothermal energy, 46, 192, 235–236, 239, 243, 249, 256, 309
geothermal heating, 272, 284
Germany, v, 12, 52–53, 76, 82, 132–135, 137, 154, 161, 170, 187, 231, 233, 244, 266, 309, 312, 316, 326–329, 369
GHG emission reduction, 64, 78
GHG emissions, ix–x, 8, 100, 116, 118, 127–128, 222
GHGs, 344
GII, 172, 174
GINI index, 345
global efforts, 18, 298
Global Reporting Initiative, 336, 346, 350, 359, 371
global representation, 141
global trend, 333
global warming, 95, 123, 247, 253, 268, 294
Global Wind Organisation, 319
goals and objectives, 204, 215
governance model, 52
government agencies, 8, 24, 46, 259, 283–284, 303
government expenditure, 172, 174, 177–178, 183, 218
government procurement policies, 259, 289
government support, 237, 287
GPP, 259, 314
Greece, 54, 60, 62, 74, 82, 90–91, 93, 131–135, 263, 273, 327–328
Green Bond Principles, 346, 359, 371
green building, 46, 284, 289, 299, 315–316, 319, 363
Green Climate Fund, 35, 321, 359
Green Deal, 51, 330, 333, 339–340, 371–372, 375
Green Employment, vii, 146, 292, 300, 303–306, 309–311, 313–318, 324–326, 329
green initiative, 145, 290, 297–300, 303–305, 334, 352–353, 355, 360
green investment, i, viii, x, 289–290, 298–299, 305–306, 312, 332–337, 339–363, 365, 367, 369–375
Green Investments, i, viii, x, 290, 298–299, 305, 333–337, 339, 341, 343, 345–347, 349–363, 365, 367, 369, 371–373, 375
Green Jobs, vii, ix–x, 77, 274–293, 295–309, 311, 313–321, 323–332
Green Jobs Concept, vii, 274
green procurement, 289, 295, 300, 302
Green Public Procurement, 217, 259, 309, 313–314, 329–331
Green Skills Agreement, 316
GRI, 336, 346, 350, 359, 371
grid behavior, 10
grid reliability, 10, 119
groundwater, 195, 346
guideline, 27, 36, 54, 92, 164, 214, 240, 262, 289, 336, 339, 346, 359, 368
GWO, 315, 319–320

H

harmonization, 43, 357
harmonize, 163, 178
harmonized, ix, 251, 359
HBS, 80–83
HDI, 246
health benefits, 31, 36, 49, 51
health evaluation, 152
health initiative, 152
heat recovery, 6, 8, 54
heating affordability, 73
HEP, 80, 82
Hidden Energy Poverty, 62, 80, 82–83, 85, 88
high costs, 4, 321
High-efficiency, 5, 7
high-energy, 73, 82
holistic approach, 48, 122, 344
Household Budget Surveys, 81
household consumer, 81, 127
household expenditure, 61, 64, 71, 81, 92
human activities, 274, 283, 293, 344
human health, 2, 294
human rights, 33, 46, 120, 140–141, 163–164, 178, 185, 188, 349
Hungary, 62–64, 74, 76, 82, 94, 132–135, 327, 369
HVAC, 6, 8, 77, 282, 284, 308
hydroelectric generator, 282, 284
hydroelectric power, 33, 264, 309
hydropower, 2, 14, 235–236, 238–239, 242, 244, 247, 252–253, 258, 271–272, 329
hypothesis, 206–207, 210–211, 224, 231, 357–358, 360, 372

I

ICMA, 346, 359
IEA, 61, 66–67, 92, 195, 247
implementing policies, 21, 25, 217, 277
implementing technologies, 5
Improving the Efficiency of Energy, vii, 5, 213
inclusive economic growth, 127, 177, 187
inclusive leadership, 140, 142, 180–181, 188
inclusivity, 147, 167–168, 170, 179, 185, 329
independent variables, 68
india, 51–52, 93, 153–155, 158–161, 170–171, 185–186, 189, 244, 263, 270, 308, 312
indicators of energy efficiency, vii, 202
indicators of green jobs, vii, 282, 284, 287, 292–293, 329
indicators set, vi, 98, 100, 102, 104, 106, 108, 110, 112, 114, 117–118, 124–125, 136
indigenous, 166
industrial facilities, 8, 282
industrial policies, 331
industrial process, 5, 7–8, 12, 18, 24, 26, 46, 191, 202, 207, 212, 287, 294
industrial revolution, 13, 148
industrial sector, vii, 51, 121, 193, 202, 206, 224–225, 229, 231
industrialization, 112, 114, 224, 235
industrialized countries, 207
inefficiency, 7, 74, 128, 195–196, 200, 209–210, 216, 226, 301, 303–304
inequality, i, v, ix, 24, 47, 55–57, 63, 65, 70, 76, 90, 115, 128, 140–145, 149–160, 162, 164–165, 170–174, 176–177, 179–189, 222, 232, 292, 294, 301, 304, 345
inexhaustible, 2, 128, 227
information dissemination, 56, 62, 64, 77–78, 90, 367
infrastructure and supply chain, 302

infrastructure development, x, 14, 27, 118–119, 129, 226, 287, 297, 321, 324, 349
infrastructure limitations, 250, 329
innovation clusters, 321–324, 329
innovation funding, 289
innovative approache, 352
innovative technologies, 12, 21, 212, 249, 320
input-output analysis, 233
institutional and governance constraints, 24
institutional barriers, 27, 182, 303
institutional impediments, 157, 181
Integrated National Energy and Climate Plan, 12, 17, 51
intellectual property, 301, 324, 329, 355, 363, 371
intermittency, ix, 237–238, 249–250, 268
international agreements, ix, 13, 18, 35, 118, 163, 278
International Capital Market Association, 346, 359
international collaboration, 33, 36, 48–49, 162
international cooperation, 21, 24–25, 34, 36, 43, 48, 120, 254, 282, 340
intersectionality, 139, 143–144, 184, 188, 190
investment capital, 21, 305
investment costs, 251, 299, 302, 305
investment decision, 339, 353
investment decisions, 290, 335–336, 339–340, 342, 368, 371
investor preferences, 352, 371
Iran, 364, 366, 372
Ireland, 59, 74, 76, 87, 131–135, 325, 327
IRENA, 51, 247
irrigation water, 345
ISO 14001, 319–320
Italy, 60, 76, 82, 91, 131–133, 135, 157, 171, 188, 266, 327–328, 369

J

JESSICA, 64
job opportunities, 139, 245, 252, 255, 300, 305, 316–318, 320–321, 347
just Transition, v, 25, 46–48, 54, 263, 315–316
just transition fund, 315–316

K

kerosene, 146
key challenges, ix, 24
key element, 1, 5, 8, 46–47, 95, 253, 258
knowledge sharing, 33, 36, 254, 298
Kyoto Protocol, 278, 336

L

labor market, 146, 154–157, 165, 170, 172, 174–179, 183, 189, 222, 289, 291, 314, 329, 345
labour, 186–187, 189
landscape, 9, 24, 90, 96, 120, 171, 178, 262–263, 272, 289, 333, 336–337, 340, 348, 357, 361, 369
Latin America, 152, 244
Latvia, 82, 86, 132–133, 135, 327
Leadership in Energy and Environmental Design, 284, 319
lean manufacturing, 6, 8
LED, 8, 14, 30, 76, 96, 145, 167, 192, 219, 260, 276, 309
LEED, 284, 319–320, 363
legacy infrastructure, 24
legal framework, 111, 142, 153, 161–163, 178, 180–181, 183, 216
legal indicators, vi, 160, 164
lifecycle assessment, 6, 299
lifestyles, 8, 60, 72, 203, 229
LIHC, 80
limitations, 27, 65–67, 145, 156, 159, 161, 166, 170, 179–180, 216, 250, 300, 329, 343
Limited access, 21–22, 24, 78, 300–301, 305
Limited Financial Resources, v, 21, 299
limited income, 72, 366
Lithuania, v, xiii, 54, 64, 74, 76, 82–89, 93, 100, 131–135, 263, 273, 327
livelihood, 158, 169, 287, 294
low-carbon energy transition, i, v, ix–x, 1, 5, 8–9, 11–12, 17, 27, 35–36, 38–39, 43–44, 48–50, 52, 54, 62, 118, 134, 262–264, 269–271
low-carbon practices, 365–366
low-income, ix, 46, 60, 63–65, 68, 71–76, 82, 88, 127, 144, 150–151, 170–171, 178, 292
Luxembourg, 82, 133–134, 325

M

machinery, 5, 7, 54, 229, 349
Malta, 60, 82, 86, 131–135, 326–327, 329
market access, 159, 182, 276, 301
market entry, 323
market expansion, 256
market failure, 195–196
market failures, ix, 200, 208, 212, 251
market mechanisms, 26, 254
market opportunities, 260, 339
materials and product innovation, 6
mathematical model, 358
middle-income, 60, 153, 159, 163, 178
Mission Innovation, 33, 321
misunderstanding, 254
mitigation efforts, 13, 17, 25, 146, 168
mitigation measures, 119, 167, 255, 307
mitigation policies, 72, 77–78, 88, 90, 263
mitigation strategies, 20, 119
mitigation targets, 118
mobility, 53, 66, 77, 234, 272, 291–293, 303, 340
modern infrastructure, 46, 246
Modernisation Programme, 89
modernization, v, 18, 45–46, 64
MTF, 69
Multi-Tier Framework, 69
multitude, 1, 203, 245, 287
multitude criteria decision-making, 245

N

N2O, 25
NABERS, 319
National Australian Built Environment Rating System, 319
National Climate Action Plans, 12
National Climate Agreement, 16–17, 53
national statistics, 73, 79, 92
natural environment, 335, 356
natural gas, 62–63, 78, 90, 227, 255, 263, 270
natural resource, 71, 147, 166, 186, 259, 344, 366

negative consequences, 68, 178, 251
negative effect, 73, 115, 173, 246–247
negative environmental impacts, 335, 353
negative impact, 27, 48, 73, 75–76, 118, 242, 250, 357
net metering, 254
Netherlands, v, 16–17, 82, 91, 131–135, 271, 326–327, 329, 369
new resources, 116
New Zealand, 59, 73
Next Generation EU, 340
NGO, 33, 163, 278, 298, 302, 330, 350, 352, 362
Nigeria, 150–151, 153–154, 159, 166, 188–189, 270
NIMBY, 56
nimbyism, 251, 305
nitrogen, 118
nitrous oxide, 25
Nordic, 155
normative indicators, vi, 164
North America, 269
Norway, v, 13–14, 53, 266–267, 364, 372
NPC, 246
nuclear, 12, 51, 97, 100, 106, 137, 192, 202, 233, 246, 263, 363

O

OECD, 14, 53, 61, 92, 99, 107, 137, 187, 189, 232, 334, 344–346, 359, 361, 366, 368–369, 371–375
operator, 250
optimize, 9–10, 121, 181, 193–194, 243, 264, 282
organic agriculture, 278
organic material, 236
organic products, 331
organizational culture, 303–304, 352, 371
outages, 10, 246

P

Pakistan, 92, 104, 107, 113, 136, 151, 170–171, 184, 366
Paris Agreement, 12–13, 15–16, 18, 33, 36, 245, 263–264, 278, 294, 298, 322, 333, 335, 337–339, 364, 366–367, 369–371, 373–375
Patent System, 325
pay gap, 140, 144, 156, 173–174, 179–180
payback period, 214, 255, 299, 354, 371
Paycheck Fairness Act, 156
penetration, ix–x, 1, 218, 231, 250, 255, 269, 271, 273, 297, 301–302
penetration of renewables, x
permanent basis, 58
pesticide use, 335
phasing out, 12, 24, 31
phenomena, 144, 167, 198, 342
photovoltaic, 89, 235–236, 242, 256, 269, 282, 284
pinch analysis, 6
PNIEC, 17
Poland, 74, 76, 82, 132–135, 155, 157, 327–328
policy alignment, 287, 289
policy and regulatory, 11–12, 24, 28–29, 49, 251, 287, 289, 293, 295, 329, 357, 371
policy and regulatory uncertainty, 24, 251
policy areas, v, 38, 48–49
policy barriers, 24
policy changes, 180, 265, 271, 339, 366
policy coherence, 26, 48, 119, 370
policy constraints, 305
policy development, 35, 43, 219, 284
policy directions, 121
Policy Documents, v, 12, 17, 48, 120, 362
policy domains, 49, 214, 367, 369
policy environment, 361, 363
policy evaluation, 216
policy execution, 149
policy Failure, vii, 216
policy flexibility, 256
policy formulation, 129, 145, 149, 179, 214
policy framework, 13, 24, 36, 48, 52, 139, 143–144, 148, 277, 298, 300, 306, 365
policy implementation, 26–27, 129–130, 216, 218–219, 262, 303
policy incentives, 340, 374
policy interventions, 22, 145, 154, 182, 207–209, 301
policy measures, 5, 83, 139, 182, 197, 204, 210, 213–216, 218, 230, 253
policy objectives, 44, 84, 255
policy package, 62
policy priorities, 281, 300
policy reforms, 160, 267, 290
policy support, 248, 300
policy tools, 41, 43, 215–216
policy uncertainty, 255
policy-making, 69, 82, 122, 198, 214, 216, 218–219, 245
policymaking, 49, 169, 220, 331, 342, 344, 370, 375
political action, 115
political regulations, vii, 214
political spheres, 141, 163
political support, ix
pollutants, 2, 118, 195, 227, 237, 240, 243, 294, 334
pollution prevention, 294
population growth, 19, 25, 345
Portugal, 60, 131–135, 267–268, 327
positive implications, 33
poverty alleviation, 24, 57–58, 60, 62, 65, 73, 75, 77–78, 88, 90, 94, 139, 189, 338
poverty indicator, v, 60, 62, 65, 70, 73, 75, 78–83, 85, 88–91, 93
poverty level, 66, 89, 252
poverty reduction, 69, 90–91, 94, 133, 140, 290
power generation, 8, 18, 26, 102, 122, 192, 227, 238–240, 242–243, 253, 255, 257, 267, 272, 374
power grid, 240, 242, 249
power plant, 1–2, 17, 54, 56, 101, 194, 202, 226–227, 236, 239, 241, 243, 248, 272
power sources, 235, 250, 252, 264, 267
Power System Security, 97
practical application, ix, 83, 97
practical execution, 164
preservation, 147, 214, 259, 293, 333–334, 340, 344, 356, 358, 371
preservation of biodiversity, 340
preventive measures, 57
price fluctuation, 109, 129, 255
primary objective, 97, 164, 197, 199, 214
principal-agent conflicts, 196
problems of security, ix
process integration, 6

process optimization, 5, 7, 33
product innovation, 6, 301
production cost, 127, 194
production processes, 8, 294
productivity decline, 227
profitability, 157, 290, 298, 352, 354, 358
promotion of innovation, 256
PSO, 65
public awareness, vii, 26, 33, 54, 245, 253–254, 261–262, 335, 358
public health, 31, 36, 49, 51, 72–73, 94, 118, 120, 128, 185, 188, 247, 259, 273, 299, 306, 314, 331
public policy, 73, 111, 186, 188, 205, 215, 366
public service, 51, 57, 65, 129, 216, 259

Q

quality of life, 36, 48, 69, 75, 93, 119, 127–128, 146, 152, 219, 292, 314, 344, 347, 371
quantitative indicators, 342
quotas, 141–142, 144, 180, 254

R

R&D, 103, 107, 109, 126, 129–130, 134, 220–223, 225, 230, 238, 260–261, 290, 301, 318, 320–325, 329, 349, 356
Ranking Tool, vi, 130
realization, 27, 164, 236
Rebound Effect, vi, 50, 195–198, 203, 231, 233–234
recognition, 49, 55–56, 96, 130, 142, 276, 293, 318, 320
recycling programs, 283, 286
reduce emissions, 12, 17–18, 25, 51, 263, 306, 320
reduce GHG emission, x, 127
reducing carbon emissions, 8, 18, 49, 196, 274, 278, 284, 287, 308
reducing emissions, 8, 13, 15, 25–26, 356
reducing energy costs, 8
reducing environmental impact, 316, 319
reducing GHG emissions, 127
reducing greenhouse, 1–2, 12–13, 25–26, 33, 43, 195, 294, 307, 333, 348
reducing waste, 8
reference, i, v–viii, 49, 89–90, 136, 184, 200, 230, 269, 330, 372
regional development, 292
regional disparities, 292
regulatory challenge, 305
regulatory changes, 337
regulatory compliance, 6, 26, 320
regulatory framework, v, 12, 24, 28–29, 43–45, 49, 129, 211, 249, 251, 254, 287, 289, 295, 302, 305, 329, 368–369, 372
regulatory uncertainty, 24, 251, 300, 305, 355, 371
reliable energy service, 22, 127
remuneration, 140–141, 156, 172, 178, 182, 257
renewable energy grants, vii, 257
renewable energy indicators, vii, 116, 244–245
renewable energy obligations, 258
renewable energy policies, 54, 233, 256, 262–263, 269
renewable energy production, 255, 335, 363
renewable energy promotion, v, 39, 261
renewable energy subsidies, 208
renewable energy usage, vii, 88, 206, 247, 343, 348, 365
renewable energy usage, vii, 88, 206, 247, 343, 348, 365
Renewable Portfolio Standards, 254, 258, 287, 309
REO, 258
replace, 194, 226
replacement, x, 1, 62, 192, 202, 299
replacing, x, 116, 192, 209, 252, 355, 371
reputational consideration, 353, 371
Research Triangle Park, 323
researchers, i, 73–76, 115, 130, 136, 150, 206, 244, 255, 282, 320, 324, 353, 358, 362
residential building, 62–64, 74–77, 81, 85, 89–91, 231, 307
residential sector, vii, 58, 60, 81, 202, 231–233
resilience and adaptation, 25–26, 287
resilience building, 282
resistance to change, 24, 303–305
reskilling, 146–147, 316
resource allocation, 129, 284, 299, 354
resource conservation, 278, 283–284, 287, 319
resource consumption, 5, 8, 26, 294, 349
resource depletion, 116, 127, 239, 274, 276, 278, 293, 295, 329
resource efficiency, 6, 8, 25, 33, 200, 294, 306–307, 322, 343, 349
resource endowment, 211, 263
resource extraction, 366
resource management, 129, 144–145, 147, 158, 165–169, 186, 283, 294, 343, 348
resource scarcity, 265
resource supply, 115
resource usage, 353
resource use, 17, 194, 294
resource utilization, 299, 345, 347
respiratory illness, 31
retrofit, 51
return on investment, 1, 238, 245, 261, 299
Rights of the Child, 141, 144
risk mitigation, 119, 255, 287
risk of poverty, 70, 74, 81, 88
risk reduction, 287
river, v, 27, 70, 172, 239, 243, 271
roadmap, 13, 17, 21, 48, 50, 270, 272, 337–338, 340, 343, 359
Romania, 131–135, 325, 327
RPS, 258, 287, 309
Rule of Law, 173–176, 183
rural area, 62–63, 67–68, 76, 144, 151, 187, 226, 238, 250, 252, 273, 292
rural communities, 292
rural development, xi, xiii, 17, 252
rural revitalization, 149, 169, 292

S

salary disparities, 142
SAW, 235, 328–329, 335–336
scalable, 240
scientific literature, ix, 71, 79, 85, 95, 115, 125, 135, 362
scientific methods, 116
secondary energy, 194, 205
self-assessment, 70
self-sufficiency, 95, 99–101, 103–104, 111, 129
service providers, 213, 258
short-term, 20, 57, 62–63, 78, 88, 90, 115, 203, 281, 304, 343
shortcoming, 85–86, 88, 194
Siksnelyte-Butkiene, I., xiii, 66, 93, 95, 137, 273

Silicon Valley, 322
Singapore, 49, 99–100, 136, 186, 317
Slovakia, 131–135, 327
Slovenia, 82, 132–135, 327–328, 369
slow down, 247, 281, 303
smart grid technology, 12, 50, 119, 192, 237
Smart grids, 1, 9–11, 43, 50, 244, 271, 321, 323
smooth transition, 316
Social and cultural shifts, 36, 49
social barriers, 56, 302
social benefits, 41, 88, 255, 287–288, 298, 314, 329, 349
social capital, 160
social equity, 33, 48–49, 118–120, 128, 254, 264, 270, 278, 287, 292–293, 305, 316, 329
social exclusion, 75, 93, 174–175, 304
social identities, 143–144
social indicator, vi, 54, 93, 119–120, 150, 153, 245–246, 273
social justice, 25, 46, 139, 142–143, 147, 149–150, 172, 177, 179, 189
social protection, 92, 173–179, 183–185, 190
social sustainability, 117, 149
societal group, 55–56
soil fertility, 294
soil health, 283
Solar Energy International, 319
solar home systems, 67
solar power, 14, 17, 237, 242, 244, 247, 249–250
Solar Professional Certification, 319
source reduction initiatives, 286
Spain, v, 17, 60, 62, 82, 131–135, 171, 188, 244, 266, 327–328
STAMP, 123–124, 130, 136–137
standards and regulations, v, 8, 40–41, 49
startups, 305, 320–323, 363
storage capacities, ix
strategic approach, 1, 182, 198
Streimikiene, D., iii–v, xi, xiii, 1, 9, 25, 52–55, 64, 68, 77, 93, 248, 258, 262–263, 272–273, 298, 330
Streimikis, J., vii, xiii, 273–274
structural changes, 182, 199, 201–202, 305
substantial quantities, 1, 3, 226
substitution, 63, 194, 202, 209, 226, 240, 264, 269
succession, 259
supply capacity, 97
supply chain, 33, 116, 119, 122, 229, 248, 269, 276–277, 281, 289, 300–302, 331–332, 336, 352, 369
supply chain disruptions, 281
supply chain initiatives, 289
supply chain management, 33, 229, 331
supply systems, 3–4, 56
susceptible, 163, 215
sustainability assessment, xi, 123, 130, 137, 283, 359
Sustainability Assessment and Measurement Principles, 123
sustainability challenges, 11, 33
sustainability criteria, 259, 320, 339
sustainability dimensions, 123–124
sustainability efforts, 36, 320, 348, 352, 362
sustainability framework, 330
sustainability goals, 48, 281–282, 287, 289–290, 298, 304, 348–349, 362
sustainability indicator, vi, 165, 271
sustainability issues, 124, 349
sustainability objectives, 48, 120, 149, 287, 304, 340
sustainability performance, 283, 339, 343, 348
sustainability principles, 290, 352
sustainable agriculture, 147, 278, 282, 289, 292, 294, 298, 334, 339, 364, 371
sustainable agriculture practices, 294
Sustainable Development Agenda, 274–275, 277
sustainable resource management, 167, 294, 348
Sweden, v, 13, 52, 82, 131–135, 154, 244, 267, 325, 327, 364system flexibility, 116
systematic approach, ix
systematic discrimination, 163

T

targets and indicators, v, ix, 18, 49
tax incentives, vii, 39, 78, 180, 214, 220, 245, 249, 256–257, 302, 309, 311–312, 314, 329
tax policy, 214
taxonomy perspective, 344, 371
technological advancement, 254, 260, 321
technological advancements, 8, 119–120, 129–130, 192, 197, 205, 208, 225, 238, 249, 299, 321–322, 324, 329, 340
technological capacity, 22
technological development, 1, 116, 119, 125, 129, 135, 200, 269
technological indicators, vi, 119–120, 169
technological risks, 354, 371
technology adoption, 27, 33, 129, 232
technology advances, 237
technology development, 51, 242, 255–256, 263, 321
technology risk, 255
technology transfer, 21, 24–25, 34, 36, 48, 208, 210, 276–277, 282, 298, 301
test, 174, 186
thermal energy, 8, 75
thermal power, 1, 202
thermodynamic, 199, 205
threshold, 64, 67–68, 70, 72–73, 80–81, 84–85, 89, 93
tidal energy, 236, 240–241, 271
timescale, 2
TOPSIS, 125, 130
total factor productivity, 375
Trade Secrets, 324–325
traditional energy sources, 237, 250
traditional knowledge, 145
traditional practices, 152
transition planning, v, 46
transparent, 42, 104, 130, 258
travel, 8, 145, 226
Turkey, 151, 185, 266–267

U

UK, iv, 13, 50, 59–60, 72–73, 91–93, 157, 180, 184, 187, 189, 213, 271, 323, 333, 371
Ukraine, 95, 108, 111, 137, 262, 266–267, 281–282
UN PRI, 346, 350, 371
undermine, 24, 27, 58, 140
unemployment, 74, 157, 174–176, 315, 317–318, 345
uneven allocation, 56
UNFCCC, 36, 50, 145, 185, 321
unit consumption, 199, 218
United Kingdom, v, 13, 52, 55, 60, 72, 74, 76, 266, 312, 369, 374

United Nations, 36, 67, 91–92, 143, 145, 149, 162, 172, 174, 263, 274, 314, 334, 338, 343–344, 346, 358–359, 371, 374–375
United Nations Principles for Responsible Investment, 346, 371
United States, 51, 97, 106, 137, 155, 157, 195, 244, 270, 273, 307–308, 311–312, 315–317, 320, 369
upfront costs, 255, 298–299
urban-rural divide, 292
urbanization, 114, 206, 209, 212, 220, 222, 226, 229–230, 374
utility bills, 59, 61, 63–64, 79–83, 86, 88–89, 126–127, 133, 195

V

Valasevičienė, R., xiii, vi
value chains, 112, 115, 137, 296–297
variability, 11, 203, 222, 250, 354
ventilation, 8, 203, 284
violence, 140–141, 143–144, 152–154, 160–162, 165, 172, 174, 184–188
Volodzkiene, L., vi, xiii, 139
vulnerabilities, 46, 57–58, 75, 117, 119–120, 145, 148–149, 167, 178, 186, 346
vulnerable population, 46, 55, 64, 76–77, 163, 294, 304

W

warmth, 80, 85, 88, 90
waste collection, 226
waste disposal, 283, 349
waste generation, 8, 286, 294, 345, 348–349
waste heat recovery, 6, 8, 54
waste products, 236
waste reduction, 17, 25, 283, 286, 294, 331, 348–349
wastewater, 283, 325, 356
water conservation, 284, 294, 319, 348
water management, xiii, 167, 253, 294, 348
water resources, 166, 283, 294, 325, 348
wave energy, 236, 241–243, 270
Weighted Energy Security Index, 115
willingness, 51, 245, 268
wind energy, 53, 94, 236–238, 242, 244, 246, 248, 250, 270, 312, 315, 319–320, 323
wind power, 236–238, 242, 244, 246–248, 250, 253–254, 256, 272
women, x, 139–174, 176–189, 292
women in leadership, 140–142, 146–147, 167–168, 180, 187
workplace discrimination, 148, 162, 178, 292
World Health Organization, 94, 153
World Trade Organization, 182
WTO, 137, 182, 185

Y

Year of Energy Efficiency, 213

Z

zero net emissions, 13
Zimbabwe, 166, 188

For Product Safety Concerns and Information please contact our EU representative GPSR@taylorandfrancis.com Taylor & Francis Verlag GmbH, Kaufingerstraße 24, 80331 München, Germany